Assessing Animal Welfare

Science for Animal Welfare

Science for Animal Welfare (formerly known as the Universities Federation for Animal Welfare, UFAW) is an international independent, scientific and educational animal welfare charity and membership society.

Our aims are to discover what matters to animals, develop scientific solutions to animal welfare problems, and disseminate evidence-based animal welfare information by:

- funding scientific research
- supporting the careers of animal welfare scientists
- disseminating animal welfare science knowledge both to experts and the wider public
- providing expert advice to government departments and other bodies, helping to draft and amend laws and guidelines

You can help make even bigger strides in animal welfare science by donating to, leaving a legacy for, or becoming a member of *Science for Animal Welfare.*

For further information about *Science for Animal Welfare* and how you can help promote and support our work, please visit our website:

ScienceforAnimalWelfare.org

The aim with the Animal Welfare book series is to promote interest and facilitate debate on the subject of animal welfare, whilst disseminating information relevant to improving the welfare of kept animals and those affected by human activities in the wild. The books in this series are the works of their authors, and the views they express do not necessarily reflect the views of *Science for Animal Welfare.*

Assessing Animal Welfare

A Guide to the Valid Use of Indicators of Affective States

Edited by
Georgia J. Mason
University of Guelph, Canada
Birte L. Nielsen
Science for Animal Welfare, UK
Michael T. Mendl
University of Bristol, UK

WILEY

Science for Animal Welfare

This edition first published 2026

Registered Offices
John Wiley & Sons, Inc., 111 River Street, Hoboken, NJ 07030, USA
John Wiley & Sons Ltd, New Era House, 8 Oldlands Way, Bognor Regis, West Sussex, PO22 9NQ, UK

For details of our global editorial offices, customer services, and more information about Wiley products visit us at www.wiley.com.

The manufacturer's authorized representative according to the EU General Product Safety Regulation is Wiley-VCH GmbH, Boschstr. 12, 69469 Weinheim, Germany, e-mail: Product_Safety@wiley.com.

Wiley also publishes its books in a variety of electronic formats and by print-on-demand. Some content that appears in standard print versions of this book may not be available in other formats.

Library of Congress Cataloging-in-Publication Data Applied for
Paperback ISBN: 9781394182190

Cover Design: Wiley
Cover Images: © Science for Animal Welfare

Set in 9.5/12pt Sabon by Straive, Pondicherry, India

Printed and bound by CPI Group (UK) Ltd, Croydon CR0 4YY

C9781394182190_020726

Contents

List of Contributors

Melissa Bateson
Faculty of Medical Sciences
Newcastle University
Newcastle upon Tyne
UK

Elodie F. Briefer
Behavioural Ecology Group
Section for Ecology and Evolution
Department of Biology
University of Copenhagen
Copenhagen
Denmark

Thomas Ede
School of Veterinary Medicine
University of Pennsylvania
Philadelphia
Pennsylvania
USA

Neil P. Evans
School of Biodiversity
One Health and Veterinary Medicine
University of Glasgow
Glasgow
UK

Carole Fureix
Bristol Veterinary School
University of Bristol
Bristol
UK

Suzanne D.E. Held
Bristol Veterinary School
University of Bristol
Bristol
UK

Lindsey Kitchenham
Department of Integrative Biology & Campbell
Centre for the Study of Animal Welfare
University of Guelph
Guelph
Ontario
Canada

Georgia J. Mason
Department of Integrative Biology & Campbell
Centre for the Study of Animal Welfare
University of Guelph
Guelph
Ontario
Canada

Dorothy E.F. McKeegan
School of Biodiversity
One Health and Veterinary Medicine
University of Glasgow
Glasgow
UK

Rebecca K. Meagher
Department of Animal Science and
Aquaculture
Dalhousie University
Truro
Nova Scotia
Canada

Michael T. Mendl
Bristol Veterinary School
University of Bristol
Bristol
UK

Lance J. Miller
Brookfield Zoo Chicago
Brookfield
Illinois
USA

Vikki Neville
Bristol Veterinary School
University of Bristol
Bristol
UK

Birte L. Nielsen
Science for Animal Welfare
Hertfordshire
UK

Elizabeth S. Paul
Bristol Veterinary School
University of Bristol
Bristol
UK

Colline Poirier
Faculty of Medical Sciences
Newcastle University
Newcastle upon Tyne
UK

Marsha L. Reijgwart
Animals in Science & Society
Faculty of Veterinary Medicine
Utrecht University
Utrecht
The Netherlands

Luise A. Seeker
Department of Bioengineering
Stanford University
Stanford
California
USA

Avelyne S. Villain
Behavioural Ecology Group
Section for Ecology and Evolution
Department of Biology
University of Copenhagen
Copenhagen
Denmark

GenPhySE
Université de Toulouse, INRAE, ENVT
Castanet-Tolosan
France

Claudia M. Vinke
Animals in Science & Society
Faculty of Veterinary Medicine
Utrecht University
Utrecht
The Netherlands

Jessica C. Whitham
Brookfield Zoo Chicago
Brookfield
Illinois
USA

Endorsements

"This exciting book addresses the most complex and challenging questions head-on, providing the conceptual tools and empirical benchmarks necessary to stimulate a great leap forwards in animal welfare science."

Professor Christine Nicol, Royal Veterinary College, UK

"… a very timely book on a subject of emerging and highest importance in animal welfare science… It is written in an easily accessible, yet scholarly thorough language and will therefore be of great interest to anyone concerned about animal welfare, laypeople and scientists alike."

Professor emeritus Per Jensen, Linköping University, Sweden

"Throughout the book it is emphasized that affective states should be studied rigorously before drawing conclusions, and the authors provide guidance through text examples and illustrations on how to achieve this. Overall, this book is a must-read for animal welfare students and researchers regardless of their career stage."

Dr Irene Camerlink, Polish Academy of Sciences, Poland

"An essential resource that will shape how animal welfare is taught, studied, and interpreted for years to come."

Dr Carly Moody, UC Davis, California, USA

"This urgently needed book addresses a core ongoing challenge in animal welfare science: how to validate indicators of animal affective states. It brings together clear explanation of the theoretical background with a comprehensive review of current indicators, to help practitioners choose the best indicator (or set of indicators) for their context. It's accessible even to non-specialists, with effective use of examples and diagrams throughout. An essential resource for anyone working in animal welfare."

Dr Heather Browning, University of Southampton, UK

"…not just scientifically authoritative but also an unapologetic, elegant, accessible and joyful portrayal of the core work of those dedicated to understanding and improving the welfare of other animals."

Professor Ngaio Beausoleil, Massey University, New Zealand

Acknowledgements

Heartfelt thanks to all the authors of this book for making it so, so much better than I ever envisaged; to Mike for being brilliant and for sticking with this project for some 18 years (and his partner, Liz Paul, and family for regularly letting me stay for days and spread paper everywhere); to Birte for ensuring this long journey actually had an ending; and to the 100+ Campbell Centre for the Study of Animal Welfare graduate students who, over about a decade, took my 'Assessing affective states' class (which really was not that good at the start; thanks for letting me try out various 'heuristic graphs' on you).

Thanks too to Pat Bateson for encouraging my first forays in animal welfare; to Marian Dawkins for her inspirational (albeit sometimes terrifying) commitment to absolute rigour and some lovely lunches; to Ian Duncan for his persuasive commitment to the central role of affective states (and some good laughs); to my own brilliant graduate students and postdocs, especially those working on this kind of topic while in my lab (e.g. Jamie Ahloy-Dallaire, Prathipa Anandarajan, Charlotte Burn, Jessie Cait, Jenna Cheal, Jonathan Cooper, Lauren Dawson, Maria Díez-León, Jeanette Kroshko, Naomi Latham, Carole Fureix, Michelle Lavery, Becky Meagher, Aileen MacLellan, Andrea Polanco, Misha Ross, Mike Walker and Harriet Warburton); to every philosopher who has ever challenged me; to my partner Jonathan Newman for his love and support (and enduring the craziness this project sometimes caused); and to Sophie, Mouse, Luke and Sylvie for showing me that cats can smile.

Georgia J. Mason, February 2026

This book started life nearly two decades ago. Special thanks to Georgia, its catalyst and my accomplice during its tortuous journey to maturity, and to Birte for resurrecting it from our hard drives and guiding it to completion. Many thanks to Pat Bateson for giving me the opportunity to study for a PhD and begin a career in behavioural biology; to Don Broom for introducing me to the exciting and challenging field of animal welfare science, and for making us think about how to define things, essential for avoiding all those crossed wires; and to others from the 'early days' – Alistair Lawrence, Christine Nicol, Francoise Wemelsfelder and Adroaldo Zanella – for sharing their knowledge, insights and skills, from the philosophical to the practical (setting up a pig research unit!), and for allowing me to experience life in different research labs.

Thanks to all the brilliant, dedicated and kind-hearted colleagues and students, including the chapter authors who produced such excellent work and tolerated our editorial intrusions; to all members of the Bristol University Animal Welfare and Behaviour Group, especially the graduate students and postdocs who have worked with me on affective states (e.g. Sharyn Bistre, Oliver Burman, Nick Cherbanich, Alastair Cockburn, Molly Davidson, Amanda Deakin, Louise Evans, Emma Harding, Helena Hale, Jillian Hendricks, Justyna Hinchcliffe, Aurelie Jolivald, Sam Jones, Sarah Kappel, Vikki Neville, Richard Parker, Marco Ramirez, Poppy Statham, Amelia St John Wallis, Helena Telkanranta, Ralph Thompson, Anna Trevarthen) and, of course, to their co-supervisors, including Emily Blackwell, Peter Dayan, Carole Fureix, Iain Gilchrist, Suzanne Held, Ben Lecorps, Emma Robinson and Nikki Rooney (apologies to any I have missed). Finally, heartfelt gratitude to Liz Paul – my partner in crime in trying to understand and assess animal affective states – for providing her deep knowledge of the psychology of emotion from those stacks of papers, discussing ideas and concepts at all times of the day and sharing the ups and downs of life.

Michael T. Mendl, February 2026

It has been a pleasure and an honour to work with so many brilliant researchers in the creation of this book. Disseminating evidence-based animal welfare information is one of the cornerstones of *Science for Animal Welfare*, and this book will do a lot of heavy lifting to enable more accurate, transparent and scientifically robust assessments of animal affect. A warm thank you goes to all the authors who have contributed; unfortunately, writing textbooks is not something that is prioritised by research institutes, and I know many of you have been writing this in your spare (ha!) time. I am someone who has changed career from active research to working in an international scientific and educational animal welfare and membership society. One of the many positive experiences has been working on this book with Georgia and Mike, who have taught me more about affective states and science communication than any number of courses could have done. I would also like to thank my employer, *Science for Animal Welfare*, especially my boss, Huw Golledge, who has allowed me to spend time on this project, which (of course!) took longer than anticipated. Finally, I would like to thank Elinor for drawing all the heuristic graphs, and many of the figures and animals included in various chapters, as well as the front cover. She stepped up at times when we changed our minds at short notice. Overall, this book project has been a challenging, yet fun and educational ride, and I really, *really* hope the content will be used widely in the teaching of animal welfare assessment.

Birte L. Nielsen, March 2026

Animal Welfare and Affective States

1

Michael T. Mendl[1], Elizabeth S. Paul[1], and Georgia J. Mason[2]

[1]Bristol Veterinary School, University of Bristol, Bristol, UK
[2]Department of Integrative Biology & Campbell Centre for the Study of Animal Welfare, University of Guelph, Guelph, Ontario, Canada

Introduction: the central role of positive and negative feelings in animal welfare

> "*Let us not mince words: animal welfare involves the subjective feelings of animals.*"
> (Dawkins 1990)

In this book, we follow the view that Marian Dawkins so succinctly summarised over 30 years ago; we are concerned about animal welfare because we are concerned that animals have *subjective experiences* akin to suffering. Take a look at the cheeseburger shown in Figure 1.1. If animal welfare troubles you, you'll probably be worried about the welfare of the organisms that provided the patty and cheese but not about those that made up the bread and garnish. Why is this? All the organisms were likely reared under unnatural conditions, their growth and physiology were adjusted through the use of chemicals (and by genetic selection and modification), and their lives were cut short when 'harvested'. Yet although they all had a challenging time, this matters to us when the organisms were animals but not when they were plants, because most of us believe that plants, lacking a nervous system, cannot consciously experience suffering[1]. The capacity to suffer is therefore core to our concept of animal welfare. More broadly it is the capacity for *sentience*, the ability to subjectively experience *feelings*, both positive and negative.

---

[1] Some researchers argue, controversially, that plants may have a form of conscious experience (Segundo-Ortin and Calvo 2023 and commentaries therein), or that nervous systems are not required for an organism to be sentient (Reber and Baluska 2021). This is a live debate, and the consensus may change with time but a telling conclusion from Segundo-Ortin and Calvo (2023) is that if plant sentience were to be established, the ethical treatment of plants would then need to be considered; sentience is thus key to welfare concerns.

Figure 1.1 The cheeseburger of welfare concerns. Source: Created using Canva.

Identifying which species are sentient is a challenging enterprise, and one we return to at the end of the chapter. However, there is a broad consensus that, assuming that a species is sentient, it is the stuff of sentience – *affective states* – that is the determinant of animal welfare. Affective states are *valenced* mental experiences – ones that are positive, pleasant and rewarding, or negative, unpleasant and punishing. They include *emotions* and *moods* such as fear, happiness, sadness and anger. They also include the affective components of sensations such as pain, thirst, hunger and sickness, which are sometimes referred to as *homeostatic emotions* (Craig 2003), *homeostatic feelings* (Damasio and Damasio 2022) or *primordial emotions* (Denton et al. 2009), and which reflect interoceptive evaluation of the state of the body. To assess animal welfare, we must therefore be able to assess these states. And to assess affective states, we must further clarify what they are. This is the focus of the current chapter.

But first, why are affective states so central to animal welfare? It has been said of such experiences, at least for humans, "*No aspect of our mental life is more important … . They are what make life worth living, or sometimes ending*" (Scarantino and De Sousa 2021). The idea that these subjective states also play a pivotal role in other species' well-being has a long history, that includes Jeremy Bentham's (1789) famous quotation "*The question is not, Can they reason? nor, Can they talk? but, Can they suffer?*"; Russell and Burch's (1959, p. 23) aspiration to "*aim at well-being rather than the mere absence of distress*"; UFAW's original mission "*to promote humane behavior towards wild and domestic animals … so as to reduce the sum total pain and fear inflicted on animals by man*", the 1965 Brambell Committee's concerns for "*the physical and mental well-being of the animal*"; and today's Five Domains' framework, with its goal of "*an ultimate assessment of the overall welfare state of the animals, understood in terms of what they (are) likely to experience subjectively*" (Mellor et al. 2020). Minimising negative subjective states such as fear or pain, and promoting positive subjective states such as contentment, is thus seen as synonymous with improving the quality of animals' lives (see also Duncan 1993; Mason and Mendl 1993; Dawkins 1998;

Rushen 2003; Mendl and Paul 2004; Broom 2007; Kirkwood 2007; Mason and Veasey 2010; Browning 2020, for similar views of animal welfare).

This is also evident in real-world guidelines and legislation. Estimating the severity of laboratory animals' negative states and suffering is a crucial part of the ethical review process for much animal-based research, and ensuring good well-being is core to many farm assurance and food labelling schemes. Furthermore, the capacity for *sentience* is now enshrined in legislation in several countries as the key factor underpinning our animal welfare obligations (e.g. EU Treaty of Amsterdam (EU 1997); New Zealand Animal Welfare Act (New Zealand Government 1999) and UK Animal Welfare (Sentience) Act (UK Government 2022)). It is also implicit in the way that many animal species receive levels of protection that plants do not (e.g. Canada has two federally-influential animal care councils, the Canadian Council on Animal Care and National Farm Animal Care Council, but no equivalents for plants). This feelings-centred view of animal welfare is thus widespread and conventional. It relies on two assumptions, additional to the animals in question being sentient: first, that positive and negative feelings matter to animals, and therefore have ethical weight (see Wittgenstein 1953; Singer 1976; Midgley 1984); and second, that we can detect or measure them in some way, so that welfare can be assessed. The latter, the process of objective assessment, is the focus of this book.

To begin, this opening chapter will delve into what affective states actually are. First, we consider two main ways in which they have been conceptualised: (i) as a collection of discrete, modular states such as 'fear', 'joy', 'jealousy' or 'anger', and (ii) as states that vary along two or more underlying dimensional characteristics, including the key property of *valence*. Next, we describe different temporal categories of affective state, spanning short-term 'emotions', longer-term 'moods' and 'affective disorders' that may last for many months to years, and overall 'cumulative affective experience'. We also briefly consider that different individuals may have different 'affective personalities'. It is important to note that these perspectives and categories are all *constructs*: conceptual ways of thinking about the organisation and structure of affective states, rather than hard-and-fast biological entities, and they may overlap with each other.

We end this chapter by outlining why these states are so challenging to investigate in animals, and why their subjective components are, at present, essentially unmeasurable. Nevertheless, by developing and using a logical inference framework, measurable indicators of affect can be identified and validated as tools for assessing welfare. In Chapter 2, we introduce such a framework.

Affective states: their natures, temporal characteristics, functions and long-term consequences

In this section, we give more detail about the constructs identified earlier and also consider their potential functions and long-term consequences.

Affective state concepts: discrete and dimensional views

Because affective states are human constructs derived from our own subjective experience of emotions, moods and other valenced feelings, much of what we know about them comes from research with humans, where linguistic self-report is used as the gold standard indicator of felt experience (e.g. Russell 2003; Mauss and Robinson 2009; Mendl et al. 2022). This research often identifies two distinct ways of conceptualising affective states: either as discrete entities (such as 'fear', 'sadness' or 'joy'), or as states that vary continuously along a few simple dimensions (such as *valence* – the positive or negative nature of a state). The merits and demerits of these two views attract much discussion in human emotion science, and understanding them is helpful for welfare scientists too (e.g. Izard 2007; Panksepp 2007; Adolphs 2017; Barrett 2017; van Heijst et al. 2023).

Discrete emotion theories

Discrete or 'basic' emotion theories propose that distinct emotions are elicited by different situations and characterised by fundamentally different responses (e.g. facial expressions) and unique 'feels' (Ekman 1992; Panksepp 1998). Furthermore, many of these theories propose that such states are universal across humans and have their origins in adaptive suites of behaviours, underpinned by distinct neurobiological circuits or systems (e.g. those located in the amygdala, LeDoux 1996), that can also be seen in many other (especially mammalian) species. For animal welfare researchers, this has an obvious intuitive appeal since it allows direct mapping of human and non-human states, and captures how different discrete emotions have different causes. But it has some disadvantages too. One is that the emotional and mood labels used in our own culture may limit what we think to look for and assess in other animals. English speakers, for example, may simply overlook the possibilities of *uitbuiken*, the Dutch concept of feeling pleasantly full after a meal, or the Japanese *shinrin-yoku*, the calming experience of being in a forest. More worryingly, it risks unjustified anthropomorphism: assumptions that other species are like us even when they have evolved in sensory or social worlds very different from our own. It is probably absurd to look for 'jealousy' or 'grief' in non-social species, for example, while conversely some species may have states so dissimilar from our own that we cannot imagine them. What does it feel like to be a lion who has just killed and eaten a rival's cubs, for example, some affectively positive combination of uitbuiken and lust as he waits for the lioness to come into oestrous? As Wittgenstein (1953) put it, "*if a lion could talk, we would not understand him*". And this challenge is even greater for animals whose sensory worlds and abilities diverge even more considerably from our own. Thomas Nagel famously asked "*What is it like to be a bat?*", in the title of his paper discussing the unknowability of alien minds (Nagel 1974). Perhaps bats lack some of the specific emotions that we have (such as guilt) but feel their own unique negative emotions if their sonars are bombarded with very strong ultrasound, or their own unique positive emotions if they capture a particularly agile insect.

Affective dimensions

The second, alternative conception of affective states sidesteps this issue by treating them as varying continuously along some simple, underlying dimensions. The prominent 'core affect' model, resulting from statistical analyses (e.g. principal components analysis [PCA]) of perceived differences, similarities and temporal associations between human emotion words and expressions, identifies two key dimensions: (i) *valence*, the degree to which a state is positive or negative, and (ii) *arousal* – how activating or arousing the state is, from soporific to frenetic (e.g. Russell 2003). The 'location' of different emotional states in this two-dimensional *core affect space* thus depends on their valence and arousal characteristics. In Figure 1.2, we show examples of (human) discrete emotions arranged according to the levels of arousal and valance that are likely associated with them. Their distribution in our figure differs from the classic 'circumplex' in which they are arranged, according to PCA statistical models, in a circle around the *edge* of core affect space (e.g. Russell and Pratt 1980; Russell and Barrett 1999). However, our figure follows the recognition that, at any one time, an individual's affective state reflects their current combination of valence and arousal and can occupy *any* area of the space. Additionally, it is suggested that the further away the current state is from the centre of core affect space, the more *intense* the accompanying emotional feeling (Russell 2003; Russell and Barrett 1999). Hence, 'fearful' is further from the centre than 'distressed', as is 'depressed' relative to 'sad', and 'elated' compared to 'happy'. Figure 1.2 thus implies that arousal and valence vary independently from each other; they are orthogonal. Nevertheless, as we will see in later chapters (e.g. vocalisations in Chapter 4 and cortisol in Chapter 5), changes in some indicators of affective state appear to reveal an underlying correlation between arousal and valence. For example, increases in the release of stress hormones (e.g. cortisol, see Chapter 5) as a situation becomes more challenging or dangerous reflect elevations in arousal plus a strengthening of valence and, in humans at least, an increasingly intense emotional feeling (e.g. increasing fear). So, for some markers and affective states, arousal, valence and emotional intensity may covary. In these cases, levels of arousal may be a useful proxy marker of valence strength.

Some argue that the core affect model best represents the structure of and relationships between different human affective states; Mauss and Robinson (2009), for instance, said "*it is quite clear that dimensions such as valence and arousal ... capture the lion's share of variance*" (see also Lindquist et al. 2012; although for counters, see Kragel and Labar 2016; Vytal and Hamann 2010). Others further argue that the combination of core affect with sensory, perceptual and interoceptive information enables humans to construct a potentially limitless set of feelings (e.g. Barrett 2017). This so-called 'constructionist' perspective is in marked

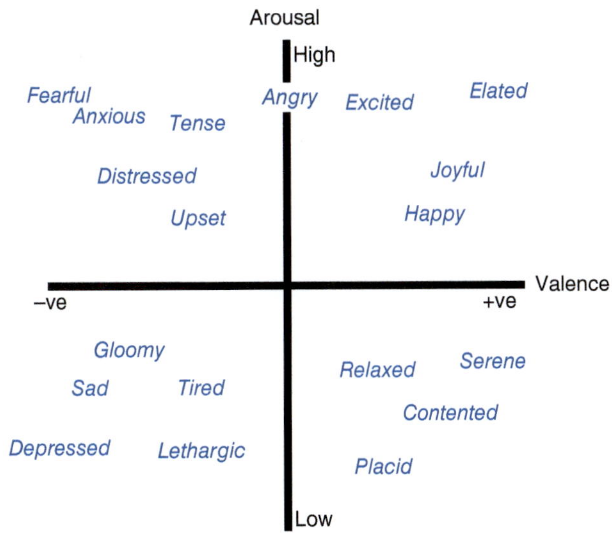

Figure 1.2 A dimensional core affect model. According to this model, valence and arousal, or activation, are the two main dimensions along which affective states can be separated. Human emotion words in italics illustrate where different types of affective state might lie in this 'core affect' space depending on the level of arousal and strength of valence that characterise them. Source: Adapted from Russell (2003).

contrast to the idea of a limited set of discrete emotions that is experienced by all members of a species and supports a view of affective states as primarily culturally determined, and having 'fuzzy' boundaries (Russell 2003), with gradual transitions between related states.

Other statistically-generated dimensional models, whilst yielding the same basic affective space, emphasise the importance of two dimensions lying at roughly 45° from the valence and arousal axes (Russell and Barrett 1999; Watson et al. 1999; Yik et al. 1999; Carver and Harmon-Jones 2009; see also Burgdorf and Panksepp 2006; Knutson and Greer 2008). These have been given labels such as *Positive Activation* and *Negative Activation* (Watson et al. 1999) or *Promotion* and *Prevention* (Higgins 1998), and linked to neurobehavioural systems underpinning two evolutionarily ancient imperatives – the need to acquire resources for survival (*reward acquisition*), and the need to avoid harm (*punishment avoidance*; Gray 1987; Carver 2001; Corr 2008; Norris et al. 2010; Norman et al. 2011). Neurobiological substrates of reward acquisition systems in mammals include mesolimbic dopaminergic, opioidergic and endocannabinoid systems acting at various brain locations (e.g. nucleus accumbens, ventral striatum, insular and orbitofrontal cortices) during anticipation or presence of reward, and are associated with high-arousal positive affect (e.g. Schultz et al. 1997; Knutson and Cooper 2005; Nguyen et al. 2021). Reduced activation of these systems has also been associated with low-arousal negative states such as depression (e.g. Forbes et al. 2009). Similarly, associations have been found, again at least in mammals, between anticipation or presence of punishment and activity in amygdala insula and anterior cingulate cortex, and between anticipation of relief and activation in the amygdala and substantia nigra (Seymour et al. 2005 2007; Carlson et al. 2011).

The activity of reward acquisition systems thus appears to vary according to presence/anticipation and absence/loss of reward and is associated with, respectively, high-arousal positive affect and low-arousal negative affect. Likewise, the activity of punishment avoidance systems varies with presence/anticipation and absence/avoidance of punishment, and is associated with, respectively, high-arousal negative affect and low-arousal positive affect (Watson et al. 1999). This is illustrated in Figure 1.3 which shows how the presence or absence of rewarding and punishing events can be mapped to locations in core affect space. Importantly, these neurobiologically-focused views also illustrate hypothesised links between affective arousal and valence on the one hand, and the evolutionarily important activities of acquiring resources and avoiding harm on the other. They may also underlie the positive associations between arousal and valence observed for some affective states and indicators as discussed earlier.

The link between reward, punishment and location in core affect space also maps to Rolls' (2014) definition of emotional states as "*states elicited by rewards and punishers*", where rewards are things for which an animal will work and punishers are things that it will work to escape or avoid (see also Mendl and Paul 2020). This is an example of an *operational definition* in which a concept (emotional state) is defined in terms of how it can be identified and/or measured, thus rendering it open to experimental investigation. This operational definition allows rewards and punishers to be identified empirically according to the behavioural work that animals do to, respectively, access or avoid them. Rewards and punishers identified in this way can then be used to induce emotional states as defined: delivery of rewards *and* removal of punishers gives rise to positive affective states, while delivery of punishers *and* removal of rewards gives rise to negative states. Anticipating such events can also elicit affective states (Rolls 2014; Mendl and Paul 2020; see Figure 1.3). Later in the book, we will see how this dovetails with two tests (Tests 2 and 3 in Chapter 2) that can be used to validate indicators of animal affective states.

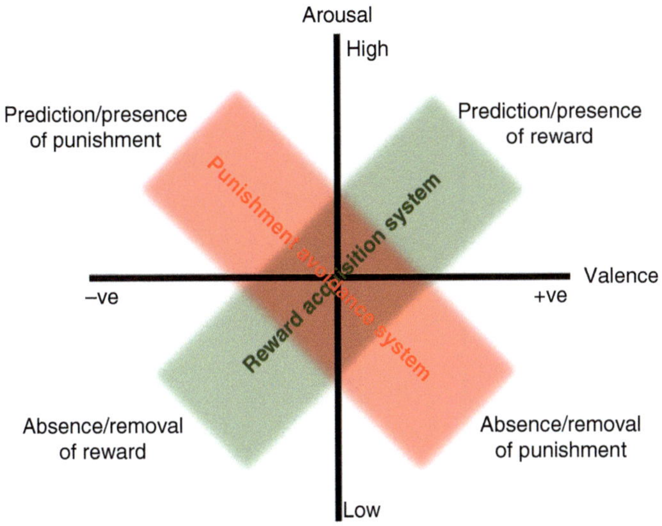

Figure 1.3 Presence/absence of reward/punishment and location in core affect space. The green and red bars represent axes along which affective states change according to the activity of neurobehavioural systems underpinning evolutionarily ancient imperatives: acquiring resources and avoiding harm.

Dimensional views have distinct advantages for animal welfare researchers, not least because they avoid the risks of anthropomorphism inherent in translating human-specific discrete emotion terms to other species (LeDoux 2017). They acknowledge that animal affective states are influenced by their sensory and perceptual worlds, and hence may be very different from our own (e.g. Bliss-Moreau 2017). And they also focus our attention on the attribute most likely to matter to animals themselves: the extent of positive or negative valence (e.g. Duncan 2006). However, these views have disadvantages too, including that states like pain and fear might be treated as similar, since their locations in two-dimensional 'core affect space' may overlap, when of course their causes and outcomes are very different.

In this book, we will use the dimensional approach, and particularly the concept of valence, to build a framework for validating animal affective states. But first we consider how affective states can also be distinguished according to their temporal characteristics. In doing so, we identify putative functions and consequences of these different affective categories.

Affective state concepts: temporal categories of affect

Affective states change across time, as illustrated in Figure 1.4, for core affect. An additional issue in the categorisation of affective states therefore relates to their duration. Here we consider the distinction between short-term emotions, longer-term moods, affective disorders, affective personality and cumulative affective experience across time. In doing so, we identify causes and putative functions of these different temporal categories.

Short-term emotions, their causes and functions

The terms 'emotion' and 'emotional state' are often used colloquially to denote a range of affective phenomena. However, 'emotion' also has a specific, narrower meaning in affective science, referring to short-term affective states that are closely tied to rewarding or punishing external or internal events that human subjects can typically pinpoint, the latter including

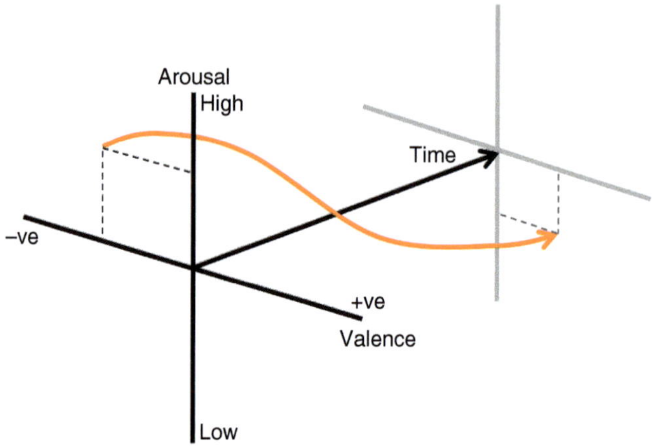

Figure 1.4 Affective states have a temporal component. The figure shows a hypothetical trajectory of core affect for an individual moving from a high-arousal, negatively valenced state to a low-arousal, positively valenced state, for example, as a result of the successful avoidance of a predator.

bodily signals and recall of previous experiences. 'Emotions' are thus often said to be 'about' something – to have a specific cause or focus (e.g. *I am delighted by X*, *scared of Y*, *thrilled about Z* or *pained by this burn*) – and are consequently fairly short-lived. Their duration is typically considered to be seconds to minutes and, occasionally, hours (Oatley and Johnson-Laird 2014).

The generation of emotions usually involves some interpretation or *appraisal* of the causal event (Schachter and Singer 1962; Scherer 1999). Some rewards and punishers are stimuli that would have been reliably good or bad for ancestral fitness (e.g. attractive mates or dangerous predators, respectively; e.g. Rolls 2014). For these, it is likely that all members of a species will undertake similar appraisals and show similar emotional responses. They will likely also experience similar homeostatic or 'primordial emotions': affective components of sensations (e.g. pain caused by injury, nausea caused by sickness) or motivations (e.g. intense thirst caused by disruptions to homeostasis). But some stimuli can be rewards to some individuals and punishers to others (e.g. licorice, Marmite or the Japanese food natto), or rewards in one situation and punishers in another (e.g. iced tea on a hot day or a cold day). New punishers and threats can also be learned, for instance by classical conditioning or social transmission. Human examples might include joy when your team scores a goal, replaced by eye-rolling disappointment or anger when the opposition equalises minutes later, or excitement when you hear fireworks, replaced by fear as you realise they are gunshots. Learning from individual experience can result in the same event generating quite different emotions in different people. And again, thanks to learning, emotions in humans can also be elicited by simply hearing good or bad news, or remembering past negative or positive events.

Ekman and Cordaro (2011) argued that "*Each … emotion prompts us in a direction that, in the course of our evolution, has done better than other solutions in recurring circumstances that are relevant to our goals*". But what *are* the adaptive functions of emotions? Emotions that involve facial expressions can be important for enhancing or minimizing the acquisition of sensory information; for example, the eyes and nostrils are widened in fear and narrowed in disgust (e.g. Susskind et al. 2008). They can also function for communicating with conspecifics (something returned to in Chapter 4 on vocal signals). But more broadly, *all* emotions appear to be involved in appropriate decisions to approach or avoid, choose between diverse options and drive reinforcement learning that shapes future behavioural responses (Mendl and Paul 2020). These three functions are described next.

"*Some argue that approach and avoidance are more or less synonymous with positive and negative emotional states*", proposed Mauss and Robinson (2009). This chimes with the views of those who emphasise the role of homeostatic or primordial emotions such as hunger, thirst and pain in motivating homeostatic responses including searching for food or water and avoiding noxious stimuli (Craig 2003). Looking first at positive emotions, humans have a tendency to view even objectively neutral stimuli as mildly positive and to expect positive things to happen (Ito and Cacioppo 2005; Sharot 2011), perhaps because we are a naturally curious, exploratory species. However, stimuli that are *strongly* liked (as opposed to only weakly liked) typically attract the fastest approach, most attention, greatest consumption and greatest willingness to bear costs to obtain access (e.g. Cabanac 1992; Pessiglione et al. 2007; Anderson and Adolphs 2014). That such effects are adaptive is nicely illustrated by the way that emotional responses to certain stimuli change with the body's homeostatic need for them. Thus, warmth is judged as more emotionally positive when we are cold, motivating us to seek shelter when needed (Cabanac 1992). Similarly, food, water and salt are judged as more delicious and pleasurable when our bodies are deprived of them, again helping motivate the restoration of homeostasis (Robinson and Berridge 2013; Rolls 2014; Saker et al. 2014). To paraphrase Cabanac (1992), "*What's pleasant is useful*".

Negative emotions, and the avoidance behaviour they trigger, are also adaptive. The affective component of pain, for instance, helps us avoid damaging tissue injury (Eccleston and Crombez 1999), while warmth is judged as negative and unpleasant when we are already too hot, promoting adaptive avoidance that helps with thermoregulation (Cabanac 1992; see Chapters 2 and 5). Ancestrally important signs of danger (e.g. spiders or snakes) are also highly motivating, often with very negative affective impacts, including potent abilities to attract our attention and generate revulsion. This can even occur despite little modern-day relevance: as Nettle (2009) and many others have pointed out, today people still tend to fear things like snakes and spiders even though electric shocks and road accidents are now a much bigger threat for most modern humans. The evolutionary imperative of avoiding harm is likely why, at least in humans, negative emotions are often more rapidly triggered, stronger, longer-lasting and more diverse than positive ones (Baumeister et al. 2001; Ito and Cacioppo 2005). Such motivational correlates of emotion are also what makes assessing preference and avoidance such a valuable tool in animal welfare research (see Chapters 2 and 3), and when inferring positive or negative states during the construct validation of indicators (as discussed in the next chapter).

Cabanac (1992) proposed that valence, as a scalable sense of pleasure or displeasure, functions as a "common currency" for decision-making. This would operate when obtaining a resource comes with a cost (such that approach only occurs when the positive affect offered by a resource outweighs the negative affect involved in obtaining it). It would also enable adaptive choices between diverse options offering different costs and benefits such as deciding whether to risk cold and predation in order to eat, versus staying safe and warm but unnourished at home (Mendl and Paul 2020). Some human data support this hypothesis: emotional 'gut feel' choices between options appear to reflect an internal weighing up of their diverse pros and cons. Complex decisions – such as which apartment to rent or which roommate to choose – thus can operate particularly efficiently when rapid, feelings-based choices are made (e.g. Mikels et al. 2011; Dijksterhuis 2004).

As a final, and perhaps most important, function, emotions influence learning and hence future actions. Maximising long-term rewards over punishments, and hence positive over negative states, is at the heart of reinforcement learning and decision-making (Mendl and Paul 2020). Associative learning thus occurs rapidly when it results in acquisition of very pleasant *outcomes* – powerful rewards – or avoidance of very unpleasant, punishing ones (Panksepp 2005; Rolls 2013), especially if these are surprising and/or unexpectedly pleasant or unpleasant. Similarly, ancestrally important *cues*, such as the shape of snakes, are particularly efficient at promoting learning about the negative consequences that can follow them (e.g. Tomarken et al. 1989; so-called *pre-prepared* fear learning: Öhman and Mineka 2001). These cues and outcomes do not even need to be close together in time for learning to occur. For example, the adaptive avoidance of contaminated food can be learnt with just one single pairing of a novel food/flavour with the later punishing experience of nausea (Garcia et al. 1955).

Long-term moods, their causes and functions

Moods are longer-lasting affective states that often do not, from the human subject's perspective, have an obvious immediate cause – they seem unattached to a specific event and are sometimes described as 'free-floating' (Beedie et al. 2005; Mendl et al. 2010, 2022; Nettle and Bateson 2012;). Examples include feelings of contentment that are not linked with any one particular event, or senses of doom, gloom or sadness. Lingering sickness-related malaise, or lassitude, is also a mood-like state that appears to be related to immune responses to infection (Schrock et al. 2020; Bucks et al. 2008) and hence is an example of a homeostatic emotional feeling linked to interoceptive evaluation of bodily function. Moods are usually considered to last from hours to days, sometimes weeks (Oatley and Johnson-Laird 2014), thus forming fuzzy temporal boundaries with the concepts of 'Emotions' (above) and 'Affective Disorders' (below), and illustrating the overlapping nature of these constructs.

Many moods seem to result from the accumulation and integration of previous shorter-term emotional experiences. For example, minor and major *stressful life events* (e.g. job loss, bereavement) can give rise to negative emotions of varying intensities and durations (e.g. Anderson and Adolphs 2014), which at some point seem to accumulate, fuse and persist as longer-term moods that are detached from their original causes (Beedie et al. 2005). Moods can also change emotion dispositions. For example, negative affective states may increase the likelihood of experiencing negative or less positive emotions by, for example, lowering thresholds to experience negative affect or biasing attention to negative stimuli (e.g. Mathews and Macleod 1994; Benning and Oumeziane 2017, Clarkson et al. 2020; see also Larsen and Ketelaar 1991). Moods also alter the appraisal of events (and thus the short-term emotions that arise from these events), such that there can be bidirectional, mutually potentiating relationships between short-term emotions and longer-term mood states (Mendl et al. 2010; Mendl and Paul 2020).

The mechanisms by which emotional experiences shape long-term moods are far from elucidated. However, positive and negative moods may be influenced differentially (e.g. Watson et al. 1999). For example, Clark and Watson (1988) found that self-reported positive moods were associated with pleasant social interactions and other positive daily life events, while negative, anxious moods were instead more often predicted by health problems (e.g. headaches, gastrointestinal problems, sickness such as colds or flu). In a similar study of people in the workplace, Miner et al. (2005) found that, while positive events were more common, negative events had a considerably stronger influence on ongoing mood. Further evidence that experiencing negative emotions predicts the emergence of negative moods comes from research on the mood disorders discussed in the next section (e.g. Tennant 2002), although the picture here is complex – factors like personality, socio-economic status and gender can moderate the link between stressful life events, emotions, moods and mood disorders (Hankin and Abramson 2001; Hyde et al. 2008; Sullivan et al. 2000).

Figure 1.5 (adapted from Mendl and Paul 2017; see also Webb et al. 2019) shows a simple conceptual model of affective valence incorporating emotion, mood and also affective disorder. The length of the *x*-axis represents minutes to hours. Rewarding or punishing events generate short-term emotions indicated by spikes in affective valence (black line) which return to a longer-term underlying mood state (wide grey line). The size and duration of spikes reflect the impact of the affective events. Mood is influenced by cumulative experience of short-term emotions, becoming more positive after positive emotional responses and more negative after negative responses, but with a tendency to regress towards a 'baseline' state (dashed horizontal line). This state may vary between, and be characteristic of, individuals but, at population level, it tends to be slightly positive (Sharot 2011). The grey mood line interpolates mood trajectory between emotional event spikes. As discussed later, negatively-valenced affective disorders may start to develop after repeated or particularly intense negative experiences. These may reflect qualitative changes in the relationship between experience and emotions – such that, for example, there is decreased sensitivity to reward and/or increased sensitivity to punishment, leading to a downward trajectory of mood. The wide blue line indicates the onset of such changes. Whether there is an absolute neutral valence point on the *y*-axis, above which states are qualitatively positive and below which they are negative, or whether the *y*-axis represents relative valence, has received little discussion in the animal literature. Some human psychology researchers argue that neutral affect does exist, is qualitatively distinct from positive and negative affect, and may function to promote inactivity (Gasper 2023). For simplicity, Figure 1.5 does not show any difference between the impact of strongly rewarding or strongly punishing events on mood. And effects of mood on appraisal of and reactivity to rewarding or punishing events are illustrated only for the blue affective disorder line; in this example of a negatively valenced affective disorder, negative events generate relatively stronger emotions than do positive events (e.g. anhedonia during depression).

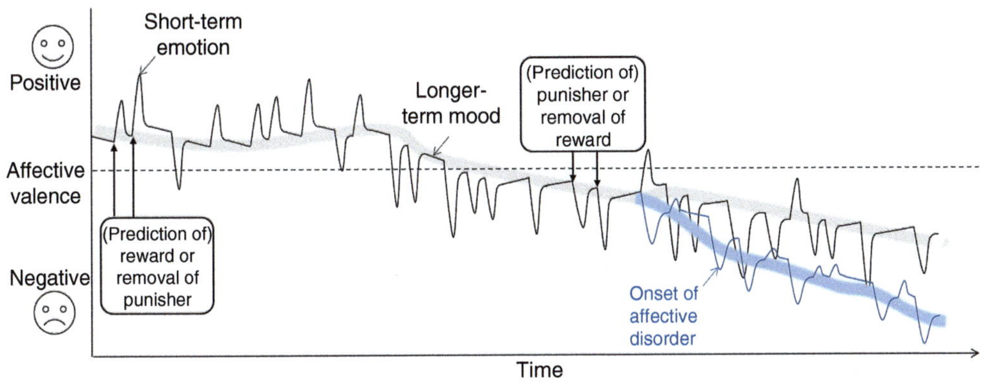

Figure 1.5 A simple conceptual model of temporal changes in affective valence resulting from the interplay between emotions and moods and including the onset of negatively-valenced affective disorder (see text for details).

Functionally, moods play a role in many aspects of cognition and decision-making, but unlike those associated with object-focused emotions, these effects mostly concern broad classes of stimuli rather than specific ones. The positive or negative nature of a mood can influence decisions made about a whole range of daily events (e.g. Schwarz and Clore 1983; Forgas 1995; Loewenstein et al. 2001; Mendl and Paul 2020). For example, people in negative moods show increased attention to threatening stimuli (Bar-Haim et al. 2007), make more pessimistic judgements about the future (Strunk et al. 2006) and more negative judgements about a range of ambiguous stimuli (for reviews, see Blanchette and Richards 2010; Paul et al. 2005). Unpleasant though they are to experience, such negative moods almost certainly have (and/or have had in the past) greater evolutionary significance than positive ones, helping us to avoid threats and minimise energetic costs (Nettle and Bateson 2012). For example, anxious moods, which involve great behavioural activation and rapid shifts of attention towards potentially threatening stimuli, may have evolved because being both fast and decisive is important to successfully avoid dangers such as predator attack (Nesse 2000, 2005). Moods seem to mediate these effects via a variety of cognitive processes including altered attention, memory and decision-making (Paul et al. 2005; Mendl and Paul, 2020), as considered further in Chapter 7 on judgement biases.

Affective disorders, their causes and possible functions

People suffering from *affective disorders* such as generalised anxiety disorder, panic disorders, post-traumatic stress disorder and major depression experience prolonged, disproportionately negative affective states. These states are longer lasting, and tougher-to-reverse, than everyday negative moods; in fact, they can last several weeks, months or even years (e.g. Oatley and Johnson-Laird 2014). In many cases, they co-occur (e.g. clinical anxiety and depression are particularly likely to be comorbid; Saha et al. 2021) and may also involve many non-affective elements which contribute to their debilitating nature (e.g. in depression or anxiety: insomnia or hypersomnia, problems concentrating, weight loss or weight gain, cognitive impairments and 'fight or flight' physiological responses like racing hearts: MacLellan et al. 2021). In humans, depression and anxiety appear to be associated with changes in the processing of rewards and punishments. A recent meta-analysis (Katz et al. 2020) for example indicates increased sensitivity to punishment in both depression and anxiety, and decreased reactivity to reward (i.e. anhedonia – see *Diagnostic and Statistical Manual of Mental Disorders,* American Psychiatric Association 2013]) in depression. Relatedly, Bishop and Gagne (2018) argued that reduced estimation of the likelihood and value of rewarding events and increased estimation of the likelihood and value of punishment are key components of depression and anxiety respectively which, in turn, impact on decision-making. As briefly mentioned earlier, we attempt to illustrate these changes in reward and punishment processing in Figure 1.5 (blue lines). This shows that during (the onset of) negatively valenced affective disorders, rewarding events may generate a smaller positive emotion peak whilst punishing events generate a larger negative peak, leading to a downward trajectory of mood (see Chapter 9 on reactivity to reward).

Affective disorders in humans are generally more likely in individuals who have faced childhood adversity (Brown et al. 1999; Heim et al. 2008; Colman and Ataullahjan 2010; Nanni et al. 2012) and also dealt with numerous unpleasant challenges throughout their lives (Tennant 2002; Hammen 2005). Thus, as well as involving current negative affect, they provide indirect evidence of past poor well-being. It is therefore not surprising that, according to many medical perspectives, these affective disorders are counter-productive, non-functional and pathological (e.g. Drevets et al. 2008). However, an alternative view is that these disorders are functional in some way, or were to our ancestors, at least for those individuals experiencing milder (non-lethal) forms (e.g. McNamara and Buchanan 2005). For example, a recent meta-analysis has shown that depressive disorders are associated with enhanced learning about punishment (Pike and Robinson 2022), supporting ideas that depression can function to enhance focus on the analysis of challenging problems (Andrews and Thomson 2009). Another possibility is that depression functions to inhibit potentially risky actions in a dangerous environment and to encourage inactivity that may allow the individual to 'ride out' difficult times (Nesse 2000). So, while often pathological in the modern human context, depression could have helped our ancestors survive in social or physical environments where punishments were common and severe. Likewise, temporary extreme anxiety or panic, although highly unpleasant to the sufferer, could be a small price to pay if the 'fight or flight' responses typifying them help protect an individual from danger (Nesse 2005).

Affective personality

We briefly note here that many human and animal personality characteristics (e.g. 'shyness', 'boldness', 'neuroticism', 'optimism', 'fearfulness', 'aggressiveness' and 'emotionality' itself) appear to have an affective basis and may reflect underlying genetic or epigenetic predispositions to exhibit particular affective behaviours and states (e.g. Larsen and Ketelaar 1991; Gosling and John 1999; Diener et al. 2003). From a dimensional perspective, the propensities to show stable differences in approach and withdrawal tendencies, or to be more or less sensitive to rewards and punishments, have been identified as important personality traits underlying vulnerability to disorders such as anxiety and depression (Davidson 1998; Corr 2008). These traits likely last for years or are lifelong (Oatley and Johnson-Laird 2014), and may equate to a 'baseline' (e.g. dashed horizontal line in Figure 1.5), towards which an animal's affective state will drift when not experiencing short-term emotion-inducing events or longer-term affective disorders. We will not dwell on this large topic area, but suffice to say that such predispositions can act alongside the impact of experience to influence an individual's affective states, disorders and hence their welfare (e.g. Diener et al. 2003).

Cumulative affective experience across time

Cumulative affective experience can be thought of as the summed experience of emotions, moods and affective disorders across a whole lifespan, or a specified timescale; and they manifest in the way that affective states of all kinds can have long-lasting consequences on behaviour, cognition and physiology. These effects have been most extensively studied in humans experiencing negative emotional states, in research focusing on what is variously termed *cumulative adversity*, *lifetime stress exposure* or *cumulative stress*: the combination of chronic difficulties and acute challenges that plague everyone to greater or lesser extents. These studies find that early, repeated or prolonged negative affect often accelerates the processes of ageing (e.g. Shields and Slavich 2017; Acer et al. 2020), reflecting underlying autonomic and cellular changes (e.g. reduced telomere length: see Chapter 12) that are collectively often called *allostatic load* (McEwen and Stellar 1993; McEwen 1998). These changes can also have harmful effects on ongoing physical health, including the increased likelihood and severity of cancer, stroke and cardiovascular disease, plus elevated all-cause mortality (increased chances of dying from a wide range of causes; e.g. Lampert et al. 2016; Shields and Slavich 2017; Cohen et al. 2019): effects that are mirrored in other animal species (e.g. Cavigelli and Mcclintock 2003; Walker et al. 2012; Cait et al. 2022). Early, repeated or prolonged negative affect can have harmful effects on *mental* health too, exacerbating not only the affective disorders already discussed (see e.g. Coelho et al. 2021; Foubert et al. 2021), but also conditions characterised by abnormal repetitive actions such as obsessive-compulsive disorder (e.g. Cromer et al. 2007) and trichotillomania (a hair-plucking disorder; e.g. Özten et al. 2015). And as we will discover in Chapter 11, this has important implications for understanding repetitive stereotypic behaviours in animals.

In humans, many of the lasting consequences of negative affect are believed to be mediated via three processes that are not mutually exclusive (e.g. Shields and Slavich 2017; Cohen et al. 2019; Yarrington et al. 2023). The first involves changes in *appraisal* (as described earlier), with people exposed to aversive events altering their anticipation, valuation and processing of rewarding and punishing events such that successive stressors become increasingly impactful. The second involves exposure to, or responses to, one type of adversity making other stressors more likely (e.g. divorce precipitating financial problems or loneliness). The third involves structural and functional consequences (such as alterations in skin, hair, blood vessels, endocrine systems and brain) that are relatively slow to reverse, such that progressive changes accumulate over time. Poirier illustrates this in her chapter on hippocampal changes within the brain (see Chapter 13), and we can sometimes see such effects visually too, as in the case of stress-induced grey hairs in mice or in humans above a certain age (Zhang et al. 2020; Rosenberg et al. 2021). Such somatic damage may in turn cause further declines in affective state (see Chapter 12).

So far, we have focused on cumulative negative states only. But what of net cumulative affective experience, taking positive states into account? For animals, any indicator that could track cumulative affective experience would be enormously useful for assessing their lifetime welfare or overall quality of life. It might allow the net cumulative impacts of entire husbandry systems to be evaluated, and be used to inform decisions about when to euthanise pets or retire research animals (see Yeates 2011; Bateson and Poirier 2019; Poirier et al. 2019). For some outcomes in humans (e.g. aspects of self-reported depressed mood), positive experiences seem to have rather little impact (e.g. Frank et al. 1996; Coelho et al. 2021). Other studies are less downbeat, finding that affectively positive events can mitigate the effects of negative ones (e.g. on self-reported general distress, albeit not anhedonia; Yarrington et al. 2023), and that mental attitudes and physical activities that may not be affective *per se* (e.g. healthy diets, exercise, positive appraisals; Puterman et al. 2015; Riepenhausen et al. 2022; see also Chapter 12) can have similar effects. Unfortunately, effects of cumulative affective experience are far less well-studied than the effects of cumulative adversity alone, particularly for variables that can be measured in animals. However, Chapters 12 and 13 present promising evidence that telomere length and certain hippocampal biomarkers have potential here.

Summary

In this chapter, we have seen that affective states can be conceptualised according to their structure and temporal characteristics and that different categories (emotions, moods, affective disorders, cumulative affective experience) have different causes and putative functions. When considering cumulative affective experience, we also touched on the issue of how we might assess these states and saw that a range of measures (e.g. behavioural, physical, cognitive, neurophysiological, physical) can be used. These indicators are considered to be 'components' of affective states which, unlike the conscious subjective component, are amenable to direct measurement (Scherer 1984). Figure 1.6 brings this all together and illustrates how short- and longer-term affective states (valence) are related and how

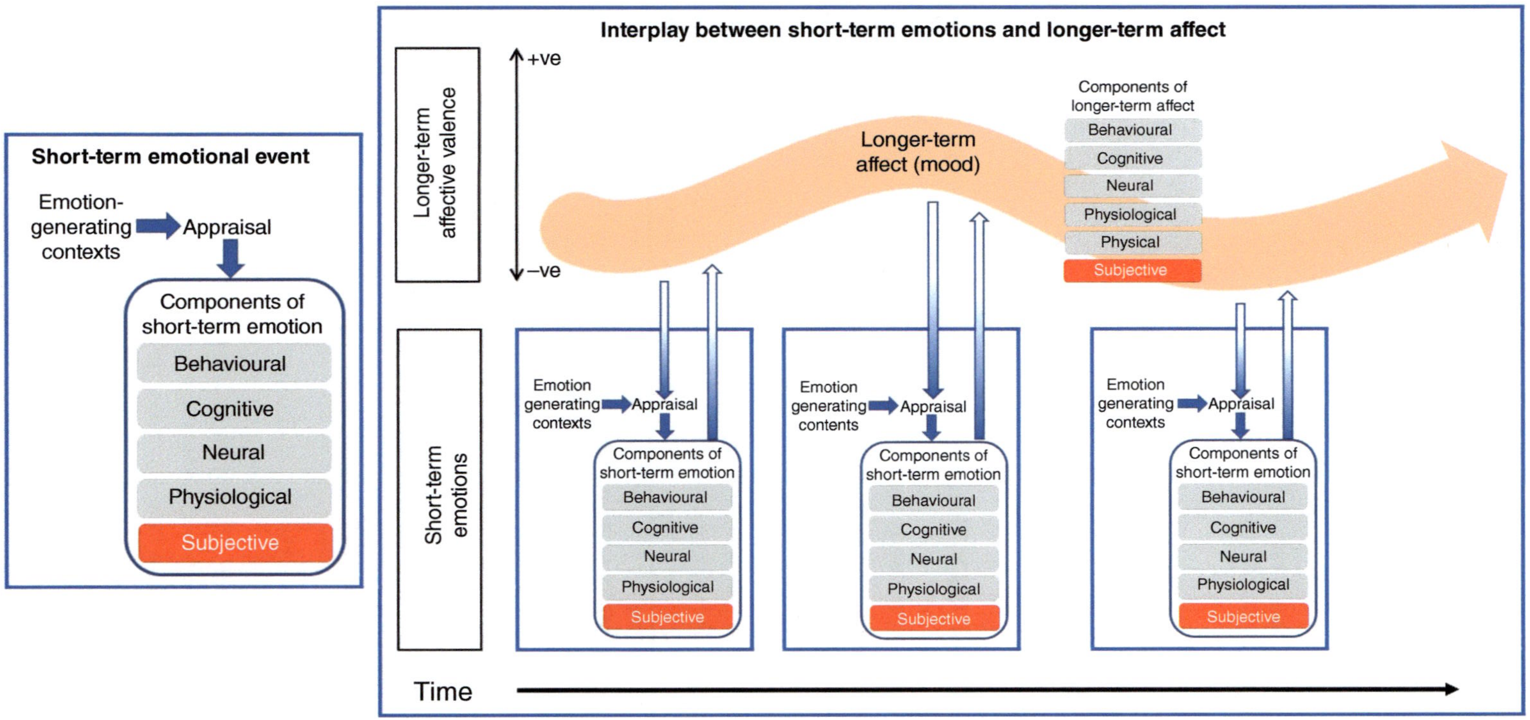

Figure 1.6 Short-term emotion, longer-term affect and the componential view of affective states. Left-hand panel: emotion-generating contexts elicit appraisal and interpretation of the situation and lead to the resulting short-term emotion including, in sentient species, a subjective affective experience. This is what we are interested in from an animal welfare perspective, but it needs to be inferred from changes in other measurable components (also known as 'welfare indicators'; see Chapter 2) of the affective state, including behavioural, cognitive, neural and physiological markers. Right-hand panel: cumulative experience of short-term emotions leads to fluctuating 'background' mood states (and, in some cases, longer-term affective disorders) which in turn influence appraisal processes. The lower half of the panel shows a sequence of three emotion-generating events across time, appraisal of each of which is influenced by background mood (upper half of panel) which in turn is then influenced by the resulting short-term emotion. Longer-term affective states comprise similar components to emotions but may also include physical indicators that change more slowly such as morbidity and mortality.

they can be assessed by measuring these components. The final section of this chapter briefly introduces the challenges inherent in trying to use these measurable changes to make inferences about what matters most from a welfare perspective, an animal's subjective experience of their affective state: how positively or negatively they feel. It is these challenges that this book aims to tackle.

Assessing animal affective states: the other minds problem and challenges faced by welfare scientists

In this chapter, we have seen that concerns about subjective affective states are at the heart of animal welfare (e.g. Mendl and Paul 2017). To better understand welfare, we therefore need to better understand and assess these states. This makes welfare science not only a fascinating field, but also a very challenging one.

That such internal, subjective experiences are private and inaccessible to measurement in others – the 'Other Minds' problem (e.g. Dennett 1996; Dawkins 2000; Farah 2008; Godfrey-Smith 2017; Avramides 2023) – is perhaps the biggest challenge facing any scientist interested in animal affective states (Paul et al. 2020). Historically, it explains why studying animal emotion has sometimes been regarded as "taboo" and not a proper topic for scientific research (e.g. Midgley 1984; LeDoux 1996; De Waal 2011). And today, it means that we still do not know where the boundaries of sentience are (Birch 2024) – that is, which non-human species (or developmental stages) *consciously* experience affective states. With our current knowledge, we just have to assume that these states are present in a variety of animals. This assumption makes biological sense given evolutionary homologies, and behavioural and cognitive similarities, between us and many other species. For example, in 2012, a group of consciousness researchers, including cognitive and neuroscientists, signed the Cambridge Declaration on Consciousness (Low et al. 2012), concluding that "*...the weight of evidence indicates that humans are not unique in possessing the neurological substrates that generate consciousness. Nonhuman animals, including all mammals and birds, and many other creatures, including octopuses, also possess these neurological substrates*". And in 2024, scientists, philosophers and other signatories of the New York Declaration on Animal Consciousness (Andrews et al. 2024) contended that "*empirical evidence indicates at least a realistic possibility of conscious experience in all vertebrates (including reptiles, amphibians, and fishes) and many invertebrates*". Such proposals make ethical sense as a 'precautionary principle', since to mistakenly treat sentient beings as unfeeling could cause great harm (e.g. Bradshaw 1998; Birch 2024). Nevertheless, they are not yet proven facts, which is why this book largely focuses on the postnatal birds and mammals for whom sentience assumptions are currently least controversial.

But the distribution of sentience is not the only challenge raised by the Other Minds problem. A second is that, as we noted earlier, even when we are comfortable assuming that sentience is present, we cannot know what different animals' specific affective experiences are like. As in our imagined examples of bats and lions, some animals' affective states may well be very different from our own: species a long way from us evolutionarily, or very different from us in their social or sensory worlds, are likely to have types of subjective experience that we lack, and to lack ones that we have. This is one key reason why this book largely takes a simple, dimensional view of affect, focusing on valence as a broad reference point for emotions in general: how positive or negative an affective state is, and therefore, how good or bad an animal's welfare can be said to be.

The third challenge arising from the Other Minds problem is that our job as welfare scientists is therefore to measure the unmeasurable! We can, however, measure in animals the changes in behaviour, cognition, physiology and health that, as we have seen in the preceding sections, accompany valenced subjective experiences (feelings). The challenge then is to infer reasonable conclusions from these measures about animals' subjective affective states. The next chapter covers how to do this logically and well.

References

Acer, E., Erdogan, H.K., Kocatürk, E., Saracoglu, Z.N., Alatas, O. and Bilgin, M. (2020) *Evaluation of oxidative stress and psychoemotional status in premature hair graying. Journal of Cosmetic Dermatology*, 19, 3403–3407. https://doi.org/10.1111/jocd.13428

Adolphs, R. (2017) *How should neuroscience study emotions? By distinguishing emotion states, concepts, and experiences. Social Cognitive and Affective Neuroscience*, 12, 24–31. https://doi.org/10.1093/scan/nsw153

American Psychiatric Association (2013) *Diagnostic and Statistical Manual of Mental Disorders (DSM-5)*, 5th ed. American Psychiatric Publishing, Arlington, VA, USA.

Anderson, D.J. and Adolphs, R. (2014) *A framework for studying emotions across species. Cell*, 157, 187–200. https://doi.org/10.1016/j.cell.2014.03.003

Andrews, K., Birch, J., Sebo, J. and Sims, T. (2024) *Background to the New York Declaration on Animal Consciousness*. Available at: https://sites.google.com/nyu.edu/nydeclaration/declaration

Andrews, P.W. and Thomson, J.A. (2009) *The bright side of being blue: Depression as an adaptation for analyzing complex problems. Psychological Review*, 116, 620–654. https://doi.org/10.1037/a0016242

Avramides, A. (2023) Other minds. In Zalta, E. N., Nodeelman, U. (eds) *The Stanford Encyclopedia of Philosophy*. Available at: https://plato.stanford.edu/archives/win2023/entries/other-minds/

Bar-Haim, Y., Lamy, D., Pergamin, L., Bakermans-Kranenburg, M.J. and Van Ijendoorn, M.H. (2007) *Threat-related attentional bias in anxious and nonanxious individuals: A meta-analytic study. Psychological Bulletin*, 133, 1–24. https://doi.org/10.1037/0033-2909.133.1.1

Barrett, L.F. (2017) *The theory of constructed emotion: An active inference account of interoception and categorization. Social Cognitive and Affective Neuroscience*, 12, 1–23. https://doi.org/10.1093/scan/nsw154

Bateson, M. and Poirier, C. (2019) *Can biomarkers of biological age be used to assess cumulative lifetime experience? Animal Welfare*, 28, 41–56. https://doi.org/10.7120/09627286.28.1.041

Baumeister, R.F., Bratslavsky, E., Finkenhauer, C. and Vohs, K.D. (2001) *Bad is stronger than good. Review of General Psychology*, 5, 323–370. https://doi.org/10.1037/1089-2680.5.4.323

Beedie, C.J., Terry, P.C. and Lane, A.M. (2005) *Distinctions between emotion and mood. Cognition and Emotion*, 19, 847–878. https://doi.org/10.1080/02699930541000057

Benning, S.D. and Oumeziane, B.A. (2017) *Reduced positive emotion and underarousal are uniquely associated with subclinical depression symptoms: Evidence from psychophysiology, self-report, and symptom clusters. Psychophysiology*, 54, 1010–1030. https://doi.org/10.1111/psyp.12853

Bentham, J. (1789) *An Introduction to The Principles of Morals and Legislation*. T. Payne and Son, London.

Birch, J. (2024) *The Edge of Sentience. Risk and Precaution in Humans, Other Animals and AI*. Oxford University Press, Oxford.

Bishop, S.J. and Gagne, C. (2018) *Anxiety, depression, and decision making: a computational perspective. Annual Review of Neuroscience*, 41, 371–388. https://doi.org/10.1146/annurev-neuro-080317-062007

Blanchette, I. and Richards, A. (2010) *The influence of affect on higher level cognition: A review of research on interpretation, judgement, decision making and reasoning. Cognition and Emotion*, 24, 561–595. https://doi.org/10.1080/02699930903132496

Bliss-Moreau, E. (2017) *Constructing nonhuman animal emotion. Current Opinion in Psychology*, 17, 184–188. https://doi.org/10.1016/j.copsyc.2017.07.011

Bradshaw, R.H. (1998) *Consciousness in non-human animals: Adopting the precautionary principle. Journal of Consciousness Studies*, 5, 108–114.

Broom, D.M. (2007) *Quality of life means welfare: how is it related to other concepts and assessed? Animal Welfare*, 16, 45–53. https://doi.org/10.1017/s0962728600031729

Brown, J., Cohen, P., Johnson, J.G. and Smailes, E.M. (1999) *Childhood abuse and neglect: Specificity of effects on adolescent and young adult depression and suicidality. Journal of the American Academy of Child and Adolescent Psychiatry*, 38, 1490–1496. https://doi.org/10.1097/00004583-199912000-00009

Browning, H. (2020) *The natural behavior debate: Two conceptions of animal welfare. Journal of Applied Animal Welfare Science*, 23, 325–337. https://doi.org/10.1080/10888705.2019.1672552

Bucks, R.S., Gidron, Y., Harris, P., Teeling, J., Wesnes, K.A. and Perry, V.H. (2008) *Selective effects of upper respiratory tract infection on cognition, mood and emotion processing: A prospective study. Brain Behavior and Immunity*, 22, 399–407. https://doi.org/10.1016/j.bbi.2007.09.005

Burgdorf, J. and Panksepp, J. (2006) *The neurobiology of positive emotions. Neuroscience and Biobehavioral Reviews*, 30, 173–187. https://doi.org/10.1016/j.neubiorev.2005.06.001

Cabanac, M. (1992) *Pleasure – the common currency. Journal of Theoretical Biology*, 155, 173–200. https://doi.org/10.1016/s0022-5193(05)80594-6

Cait, J., Cait, A., Scott, R.W., Winder, C.B. and Mason, G.J. (2022) *Conventional laboratory housing increases morbidity and mortality in research rodents: Results of a meta-analysis. BMC Biology*, 20. https://doi.org/10.1186/s12915-021-01184-0

Carlson, J.M., Greenberg, T., Rubin, D. and Mujica-Parodi, L.R. (2011) *Feeling anxious: Anticipatory amygdalo-insular response predicts the feeling of anxious anticipation. Social Cognitive and Affective Neuroscience*, 6, 74–81. https://doi.org/10.1093/scan/nsq017

Carver, C.S. (2001) *Affect and the functional bases of behavior: On the dimensional structure of affective experience. Personality and Social Psychology Review*, 5, 345–356. https://doi.org/10.1207/s15327957pspr0504_4

Carver, C.S. and Harmon-Jones, E. (2009) *Anger is an approach-related affect: Evidence and implications. Psychological Bulletin*, 135, 183–204. https://doi.org/10.1037/a0013965

Cavigelli, S.A. and Mcclintock, M.K. (2003) *Fear of novelty in infant rats predicts adult corticosterone dynamics and an early death. Proceedings of the National Academy of Sciences of the United States of America*, 100, 16131–16136. https://doi.org/10.1073/pnas.2535721100

Clark, L.A. and Watson, D. (1988) *Mood and the mundane – relations between daily life events and self-reported mood. Journal of Personality and Social Psychology*, 54, 296–308. https://doi.org/10.1037//0022-3514.54.2.296

Clarkson, J.M., Leach, M.C., Flecknell, P.A. and Rowe, C. (2020) *Negative mood affects the expression of negative but not positive emotions in mice. Proceedings of the Royal Society B-Biological Sciences*, 287, 20201636. https://doi.org/10.1098/rspb.2020.1636

Coelho, H.L., Trindade, I.A., Mendes, A.L. and Ferreira, C. (2021) *The mediating role of shame and fear of compassion on the relationship between major life events and depressive symptoms. Current Psychology*, 40, 4553–4562. https://doi.org/10.1007/s12144-019-00396-6

Cohen, S., Murphy, M.L.M. and Prather, A.A. (2019) *Ten surprising facts about stressful life events and disease risk. Annual Review of Psychology*, 70, 577–597. https://doi.org/10.1146/annurev-psych-010418-102857

Colman, I. and Ataullahjan, A. (2010) *Life course perspectives on the epidemiology of depression. The Canadian Journal of Psychiatry*, 55, 622–632. https://doi.org/10.1177/070674371005501002

Corr, P. J. (ed.) (2008) *The Reinforcement Sensitivity Theory of Personality*. Cambridge University Press, Cambridge.

Craig, A.D.B. (2003) *A new view of pain as a homeostatic emotion. Trends in Neurosciences*, 26, 303–307. https://doi.org/10.1016/s0166-2236(03)00123-1

Cromer, K.R., Schmidt, N.B. and Murphy, D.L. (2007) *An investigation of traumatic life events and obsessive-compulsive disorder. Behaviour Research and Therapy*, 45, 1683–1691. https://doi.org/10.1016/j.brat.2006.08.018

Damasio, A. and Damasio, H. (2022) *Homeostatic feelings and the biology of consciousness. Brain*, 145, 2231–2235. https://doi.org/10.1093/brain/awac194

Davidson, R.J. (1998) *Affective style and affective disorders: Perspectives from affective neuroscience. Cognition and Emotion*, 12, 307–330. https://doi.org/10.1080/026999398379628

Dawkins, M.S. (1990) *From an animals point of view – motivation, fitness, and animal welfare. Behavioral and Brain Sciences*, 13(1), 1–61. https://doi.org/10.1017/s0140525x00077104

Dawkins, M.S. (1998) *Through Our Eyes Only. The Search for Animal Consciousness*. Oxford University Press, Oxford.

Dawkins, M.S. (2000) *Animal minds and animal emotions. American Zoologist*, 40, 883–888. https://doi.org/10.1093/icb/40.6.883

De Waal, F.B.M. (2011) *What is an animal emotion? Annals of the New York Academy of Sciences*, 1224, 191–206. https://doi.org/10.1111/j.1749-6632.2010.05912.x

Dennett, D.C. (1996) *Kinds of Minds. Towards and Understanding of Consciousness*. Basic Books, New York.

Denton, D.A., Mckinley, M.J., Farrell, M. and Egan, G.F. (2009) *The role of primordial emotions in the evolutionary origin of consciousness. Consciousness and Cognition*, 18, 500–514. https://doi.org/10.1016/j.concog.2008.06.009

Diener, E., Oishi, S. and Lucas, R.E. (2003) *Personality, culture, and subjective well-being: Emotional and cognitive evaluations of life. Annual Review of Psychology*, 54, 403–425. https://doi.org/10.1146/annurev.psych.54.101601.145056

Dijksterhuis, A. (2004) *Think different: The merits of unconscious thought in preference development and decision making. Journal of Personality and Social Psychology*, 87, 586–598. https://doi.org/10.1037/0022-3514.87.5.586

Drevets, W.C., Price, J.L. and Furey, M.L. (2008) *Brain structural and functional abnormalities in mood disorders: Implications for neurocircuitry models of depression. Brain Structure and Function*, 213, 93–118. https://doi.org/10.1007/s00429-008-0189-x

Duncan, I.J.H. (1993) *Welfare is to do with what animals feel. Journal of Agricultural and Environmental Ethics*, 6, 8–14.

Duncan, I.J.H. (2006) *The changing concept of animal sentience. Applied Animal Behaviour Science*, 100, 11–19. https://doi.org/10.1016/j.applanim.2006.04.011

Eccleston, C. and Crombez, G. (1999) *Pain demands attention: A cognitive-affective model of the interruptive function of pain. Psychological Bulletin*, 125, 356–366. https://doi.org/10.1037/0033-2909.125.3.356

Ekman, P. (1992) *An argument for basic emotions. Cognition and Emotion*, 6, 169–200. https://doi.org/10.1080/02699939208411068

Ekman, P. and Cordaro, D. (2011) *What is meant by calling emotions basic. Emotion Review*, 3, 364–370. https://doi.org/10.1177/1754073911410740

EU (1997) Treaty of Amsterdam. Luxembourg: Office for Publications of the European Communities. Available at: https://eur-lex.europa.eu/eli/treaty/tec_2006/pro_33/oj/eng

Farah, M.J. (2008) *Neuroethics and the problem of other minds: implications of neuroscience for the moral status of brain-damaged patients and nonhuman animals. Neuroethics*, 1, 9–18. https://doi.org/10.1007/s12152-008-9006-8

Forbes, E.E., Hariri, A.R., Martin, S.L., Silk, J.S., Moyles, D.L., Fisher, P.M., Brown, S.M., Ryan, N.D., Birmaher, B., Axelson, D.A. and Dahl, R.E. (2009) *Altered striatal activation predicting real-world positive affect in adolescent major depressive disorder. American Journal of Psychiatry*, 166, 64–73. https://doi.org/10.1176/appi.ajp.2008.07081336

Forgas, J.P. (1995) *Mood and judgment – the affect infusion model (AIM). Psychological Bulletin*, 117, 39–66. https://doi.org/10.1037/0033-2909.117.1.39

Foubert, L., Noël, Y., Spahr, C.M. and Slavich, G.M. (2021) *Beyond WEIRD: Associations between socioeconomic status, gender, lifetime stress exposure, and depression in Madagascar. Journal of Clinical Psychology*, 77, 1644–1665. https://doi.org/10.1002/jclp.23131

Frank, E., Tu, X.M., Anderson, B., Reynolds, C.F., Karp, J.F., Mayo, A., Ritenour, A. and Kupfer, D.J. (1996) *Effects of positive and negative life events on time to depression onset: An analysis of additivity and timing. Psychological Medicine*, 26, 613–626. https://doi.org/10.1017/s0033291700035686

Garcia, J., Kimeldorf, D.J. and Koelling, R.A. (1955) *Conditioned aversion to saccharin resulting from exposure to gamma radiation. Science*, 122, 157–158. https://doi.org/10.1126/science.122.3160.157

Gasper, K. (2023) *A case for neutrality: why neutral affect is critical for advancing affective science. Affective Science*, 4, 458–462. https://doi.org/10.1007/s42761-023-00214-0

Godfrey-Smith, P. (2017) Other Minds. The Octopus and The Evolution of Intelligent Life, Dublin, Harper Collins.

Gosling, S.D. and John, O.P. (1999) *Personality dimensions in nonhuman animals: A cross-species review. Current Directions in Psychological Science*, 8, 69–75. https://doi.org/10.1111/1467-8721.00017

Gray, J.A. (1987) *The Psychology of Fear and Stress*, 2nd ed. Cambridge University Press, Cambridge.

Hammen, C. (2005) *Stress and depression. Annual Review of Clinical Psychology*, 1, 293–391. https://doi.org/10.1146/annurev.clinpsy.1.102803.143938

Hankin, B.L. and Abramson, L.Y. (2001) *Development of gender differences in depression: An elaborated cognitive vulnerability-transactional stress theory. Psychological Bulletin*, 127, 773–796. https://doi.org/10.1037/0033-2909.127.6.773

Heim, C., Newport, D.J., Mletzko, T., Miller, A.H. and Hemeroff, C.B. (2008) *The link between childhood trauma and depression: Insights from HPA axis studies in humans. Psychoneuroendocrinology*, 33, 693–710. https://doi.org/10.1016/j.psyneuen.2008.03.008

van Heijst, K., Kret, M. and Ploeger, A. (2023) *Basic emotions or constructed emotions: insights from taking an evolutionary perspective. Perspectives on Psychological Science*. https://doi.org/10.1177/17456916231205186

Higgins, E.T. (1998) *Promotion and prevention: Regulatory focus as a motivational principle. Advances in Experimental Social Psychology*, 30, 1–46. https://doi.org/10.1016/s0065-2601(08)60381-0

Hyde, J.S., Mezulis, A.H. and Abramson, L.Y. (2008) *The ABCs of depression: Integrating affective, biological, and cognitive models to explain the emergence of the gender difference in depression. Psychological Review*, 115, 291–313. https://doi.org/10.1037/0033-295x.115.2.291

Ito, T.A. and Cacioppo, J.T. (2005) *Variations on a human universal: Individual differences in positivity offset and negativity bias. Cognition and Emotion*, 19, 1–26. https://doi.org/10.1080/02699930441000120

Izard, C.E. (2007) *Basic emotions, natural kinds, emotion schemas, and a new paradigm. Perspectives on Psychological Science*, 2, 260–280. https://doi.org/10.1111/j.1745-6916.2007.00044.x

Katz, B.A., Matanky, K., Aviram, G. and Yovel, I. (2020) *Reinforcement sensitivity, depression and anxiety: A meta-analysis and meta-analytic structural equation model. Clinical Psychology Review*, 77, 101842. https://doi.org/10.1016/j.cpr.2020.101842

Kirkwood, J.K. (2007) *Quality of life: The heart of the matter. Animal Welfare*, 16, 3–7. https://doi.org/10.1017/s0962728600031663

Knutson, B. and Cooper, J.C. (2005) *Functional magnetic resonance imaging of reward prediction. Current Opinion in Neurology*, 18, 411–417. https://doi.org/10.1097/01.wco.0000173463.24758.f6

Knutson, B. and Greer, S.M. (2008) *Anticipatory affect: neural correlates and consequences for choice. Philosophical Transactions of the Royal Society B: Biological Sciences*, 363, 3771–3786. https://doi.org/10.1098/rstb.2008.0155

Kragel, P.A. and Labar, K.S. (2016) *Decoding the nature of emotion in the brain. Trends in Cognitive Sciences*, 20, 444–455. https://doi.org/10.1016/j.tics.2016.03.011

Lampert, R., Tuit, K., Hong, K.I., Donovan, T., Lee, F. and Sinha, R. (2016) *Cumulative stress and autonomic dysregulation in a community sample. Stress-the International Journal on the Biology of Stress*, 19, 269–279. https://doi.org/10.1080/10253890.2016.1174847

Larsen, R.J. and Ketelaar, T. (1991) *Personality and susceptibility to positive and negative emotional states. Journal of Personality and Social Psychology*, 61, 132–140. https://doi.org/10.1037//0022-3514.61.1.132

LeDoux, J. (1996) *The Emotional Brain*. Simon and Schuster, New York.

LeDoux, J.E. (2017) *Semantics, surplus meaning, and the science of fear. Trends in Cognitive Sciences*, 21, 303–306. https://doi.org/10.1016/j.tics.2017.02.004

Lindquist, K.A., Wager, T.D., Kober, H., Bliss-Moreau, E. and Barrett, L.F. (2012) *The brain basis of emotion: A meta-analytic review. Behavioral and Brain Sciences*, 35, 121–143. https://doi.org/10.1017/s0140525x11000446

Loewenstein, G.F., Weber, E.U., Hsee, C.K. and Welch, N. (2001) *Risk as feelings. Psychological Bulletin*, 127, 267–286. https://doi.org/10.1037/0033-2909.127.2.267

Low, P., Panksepp, J., Reiss, D., Edelman, D., van Swinderen, B. and Koch, C. (2012) The Cambridge Declaration on Consciousness. *In: Proceedings of the Francis Crick Memorial Conference*, July 7 2012 Churchill College, Cambridge University, United Kingdom.

MacLellan, A., Fureix, C., Polanco, A. and Mason, G. (2021) *Can animals develop depression? An overview and assessment of 'depression-like' states. Behaviour*, 158, 1303–1353. https://doi.org/10.1163/1568539x-bja10132

Mason, G. and Mendl, M. (1993) *Why is there no simple way of measuring animal welfare? Animal Welfare*, 2, 301–319. https://doi.org/10.1017/s0962728600016092

Mason, G.J. and Veasey, J.S. (2010) *How should the psychological well-being of zoo elephants be objectively investigated? Zoo Biology*, 29, 237–255. https://doi.org/10.1002/zoo.20256

Mathews, A. and Macleod, C. (1994) *Cognitive approaches to emotion and emotional disorders. Annual Review of Psychology*, 45, 25–50. https://doi.org/10.1146/annurev.ps.45.020194.000325

Mauss, I. and Robinson, M. (2009) *Measures of emotion: A review. Cognition and Emotion*, 23, 209–237. https://doi.org/10.1080/02699930802204677

McEwen, B.S. (1998) *Stress, adaptation, and disease – allostasis and allostatic load. Annals of the New York Academy of Sciences*, 840, 33–44. https://doi.org/10.1111/j.1749-6632.1998.tb09546.x

McEwen, B.S. and Stellar, E. (1993) *Stress and the individual – mechanisms leading to disease. Archives of Internal Medicine*, 153, 2093–2101. https://doi.org/10.1001/archinte.1993.00410180039004

McNamara, J.M. and Buchanan, K.L. (2005) *Stress, resource allocation, and mortality. Behavioral Ecology*, 16, 1008–1017. https://doi.org/10.1093/beheco/ari087

Mellor, D.J., Beausoleil, N.J., Littlewood, K.E., Mclean, A.N., Mcgreevy, P.D., Jones, B. and Wilkins, C. (2020) *The 2020 Five Domains Model: including human-animal interactions in assessments of animal welfare. Animals*, 10, 1870. https://doi.org/10.3390/ani10101870

Mendl, M., Burman, O.H.P. and Paul, E.S. (2010) *An integrative and functional framework for the study of animal emotion and mood. Proceedings of the Royal Society B: Biological Sciences*, 277, 2895. https://doi.org/10.1098/rspb.2010.0303–2904.

Mendl, M., Neville, V. and Paul, E.S. (2022) *Bridging the gap: human emotions and animal emotions. Affective Science*, 3, 703–712. https://doi.org/10.1007/s42761-022-00125-6

Mendl, M. and Paul, E.S. (2004) *Consciousness, emotion and animal welfare: insights from cognitive science. Animal Welfare*, 13, S17–S25. https://doi.org/10.1017/s0962728600014330

Mendl, M. and Paul, E.S. (2017) *Getting to the Heart of Animal Welfare. The Study of Animal Emotion.* Stichting Animales, Utrecht.

Mendl, M. and Paul, E.S. (2020) *Animal affect and decision-making. Neuroscience and Biobehavioral Reviews*, 112, 144–163. https://doi.org/10.1016/j.neubiorev.2020.01.025

Midgley, M. (1984) *Animals and Why They Matter.* University of Georgia Press, Athens.

Mikels, J.A., Maglio, S.J., Reed, A.E. and Kaplowitz, L.J. (2011) *Should I go with my gut? Investigating the benefits of emotion-focused decision making. Emotion*, 11, 743–753. https://doi.org/10.1037/a0023986

Miner, A.G., Glomb, T.M. and Hulin, C. (2005) *Experience sampling mood and its correlates at work. Journal of Occupational and Organizational Psychology*, 78, 171–193. https://doi.org/10.1348/096317905x40105

Nagel, T. (1974) *What is it like to be a bat? The Philosophical Review*, 83, 435–450. https://doi.org/10.4159/harvard.9780674594623.c15

Nanni, V., Uher, R. and Danese, A. (2012) *Childhood maltreatment predicts unfavorable course of illness and treatment outcome in depression: A meta-analysis. American Journal of Psychiatry*, 169, 141–151. https://doi.org/10.1176/appi.ajp.2011.11020335

Nesse, R.M. (2000) *Is depression an adaptation? Archives of General Psychiatry*, 57, 14–20. https://doi.org/10.1001/archpsyc.57.1.14

Nesse, R.M. (2005) *Natural selection and the regulation of defenses – a signal detection analysis of the smoke detector principle. Evolution and Human Behavior*, 26, 88–105. https://doi.org/10.1016/j.evolhumbehav.2004.08.002

Nettle, D. (2009) *Evolution and Genetics for Psychology.* Oxford University Press, Oxford.

Nettle, D. and Bateson, M. (2012) *The evolutionary origins of mood and its disorders. Current Biology*, 22, R712–R721. https://doi.org/10.1016/j.cub.2012.06.020

New Zealand Government (1999) *Animal Welfare Act 1999*. Available at: https://www.legislation.govt.nz/act/public/1999/0142/latest/DLM49664.html?search=ts_act%40bill%40regulation%40deemedreg_animal+welfare_resel_25_a&p=1

Nguyen, D., Naffziger, E.E. and Berridge, K.C. (2021) *Positive affect: nature and brain bases of liking and wanting. Current Opinion in Behavioral Sciences*, 39, 72–78. https://doi.org/10.1016/j.cobeha.2021.02.013

Norman, G.J., Norris, C.J., Gollan, J., Ito, T.A., Hawkley, L.C., Larsen, J.T., Cacioppo, J.T. and Berntson, G.G. (2011) *Current emotion research in psychophysiology: the neurobiology of evaluative bivalence. Emotion Review*, 3, 349–359. https://doi.org/10.1177/1754073911402403

Norris, C.J., Gollan, J., Berntson, G.G. and Cacioppo, J.T. (2010) *The current status of research on the structure of evaluative space. Biological Psychology*, 84, 422–436. https://doi.org/10.1016/j.biopsycho.2010.03.011

Oatley, K. and Johnson-Laird, P.N. (2014) *Cognitive approaches to emotions. Trends in Cognitive Sciences*, 18, 134–140. https://doi.org/10.1016/j.tics.2013.12.004

Öhman, A. and Mineka, S. (2001) *Fears, phobias, and preparedness: Toward an evolved module of fear and fear learning. Psychological Review*, 108, 483–522. https://doi.org/10.1037/0033-295x.108.3.483

Özten, E., Sayar, G.H., Eryilmaz, G., Kagan, G., Isik, S. and Karamustafalioglu, O. (2015) *The relationship of psychological trauma with trichotillomania and skin picking. Neuropsychiatric Disease and Treatment*, 11, 1203–1210. https://doi.org/10.2147/ndt.s79554

Panksepp, J. (1998) *Affective Neuroscience*. Oxford University Press, Oxford.

Panksepp, J. (2005) *Affective consciousness: Core emotional feelings in animals and humans. Consciousness and Cognition*, 14, 30–80. https://doi.org/10.1016/j.concog.2004.10.004

Panksepp, J. (2007) *Neurologizing the psychology of affects how appraisal-based constructivism and basic emotion theory can coexist. Perspectives on Psychological Science*, 2, 281–296. https://doi.org/10.1111/j.1745-6916.2007.00045.x

Paul, E.S., Harding, E.J. and Mendl, M. (2005) *Measuring emotional processes in animals: the utility of a cognitive approach. Neuroscience and Biobehavioral Reviews*, 29, 469–491. https://doi.org/10.1016/j.neubiorev.2005.01.002

Paul, E.S., Sher, S., Tamietto, M., Winkielman, P. and Mendl, M.T. (2020) *Towards a comparative science of emotion: Affect and consciousness in humans and animals. Neuroscience and Biobehavioral Reviews*, 108, 749–770. https://doi.org/10.1016/j.neubiorev.2019.11.014

Pessiglione, M., Schmidt, L., Draganski, B., Kalisch, R., Lau, H., Dolan, R.J. and Frith, C.D. (2007) *How the brain translates money into force: A neuroimaging study of subliminal motivation. Science*, 316, 904–906. https://doi.org/10.1126/science.1140459

Pike, A.C. and Robinson, O.J. (2022) *Reinforcement learning in patients with mood and anxiety disorders vs control individuals a systematic review and meta-analysis. JAMA Psychiatry*, 79, 313–322. https://doi.org/10.1001/jamapsychiatry.2022.0051

Poirier, C., Bateson, M., Gualtieri, F., Armstrong, E.A., Laws, G.C., Boswell, T. and Smulders, T.V. (2019) *Validation of hippocampal biomarkers of cumulative affective experience. Neuroscience and Biobehavioral Reviews*, 101, 113–121. https://doi.org/10.1016/j.neubiorev.2019.03.024

Puterman, E., Lin, J., Krauss, J., Blackburn, E.H. and Epel, E.S. (2015) *Determinants of telomere attrition over 1 year in healthy older women: stress and health behaviors matter. Molecular Psychiatry*, 20, 529–535. https://doi.org/10.1038/mp.2014.70

Reber, A.S. and Baluska, F. (2021) *Cognition in some surprising places. Biochemical and Biophysical Research Communications*, 564, 150–157. https://doi.org/10.1016/j.bbrc.2020.08.115

Riepenhausen, A., Wackerhagen, C., Reppmann, Z.C., Deter, H.C., Kalisch, R., Veer, I.M. and Walter, H. (2022) *Positive cognitive reappraisal in stress resilience, mental health, and well-being: A comprehensive systematic review. Emotion Review*, 14, 310–331. https://doi.org/10.1177/17540739221114642

Robinson, M.J.F. and Berridge, K.C. (2013) *Instant transformation of learned repulsion into motivational "wanting". Current Biology*, 23, 282–289. https://doi.org/10.1016/j.cub.2013.01.016

Rolls, E.T. (2013) *What are emotional states, and why do we have them? Emotion Review*, 5, 241–247. https://doi.org/10.1177/1754073913477514

Rolls, E.T. (2014) *Emotion and Decision-Making Explained*. Oxford University Press, Oxford.

Rosenberg, A.M., Rausser, S., Ren, J.T., Mosharov, E.V., Sturm, G., Ogden, R.T., Patel, P., Soni, R.K., Lacefield, C., Tobin, D.J., Paus, R. and Picard, M. (2021) *Quantitative mapping of human hair greying and reversal in relation to life stress. eLife*, 10, e67437. https://doi.org/10.7554/eLife.67437

Rushen, J. (2003) *Changing concepts of farm animal welfare: bridging the gap between applied and basic research. Applied Animal Behaviour Science*, 81, 199–214. https://doi.org/10.1016/S0168-1591(02)00281-2

Russell, J.A. (2003) *Core affect and the psychological construction of emotion. Psychological Review*, 110, 145–172. https://doi.org/10.1037/0033-295x.110.1.145

Russell, J.A. and Barrett, L.F. (1999) *Core affect, prototypical emotional episodes, and other things called emotion: Dissecting the elephant. Journal of Personality and Social Psychology*, 76, 805–819. https://doi.org/10.1037//0022-3514.76.5.805

Russell, J.A. and Pratt, G. (1980) *A description of the affective quality attributed to environments. Journal of Personality and Social Psychology*, 38, 311–322. https://doi.org/10.1037/0022-3514.38.2.311

Russell, W.M.S. and Burch, R.L. (1959) *The Principles of Humane Experimental Technique*. Methuen and Co. Ltd, London.

Saha, S., Lim, C.C.W., Cannon, D.L., Burton, L., Bremner, M., Cosgrove, P., Huo, Y. and Mcgrath, J.J. (2021) *Co-morbidity between mood and anxiety disorders: A systematic review and meta-analysis. Depression and Anxiety*, 38, 286–306. https://doi.org/10.1002/da.23113

Saker, P., Farrell, M.J., Adib, F.R.M., Egan, G.F., Mckinley, M.J. and Denton, D.A. (2014) *Regional brain responses associated with drinking water during thirst and after its satiation. Proceedings of the National Academy of Sciences of the United States of America*, 111, 5379–5384. https://doi.org/10.1073/pnas.1403382111

Scarantino, A. and De Sousa, R. (2021). Emotion. In: Zalta, E.N. (ed.) *The Stanford Encyclopedia of Philosophy*. Available at: https://plato.stanford.edu/archives/sum2021/entries/emotion/

Schachter, S. and Singer, J. (1962) *Cognitive, social, and physiological determinants of emotional state. Psychological Review*, 69, 379–399. https://doi.org/10.1037/h0046234

Scherer, K. (1984) On the nature and function of emotion: A component process approach. In Scherer, K. R., Ekman, P. (eds) *Approaches to emotion*. Psychology Press, New York, pp. 293–317.

Scherer, K.R. (1999) Appraisal theory. In Dagleish, T., Power, M. (eds) *Handbook of Cognition and Emotion*. John Wiley and Sons Ltd, Chichester, England.

Schrock, J.M., Snodgrass, J.J. and Sugiyama, L.S. (2020) *Lassitude: The emotion of being sick. Evolution and Human Behavior*, 41, 44–57. https://doi.org/10.1016/j.evolhumbehav.2019.09.002

Schultz, W., Dayan, P. and Montague, P.R. (1997) *A neural substrate of prediction and reward. Science*, 275, 1593–1599. https://doi.org/10.1126/science.275.5306.1593

Schwarz, N. and Clore, G. (1983) *Mood, misattribution, and judgements of well-being – informative and directive functions of affective states. Journal of Personality and Social Psychology*, 45, 513–523. https://doi.org/10.1037/0022-3514.45.3.513

Segundo-Ortin, M. and Calvo, P. (2023) *Plant sentience? Between romanticism and denial: science. Animal Sentience*, 33(1). doi: 10.51291/2377-7478.1772

Seymour, B., O'doherty, J.P., Koltzenburg, M., Wiech, K., Frackowiak, R., Friston, K.J. and Dolan, R. (2005) *Opponent appetitive-aversive neural processes underlie predictive learning of pain relief. Nature Neuroscience*, 8, 1234–1240. https://doi.org/10.1038/nn1527

Seymour, B., Singer, T. and Dolan, R. (2007) *The neurobiology of punishment. Nature Reviews Neuroscience*, 8, 300–311. https://doi.org/10.1038/nrn2119

Sharot, T. (2011) *The optimism bias. Current Biology*, 21, R941–R945. https://doi.org/10.1016/j.cub.2011.10.030

Shields, G.S. and Slavich, G.M. (2017) *Lifetime stress exposure and health: a review of contemporary assessment methods and biological mechanisms. Social and Personality Psychology Compass*, 11(8), e12335. https://doi.org/10.1111/spc3.12335

Singer, P. (1976) *Animal Liberation: A New Ethics for our Treatment of Animals*. Cape, London.

Strunk, D.R., Lopez, H. and Derubeis, R.J. (2006) *Depressive symptoms are associated with unrealistic negative predictions of future life events. Behaviour Research and Therapy*, 44, 861–882. https://doi.org/10.1016/j.brat.2005.07.001

Sullivan, P.F., Neale, M.C. and Kendler, K.S. (2000) *Genetic epidemiology of major depression: Review and meta-analysis. American Journal of Psychiatry*, 157, 1552–1562. https://doi.org/10.1176/appi.ajp.157.10.1552

Susskind, J.M., Lee, D.H., Cusi, A., Feiman, R., Grabski, W. and Anderson, A.K. (2008) *Expressing fear enhances sensory acquisition. Nature Neuroscience*, 11, 843–850. https://doi.org/10.1038/nn.2138

Tennant, C. (2002) *Life events, stress and depression: a review of recent findings. Australian and New Zealand Journal of Psychiatry*, 36, 173–182. https://doi.org/10.1046/j.1440-1614.2002.01007.x

Tomarken, A.J., Cook, M. and Mineka, S. (1989) *Fear-relevant selective associations and covariation bias. Journal of Abnormal Psychology*, 98, 381–394. https://doi.org/10.1037//0021-843x.98.4.381

Vytal, K. and Hamann, S. (2010) *Neuroimaging support for discrete neural correlates of basic emotions: a voxel-based meta-analysis. Journal of Cognitive Neuroscience*, 22, 2864–2885. https://doi.org/10.1162/jocn.2009.21366

UK Government (2022). *Animal Welfare (Sentience) Act 2022*. Available at: https://www.legislation.gov.uk/ukpga/2022/22

Walker, M.D., Duggan, G., Roulston, N., Van Slack, A. and Mason, G. (2012) *Negative affective states and their effects on morbidity, mortality and longevity. Animal Welfare*, 21, 497–509. https://doi.org/10.7120/09627286.21.4.497

Watson, D., Wiese, D., Vaidya, J. and Tellegen, A. (1999) *The two general activation systems of affect: Structural findings, evolutionary considerations, and psychobiological evidence. Journal of Personality and Social Psychology*, 76, 820–838. https://doi.org/10.1037/0022-3514.76.5.820

Webb, L.E., Veenhoven, R., Harfeld, J.L. and Jensen, M.B. (2019) *What is animal happiness? Annals of the New York Academy of Sciences*, 1438, 62–76. https://doi.org/10.1111/nyas.13983

Wittgenstein, L. (1953) *Philosophical Investigations*. Blackwell, Oxford.

Yarrington, J.S., Metts, A.V., Zinbarg, R.E., Nusslock, R., Wolitzky-Taylor, K., Hammen, C.L., Kelley, N.J., Bookheimer, S. and Craske, M.G. (2023) *The role of positive and negative aspects of life events in depressive and anxiety symptoms. Clinical Psychological Science*, 11, 910–920. https://doi.org/10.1177/21677026221141654

Yeates, J.W. (2011) *Is 'a life worth living' a concept worth having? Animal Welfare*, 20, 397–406. https://doi.org/10.1017/s0962728600002955

Yik, M.S.M., Russell, J.A. and Barrett, L.F. (1999) *Structure of self-reported current affect: Integration and beyond. Journal of Personality and Social Psychology*, 77, 600–619. https://doi.org/10.1037/0022-3514.77.3.600

Zhang, B., Ma, S., Rachmin, I., He, M., Baral, P., Choi, S., Gonçalves, W.A., Shwartz, Y., Fast, E.M., Su, Y.Q., Zon, L.I., Regev, A., Buenrostro, J.D., Cunha, T.M., Chiu, I.M., Fisher, D.E. and Hsu, Y.C. (2020) *Hyperactivation of sympathetic nerves drives depletion of melanocyte stem cells. Nature*, 577, 676–681. https://doi.org/10.1038/s41586-020-1935-3

Measuring the Unmeasurable: The Construct Validation of Affective State Indicators

2

Georgia J. Mason[1] and Michael T. Mendl[2]

[1]Department of Integrative Biology & Campbell Centre for the Study of Animal Welfare, University of Guelph, Guelph, Ontario, Canada
[2]Bristol Veterinary School, University of Bristol, Bristol, UK

Introduction

Chapter 1 reviewed the constructs known as 'affective states', including how they are caused by fitness-relevant stimuli from the body and external world, and how their subjective, valenced components are private, yet accompanied by measurable changes in behaviour, cognition, physiology and potentially even health. Building on this, here we introduce the book's recurring theme: identifying such changes for use as accurate proxies (or *indicators*) of animal feelings. Section 1 reviews how to identify such variables, via five well-accepted tests for *construct validation.* Section 2 digs into two further attributes of useful, valid indicators: *responsiveness* and *selectivity*. **Responsiveness** reflects an indicator's ability to sensitively react to all relevant changes in affect (even small, subtle ones): important because without this, we will fail to detect some changes in affective states even when they are present. **Selectivity** reflects the specificity of an indicator to the particular affective states we wish to assess: equally important because without this, we risk mistakenly inferring changes in affective states that are *not* really present.

A perfect indicator of affect would thus be highly responsive to all changes in the states we wish to assess (never missing any) *and* highly selective to them (never responding to anything else). Now, such an indicator does not exist, and probably never will. But the good news is that through careful validation, we can still appreciate the strengths and weaknesses of different indicators *a priori*, so as to choose the best (or best combination) for any given welfare assessment task and interpret it wisely. To quote Dawkins (1980), this is like "*Be[ing] aware of the limitations of your materials before you start building a house*". Section 2 shows how to do this, aided with examples and some heuristic figures – graphs designed to illustrate how affect indicators behave or misbehave – that will be used throughout the book.

As we wrote this chapter, we were inspired by many thoughtful previous publications on how to make accurate inferences about animals' affective states, not just from welfare scientists but also neuroscientists, veterinary researchers and others: publications we pull together for you here. And we found many great examples of good construct validation in animal welfare science (something our co-authors demonstrate throughout the book, too). Yet despite this, we noticed that construct validation is sometimes skipped by animal welfare researchers. This could reflect pressures to publish, as well as the urgency of animal welfare problems (and sometimes perhaps too the seductive appeal of the loaded terms that can appear in welfare research, such as "stress hormones", "enrichment" and "optimism"). Omitting proper construct validation, however, risks welfare research being confusing to read, unconvincing to sceptics, and worst of all, perhaps actually wrong about the very animals it aims to help (something also argued powerfully by Watters et al. 2021). We revisit these important topics at the end of this chapter, and again in Chapter 14. But here, we focus on what good construct validation looks like. Done well, this equips scientists for the enormous challenge of assessing states in animals that cannot be measured directly, and allows them to be objective about assessing the subjective.

Section 1: Construct validation for welfare researchers

Introducing the construct validation of indicators of animal affect

Construct validation is the careful, logical process of checking that data collected to evaluate something conceptual (such as a hypothesised internal process or state) do indeed *accurately* reflect the concept of interest. It maps onto what Browning (2022a) terms *correctness*: "*the degree to which the measure will give the right results . . . numbers that really do reflect the welfare as experienced by the animals*". Construct validation involves carefully considering questions like "What other kinds of evidence can you bring to support your interpretation?" (Cronbach and Meehl 1955). More formally, the hypothesis '*these data accurately reflect construct X*', where X is, say, severe negative affect, should be able to generate clear predictions (for instance about group differences in a particular variable, or about how certain measures covary). Testing these predictions is the process of construct validation. If this validation is successful, the choice and use of a particular indicator will land *accurately* on the metaphorical bullseye: the affective state of interest (see Figure 2.1).

Let's illustrate with an imaginary scenario. Some intelligent Martians suspect that upward-curled mouths in humans indicate positive affect, and that downward-curled mouths indicate the opposite: a hypothesis represented by Figure 2.2: the first of the heuristic graphs developed here (and throughout the book; see Box 2.1 for details). What predictions might this hypothesis make, that allow the Martians to test their idea? The most obvious might be the following: *Prediction 1: humans with upward-curling mouths self-report feeling positive, while humans with downward-curled mouths self-report feeling negative.* But, having read Chapter 1, our alien experimenters also propose the following: *Prediction 2: humans will seek out stimuli that elicit upwards mouth curvature, but avoid stimuli that elicit downwards mouth curvature*; *Prediction 3: stimuli that would ancestrally have promoted human fitness (e.g. healthy, receptive mates) cause upwards mouth curvature, while those that would decrease fitness (like bodily injury, or infant deaths) cause downwards mouth curvature.* And having acquired some knowledge of both human writing and human neurophysiology, the aliens might also propose *Prediction 4: humans given drugs that they label "for pain relief" or "to treat anxiety" will be less likely to show downward curved mouths.* The

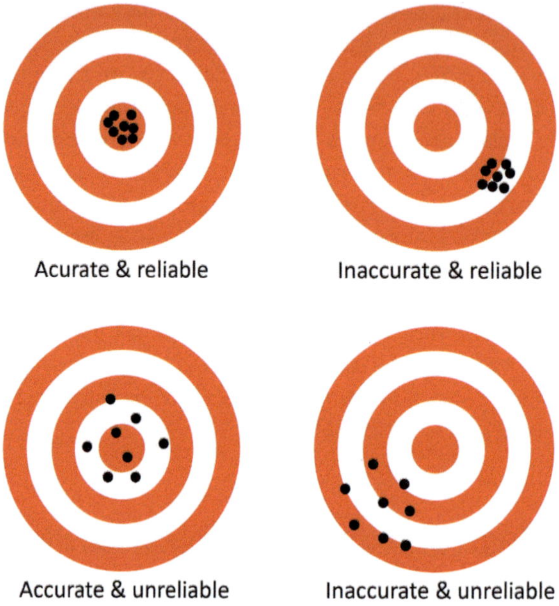

Figure 2.1 In these targets, the bullseye at the centre is the affective state we wish to assess (the true state of the animal, or what Mendl et al. 2022 term the 'ground truth'), and the black dots represent data from four potential indicators of affective state. The two indicators on the left are accurate ones: their average values land on the bullseye. Both thus have good construct validity as indicators of this state (though the one at the top is more reliable – i.e. consistent and repeatable – than the one below). The variables on the right, in contrast, lack construct validity: they are inaccurate indicators (even if highly repeatable, as in the reliably inaccurate indicator at the top). Source: Adapted from Precision and Accuracy (2022).

Martians are keen to test more than one prediction because they recognise that none is perfect: each relies on a particular assumption that may or may not be correct (Prediction 1, for example, assumes that people report their feelings truthfully). And the more predictions supported by the data, the more confident can our aliens be that mouth curvature *is* a good proxy for assessing human affective state (Note: if Figure. 2.2 has you anticipating ways in which the aliens might actually be misled by human smiles, don't worry: we revisit this later!).

Similar processes of clear, defensible logic should underlie the construct validation of indicators of animal affect. Furthermore, well-accepted methods for doing this *are* already described in animal welfare publications (e.g. Dawkins 1980; Stafleu et al. 1992; Browning 2023), including some of our own (Mason and Mendl 1993; Mason and Veasey 2010; Walker et al. 2012; Ahloy-Dallaire et al. 2018; Mendl et al. 2022). Here, we pull these together and integrate them with relevant work from other fields, especially: studies of the psychology of human and animal emotion (e.g. Panksepp 2005; Mauss and Robinson 2009; de Waal 2011); veterinary research on the affective impacts of health problems and their treatment (e.g. Kenward et al. 2015; Evangelista et al. 2019) and biomedical experiments aiming to validate lab animal "models" of pain, anxiety, nausea and other affective states clinically important to humans (e.g. Willner 1986; Belzung and Greibel 2001; Andrews and Horn 2006; McNaughton and Zangrossi Jr. 2008; Belzung and Lemoine 2011).

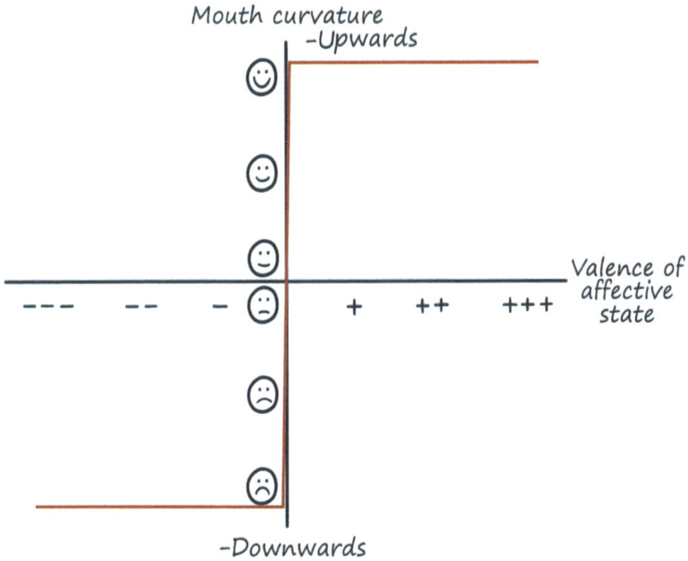

Figure 2.2 A hypothesised simple relationship between human mouth curvature and affect. The *x*-axis indicates the affective valence of the subject, from very negative (– – –) to very positive (+ + +). The *y*-axis represents the state of the indicator that we are considering, here spanning the range from a very downwardly curved mouth to a very upwardly curved mouth (see Box 2.1 for details). The red line indicates a hypothesised relationship between the affect indicator and true affective valence. Here, our Martians are thus hypothesising that the direction of mouth curvature simply reflects whether the human is in a positive or negative state (no matter how subtle). They now want to investigate whether this pattern is correct.

Box 2.1 A short note about our heuristic graphs

Heuristic graphs like Figures 2.2 and 2.4 will be recurring tools throughout the book. It is therefore timely to explain their structure.

First, the value of our measurable variable of interest, the potential indicator (be it behavioural or physiological or something else) will always be on the vertical axis. For some indicators, this vertical axis starts at the intersection with the horizontal valence axis, as no negative indicator values are possible (e.g. cortisol or vocalisations). Correspondingly, the target we are interested in – affective valence, along with intensity of this felt state – is on the horizontal axis. This layout is inspired by typical dose–response curves, as well as by the standard calibration curves used in some assay and instrumentation work. Overall, this depiction thus reflects the focus of this book: how well an indicator is able to capture changes in underlying affective valence.

Second, we depict valence on a cardinal scale (see Browning 2022b), where two units have twice the magnitude of one (e.g. to be at '+ +' on the *x* axis is twice as pleasant as being at '+'). Ideally, it would also be on a 'ratio scale,", such that there is a known *zero point* of neutral affect (although where this is for animals can be, as discussed later in this chapter, hard to assess). Third and relatedly, since subjective experience can never be measured directly, locating animal subjects precisely along the horizontal valence axis will always be a judgment call (albeit one aided by the validation tests we discuss next: Tests 2–5 on the species in question, and Test 1 using human-generated data).

Together, these approaches to construct validation echo the Martians' four tests mentioned earlier, and also generate a fifth one, yielding five key validation tests for a potential indicator to pass or fail. (These five tests will be referred back to in all subsequent chapters of this book, so if you like bookmarks, please insert one here.) The five tests are widely applicable to different situations and species; they are logical and complementary (each testing a different prediction of the underlying hypothesis that the variable of interest correlates with affect); and they can be summarised as follows:

- **Test 1:** *Does the variable of interest accurately track self-reported affective states in humans?*
- **Test 2:** *Does the variable of interest reliably change in animals exposed to stimuli that they prefer/avoid?*
- **Test 3:** *Does the variable of interest reliably change in animals exposed to ancestrally fitness-boosting or fitness-harming situations?*
- **Test 4:** *Does the variable of interest reliably change in animals given affect-modulating drugs?*
- **Test 5:** *Does the variable of interest reliably covary with pre-validated affect indicators?*

None of these tests is perfect: they have different pros and cons, largely because they rely on different assumptions. Subjecting a potential indicator to different validatory tests can therefore increase our confidence that any conclusions are robust: just as for the clever Martians studying humans' faces, the more tests used to validate a putative indicator for animals (and the more it passes), the surer we can be of validity. To illustrate the application of these tests, Table 2.1 uses the example of '*Variable Y*': an imaginary potential indicator of negative affect in cats.

Now, on their own, these five tests may not tell us everything we want to know (even if they successfully validate something like Variable Y for cats), their limitations being:

- They do not demonstrate animal sentience: they take this as a given (the whole process resting on the crucial underlying *assumption* that our species of interest does experience valenced subjective states).
- They do not address an indicator's reliability or repeatability (see Figure 2.1), nor its practical feasibility.
- Without extra thought, they cannot reveal whether a state is a short-term emotion, a longer-term mood or a sign of cumulative affective experience, let alone a state's specific 'feel' (thus Tests 1–5 could not automatically determine whether Variable Y reflects a specific feline emotion like fear, for instance, or the longer-term cumulative effects on a cat of having a tough life).
- And without extra work, they may not tell us much about a valid indicator's responsiveness or selectivity: the key properties that will be reviewed in Section 2.

Nevertheless, these tests are still *essential* for determining whether potential indicators of animal affect do indeed reflect the states we are trying to assess. We therefore look at each test in detail next, before outlining how they can be best used (including being refined and combined to help identify more specific affective states like discrete emotions).

Table 2.1 Testing the hypothesis that Variable Y indicates negative affect in domestic cats (*Felis catus*).

Predictions of this hypothesis[a]	Assumptions that each prediction relies on (*in addition to the assumption that cats are sentient*)
Variable Y passes Test 1: Humans in states of negative affect will show high levels of Variable Y, while humans in states of positive affect will show low levels/nil.	Humans self-report their affective states truthfully. Responses are biologically homologous across species (human and cat).
Variable Y passes Test 2: Cats in aversive situations (ones they would avoid if they could) will show high levels of Variable Y, while cats in preferred situations will show low levels/nil.	Negative states function to motivate avoidance, while positive states function to motivate seeking and approach. Approach indicates relatively positive affect rather than a problem-solving/predator-inspection type response (see later).
Variable Y passes Test 3: Cats in situations that would threaten fitness in their ancestors will show high levels of Variable Y, while cats in situations that would promote fitness in their ancestors will show low levels/nil.	Affective states evolved to promote fitness (such that e.g. fitness-threatening situations induce negative affect that motivate avoidance). We know what would have promoted or harmed fitness in cats' wild ancestors. Domestic cats' brains retain the legacy of these past selection pressures.
Variable Y passes Test 4: Cats given drugs that increase negative affect will show increases in Variable Y, while cats given drugs that decrease negative affect or promote positive affect will show low levels/nil.	We know the affective consequences of various drugs on cats.
Variable Y passes Test 5: Cats displaying other validated signs of negative affect will show increased levels of Variable Y, while cats displaying signs of positive affect will show low levels/nil.	We have other well-validated indicators of affective state for cats. These indicators reflect affective states with similar time courses (e.g. emotions versus moods) and similar ceiling/floor effects (see Section 2).

[a] NB. In our example, the prediction is that Variable Y will increase with increasing negativity of the animal's affective state, but of course some indicators may *decrease* with increasing negativity.

Test 1: Does a variable of interest accurately track self-reported affective states in humans?

Test 1 uses humans as models for understanding other species, by assessing which variables (e.g. changes in cognition, immunology, and hormone levels) track self-reported affective states in people. It then treats these changes as indicators of similar states in other species. A potential indicator thus passes Test 1 if, in humans, it covaries with the state we wish to assess in non-human animals.

This approach is very common. Many indicators of nausea in dogs, for instance (from behavioural signs like hyper-salivation and groaning, to blood biomarkers like high levels of Substance P), are derived from studies of human nausea (Kenward et al. 2015), and the following chapters show many further instances. For example, glucocorticoid release in humans self-reporting negative states helped inspire Hans Selye to conduct his animal research on the hypothalamic–pituitary–adrenal (HPA) axis, and this is one reason why these hormones are often used in animal welfare assessment today (see Chapter 5). Likewise, humans with depression often report a loss of pleasure, no longer enjoying things they used to; tests for this 'anhedonia' have therefore been developed for animals to infer similar depressive states in them (e.g. Willner 1997, MacLellan et al. 2022; see Chapter 9).

Test 1 has many advantages. It capitalises on the huge amount of research on human affect: experimental studies inducing states in the lab, along with opportunistic studies of people experiencing affective disorders (e.g. clinical anxiety) and negative events (from divorce and bereavement, to extreme traumas like the Holocaust). Furthermore, we can be pretty certain of the valence and intensity of these subjects' affective states, thanks to their verbal self-report and our own abilities to empathise. Thus, looking at Figure 2.2, we can locate our human subjects accurately on the *X* axis, confident as to whether they lie to the right or left of neutral, and to what extent.

Test 1 does have some disadvantages, however. First, it relies on biological homology – the idea that affective states have similar correlates in humans and other species due to shared evolutionary history. This assumption is fair for many evolutionarily ancient responses, but can break down for evolutionarily newer ones (like species-typical signals) and with increasing phylogenetic distance between ourselves and the species of interest. Thus, to revisit the cats in Table 2.1, Test 1 could never validate ear retraction as indicators of feline affect, for example, because we humans – being largely unable to move our ears – simply never use these signals. Test 1 could even be anthropomorphically misused, drawing parallels between responses that look superficially similar to certain human ones, yet which actually differ greatly in affective significance (e.g. apparently disgusted-looking nose-wrinkling in cats, when they are merely sampling odours using 'flehmen' responses). Another disadvantage is that there are some limitations in what we know about humans. In particular, the extreme combinations of challenge experienced by certain animals on intensive farms or used in research (e.g. losing the mother early in infancy, followed by long-term barren housing and intermittent painful procedures) have never been studied in humans: a data gap relevant for some potential indicators (e.g. stereotypic behaviour, see Chapter 11).

As we see next, the other four validation methods avoid Test 1's extrapolation of indicators from humans to other animals. Tests 2–5 thus focus *only* on changes observed in the animal species of interest, using cats to understand feline signs of affect, for instance.

Test 2: Does a variable of interest change in animals exposed to stimuli that they prefer/avoid?

In **Test 2**, researchers identify stimuli that animals will approach or work for (Chapter 1's, 'rewards'), and/or aversive stimuli that animals will avoid, or work to prevent or escape from ('punishers'). Animals provided with these rewards are then assumed to be in comparatively more positive states than animals exposed to the identified punishers (assumed to be in comparatively more negative ones). Changes in the behaviour, physiology, cognition and health of such subjects can then be assessed, with those reliably present (e.g. seen across a range of rewards or punishers) passing Test 2.

To outline a few good examples, some researchers have manipulated food availability so as to compare animals given as much food as they prefer with animals given much less. In broiler breeders and calves, for instance, both aggression towards conspecifics and foraging-related oral movements (redirected pecking; non-nutritive sucking) were found to increase in the food-denied subjects, consistent with these responses indicating more negative affect (e.g. Vieira et al. 2008, D'Eath et al. 2009). Other researchers have exposed animals to more or less preferred environments. Blaser and Rosemberg (2012), for instance, assessed zebrafishes' preferences for various tank types, finding that they preferred deep water and black walls over shallow water and white walls. Fish in the least preferred tank types were also found to show behavioural differences from those in the most preferred – more immobility, and thigmotaxis or 'wall-clinging' – so flagging these as potential indicators of more negative affect. As a final, similar illustration, compared with hens in less preferred types of pens, laying hens housed in conditions they prefer showed changes that included reductions in body temperature, blood glucose levels, faecal water content and smaller autonomic and behavioural responses to startling novel stimuli (Nicol et al. 2009, 2011; Ross et al. 2019, 2020): all consistent with these changes being valid indicators of relatively positive affect for chickens.

The main advantage of Test 2 is that it is theory-driven, building on the apparent proximate functions of affect covered in Chapter 1 (e.g. Rolls 2005, 2014; Mendl and Paul 2020; Dawkins 2023; see also Chapter 3). Like Tests 3–5, it is also freed from the limitations of Test 1, allowing the validation of species-unique indicators. However, it also has disadvantages. First, it can only tell us about animals' *relative* affective states (with preferred situations moving them rightwards along the *x* valence axis: see Figure 2.3): not their *absolute* states. That is because, even if animals given preferred resources are usually in more positive states than animals denied these resources, we cannot know where exactly to locate them on the valence scale (see Figure 2.3). Are the former in absolutely positive states (to the right of the origin), while the latter are in absolutely negative ones? Or are they respectively positive and neutral? Neutral and negative? Or even just *less* negative compared to *more* negative? Such uncertainty does not matter if we simply want validated indicators for improving animal welfare, but it *does* if we want something more absolute (such as validated indicators of truly positive affect).[1]

Test 2's second disadvantage of is that sought-after resources might have unintended consequences that, in the longer-term, harm welfare (e.g. sugar rots teeth, opiates cause addiction). This means that unless preference tests are run over enough time for animals to learn to regulate their choices according to longer-term consequences (see e.g. Mason

[1] As we noted in Chapter 1, whether there is a truly 'neutral' point on the valence axis has received limited attention, even in studies of humans (Gasper 2023), but our points remain relevant, especially in the light of current interest in 'positive animal welfare' (Lawrence et al. 2019; Rault et al. 2025),

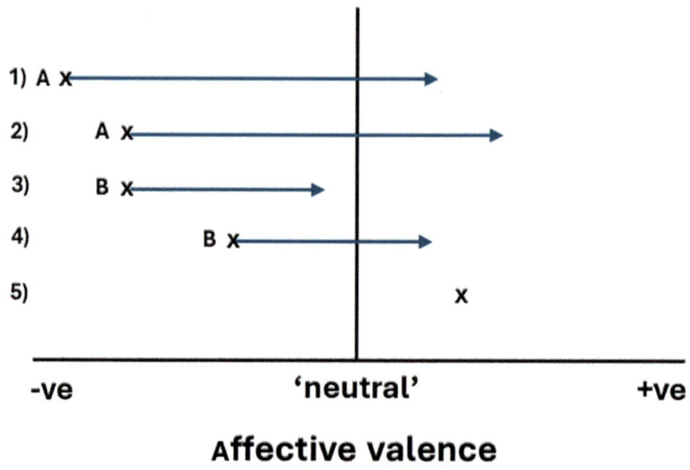

Figure 2.3 Hypothetical effects (examples 1–5) of presenting preferred stimuli (A or B), showing how individuals move from an 'initial' affective valence (x) to a more positive one (the head of each arrow). We may hope or assume that preferred stimuli A and B induce absolutely positive affective states, as in examples 1, 2 and 4 (where the animal is moved to the right of the 'neutral' point on the *x*-axis). But this may not always be the case. For example, the stimuli may not be beneficial enough (compare examples 2 and 3, where stimulus A is preferred enough to generate an absolute positive state but stimulus B is not); and/or the individual's initial valence may be too negative (compare examples 3 and 4 where stimulus B induces an absolutely positive state only in case 4). Furthermore, control animals already in positive initial states, as in example 5, will not be in negative states even if denied the preferred resource: they will merely be less positive than they otherwise might have been. Thus, access to preferred stimuli typically does move animals rightwards along the valence axis, but does not necessarily move them from absolutely negative states to absolutely positive ones.

et al. 2001; Duncan and Young 2004), Test 2 may be better at validating indicators of short-term emotions than longer-term affective states.

The third and final potential disadvantage is that approach behaviours occasionally do not reflect emotions in the ways we assume (as Chapter 3 on preference responses discusses further). Faced with predators, for instance, animals might approach them, *not* because they find predators rewarding, but because (in the non-preferred situation of having predators close by) inspecting predators to gain information is adaptive.

Test 3: Does a variable of interest reliably change in animals exposed to ancestrally fitness-boosting or fitness-harming situations?

Test 3 rests on assumptions that ancestrally, animals evolved to feel negative affect if exposed to treatments that would harm their fitness (see Chapter 1), and that the brains of our subjects still reflect this legacy today. It involves assessing which variables change in animals exposed to ancestrally fitness-boosting or fitness-harming situations. Potential indicators of affect pass Test 3 if they are influenced by these threats to, or promoters of, ancestral fitness.

As one excellent example, Weary and Fraser (1995; 1997) used a knowledge of the fitness benefits of maternal care to reason that underweight, food-deprived or cold piglets would be particularly fitness-threatened by maternal loss (thanks to their increased needs for warmth and milk). When temporarily separated from their mothers, compared to well-fed, warm peers such piglets called more, and used longer, more high-pitched vocalisations: sounds, the authors argued, that therefore accurately signalled these infants' negative states. Würbel and Stauffacher (1997) similarly considered potential fitness implications for wild house mice when inferring the affective significance for young lab mice of being removed permanently from their mothers (termed "*weaning*" in lab animal breeding units). Reasoning that "*because of a higher risk to fitness, both [early] weaning and lower body weight at standard weaning age should be associated with higher stress*", they studied pups removed prematurely from their mothers (either early or at light body weights), for comparison with controls. The prematurely removed pups showed greater corticosterone responses to this separation and went on to develop more stereotypic bar-mouthing, providing evidence for the validity of both responses as indicators of negative affect (topics revisited in Chapters 5 and 11).

One advantage of Test 3 is that, again, it is theory-driven, building on the apparent ultimate functions of affect. Test 3 is thus useful for wild animals and/or species with ancestors whose natural lives we understand. Test 3 can also allow us to learn about indicators from accidental harms to animals (e.g. wounding or disease), or challenging experiences that animals undergo in agricultural operations or research laboratories (e.g. early mother–infant separation, or sometimes deliberate tissue injury). Finally, Test 3 can sometimes help us decide where our animals are located on the valence x axis, not just relatively but absolutely. That's because some experiences would *always*, no matter the degree or context, threaten fitness in the wild (e.g. infection, predation or injury). These should thus not just move animals leftwards along a horizontal valence axis (if we picture Figure 2.2 again), but also always induce *absolutely* negative affect.

However, Test 3 cannot always give us such certainty about absolute valence, for stimuli whose impact on fitness varies with degree. Food availability, for example, can vary from very good to very bad for fitness and thence affective state, and yet exactly where it causes absolutely positive, neutral or absolutely negative valence might be hard to pinpoint (analogous to the problem illustrated in Figure 2.3). Furthermore, Test 3 is problematic if its underlying assumptions cannot be met, for instance if we are unsure what would have influenced ancestral fitness, or our subjects are so domesticated that long-distant evolutionary pressures no longer shape what their brains treat as rewards or punishers. If using Test 3 to try to validate affect indicators in dogs, for example, it would seem rash to treat all factors good for wolf fitness (like being able to hunt deer), as causes of positive affect in today's chihuahuas. Likewise, Test 3 can break down for evolutionarily new stimuli. To illustrate, animals often find veterinary treatments imposed by humans (e.g. inoculations) aversive, even if highly beneficial for health and survival. This contradiction reflects that inoculations are too evolutionarily recent for animals to have adapted to find them affectively positive (instead eliciting evolutionarily old – and ancestrally adaptive – negative responses to tissue damage). Similarly, some lethal situations (like gaseous environments laden with carbon monoxide or lacking in oxygen) will not induce aversion, despite their danger, in species that have not evolved to detect such environments (see e.g. Raj and Gregory 1995; Raj and Mason 1999). In cases like these, where animals find aversive evolutionarily new stimuli that are actually good for their current fitness, or fail to avoid ones that are bad for their current fitness (i.e. where Tests 2 and 3 would make different assumptions about underlying affect), Test 3 is not a good way to validate indicators, and Test 2 should be used instead.

Test 4: Does a variable of interest reliably change in animals given affect-modulating drugs?

Test 4 takes a pharmacological approach, manipulating affective state with drugs to then reveal which measurable variables change in dosed subjects. Potential indicators of affect then pass Test 4 if they are altered by such drugs in consistent ways. Test 4 has been used widely, including in the biomedical world to validate rodent signs of anxiety. For example, an early study by Pellow et al. (1985) on how anti-anxiety drugs affect rat behaviour in an Elevated Plus Maze showed that administering these increases the time that animals spend in the two more aversive open maze arms relative to the two enclosed arms. Analgesic drugs can also validate putative signs of pain (Weary et al. 2006), if those signs decrease or disappear when analgesics are administered. Gleerup et al. (2015) for instance provide one nice example, identifying tooth grinding and head pressing as signs of pain in dairy cows, while Keating et al. (2012) conducted similar research on rabbits, and (amongst others) Evangelista et al. (2019) on cats, to validate facial expressions indicating pain. Further examples of the use of Test 4 include several in the chapters on vocalisations and play (Chapters 4 and 8).

This approach can be elegant and persuasive. It does, however, require knowledge of how a drug influences animals' affective states. This may be inferred from its subjective effects in humans, especially if the mode of action in our species of interest is the same, but beyond such assumptions of homology, additional checks can also be made. One approach is to assess whether animals' willingness to self-administer the drug is consistent with it having the expected valence (e.g. with putative analgesics being sought out by injured animals likely to be in pain), while another is that the drug changes other welfare indicators in predicted ways: the approach taken by Test 5. This approach might also help us with another challenge: that once again (see Figure 2.3), even when we're confident that a drug induces relatively positive affect (so improving welfare), it can be hard to know if it causes an *absolutely* positive state.

Test 5: Does a variable of interest reliably covary with pre-validated affect indicators?

Test 5 involves assessing which variables covary with indicators of affect that have already been validated in our species of interest. A new potential indicator of affect thus passes this test if it covaries with existing valid indicators of states of the same valence.

Several studies, for example, present evidence for the validity of 'grimace' scales for scoring pain by investigating whether and how they correlate with other pre-validated welfare indicators, such as tooth-grinding, lameness and inactive hunched postures in sheep (Häger et al. 2017; Paterson et al. 2023). A related approach, applied at the group level, uses pre-validated indicators to identify manipulations that alter affect and then uses those manipulations to validate new indicators. Thus, some researchers interested in indicators of pain have not only used analgesics to manipulate subjects' affective states (see Test 4), but have also, before drawing conclusions about the validity of any *new* indicators, simultaneously checked that these drugs alter *already*-validated indicators in expected ways (e.g. Viscardi et al. 2017; Paterson et al. 2023). Likewise, in the first study of animal judgement biases, Harding et al. (2004) used isolation plus a treatment based on one traditionally called '*Chronic Mild Stress*', to induce negative affective states in rats. They could be confident that these treatments were negative because many previous studies (e.g. Willner 2017) had found them to produce anhedonia and other changes indicating poor welfare.

Test 5's advantages include the efficient leveraging of previous welfare research. In some instances, Test 5 might also allow the more precise location of our test subjects on the valence axis, if the pre-validated indicators used are specific to absolutely positive or negative states (as is likely for some signals, for instance, or for abnormal repetitive behaviours [see Chapter 11]). However, the validity of this approach of course critically relies on the accuracy of the pre-validated indicators.

Using multiple validatory tests for robust conclusions, and to identify indicators of specific states

Many studies or bodies of work rely on multiple validatory experiments and approaches, not just one. This is important because the validation of a new indicator is greatly strengthened, leading to more defensible conclusions, by (i) the accumulation of complementary supportive findings via different tests (as already mentioned), and also (ii) replication. Replication is important because, as we discuss later, false leads/positives or false nulls/negatives (e.g. Type I or Type II errors) may undermine the conclusions that can be drawn from a single study. To illustrate, Nicol et al. (2009) found that several behavioural and physiological indicators varied between chickens exposed to more versus less preferred housing conditions (so 'passing' Test 1); however, in a partial replication of the study, Nicol et al. (2011) obtained similar findings only for some of these markers (e.g. low faecal water content). We should therefore be wary of concluding from one single study that passing a single test validates an indicator (or that failing a single test invalidates it).

To illustrate the accumulation of complementary lines of evidence for indicator validity, a couple of examples from our own work include that Harding et al. (2004) did not just use Test 5, as described earlier, but also Test 1: knowledge that humans in negative affective states often display negative judgment biases. Likewise, Ross et al. (2019) did not just use Test 2 to validate startle response magnitudes as signs of negative affect in hens; they too also used Test 1: evidence that humans in negative states often show the same 'jumpiness'. And in their exemplary investigation of eye-widening responses in dairy cattle, Sandem et al. (2002, 2004, 2006a, 2006b) and Sandem and Braastad (2005) used *four* methodologies to validate this indicator of negative affect, finding that it passes Test 2 (e.g. responding to an opening umbrella that triggered avoidance), Test 3 (increasing in mothers having their young calves removed), Test 4 (reducing in response to diazepam), and Test 5 (correlating with aggressive butting behaviour in hungry, food-frustrated animals). Furthermore, this team's conclusions have since been supported by Test 1, via work on wide-eyed facial expressions in fearful humans (e.g. Susskind et al. 2008).

Using more than one validatory approach has a further advantage beyond robustness: it can help pinpoint the degree of positive or negative valence, and perhaps even the discrete affective state that the animal is experiencing. Thus if we want to assess whether animals have not just *relatively* good welfare compared to others, but absolutely positive states (the challenge further discussed in Chapters 8 and 9 on play and reward reactivity), we would need to identify indicators that are *unique* to humans in positive states, never occurring during negative ones (to apply Test 1); *never* seen in animals experiencing clear threats to fitness (to apply Test 3); and/or *only* occur in animals dosed with recreational drugs that cause positive affect (to apply Test 4). Likewise, we may wish to assess whether animals have, not just *relatively* poor welfare compared to others, but instead *absolutely* negative states, and perhaps even ones above some threshold (e.g. requiring a "severe" categorisation for research animals; or triggering a euthanasia decision) (Mason, in press). To do this, we would need validated indicators that *only* occur in humans who are suffering greatly (to apply Test 1); that only occur in animals experiencing great threats to fitness (to apply Test 3) and/or that weak analgesics or anxiolytics are not powerful enough to abolish (to apply Test 4).

We might even want yet more precision still, to pinpoint particular sub-types of affect. In Chapter 1, we learned how discrete affective states have distinct causes, and correlate with particular behavioural responses. Here, combinations of Tests 1–4 are therefore essential for any researcher interested in operationalising and distinguishing between such specific states as pain, fear and boredom: emotions that may be equally negative and bad for welfare, and yet which feel and manifest quite differently. For example, animal pain might be operationally defined as a negative affective response to tissue injury that causes avoidance: thus, a correlate of exposure, not to *any* aversive stimulus, but specifically to an aversive stimulus that causes actual physical harm (in a modified, more restrictive version of Test 2). Whether responses to such harms are reduced by known analgesic drugs can then be investigated (thus not reduced by *any* drugs that reduce negative affect, but specifically pain-reducing ones, in a modified version of Test 4). We might also expect pain to cause facial grimacing, 'guarding' behaviours (e.g. limping) and the self-selection of analgesics, based on how humans respond specifically to pain (in a modified version of Test 1).

Section 2: Understanding how indicators of affect can misbehave: issues with responsiveness and selectivity

Tests 1–5 thus allow us to assess the validity of indicators of animal affect. But although these tests are necessary for basic construct validation, on their own they are typically not sufficient for understanding the strengths and weakness of an indicator: they must be built on to fully reveal an indicator's properties, especially its *responsiveness* to and *selectivity* for affect (as defined in the Introduction).

To illustrate, let's return to our Martian experimenters who learned that humans have things we call expressions on the things we call faces. Figure 2.2 represents their initial guide to how affective states relate to mouth curvature. It's easy for us to see how crude this figure is: how limited is its value for human affect assessment. One obvious flaw is that it incorrectly assumes just two simple categorical responses: a big smile during all positive states (no matter how subtle those states really are), and equally strong downward mouth curvature during all negative states (again no matter how mild). Any Martian believing that this figure would then be oblivious to all changes in human affect except those strong enough to cause really large changes in mouth curvature. This would then make them prone to *false null* errors: failures (null findings) to detect certain states even when they're present (here mildly positive or negative ones that only cause small changes in mouth shape). (Some fields would call these 'false negative' conclusions, but we avoid this phrase to reduce confusion with negative affect/ negative valence.)

Luckily, some human research assistants help the Martians understand that peoples' facial expressions are more graded than Figure 2.2 depicts – changing in incremental degrees as valence changes in intensity, and yet not doing so infinitely due to ceiling and floor effects (beyond which mouth curvature cannot become more extreme, no matter how intense the affective state). Drawing the more realistic curve of Figure 2.4a, this now reveals to the Martians how mouth curvature is a more useful indicator than Figure 2.2 suggested, being responsive to varying degrees or intensities of valence.

However, it is clear now that at extremes of affect, mouth curvature becomes a rather blunt instrument (unable to differentiate between different degrees of very positive affect, for example: see the x_1 data points in Figure 2.4a). Furthermore, the Martians learn that mouth curvature can also fail to respond to altered affective state in other ways. Positive emotions

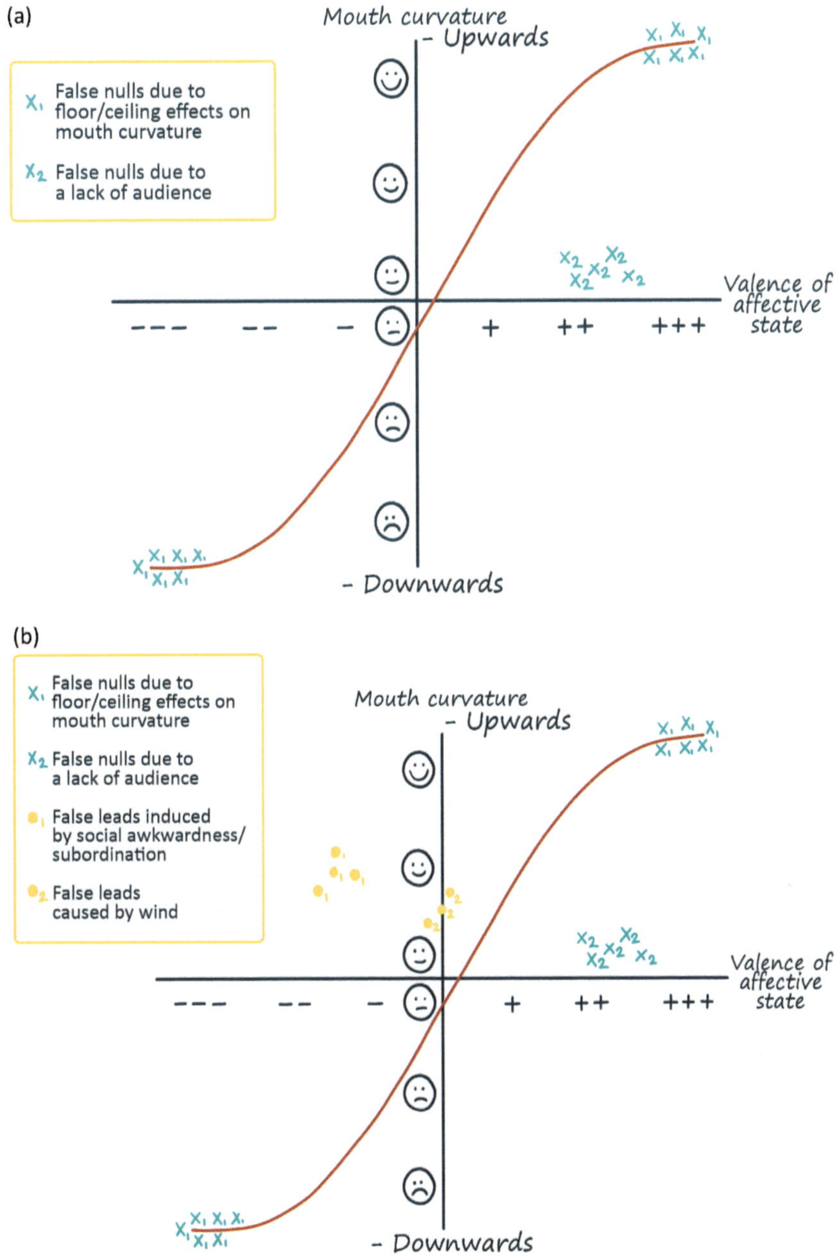

Figure 2.4 (a) A more accurate hypothesised relationship between human mouth curvature and affect (created by the aliens' human research assistants), showing its good responsiveness to varying degrees of affect, but also where using it will lead to false null results (failures to detect altered affect). Two types of false null are shown here. (b) An even more accurate depiction of the relationship between human mouth curvature and affect, showing further ways in which it can mislead.

do not always cause smiling, for instance: smiling may not happen if there is no audience because it is a social signal (Mauss and Robinson2009, Kret et al. 2020; see the x_2 data points in Figure 2.4a). Longer-term positive moods are also not typically accompanied by persistent non-stop broad grins. These cases of poor *responsiveness* to changes in affect, shown by all the crosses in Figure 2.4a, again risk false **nulls** in the Martians' assessment about human states: failures both to detect states and to discriminate between different levels of valence.

And the poor Martians have yet more to learn. Their human assistants reveal that babies with wind sometimes merely look as though they are smiling, and that people can smile in unpleasantly awkward social situations, despite feeling negative affect (e.g. Williams et al. 2020). In other words, an upwards curling mouth is not always *selective* or specific to positive states: its presence can sometimes mislead, because it can sometimes occur in neutral or even negative states too (Figure 2.4b).

What a mess! And yet of course this aspect of facial expression *is* useful for inferring human affect: we humans do it all the time, helped by keeping context in mind and supplementing it with additional indicators. This illustrates the type of nuance we must apply to indicators of animal affect in order to use them well. Construct validation must therefore involve fully understanding how our indicators of animal affect can still be imperfect, even if generally valid.

Understanding poor responsiveness in animal affect indicators: How they can generate false null results

A responsive indicator is thus one that changes in predictable ways as valence intensity changes, and responsiveness is highest in those indicators that are able to detect the smallest changes in valence. Responsiveness can be determined by seeing how the indicator varies in humans self-reporting different intensities of an affective state (to utilise Test 1; e.g. Chapter 5's examples for cortisol responses in stressed humans); in animals exposed to relevant stimuli that differ in how strongly they are preferred or avoided (to utilise Test 2; see Schwartzkopf-Genswein et al. 1997 and Core et al. 2009 for examples); in animals exposed to relevant stimuli differing in their degrees of benefit or threat to ancestral fitness, or aversiveness/degrees of threat to ancestral fitness (to utilise Test 3, Würbel and Stauffacher 1997 providing a nice example, as do Molony et al. 2002); in animals exposed to different doses of drugs that influence this affective state (to utilise Test 4; see e.g. Moloney et al. 2002); and in animals varying in their display of other validated welfare indicators (to use Test 5; e.g. 50 kHz calls and 'affective bias' in rats [Hinchcliffe et al. 2020]). This is summarised for an imaginary indicator of positive affect in Figure 2.5.

A perfect affect indicator would be very responsive to all changes in subjective experience, across both positive and negative valence. Ideally, it would also be equally responsive across the full range of intensities (a property called *linearity* in some fields [e.g. McEnroe et al. 2020]). Thus, a perfect indicator would look like the two variables in Figure 2.6, varying in graded, incremental and proportionate ways across the full range of valence from very negative to very positive. We could then use this indicator to infer not just the presence of an affective state, but also its intensity; and furthermore, we could do so precisely, making fine distinctions even when underlying changes in affect are subtle. Although some may come close over certain ranges, there are no such perfect indicators for animals (the closest, for humans, might be verbal self-report; e.g. Mauss and Robinson 2009). As we will review next, indicators of animal affect never have perfect responsiveness due to a mix of measurement issues (where our tools for quantifying indicators are not perfect), and biological issues that often reflect the indicator's evolved function.

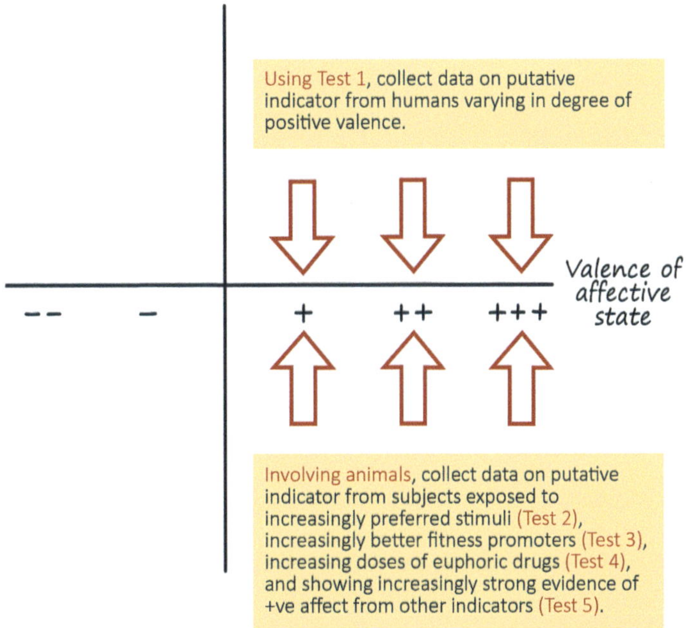

Figure 2.5 Full construct validation involves assessing how responsive an indicator is to affective states varying in intensity (here of positive valence). This process assesses the ability of the indicator to distinguish between even small changes in valence over the full range of interest.

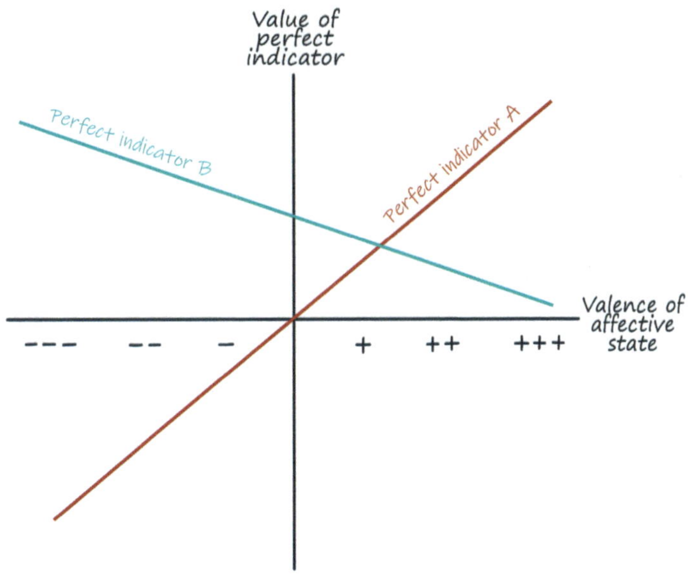

Figure 2.6 Two examples of (imaginary) perfect indicators of affect with great responsiveness across the full range of valence.

Poor responsiveness can manifest in **six main ways:**

i) The indicator reflects only the presence or absence of a valenced state: not its intensity.
ii) The indicator is subject to floor or ceiling effects.
iii) The indicator shows only small effect sizes.
iv) The indicator has threshold effects.
v) The indicator only responds to a subset of valenced experiences, not all of them.
vi) The indicator only changes during certain contexts.

Each of these is expanded below. Understanding and anticipating poor indicator responsiveness is very useful for appreciating when an apparent lack of effect (e.g. in welfare studies comparing different treatments), probably reflects false nulls. It is also useful for selecting the best measures *a priori* when designing studies, thereby reducing the chance of such problems cropping up.

i) *The indicator reflects only the presence or absence of a valenced state*
In this first form of low responsiveness, the indicator is merely the marker of *the presence or absence* of negative or positive states, but reveals nothing about their intensity (just as in our Martians had imagined for human mouth shapes in Figure 2.2). Such markers would be very useful if that's all we need to know, but are limited if we need finer-scaled information.

ii) *Floor or ceiling effects*
A second, common form of poor responsiveness is a *floor or ceiling effect*, where an indicator reaches a maximum or minimum beyond which it cannot change further. Beyond this point, further changes in affect are therefore undetectable using this indicator (Figure 2.4a,b illustrating this for human mouths). Many examples of floor and ceiling effects will occur in the chapters that follow. Furthermore, as indicators *approach* a floor or ceiling, they might also yield small effect sizes as we describe next.

iii) *Small effect sizes lead to Type II errors in statistical tests*
A related form of poor responsiveness is when indicators change only slightly in response to a given change in affect (see Figure 2.7a,b), including cases where responsiveness declines as a floor or ceiling is neared. If the indicator also shows high variability, such changes will be harder to detect statistically, making the indicator prone to Type II errors (false nulls). In a chicken housing study mentioned earlier (Ross et al. 2019), for example, birds in their preferred pens showed both positive judgement biases compared to hens housed in more barren environments and reduced startle responses. However, the effect size (Cohen's D) was only 0.31 ('small') for the former, while 2.4 ('very large') for the latter. This means that statistically significant (but subtle) housing effects on judgement biases would be missed in hens unless dozens of subjects were studied (an issue revisited in Chapter 7).

iv) *Threshold effects (detection limits)*
Some indicators only change if affective states are above a certain intensity or prolonged for a certain duration. Dawkins (1980) warned, for instance, that "*disturbances of physical health . . . are generally agreed to be signs of severe or prolonged suffering. But they may be relatively insensitive indicators of short-term or less acute suffering*". Like other indicators of cumulative affective experience (see Chapters 11–13), these would therefore be inappropriately blunt instruments for assessing the welfare impacts of different gauges of injection needle, say, or

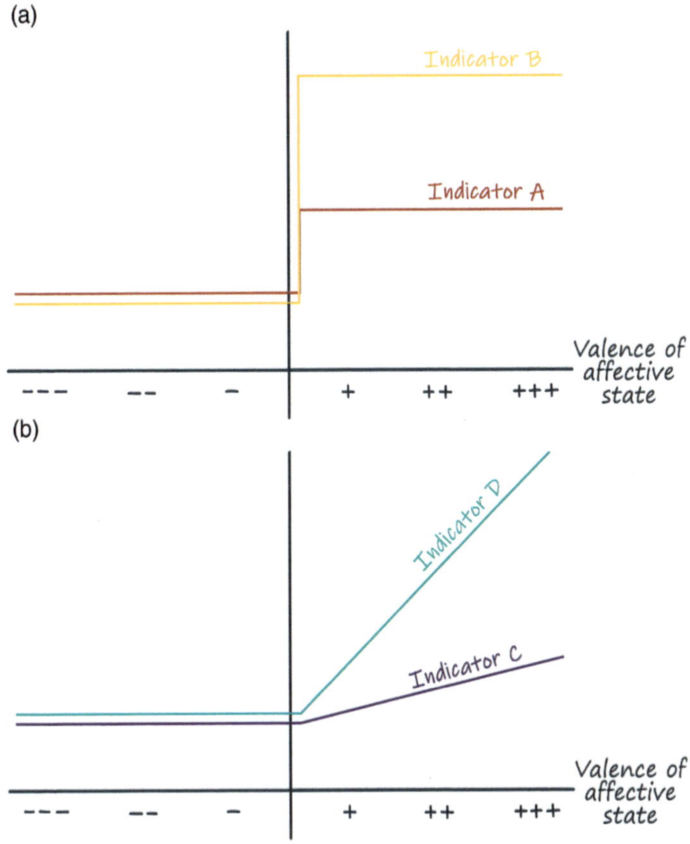

Figure 2.7 Forms of poor responsiveness that risk false nulls due to Type II errors: (a) two indicators that simply respond to the presence/absence of positive affect. Indicator B changes more than does A; assuming similar variance, this will mean smaller effect sizes for indicator A than B, and so a greater tendency to Type II errors when statistical tests are run. This reduced responsiveness is akin to low 'sensitivity' as used in epidemiology and signal detection fields; (b) two indicators that instead change in a graded, incremental way with increasing positive affect. Indicator D has a steeper slope than C; assuming similar variance, this will mean smaller effect sizes for indicator C than D, and so again a greater tendency to get Type II errors when statistical tests are run. This reduced responsiveness is akin to low 'sensitivity' as used in chemistry and some health or biochemical fields (e.g. Liang et al. 1985; Gooding 2019), to low 'precision' in others (e.g. Liang and Jette 1981), and to low 'responsivity' as used in electronics. Source: Adapted from Liang et al. (1985), Gooding (2019), Liang and Jette (1981).

other minor, transient challenges. This type of poor responsiveness thus involves threshold effects (as illustrated by Figure 2.8), or what would be termed *detection limits* in some fields.

v) *The indicator only responds to a subset of valenced experiences, not all of them*

A somewhat related form of poor responsiveness is that many indicators only respond to a subset of valenced experiences, not all. After all, as discussed earlier, humans only smile during some positive emotions, not all positive states. Indicators thus might only respond if the subjective experience is an emotion, or a mood, or more prolonged still (just like the physical

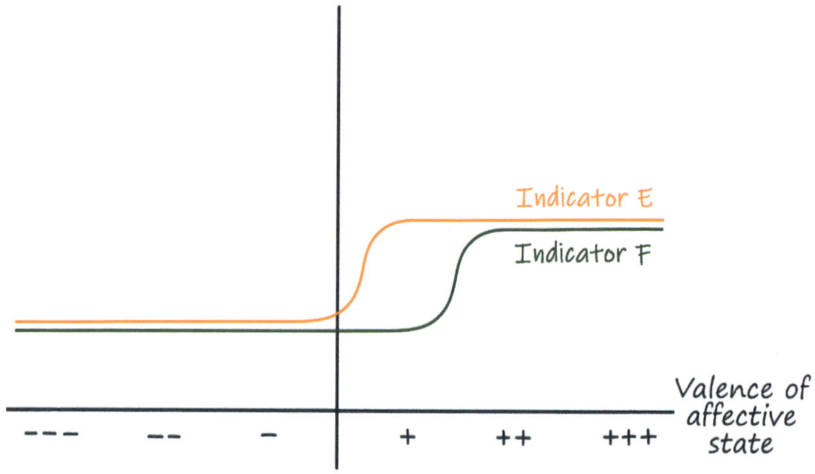

Figure 2.8 A form of low responsiveness reflecting threshold effects: indicator F is less good than indicator E at detecting very low levels of positive affect because it only responds above a certain threshold.

health example above); or they might only respond to certain discrete states, like our pain versus fear examples in the previous section.

vi) *The indicator only responds during particular contexts*
The final form of low responsiveness relates to contextual effects: some indicators might only occur during specific situations. After all, as we have discussed, humans smile more if there is an audience. Whether or not animals flee from danger is another example: fleeing is an obvious sign of aversion, but in some cases (e.g. the predator is very close and/or very fast), hiding or freezing might be more adaptive and so displayed instead (e.g. Mobbs et al. 2007). And whether corticosteroids or heart rates increase during negative emotions will depend on whether active physical responses (the types of activity these physiological responses are designed to support) are adaptive (see Chapters 5 and 6).

Understanding poor selectivity in animal affect indicators: How they can generate false leads

Selectivity is the second important property of indicators: the degree to which "*indicators are measuring the intended target state, and not some other*" (Browning 2023). Like *specificity* in epidemiology and signal detection theory, a perfect indicator is highly selective: uniquely responsive to changes in affective valence (or one sub-type if that is our focus). A non-selective indicator, in contrast, responds to other phenomena too, and so risks what some fields would call 'false positive' conclusions, and what we (to avoid confusion with positive valence) call 'false leads'. For example, earlier we saw that upwards mouth curvature in humans is not specific to positive affect, such that our Martian experimenters might be misled by the appeasing smiles of adults in uncomfortable social situations (see Figure 2.4b), mistakenly concluding that positive affect is present when it is not. Likewise, in animal welfare research, many indicators have problems with selectivity. Acute increases in glucocorticoids, for example, have been known since the early research of Selye to reflect both 'distress' (negative) and 'eustress' (positive): respectively *negative* responses to punishers and *positive* ones to rewards (see Chapter 5).

Just as with imperfect responsiveness, however, imperfect selectivity need not be a problem as long as we understand it in advance and can mitigate its effects, perhaps via well-chosen experimental controls or other techniques. Understanding selectivity is therefore also very helpful for interpreting research results, and for choosing the best *combinations* of measures for answering a given welfare question when designing a project. To do this, understanding selectivity means asking of any valid indicator, does it *only* increase in response to the valanced states one thinks (or hopes) it does? Or does it change similarly during non-affective states as well, or even during states of the opposite valence? Investigating this is a little harder than investigating responsiveness, but we suggest the following as the main causes of poor selectivity:

i) The indicator reflects arousal and intensity, but not the direction of valence.
ii) The indicator is influenced by other factors unrelated to affect.
iii) The indicator reflects the direction of valence, yet we are only interested in one discrete affective state (e.g. fear, but not pain or boredom).
iv) The indicator can increase or decrease during a particular state, such that its levels provide ambiguous information about what state the individual is in (because high values could mean either relatively good welfare *or* relatively bad welfare).

In the following, we expand on each of these.

i) *When indicators reflect arousal and intensity, but not the direction of valence*
One common reason for non-selectivity is that an indicator actually reflects arousal (as we will see for acute HPA and sympatho-adrenal responses in Chapters 5 and 6), increasing with general physiological activation and/or physical activity during both positive *and* negative states. For high-arousal affective states, such indicators therefore tend to show V- or U-shaped relationships with the intensity of valence (or potentially n-shaped), giving their strongest measurements during both high intense positive *and* negative states (see Figure 2.9), and, importantly, their lowest values at neutral (such that the V, U or n is symmetrical around the origin in this graphic).

Such effects make it always useful to check whether a presumed indicator of affective state responds similarly to high-arousal positive *and* negative states. This can be done by referring to the human literature (to apply Test 1); by observing animals when experiencing preferred as well as aversive stimuli, especially ones that elevate physical activity (to apply Test 2); by exposing animals to high-arousal promoters of ancestral fitness as well as to high-arousal threats (to apply Test 3); and so on.

If an indicator *does* increase in high-arousal states of both valences, then understanding this will help us not be misled when high levels reflect a differently valenced state to the one we are aiming to assess. Such indicators also need to be complemented with measures which *are* valence-specific (such as many of the signals of Chapter 4), and so indicate which 'side of neutral' we are on. Armed with this extra information, these indicators can then be used validly and well (as we will see in Chapters 5 and 6).

ii) *When indicators are influenced by other factors unrelated to affect*
Poor selectivity can also occur when indicators are changed by factors completely unrelated to affect. Examples could include temporal variables such as time of day or season, and demographic variables such as sex, age and genotype. Some indicators may also be mediated by mechanisms that do not require waking consciousness, such that they can

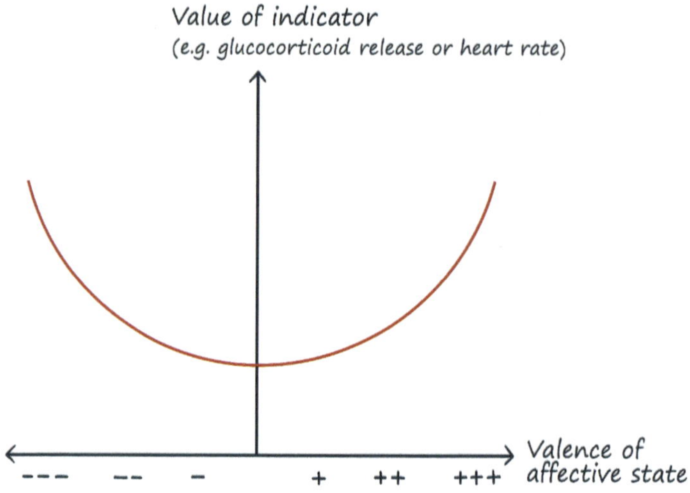

Figure 2.9 Some potential indicators have V- or U-shaped relationships with the intensity of affective valence, increasing in both positively and negatively valenced states (U-shaped responses also reflect threshold effects where small deviations from neutral affect have little impact on the indicator, i.e. at the bottom of the U). Anyone assuming that increases only meant negative affect would therefore be misled, falsely attributing negative affect to subjects who are actually in high-arousal positive states.

occur in insensible animals (e.g. those under anaesthesia) who are experiencing no affective valence at all (as we will again see for acute HPA and sympatho-adrenal responses in Chapters 5 and 6).

iii) *When indicators respond to a range of negative (or positive) states, yet we are only interested in one discrete sub-type*
Sometimes we are only interested in one discrete affective state, such as pain (if assessing animals post-operatively for example). In that case, any indicator that is responsive to other negative states will give us false lead readings if an animal is fearful, say, but our interest is in pain and only pain. (The orbital tightening and flattened ears of mice in pain, for example, are also seen in socially threatening situations: Defensor et al. 2012.)

iv) *When indicators can increase or decrease during a particular state*
This final type of poor selectivity is particularly challenging and may even rule out a potential indicator as being useful. This is when certain indicators can *both* increase *or* decrease during states of a particular valence.

In some of these cases, responses are U-shaped, V-shaped or n-shaped like we met earlier in (ii), but – crucially – they are *not* symmetrical around the neutral point. For example, during sustained adversity, corticosteroid levels can elevate, decrease or be unchanged (varying in different studies [see Chapter 5]), and Guilliams and Edwards (2010) presented this as reflecting a progression over time through different stages of chronic negative affect (Figure 2.10). Similar arguments are presented for reward reactivity in Chapter 9 (see also van der Harst and Spruijt 2007).

Other such indicators that are similarly problematic have a relationship with valence that might best be described as a 'Y on its side' (see Figure 2.11). Thus, during states of

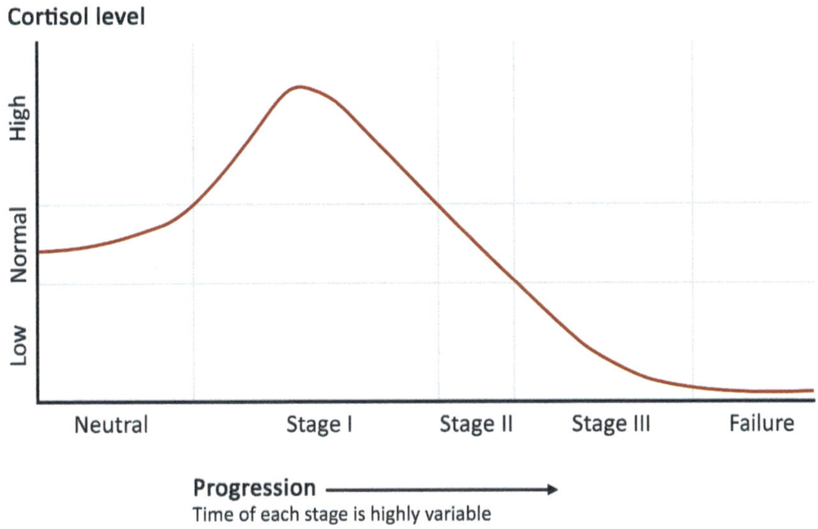

Figure 2.10 How baseline corticosteroid levels change with increasing durations of negative affect. Source: Adapted from Guilliams and Edwards (2010).

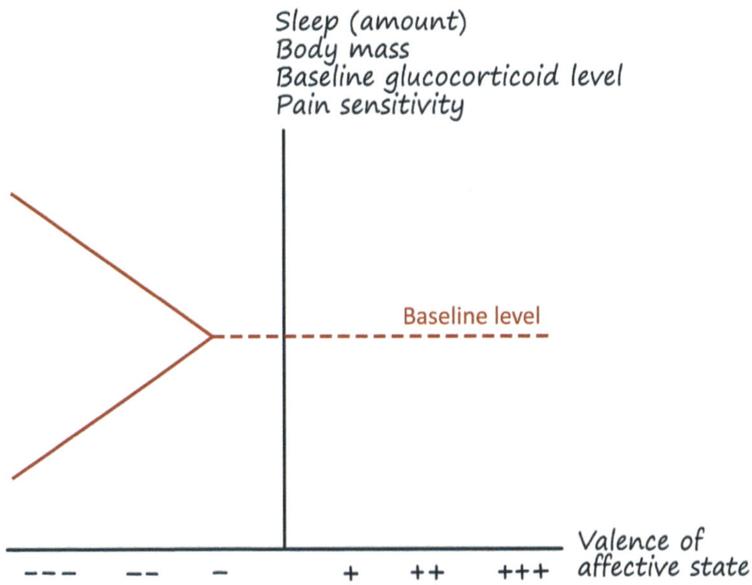

Figure 2.11 Some variables, like the ones shown here, can either increase or decrease as an affective state increases in intensity, resulting in a toppled Y-shaped relationship.

negative affect, for instance, they diverge from baseline, but either upwards or downwards, depending on the individual or the particular circumstances at play. Among humans, body weight is one such example. As Sapolsky (1998) vividly put it: "*People go through a traumatic divorce and balloon up. People go through a traumatic divorce and waste away*" (see also Andrews 2022 on birds). Similar effects occur, for instance, for sleep duration (see e.g. Pawlyk et al. 2008 on hypersomnia and insomnia), pain sensitivity (stress inducing both analgesia and pain hypersensitivity, e.g. Jørum 1988) and lying time in cattle (see Tucker et al. 2021).

All such indicators are potentially dangerous in welfare assessment because an increase in measured values is not specific to one type of affective change: it could either mean an increase in negative affect or a *decrease* in negative effect. This in turn could then lead to highly subjective, even biased interpretations of the data. Such problematic indicators could perhaps become useable if we had good *a priori* ways to identify the baseline levels evident during good welfare, and if we understood *a priori* why an increase or a decrease can occur. But on the whole, we advise against using indicators that have this property.

Conclusions and overview of the book

The complexity of affective states, combined with the Other Minds problem (Chapter 1), presents animal welfare scientists with a special challenge: assessing something that cannot be measured, only inferred. This challenge is important, potentially daunting and unavoidable: we simply must grasp it if we are to do good, accurate research that truly benefits animals and is robust to criticism. Happily, this challenge is also intellectually fascinating, shared with several other disciplines, and long-standing enough to have attracted much thoughtful scholarship. Indeed, that certain behavioural and physiological changes likely reflect animals' affective states has been recognised ever since Darwin, and how to validate such changes as indicators of affective states has attracted much thought, analysis and empirical work in animal welfare science, veterinary medical research, biomedical research and other fields.

Here we therefore integrated previous writings and experiments to collate five good validatory tests that are both complementary and widely used: assessing whether a variable tracks self-reported affective states in humans (our *Test 1*), and, focusing on subjects from the animal species whose welfare is of interest, assessing whether a variable changes in subjects (i) exposed to stimuli that they prefer/avoid (*Test 2*), (ii) exposed to ancestrally fitness-boosting or fitness-harming situations (*Test 3*), (iii) given affect-modulating drugs (*Test 4*) and/or (iv) displaying already-validated, signs of affect (*Test 5*). Used well (as modelled in this book), this process of construct validation is logical, involves clear hypotheses and predictions, and rests on explicit assumptions (e.g. around the adaptive values of affective states, and/or their homology between humans and other species). It also avoids potential pitfalls such as circular reasoning, unjustified anthropomorphism or mistaking reliability for accuracy.

Furthermore, this chapter also showed how such basic construct validation can be refined, to yield indicators that can be used with more accuracy. This can be to identify indicators of particular sub-types of affect (e.g. fear or pain), and also to understand an indicator's responsiveness and selectivity. Responsiveness ideally is high over the full range of valence. However, sometimes it is demonstrably low, and understanding this can help us avoid false nulls that would lead to false conclusions. Likewise, selectivity too ideally is high, yet sometimes is demonstrably low (indicator values changing with states other than the valence we are interested in), and understanding *this* can help avoid another error: being tricked by false leads.

Appreciating these two indicator properties is important because it does far more than just warn when false nulls or false leads are likely: it is constructively useful, suggesting which indicators are best for which 'job' (e.g. for assessing emotions versus moods versus cumulative affect), and which *combinations* of complementary indicators are likely to yield the most accurate information about animals' states. After all, animal welfare science has many excellent affect assessment tools, none of which is perfect. Such imperfections are not fatal flaws (as the following chapters demonstrate) as long as we can understand our indicators' imperfections and turn that into practical guidelines. That is why a heuristic device used throughout the book involves graphs plotting indicator values against valence (loosely based on dose–response or assay calibration curves), in ways designed to convey how various indicators can either be ideal or instead "misbehave." As Dawkins (2021) beautifully put it, "*The problem is not too few measures of welfare, but how to make sense of the many that are available*", and that is our aim in this book.

But why consider all this in such detail, and pull together such a volume, *decades* after welfare science first began? One reason (and something we revisit in Chapter 14) is a growing concern about the conduct of science that applies to welfare science too. Evidence has emerged that many published findings cannot be replicated (as especially well-documented by analysts of psychology and biomedical research), raising troubling questions about the short-cuts researchers may take when under pressure to produce papers (e.g. Tran 2024, Mason 2023). Furthermore, welfare science is arguably especially vulnerable with its challenging remit of measuring the unmeasurable. Browning recently threw down the gauntlet here, arguing that "*many of the methods currently used for estimating animal welfare. . . are subjective, unreliable and unlikely to be useful*" (Browning 2022a). And the scope for researcher subjectivity – and even bias, or at least *perceptions* of bias – is especially concerning given the ethical and political importance of animal welfare research (Browman and Skiftesvik 2011, Phillips and Petherick 2015). But a second reason is our own conviction that construct validity is not just an important problem, but also a tractable one. We obviously need indicators with good construct validity if we are to genuinely help improve animal welfare (e.g. Watters et al. 2021). But much of the hard thinking has already been done – as shown by the thoughtfulness of both the researchers we cite here and the authors of the chapters that follow.

In this vein, the following chapters therefore review how measures of behaviour, physiology and cognition have been validated as markers of affective states. Concentrating on examples from birds and mammals (for whom the assumption of sentience is most widely accepted), these chapters review indicators ranging from those that have been used for many years in animal welfare science to relatively new markers of affective state. They are grouped into those best suited for assessing emotions, those better for assessing longer-term affective states (moods) and those more reflective of the cumulative effects of affective states. These groups of course reflect a spectrum rather than hard and fast biological categories (see Chapter 1), but they do help delineate which type of welfare assessment task an indicator is likely to be best for. Those in Chapters 3–6, for instance, are most well suited for assessing short-term affective responses to particular stimuli; while those in the last group are most useful for assessing the cumulative impact on animals of the entire duration for which they are used by humans, and perhaps even for assessing whether they have had 'lives worth living' (see Yeates 2011). Each chapter exemplifies the validation approaches advocated here. And together, they provide guides for using and interpreting a range of valuable, widely used indicators, and also demonstrate what careful, thoughtful construct validation can and should look like: principles that can be applied to the many useful indicators not covered in this volume.

References

Ahloy-Dallaire, J., Espinosa, J. and Mason, G. (2018) *Play and optimal welfare: Does play indicate the presence of positive affective states? Behavioural Processes*, 156, 3–15. https://doi.org/10.1016/j.beproc.2017.11.011

Andrews, C.P. (2022) *On the use of body mass measures in severity assessment in laboratory passerine birds. Animal Welfare*, 31(3), 387–401. https://doi.org/10.71.0/09627286.31.1.013

Andrews, P.L. and Horn, C.C. (2006) *Signals for nausea and emesis: Implications for models of upper gastrointestinal diseases. Autonomic Neuroscience*, 125(1–2), 100–115. https://doi.org/10.1016/j.autneu.2006.01.008

Belzung, C. and Greibel, G. (2001) *Measuring normal and pathological anxiety-like behaviour in mice: A review. Behavioural Brain Research*, 125, 141–149. https://doi.org/10.1016/S0166-4328(01)00291-1

Belzung, C. and Lemoine, M. (2011) *Criteria of validity for animal models of psychiatric disorders: Focus on anxiety disorders and depression. Biology of Mood & Anxiety Disorders*, 1(1), 1–14. https://doi.org/10.1186/2045-5380-1-9

Blaser, R.E. and Rosemberg, D.B. (2012) *Measures of anxiety in zebrafish (Danio rerio): Dissociation of black/white preference and novel tank test. PLoS One*, 7(5), e36931. https://doi.org/10.1371/journal.pone.0036931

Browman, H.I. and Skiftesvik, A.B. (2011) *Welfare of aquatic organisms: Is there some faith-based HARKing going on here? Diseases of Aquatic Organisms*, 94(3), 255–257. https://doi.org/10.3354/dao02366

Browning, H. (2022a) *Assessing measures of animal welfare. Biology & Philosophy*, 37(4), 36. https://doi.org/10.1007/s10539-022-09862-1

Browning, H. (2022b) *The measurability of subjective animal welfare. Journal of Consciousness Studies*, 29(3–4), 150–179. https://doi.org/10.53765/20512201.29.3.150

Browning, H. (2023) *Validating indicators of subjective animal welfare. Philosophy of Science*, 90(5), 1255–1264. https://doi.org/10.1017/psa.2023.10

Core, S., Widowski, T., Mason, G. and Miller, S. (2009) *Eye white percentage as a predictor of temperament in beef cattle. Journal of Animal Science*, 87(6), 2168–2174. https://doi.org/10.2527/jas.2008-1554

Cronbach, L.J. and Meehl, P.E. (1955) *Construct validity in psychological tests. Psychological Bulletin*, 52(4), 281–302. https://doi.org/10.1037/h0040957

Dawkins, M.S. (1980) *Animal Suffering. The Science of Animal Welfare.* Chapman and Hall, London, UK.

Dawkins, M.S. (2021) *The Science of Animal Welfare: Understanding What Animals Want*, 1st ed. Oxford University Press, Oxford, UK.

Dawkins, M.S. (2023) *Farm animal welfare: Beyond "natural" behavior. Science*, 379(6630), 326–328. https://doi.org/10.1126/science.ade5437

D'Eath, R.B., Tolkamp, B.J., Kyriazakis, I. and Lawrence, A.B. (2009) *'Freedom from hunger' and preventing obesity: the animal welfare implications of reducing food quantity or quality. Animal Behaviour*, 77(2), 275–288. https://doi.org/10.1016/j.anbehav.2008.10.028

Defensor, E.B., Corley, M.J., Blanchard, R.J. and Blanchard, D.C. (2012) *Facial expressions of mice in aggressive and fearful contexts. Physiology & Behavior*, 107(5), 680–685. https://doi.org/10.1016/j.physbeh.2012.03.024

De Waal, F.B. (2011) *What is an animal emotion? Annals of the New York Academy of Sciences*, 1224(1), 191–206. https://doi.org/10.1111/j.1749-6632.2010.05912.x

Duncan, A.J. and Young, S.A. (2004) The effect of pattern of feeding on the development of food preferences in goats in response to positive and negative post-ingestive consequences. In *Poisonous Plants and Related Toxins*. CABI Publishing, Wallingford UK, pp. 526–530.

Evangelista, M.C., Watanabe, R., Leung, V.S., Monteiro, B.P., O'Toole, E., Pang, D.S. and Steagall, P.V. (2019) *Facial expressions of pain in cats: the development and validation of a Feline Grimace Scale. Scientific Reports*, 9(1), 19128. https://doi.org/10.1038/s41598-019-55693-8

Gasper, K. (2023) *A case for neutrality: why neutral affect is critical for advancing affective science. Affective Science*, 4, 458–462. https://doi.org/10.1007/s42761-023-00214-0

Gleerup, K.B., Andersen, P.H., Munksgaard, L. and Forkman, B. (2015) *Pain evaluation in dairy cattle. Applied Animal Behaviour Science*, 171, 25–32. https://doi.org/10.1016/j.applanim.2015.08.023

Gooding, J.J. (2019) *What does ultrasensitive really mean? ACS Sensors*, 4(3), 528–528. https://doi.org/10.1021/acssensors.9b00404

Guilliams, T.G. and Edwards, L. (2010) *Chronic stress and the HPA axis: Clinical assessment and therapeutic considerations. The Standard*, 9(2), 1–12.

Häger, C., Biernot, S., Buettner, M., Glage, S., Keubler, L.M., Held, N., Bleich, E.M., Otto, K., Müller, C.W., Decker, S., Talbot, S.R. and Bleich, A. (2017) *The Sheep Grimace Scale as an indicator of post-operative distress and pain in laboratory sheep. PLoS One*, 12(4), e0175839. https://doi.org/10.1371/journal.pone.0175839

Harding, E.J., Paul, E.S. and Mendl, M. (2004) *Cognitive bias and affective state. Nature*, 427(6972), 312–312. https://doi.org/10.1038/427312a

Hinchcliffe, J.K., Mendl, M. and Robinson, E.S.J. (2020) *Rat 50 kHz calls reflect graded tickling-induced positive emotion. Current Biology*, 30(18), R1034–R1035. https://doi.org/10.1016/j.cub.2020.08.038

Jørum, E. (1988) *Analgesia or hyperalgesia following stress correlates with emotional behavior in rats. Pain*, 32(3), 341–348. https://doi.org/10.1016/0304-3959(88)90046-2

Keating, S.C., Thomas, A.A., Flecknell, P.A. and Leach, M.C. (2012) *Evaluation of EMLA cream for preventing pain during tattooing of rabbits: Changes in physiological, behavioural and facial expression responses. PLoS One*, 7(9), e44437. https://doi.org/10.1371/journal.pone.0044437

Kenward, H., Pelligand, L., Savary-Bataille, K. and Elliott, J. (2015) *Nausea: Current knowledge of mechanisms, measurement and clinical impact. The Veterinary Journal*, 203(1), 36–43. https://doi.org/10.1016/j.tvjl.2014.10.007

Kret, M.E., Prochazkova, E., Sterck, E.H.M. and Clay, Z. (2020) *Emotional expressions in human and non-human great apes. Neuroscience and Biobehavioral Reviews*, 115, 378–395. https://doi.org/10.1016/j.neubiorev.2020.01.027

Lawrence, A.B., Vigors, B. and Sandøe, P. (2019) *What is so positive about positive animal welfare? A critical review of the literature. Animals*, 9, 783. https://doi.org/10.3390/ani9100783

Liang, M.H. and Jette, A.M. (1981) *Measuring functional ability in chronic arthritis. A critical review. Arthritis & Rheumatism*, 24(1), 80–86. https://doi.org/10.1002/art.1780240113

Liang, M.H., Larson, M.G., Cullen, K.E. and Schwartz, J.A. (1985) *Comparative measurement efficiency and sensitivity of five health status instruments for arthritis research. Arthritis & Rheumatism*, 28(5), 542–547. https://doi.org/10.1002/art.1780280513

MacLellan, A., Nazal, B., Young, L. and Mason, G. (2022) *Waking inactivity as a welfare indicator in laboratory mice: Investigating postures, facial expressions and depression-like states. Royal Society Open Science*, 9(11), 221083. https://doi.org/10.1098/rsos.221083

Mason, G. (in press, 2025). Assessing laboratory animal welfare: The crucial importance of construct validity. Lab Animals.

Mason, G.J. (2023) *Animal welfare research is fascinating, ethical, and useful—but how can it be more rigorous? BMC Biology*, 21(1), 302. https://doi.org/10.1186/s12915-023-01793-x

Mason, G.J., Cooper, J. and Clarebrough, C. (2001) *Frustrations of fur-farmed mink. Nature*, 410(6824), 35–36. https://doi.org/10.1038/35065157

Mason, G. and Mendl, M. (1993) *Why is there no simple way of measuring animal welfare? Animal Welfare*, 2(4), 301–319. https://doi.org/10.1017/S0962728600016092

Mason, G.J. and Veasey, J.S. (2010) *How should the psychological well-being of zoo elephants be objectively investigated? Zoo Biology*, 29(2), 237–255. https://doi.org/10.1002/zoo.20256

Mauss, I.B. and Robinson, M.D. (2009) *Measures of emotion: A review. Cognition and Emotion*, 23(2), 209–237. https://doi.org/10.1080/02699930802204677

McEnroe, R.J., Durham, A.P., Kondratovich, M.V., Johansen, J.V., Meyers, P.G., Souers, R.J. and Vaks, J.E. (2020) *Evaluation of Linearity of Quantitative Measurement Procedures*, 2nd ed. Published by Clinical and Laboratory Standards Institute. (https://clsi.org/).

McNaughton, N. and Zangrossi, H. Jr. (2008) Theoretical approaches to the modeling of anxiety in animals. In Blanchard, R. J., Blanchard, D. C., Griebel, G., Nutt, D. (eds) *Handbook of Anxiety and Fear*. Elsevier Academic Press, Amsterdam, pp. 11–27.

Mendl, M.T. and Paul, E.S. (2020) Assessing affective states in animals. In McMillan, F. (ed.) *Mental Health and Well-being in Animals*. CABI Publishing, Wallingford, UK, pp. 328–344.

Mendl, M., Neville, V. and Paul, E.S. (2022) *Bridging the gap: Human emotions and animal emotions. Affective Science*, 3(4), 703–712. https://doi.org/10.1007/s42761-022-00125-6

Mobbs, D., Petrovic, P., Marchant, J.L., Hassabis, D., Weiskopf, N., Seymour, B., Dolan, R.J. and Frith, C.D. (2007) *When fear is near: Threat imminence elicits prefrontal-periaqueductal gray shifts in humans. Science*, 317(5841), 1079–1083. https://doi.org/10.1126/science.1144298

Molony, V., Kent, J.E. and McKendrick, I.J. (2002) *Validation of a method for assessment of an acute pain in lambs. Applied Animal Behaviour Science*, 76(3), 215–238. https://doi.org/10.1016/S0168-1591(02)00014-X

Nicol, C.J., Caplen, G., Edgar, J. and Browne, W.J. (2009) *Associations between welfare indicators and environmental choice in laying hens. Animal Behaviour*, 78(2), 413–424. https://doi.org/10.1016/j.anbehav.2009.05.016

Nicol, C.J., Caplen, G., Statham, P. and Browne, W.J. (2011) *Decisions about foraging and risk trade-offs in chickens are associated with individual somatic response profiles. Animal Behaviour*, 82(2), 255–262. https://doi.org/10.1016/j.anbehav.2011.04.022

Panksepp, J. (2005) *Affective consciousness: Core emotional feelings in animals and humans. Consciousness and Cognition*, 14(1), 30–80. https://doi.org/10.1016/j.concog.2004.10.004

Paterson, E.A., O'Malley, C.I., Moody, C., Vogel, S., Authier, S. and Turner, P.V. (2023) *Development and validation of a cynomolgus macaque grimace scale for acute pain assessment. Scientific Reports*, 13(1), 3209. https://doi.org/10.1038/s41598-023-30380-x

Pawlyk, A.C., Morrison, A.R., Ross, R.J. and Brennan, F.X. (2008) *Stress-induced changes in sleep in rodents: models and mechanisms. Neuroscience & Biobehavioral Reviews*, 32(1), 99–117. https://doi.org/10.1016/j.neubiorev.2007.06.001

Pellow, S., Chopin, P., File, S.E. and Briley, M. (1985) *Validation of open: Closed arm entries in an elevated plus-maze as a measure of anxiety in the rat. Journal of Neuroscience Methods*, 14(3), 149–167. https://doi.org/10.1016/0165-0270(85)90031-7

Phillips, C.J.C. and Petherick, J.C. (2015) *The ethics of a co-regulatory model for farm animal welfare research. Journal of Agricultural and Environmental Ethics*, 28, 127–142. https://doi.org/10.1007/s10806-014-9524-9

Precision and Accuracy (2022) Available from: https://letstalkscience.ca/educational-resources/backgrounders/precision-and-accuracy, last accessed on 09 August 2025.

Raj, M. and Mason, G. (1999) *Reaction of farmed mink (Mustela vison) to argon-induced hypoxia. Veterinary Record*, 145(25), 736–737. https://doi.org/10.1136/vr.145.25.736

Raj, M. and Gregory, N.G. (1995) *Welfare implications of gas stunning pigs 1. Determination of aversion to the initial inhalation of carbon dioxide or argon. Animal Welfare*, 4, 273–280. https://doi.org/10.1017/S096272860001798X

Rault, J.-L., Bateson, M., Boissy, A., Forkman, B., Grinde, B., Gygax, L., Harfeld, J.L., Hintze, S., Keeling, L.J., Kostal, L., Lawrence, A.B., Mendl, M.T., Miele, M., Newberry, R.C., Sandoe, P.,

Spinka, M., Taylor, A.H., Webb, L.E., Whalin, L. and Jensen, M.B. (2025) *A consensus on the definition of positive animal welfare. Biology Letters*, 21, 20240382. https://doi.org/10.1098/rsbl.2024.0382

Rolls, E.T. (2005) *Emotion explained.* Oxford University Press, Oxford, UK.

Rolls, E.T. (2014) *Emotion and decision-making explained.* Oxford University Press, Oxford, UK.

Ross, M., Garland, A., Harlander-Matauschek, A., Kitchenham, L. and Mason, G. (2019) *Welfare-improving enrichments greatly reduce hens' startle responses, despite little change in judgment bias. Scientific Reports*, 9(1), 11881. https://doi.org/10.1038/s41598-019-48351-6

Ross, M., Rausch, Q., Vandenberg, B. and Mason, G. (2020) *Hens with benefits: Can environmental enrichment make chickens more resilient to stress? Physiology & Behavior*, 226, 113077. https://doi.org/10.1016/j.physbeh.2020.113077

Sandem, A.I., Braastad, B.O. and Bøe, K.E. (2002) *Eye white may indicate emotional state on a frustration–contentedness axis in dairy cows. Applied Animal Behaviour Science*, 79(1), 1–10. https://doi.org/10.1016/S0168-1591(02)00029-1

Sandem, A.I., Janczak, A.M. and Braastad, B.O. (2004) *A short note on effects of exposure to a novel stimulus (umbrella) on behaviour and percentage of eye-white in cows. Applied Animal Behaviour Science*, 89(3–4, 314), 309. https://doi.org/10.1016/j.applanim.2004.06.011

Sandem, A.I. and Braastad, B.O. (2005) *Effects of cow–calf separation on visible eye white and behaviour in dairy cows—a brief report. Applied Animal Behaviour Science*, 95, 233–239. https://doi.org/10.1016/j.applanim.2005.04.011

Sandem, A.I., Braastad, B.O. and Bakken, M. (2006a) *Behaviour and percentage eye-white in cows waiting to be fed concentrate—a brief report. Applied Animal Behaviour Science*, 97(2–4), 145–151. https://doi.org/10.1016/j.applanim.2005.08.003

Sandem, A.I., Janczak, A.M., Salte, R. and Braastad, B.O. (2006b) *The use of diazepam as a pharmacological validation of eye white as an indicator of emotional state in dairy cows. Applied Animal Behaviour Science*, 96, 177–183. https://doi.org/10.1016/j.applanim.2005.06.008

Sapolsky, R. (1998) *Why Zebras Don't Get Ulcers: An Updated Guide to Stress, Stress Related Diseases, and Coping*, 2nd ed. W.H. Freeman and Co. Ltd., New York, USA.

Schwartzkopf-Genswein, K.S., Stookey, J.M., Janzen, E.D. and McKinnon, J. (1997) *Effects of branding on weight gain, antibiotic treatment rates and subsequent handling ease in feedlot cattle. Canadian Journal of Animal Science*, 77(3), 361–367. https://doi.org/10.4141/A96-104

Stafleu, F.R., Rivas, E., Rivas, T., Vorstenbosch, J., Heeger, F.R. and Beynen, A.C. (1992) *The use of analogous reasoning for assessing discomfort in laboratory animals. Animal Welfare*, 1(2), 77–84. https://doi.org/10.1017/S0962728600014822

Susskind, J.M., Lee, D.H., Cusi, A., Feiman, R., Grabski, W. and Anderson, A.K. (2008) *Expressing fear enhances sensory acquisition. Nature Neuroscience*, 11(7), 843–850. https://doi.org/10.1038/nn.2138

Tran, N. (2024) The 'publish or perish' mentality is fuelling research paper retractions – and undermining science. The Conversation, September 23, 2024 (https://theconversation.com/the-publish-or-perish-mentality-is-fuelling-research-paper-retractions-and-undermining-science-238983?utm_source=clipboard&utm_medium=bylinecopy_url_button)

Tucker, C.B., Jensen, M.B., de Passillé, A.M., Hänninen, L. and Rushen, J. (2021) *Invited review: Lying time and the welfare of dairy cows. Journal of Dairy Science*, 104(1), 20–46. https://doi.org/10.3168/jds.2019-18074

van der Harst, J.E. and Spruijt, B.M. (2007) *Tools to measure and improve animal welfare: reward-related behaviour. Animal Welfare*, 16, 67–73. https://doi.org/10.1017/S0962728600031742

Vieira, A.D.P., Guesdon, V., De Passille, A.M., Von Keyserlingk, M.A.G. and Weary, D.M. (2008) *Behavioural indicators of hunger in dairy calves. Applied Animal Behaviour Science*, 109(2–4), 180–189. https://doi.org/10.1016/j.applanim.2007.03.006

Viscardi, A.V., Hunniford, M., Lawlis, P., Leach, M. and Turner, P.V. (2017) *Development of a Piglet Grimace Scale to evaluate piglet pain using facial expressions following castration and tail docking: a pilot study. Frontiers in Veterinary Science*, 4, 51. https://doi.org/10.3389/fvets.2017.00051

Walker, M.D., Duggan, G., Roulston, N., Van Slack, A. and Mason, G. (2012) *Negative affective states and their effects on morbidity, mortality and longevity. Animal Welfare*, 21(4), 497–509. https://doi.org/10.7120/09627286.21.4.497

Watters, J.V., Krebs, B.L. and Eschmann, C.L. (2021) *Assessing animal welfare with behavior: Onward with caution. Journal of Zoological and Botanical Gardens*, 2(1), 75–87. https://doi.org/10.3390/jzbg2010006

Weary, D.M. and Fraser, D. (1995) *Calling by domestic piglets: Reliable signals of need? Animal Behaviour*, 50(4), 1047–1055. https://doi.org/10.1016/0003-3472(95)80105-7

Weary, D.M. and Fraser, D. (1997) *Vocal response of piglets to weaning: Effect of piglet age. Applied Animal Behaviour Science*, 54(2–3), 153–160. https://doi.org/10.1016/S0168-1591(97)00066-X

Weary, D.M., Niel, L., Flower, F.C. and Fraser, D. (2006) *Identifying and preventing pain in animals. Applied Animal Behaviour Science*, 100(1–2), 64–76. https://doi.org/10.1016/j.applanim.2006.04.013

Williams, J.H., Huggins, C.F., Zupan, B., Willis, M., Van Rheenen, T.E., Sato, W., Palermo, R., Ortner, C., Krippl, M., Kret, M., Dickson, J.M., Li, C.R. and Lowe, L. (2020) *A sensorimotor control framework for understanding emotional communication and regulation. Neuroscience & Biobehavioral Reviews*, 112, 503–518. https://doi.org/10.1016/j.neubiorev.2020.02.014

Willner, P. (1986) *Validation criteria for animal models of human mental disorders: Learned helplessness as a paradigm case. Progress in Neuro-Psychopharmacology and Biological Psychiatry*, 10, 677–690. https://doi.org/10.1016/0278-5846(86)90051-5

Willner, P. (1997) *Stress and depression: Insights from animal models. Stress Medicine*, 13(4), 229–233. https://doi.org/10.1002/(sici)1099-1700(199710.13:4%3c229::aid-smi743%3e3.0.co;2-8

Willner, P. (2017) *The chronic mild stress (CMS) model of depression: History, evaluation and usage. Neurobiology of Stress*, 6, 78–93. https://doi.org/10.1016/j.ynstr.2016.08.002

Würbel, H. and Stauffacher, M. (1997) *Age and weight at weaning affect corticosterone level and development of stereotypies in ICR-mice. Animal Behaviour*, 53, 891–900. https://doi.org/10.1006/anbe.1996.0424

Yeates, J.W. (2011) *Is 'a life worth living' a concept worth having? Animal Welfare*, 20(3), 397–406. https://doi.org/10.1017/S0962728600002955

Part 1

Indicators of Emotion

Preference and Avoidance Tests

3

Marsha L. Reijgwart[1], Thomas Ede[2], and Claudia M. Vinke[1]

[1]Animals in Science & Society, Faculty of Veterinary Medicine, Utrecht University, Utrecht, The Netherlands
[2]School of Veterinary Medicine, University of Pennsylvania, Philadelphia, Pennsylvania, USA

Introduction

Preference and avoidance tests are two sides of the same coin. In their most basic form, they rely on the simple assumption that animals seek what induces positive affect and avoid what causes negative affect. Following this line of reasoning, preference and avoidance tests can be used to identify stimuli that animals (dis)like and (don't) want by observing whether the animal avoids or approaches the stimulus. Hence the focus of these tests is *affective valence*, which is arguably most relevant to animal welfare (Fraser 2009). As Marian Dawkins put it, "*Animals may not be able to talk, but they can vote with their feet and express some of what they are feeling by where they choose to go*" (Dawkins 1980). Furthermore, once preferred and avoided stimuli are identified, they can be used to validate other welfare indicators; they thus form one of the validation tests (Test 2) listed in Chapter 2. Those indicators that change reliably when animals are presented with preferred or avoided stimuli show evidence of validity as markers of, respectively, positive or negative affective valence.

In this chapter, we will illustrate the versatility and validity of preference and avoidance tests as a measure of valence, with applications in a wide range of contexts and species. Following a similar structure to other chapters in this book, we will discuss the construct validity of preference and avoidance tests, their responsiveness (i.e. how sensitive they are to changes in affect) and selectivity (i.e. the degree to which they are influenced by other variables or states unrelated to the specific type of affect we are trying to assess) and their use in practice. But first, we start with a short overview on the main principles of preference and avoidance tests and how they can be designed.

Preference and avoidance tests: a short overview

There are many methodologies for preference and avoidance paradigms. They involve presenting an array of stimuli, usually one or two at a time, and measuring where the animals spend their time or which food they eat, etc. Three important parameters that differ between different paradigms are as follows: the number of choices the animal is given (one option versus multiple options); whether appetitive or consummatory responses are assessed and whether a cost is imposed on access/avoidance and/or consumption (see Figure 3.1). It should be noted that different paradigms can be combined in one study in order to gain a more complete picture of an animal's priorities and preferences. For example, in a study by Reijgwart et al. (2016), ferrets could choose between seven different chambers that were accessible through a weighted door (motivation test). Within each chamber, the ferret could freely choose to interact with one of the three enrichment items that were in that chamber (choice test). We now briefly discuss the implications of variation in test design.

First, the number of choices that an animal is given determines whether we can infer the valence (positive or negative) of a single resource or the relative order or strength of preference/avoidance for different resources (Dawkins 1988, 1990). The number of choice options should be chosen carefully. For example, in a two-choice set-up, side-biases could influence choices made when resources are cued by their location (Adámková et al. 2017). Subjects can be pretested to quantify their inherent side-biases, and location of resources interchanged, preferably in a randomised manner, to overcome this issue. Nevertheless, a multiple-choice set-up is considered to be more efficient and responsive when assessing the value of multiple resources (see Reijgwart et al. 2018).

Second, tests can measure consummatory and/or appetitive behaviour. Consummatory behaviour can be used to assess whether an animal 'likes' something or shows aversion to it, and it involves engaging with the goal (e.g. food) of the motivational state (e.g. hunger), to reduce or satisfy the motivation via negative feedback. Examples of using consummatory behaviour as a measure of preference include determining which type of stall flooring cows prefer to lie on (Tucker et al. 2003) or which litterbox cats prefer to use (Grigg et al. 2012). Appetitive behaviours are goal directed locomotory patterns and functional actions to (dis)engage the animal with the resource (e.g. approach or avoid) before consumption can take place (e.g. rooting for food in pigs). These behaviours can thus be used to indicate how much an animal 'wants' a resource. Measures of appetitive behaviours in a preference test might include the first item that is chosen, or the latency to start interacting with the resource. Some researchers argue that '(dis)liking' is the 'true' affective reaction whilst 'wanting' is a measure of motivation which, usually, is closely linked to how much a resource is 'liked' (e.g. Berridge 2004; Berridge and Dayan 2021).

Third, we can observe the behaviour of the animals with and without imposing a cost on approach, consumption and/or avoidance. Tests that do not impose costs include simple *choice tests*, in which subjects are free to choose between different resources. Variables recorded can include the order in which resources are visited, and the number and duration of visits. These tests were the first to be used in animal welfare science to probe the animal's perspective (e.g. Hughes and Black 1973). *Conditioned place preference* (*CPP*) and *conditioned place avoidance* (*CPA*) *tests* also impose no cost. Here, the animal is trained (conditioned) to associate a location with an event (hypothesised to be pleasurable in CPP, aversive in CPA). When the animal is put in the same area again, but without exposure to the event, it is assumed that the animal will show a preference for or avoidance of the location that is associated with the pleasurable or aversive experience, respectively. CPP/CPA paradigms have been used across a wide variety of species, including fish (Wong et al. 2014), lizards (Venable et al. 2019), hermit crabs (Elwood and Appel 2009), bees (Black et al. 2018) and ants (Wystrach et al. 2020).

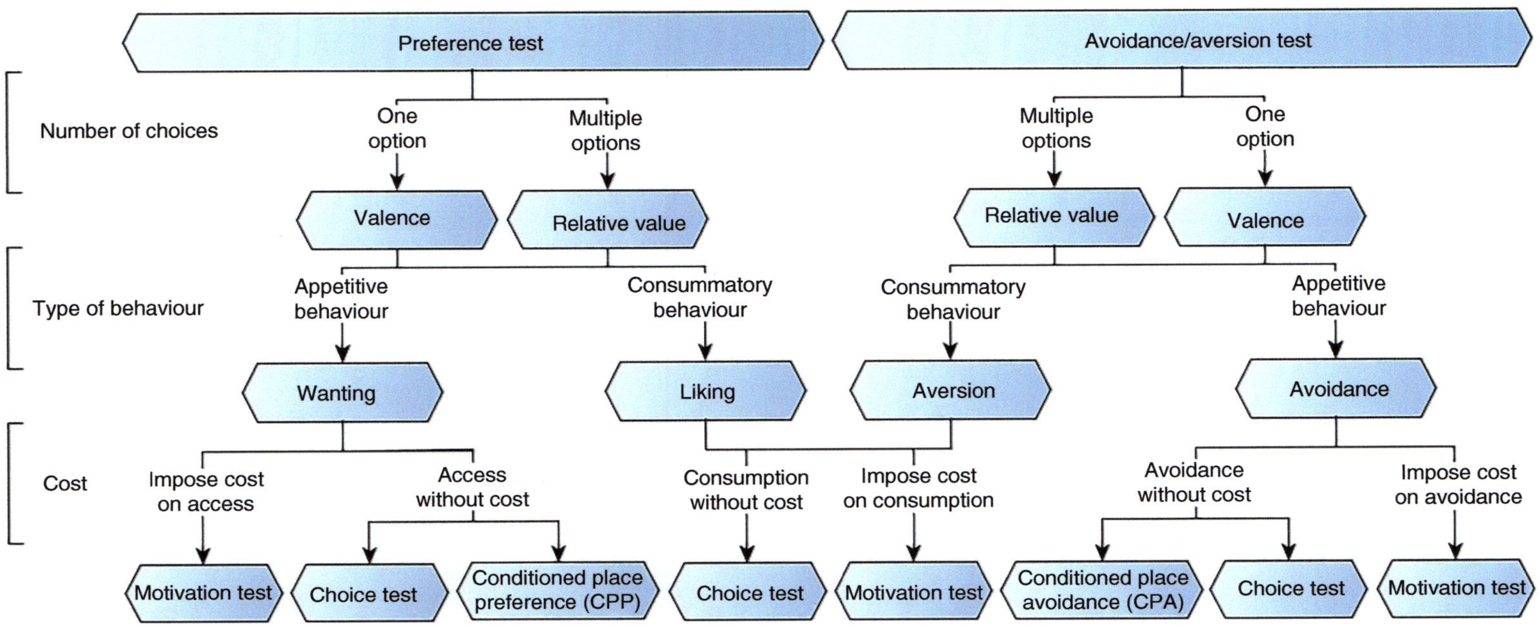

Figure 3.1 Dendrogram of preference and avoidance testing paradigms as based on three main parameters (number of choices, type of behaviour and cost). Test types are at the bottom of the figure, while the concepts that they can assess are indicated in the paths above.

Choice and CPP/CPA tests inform us about the relative value to the animal of resources or experiences. However, to gain insight into the size of this difference in perceived value, and thereby attempt to quantify absolute valuation of resources and experiences, we can impose a cost on access, consumption or avoidance in a *motivation test*. In these, the animal can be required to invest energy or time by paying costs to defend consumption or avoid or approach an anticipated resource or stimulus. This forces the animal to weigh costs and benefits in their decision-making. The outcome is assumed to reflect how much animals value (are prepared to work for) different resources.

Different types of costs can be used in these motivation tests, e.g. an animal can be required to make a specific operant response (e.g. lever pressing, nose poking), endure an aversive stimulus (e.g. noise, shock, wind, pollution) or use a more general 'budget' (e.g. spending time) within a limited time frame where they are allowed to stay with, gain access to, or get away from a resource or stimulus (Fraser and Matthews 1997). When measuring consummatory behaviour, animals' endurance of an aversive stimulus such as variable wind speed can be assessed (Faure and Lagadic 1994). For paradigms that focus on appetitive behaviour, using an operant response is more common, for example a weighted door which can be varied in weight to reflect different costs of access (Smith et al. 1990; Cooper and Appleby 2003; Galhardo et al. 2011; Olsson and Keeling 2002; Reijgwart et al. 2016; McConnachie et al. 2018; Tucker et al. 2018). As indicated earlier, motivation tests can be used to assess how aversive as well as how attractive a resource is. For example, runway tests use the speed with which an animal runs the length of a runway to arrive at a resource to infer their motivation to acquire or avoid that resource. These tests can be used to confront the animal with the decision to work, or put up a certain level of resistance, to avoid being exposed to a stimulus or event such as electro-immobilisation (E-I) or injurious handling methods (e.g. Rushen 1986; Pajor et al. 2000; see also Niel and Weary 2007; Elwood and Appel 2009).

The costs used in motivation tests provide a 'common currency' with which to quantify the value or aversiveness of different resources, allowing the construction of 'demand curves' that show how much resource is obtained as the price increases (Figure 3.2). These curves can be interpreted using various measures derived from *consumer demand studies* in human microeconomics (Dawkins 1983; Varian 1996), including consumer surplus and reservation price or maximum price paid (MPP). Consumer surplus is calculated by measuring the area under the demand curve (Varian 1996; Houston 1997; Ng 1990). MPP is calculated by measuring the highest price an individual is prepared to pay for the first visit to a resource after a period of deprivation (e.g. Cooper and Mason 2001; Mason et al. 2001; Kirkden et al. 2003).

Because any measures that rely on the *shape* of demand curves have limitations (see Houston 1997; Kirkden et al. 2003; Seaman et al. 2008), the MPP is currently a more commonly used measure of motivation. Different MPPs can be used to rank resources, and MPPs for an empty space (control condition) and essential resources (such as food) can be used as bottom and top benchmarks, respectively. Resources that rank as highly as the essential resource are regarded to be very important for the animal's well-being. For example, dairy cows were willing to pay a similar maximum price for access to pasture or use of a mechanical brush, as for fresh feed (von Keyserlingk et al. 2017; McConnachie et al. 2018), and ferrets had an equal MPP for food as for a comfortable place to sleep (Reijgwart et al. 2016, Figure 3.3).

As a final note, all of these paradigms can be presented in discrete tests or instead incorporated into animals' living conditions such that their home cage is an apparatus, the latter sometimes being called a 'closed economy' (Reijgwart et al. 2016).

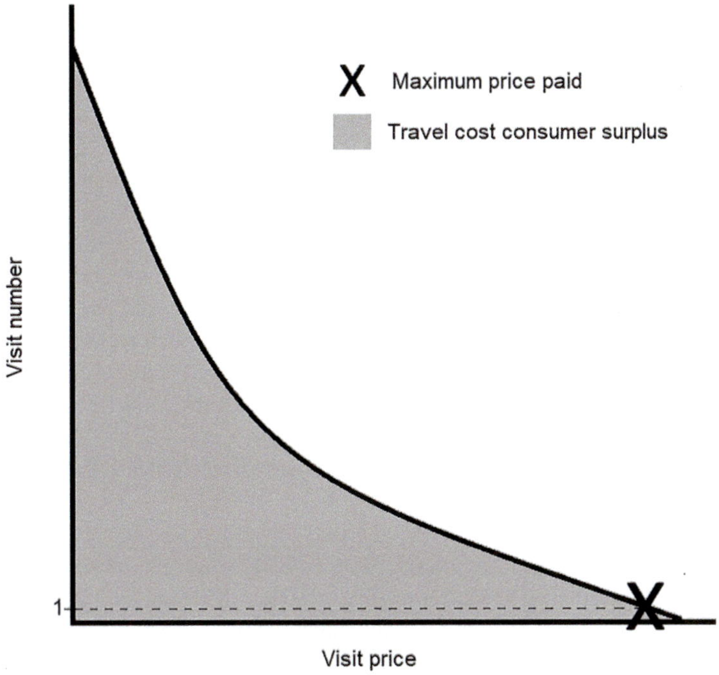

Figure 3.2 Visualisation of two types of interpretation of a demand curve: consumer surplus (area under the curve) and maximum price paid (highest price an individual is prepared to pay for a resource) as a measure of motivation. The *X*-axis represents the price of one visit to the resource, and the *Y*-axis represents the number of visits the individual pays to the resource.

Figure 3.3 Ferret resting in a hammock after having pushed a heavy door to gain access to it. Source: Photo by M.L. Reijgwart.

Basic construct validity of preference/aversion

Some argue that approach and avoidance are more or less synonymous with positive and negative emotional states (e.g. Mauss and Robinson 2009; see also Chapter 1). The assumption that negative valence is avoided, and positive valence is sought out (Rolls 2005; 2014), can be used to provide evidence that, for example, housing conditions or surgical procedures have an affective impact on animals and hence on their welfare. In this section, the construct validity of the outcomes of preference and avoidance tests as indicators of affective valence in animals is reviewed. Because preference and avoidance are integral to one (Test 2) of the five tests listed in Chapter 2, we need to validate them using some of the other tests listed. We thus discuss how emotions influence choices by humans (Test 1), whether animals avoid or approach resources that threaten or are important for survival and fitness (Test 3) and how preference and avoidance responses change in animals given affect-modulating drugs (Test 4 in Chapter 2).

Human emotional responses to stimuli predict how much they prefer or avoid them

A combination of positive (rewarding) and negative (aversive) experiences, or the expectation of these experiences, motivates individual people to behave in certain ways (Masterson and Crawford 1982; Lang et al. 1998). Emotions influence learning and future actions, and appear crucial for motivating appropriate approach or avoidance and for making adaptive choices between diverse options. People will – either consciously or unconsciously – choose the things that are expected to result in the most positive emotions (Forgas 1995; Pham 1998; Lerner and Keltner 2000; Winkielman et al. 2005). This motivating aspect of affect (pleasure/displeasure) is closely related to concepts of behavioural reinforcement and punishment (Thorndikes 1911 'Law of effect'). Maximising long-term rewards over punishments, and hence positive over negative states, is thus at the heart of reinforcement learning (Mendl and Paul 2020; also see Chapter 1).

The anticipated emotional consequences of different courses of action have also been shown to predict human choices (Cabanac 1990). In human welfare economics, the area under a demand curve (consumer surplus) is used to assess 'utility' (i.e. consumer satisfaction or perceived benefit), whereas maximum price paid (MPP) approximates the utility of having one (the first) unit of a resource versus having none at all (Houston 1997; Kirkden et al. 2003; Ng 1983).

There are many examples where negative emotions like pain or fear drive withdrawal, and positive emotions like pleasure drive approach behaviour. For example, negative affect influences a broad repertoire of behaviours, including flight behaviour triggered by fear states, banal consumer choices of not buying a bad product twice, and exaggerated avoidance of cues due to traumatic previous experiences, which may result in debilitating conditions such as agoraphobia or dental phobias (Öst 1985). Pleasure can also be a strong motivational driver for humans and has also been called 'the common currency of the brain' (Cabanac 1992). For example, it has been shown that highly preferred stimuli are better reinforcers (i.e. are more reliable in eliciting the trained response) than less preferred stimuli in children with disabilities (Piazza et al. 1996). Similarly, when self-described 'chocolate-lovers' were repeatedly offered chocolate, the reported pleasantness of the treat was a reliable predictor of how motivated the subject was to receive another piece (Small et al. 2001). Other examples of links between human affective states and preference or avoidance are provided in the following sections alongside similar animal studies.

Do animals approach or avoid stimuli that promote or threaten ancestral fitness?

From an evolutionary perspective, the capacity of organisms to identify and avoid future negative events such as painful injuries or food poisoning brings obvious selective advantages. As discussed earlier (see also Chapter 1), emotions are likely to play a critical role. Pleasure or displeasure is hypothesised to reflect the evolutionary beneficial or damaging effects of an event, guiding learning and future actions, and thereby motivating adaptive choices between diverse options and increasing survival chances. For example, stimuli that involve or threaten physical injury (such as burns, incisions or electric shock) induce strong negative affect in humans and are believed to do similarly for most animals because physical injury is bad for fitness. Preference tests can thus be validated if animals prefer or avoid stimuli that are likely to enhance or threaten ancestral fitness.

In the first learned aversion test in animal welfare research, Rushen (1986) investigated different restraining methods used for shearing sheep. Thirty-six Merino sheep were trained to run down a 7 m race, at the end of which they received different handling treatments: they were either allowed to run freely (group 'Free-run', n = 6), physically restrained (group 'Restraint', n = 10), sham electro-immobilised without current (group 'Wired-up', n = 10) or fully electro-immobilised (group 'Full E-I', n = 10) with a low-voltage current which caused paralysis through muscle contraction. The last three experimental treatments very likely caused pain and/or simulated capture by a predator, both of which reflect threats to fitness. Sheep who could run freely at the end of the race maintained a low transit time (Figure 3.4a) and were not required to be pushed down the alley by the human handler (Figure 3.4b). Sheep subjected to restraint, in contrast, increased their transit time and were required to be pushed. Such reluctance to reach the end of the race was especially marked in the sheep who were electro-immobilised. These results indicate that fitness-threatening stimuli do indeed induce aversion, and that a larger threat to fitness (assuming that pain and capture/restraint are more fitness-threatening than restraint alone) induces a stronger aversion. The study also highlights an important advantage of conditioned aversion or preference paradigms: in cases where spontaneous responses are impaired or impossible (for example, restrained or paralysed sheep), we can use the animals' memory of these events, and their subsequent preference behaviour, to assess the affective and welfare impact of the situation.

In a similar set-up, Pajor et al. (2000) evaluated injurious and hence fitness-threatening handling methods for cows using a runway. After several exposures, cows who received punishment-based handling (hitting, shouting, or electric prodding) required more time and force to be moved to the end of the race compared to cows receiving reward-based (food) or neutral (minimal interaction) treatments. Likewise, studying younger animals, de Passillé et al. (1996) found that dairy calves were more likely to avoid people who handled them using painful or stressful methods, such as nose tongs and cattle prods, than those who handled them using rewards such as milk and friendly social contact.

Aversion can develop with a single, intense event potentially interpreted as life-threatening by the animal. For example, in hot-iron disbudding an injury is incurred, tissue is damaged and, without pain control, calves struggle, frantically attempt to escape, and occasionally even faint. In a conditioned place aversion study, calves were exposed to hot-iron disbudding with a local anaesthetic and a sham procedure in visually distinct pens (Figure 3.5). After receiving both procedures, calves spent less time in the pen where they were disbudded, indicating that this damaging procedure was aversive to calves, even when treatment was provided for acute pain (Ede et al. 2019).

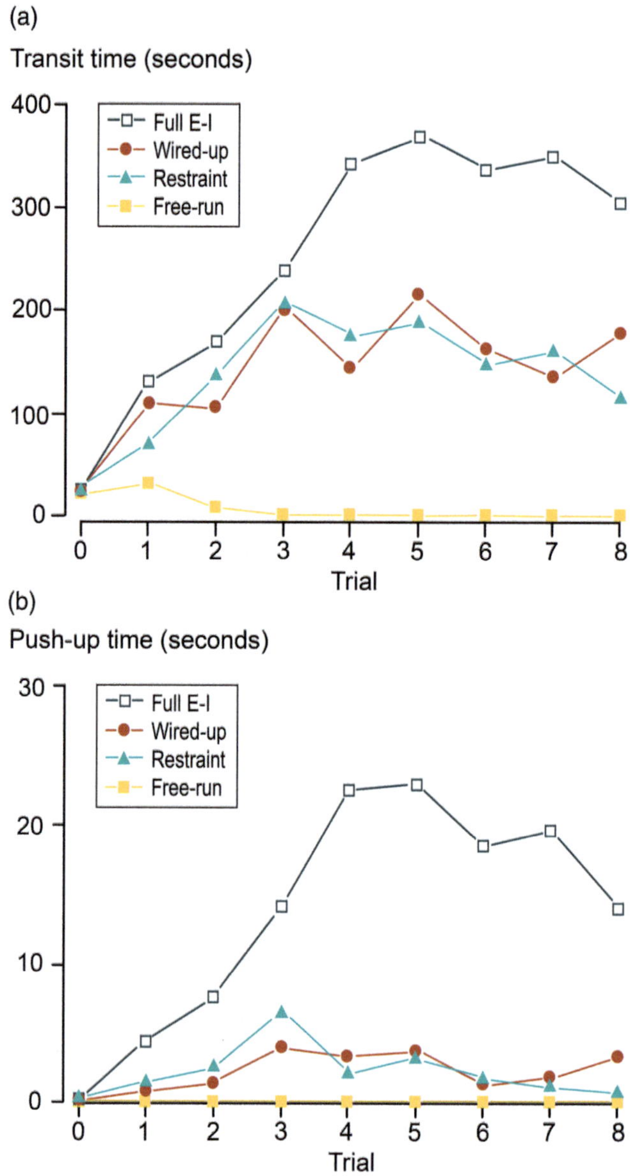

Figure 3.4 Merino sheep (n = 36) were trained to run down a 7m race, at the end of which they received different handling treatments (trial 0 is the average of six trials without any treatment for all sheep); (a) the mean transit time (in seconds) for the sheep to run down the race across trials; (b) the mean time (in seconds) needed across trials to push the animals up the four sections of the race if the end of each section was not reached within two minutes. The highest transit and push-up times observed for full electro-immobilisation (E-I) suggest this method to be the most aversive. Source: Redrawn from Rushen (1986) / with permission of Elsevier.

Figure 3.5 Conditioned place aversion (CPA) apparatus. Calves were disbudded in one pen and underwent a sham procedure in the other pen. Calves then avoided the pen where they were disbudded. Illustration by Ann Sanderson. Source: Adapted from Ede et al. (2019).

The strength of preference and avoidance may be responsive to changes in affective valence. For example, juvenile lizards who were exposed to a meal of either 100%, 50% or 0% fire ants (whose stings and venom are likely to be painful for lizards) had more intense and longer-lasting aversion responses (measured as reduced consumption) when fed the meals with the higher fire ant percentage (Venable et al. 2019). A similar graded response is seen in Rushen's (1986) study of sheep (Figure 3.4). Thus, some preference/aversion tests may be *responsive* to changes in underlying affective valence.

Stimuli that are good for fitness should, in contrast, be preferred. Many preference tests confirm this. For example, when an animal has been deprived of a resource (e.g. food or water), their motivation to gain access to that resource will be greater compared with a satiated animal (Cabanac and LaFrance 1990; Saker et al. 2014). Also, safety can often be found in numbers, which is reflected in the high motivation that many group-living prey species have for access to social partners (sheep: Dumont and Boissy 2000; calves: Holm et al. 2002, Ede et al. 2022; hens: Bradshaw 1992; zebrafish: Ogi et al. 2021). Similarly, cows are motivated to be reunited with their calves (Wenker et al. 2020), and animals who are in season are highly motivated to reach a receptive partner, both of which have a clear relation to fitness. For example, male laboratory rats will cross a shock grid in order to reach a receptive female (Warner 1927; Anderson 1938; Meyerson and Lindström 1973). This example indicates decision-making processes in which animals trade off (expected) aversive and rewarding affective states.

Do animals prefer pharmacological interventions that reduce negative affect/promote positive affect?

The ability of preference/avoidance behaviour to reflect animals' affective responses to events is supported by studies using pharmacological interventions. There are homologous modes of action of non-steroidal anti-inflammatory drugs (NSAIDs), opioids, and drugs of abuse between humans and other species. Based on human reports, these drugs can be assumed to reduce negative affect/promote positive affect in animals and can therefore be used to illustrate the validity of preference tests. When in a state of pain, human patients will self-administer analgesic drugs and report satisfaction of pain relief (Ballantyne et al. 1993).

Similar results can be found in animals, with arthritic rats or lame chickens displaying a preference for the consumption of analgesics (Colpaert et al. 1980; Colpaert et al. 2001; Danbury et al. 2000) that is not seen in control animals, strongly suggesting that these choices reflect attempts to reduce pain and negative affect.

Another example involves drugs of abuse, such as cocaine, amphetamine, and nicotine, which elicit a positive feeling in people. As would be predicted, rats and mice show CPP when conditioned with these drugs (Tzschentke 1998). Relatedly, opiate withdrawal is a well-known negative experience for humans (Schulteis et al. 2000). After rats were provided with morphine pellets for five days, withdrawal was chemically induced using naloxone, and rats showed an aversion to the place associated with this withdrawal (Mucha 1987). Non-dependent rats provided with naloxone did not display aversion, and dependent rats not provided with naloxone did not either, suggesting that aversion was specifically linked to the experience of withdrawal. The response was long-lasting, with morphine-dependent rats undergoing withdrawal displaying aversion for up to 16 weeks (Stinus et al. 2000).

Being sick can have a negative effect on our chances of survival, and nausea-induced conditioned taste aversion is a well-known phenomenon in people. Humans develop an aversion to certain flavours if they are associated with negative experiences such as food poisoning, overconsumption or allergic reactions, even if these negative feelings are not directly related to the food consumed (e.g. being induced by motion sickness, flu symptoms or chemotherapy; Batsell and Brown 1998). For instance, undergraduate students were hired to repeatedly tilt their heads and stare at a moving visual field to induce severe nausea while consuming fruit-flavoured beverages. Results showed that association with feelings of nausea decreased the students' subsequent beverage consumption (Lamon et al. 1977). Conditioned food aversion was first observed in animals in the 1950s. Rats usually display a preference for sucrose solution over tap water. When sucrose solution was associated with radiation exposure, rats displayed a strong and lasting aversion to the previously favoured solution (Garcia et al. 1955). More recently, pharmacological induction of sickness in a critically endangered species (northern quoll) generated avoidance of consuming a highly toxic invasive species (cane toad). Juvenile quolls were offered non-lethal toads laced with a nausea-inducing chemical. After release into the wild, these 'toad-smart' quolls were less likely to attack the toxic toads (and die from it) than control quolls (O'Donnell et al. 2010).

Though preference/avoidance behaviour in pharmacological studies is useful to understand decision-making and its working mechanisms, drug addiction shows the limitation for welfare studies: the most preferred is not always the best choice for survival and fitness, and may specifically affect the most dysfunctional individuals more prone to maladaptive choices and rewards (see false leads in the Selectivity section later).

A connection of preference and avoidance results predicting other welfare indicators (Test 5 in Chapter 2) can further strengthen the interpretations and conclusions.

Responsiveness and selectivity

In this section, we will discuss the responsiveness and selectivity of preference and avoidance paradigms, and list a few issues to keep in mind when designing and interpreting these tests.

Responsiveness

Preference and avoidance tests may generate false nulls (displaying poor **responsiveness**, i.e. not reflecting the affective value of the options on offer) due to ceiling effects, which arise when there is a maximum to the time or effort that an animal can devote to acquiring or avoiding a resource. For example, ferrets stopped pushing open a heavy door at the same weight (price) to gain access to food (MPP = 1450 ± 144 g) as to gain access to a hammock to sleep in (MPP = 1450 ± 120 g), even though food consumption is essential for their survival, and sleeping in a hammock when there are other safe sleeping places freely available is not (Reijgwart et al. 2016). At the MPP, ferrets were observed to make efforts to open the door at higher weights, but they seemed physically unable to do so, suggesting a ceiling effect.

Additionally, as for many indicators, there is a threshold effect: if a stimulus is affectively close to neutral for the animals, they are less likely to avoid or approach it, especially if there are other choices available that are of higher value. For example, ferrets stopped pushing a weighted door at lower weights to reach less-preferred enrichments and an empty chamber when tested in a set-up with seven chambers, compared to when tested in a set-up with three chambers (Reijgwart et al. 2018).

Learned preference or aversion paradigms draw most of their inferential strength from their cognitive component: animals learn that something is positive or negative and change their behaviour to seek or avoid it. However, the association between cue and stimulus is not always made (Domjan 1983; Smith et al. 2000). There is then a major issue when interpreting negative results: is the absence of preference or aversion due to the stimulus being judged as neutral by the animal (i.e. the test is working), or a failure in association between cue and stimulus (a false null)? Learning is more likely to occur when species-appropriate cues and actions are relevant to behavioural predispositions, response possibilities and/or senses, for example olfactory cues for dogs, auditory cues for dolphins, but not visual cues for mole rats. Even when cues are appropriate, investigators need to ensure animals' cognitive abilities and memory can be relied upon. It can be challenging for very young and old animals or subjects treated with memory-impairing drugs such as sedatives, to perform well in experiments relying on learned responses. For example, older dogs have been noted to have poorer performances in learning tasks compared to young dogs (Tapp et al. 2003), and treatment with benzodiazepines (sedative medication) affected mice in their acquisition of an avoidance response to electric shocks (Jensen et al. 1979). Learning-based preference tasks might then be inaccurate with such subjects.

A similar issue relates to the subject's perception of resource cues (visual, olfactory, auditory) at the place of payment. For example, ferrets were observed to sleep at the place of payment where they could see their conspecifics, instead of paying to get closer to them (Reijgwart et al. 2016). However, lack of payment in this example does not automatically reflect that a resource is not important. In this case, full physical interaction with the resource may not be necessary to be rewarding; just seeing or smelling others may suffice, which in some apparatuses may be possible without 'paying'. The result is thus a false null, as the test result does not reflect the affective benefits of the social contact obtained, albeit limited.

Selectivity

False leads (caused by low **selectivity**, i.e. when a response is *not* specific to affect of a particular valence, and other factors can influence it) during preference/avoidance testing may arise for a number of reasons. For example, when tests impose costs, these need to be validated to ensure the stimulus or task is indeed experienced as a cost by the animal (see *contrafreeloading* for counter examples). If they are not, then the number and duration of visits to all resources and/or the control environment remain high which, in turn, may generate the false inference that (some of) these resources are highly valued.

Alternatively, an aspect of the environment in which the resources are placed can be reinforcing for the animal. This risk is especially high when testing animals in a stimulus-poor environment, as they might interact with the device that is meant to impose a cost to seek out general stimulation (an issue revisited in 'External validity'). For example, when testing ferrets in a two-chamber consumer demand set-up using weighted doors, visits to the empty control compartment remained high despite increasing costs (Reijgwart et al. 2015). In addition to the possibility that visiting the empty cage had some value (e.g. representing checking 'territory' or being a mentally and physically enriching task), it is also conceivable that 'paying the price' became rewarding in itself, thus generating a risk of false leads.

False leads can also arise from the way some fitness-threatening stimuli may be approached instead of avoided to gain information about the resource or in an attempt to drive the stimulus away. In a laboratory setting, a high motivation to gain access to conspecifics in ferrets was found, but this behaviour is not necessarily associated with positive affect, as it could also reflect a high motivation for territorial behaviour or inspection of possible rivals (Reijgwart et al. 2016). In these cases, observation of appetitive (anticipatory) behaviour before the consumptive behaviour may give additional insight into the incentive value of the expected resource or stimulus (Spruijt et al. 2001; Vinke et al. 2006; Møe et al. 2006; Podturkin et al. 2022).

Finally, if one follows Berridge's stance that '(dis)liking' is the key measure of affect, then another source of false leads for tests using appetitive behaviour is when 'wanting' dissociates from 'liking' (e.g. drug addicts whose wanting for their drug outpaces their liking of it, leading to a false lead). Such dissociations may, however, be rare but important if testing in connection with addiction research.

Factors that may influence both responsiveness and selectivity

Design features of preference/avoidance tests can themselves give rise to false nulls and false leads. When different options are presented simultaneously, slightly less preferred items may never be chosen, or used at a lower rate, when presented together with a greatly preferred item. As a consequence, one might underestimate the absolute preference for one resource when it is paired with something that is highly valued (a responsiveness issue), or overestimate preference when it is paired with a low-value resource (a selectivity issue). For example, zebrafish preferred sand over a barren tank, but mostly ignored sand if gravel was available (Schroeder et al. 2014).

Consummatory measures (e.g. volumes of resource consumed, or utilisation/interaction time budgets) can also be inaccurate proxies for preference, if the stimuli being compared are very different in nature. For example, when comparing dilute and concentrated resources, a difference in consummatory behaviour might just reflect the different rate at which the animal is satiated. A naïve treatment of volume consumed as preference would thus underestimate the value of the concentrated resource and/or overestimate the value of the dilute resource. Likewise, when measuring consummatory behaviour, interpreting the relative time budgets for different behaviours as preference can have similar problems. If a mouse by chance chooses to sleep in one cage over the other in a free-choice preference test that is comparing resources covering different motivational systems, one may mistakenly overestimate preference for the resource contained in the chosen cage and underestimate preference for that in the other cage (when of course all the data are revealing is that sleep takes a lot of time!). In cases like these, where very divergent stimuli, resources or behaviours are being compared with each other, appetitive measures of preference are therefore far more accurate and valid than consummatory ones.

The social environment in which an animal is tested can also change the results of a motivation test. For example, when tested in groups, dominant individuals can prevent subordinates from accessing the preferred option, as observed by Schroeder et al. (2014) in zebrafish.

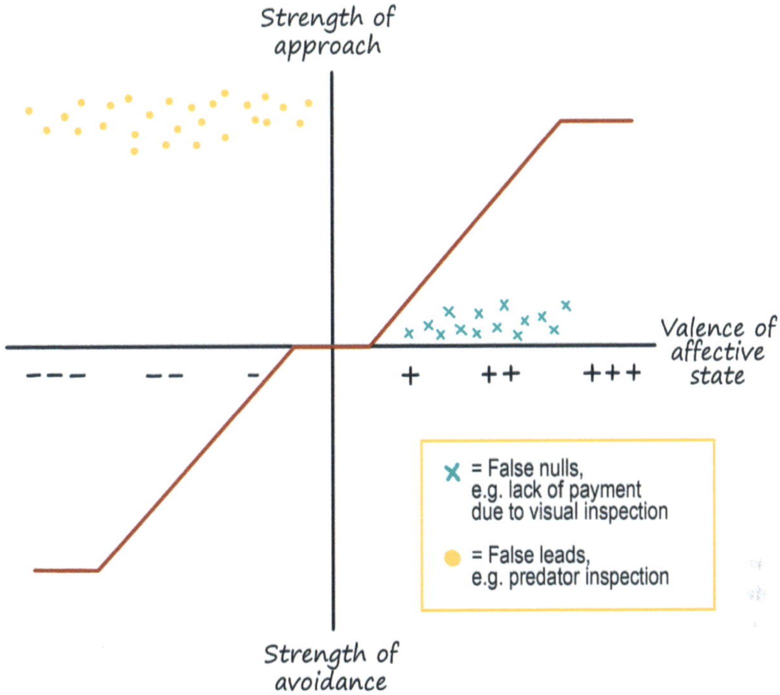

Figure 3.6 Schematic visualisation of the relationship between valence and approach/avoidance behaviour. The flat lines at the bottom left and upper right corner of the plot reflect floor and ceiling effects, respectively. The flat line in the centre reflects a threshold effect, and examples of false nulls and false leads are indicated.

This would lead to a false null for subordinates. On the other hand, social animals may influence each other's choices because they are motivated to spend time with conspecifics in the same choice compartment (Sherwin 2003). False leads or false nulls, depending on where the group settles, are especially likely if motivation for social cohesion outweighs that for resource access. Although wanting something because conspecifics are interacting with it can have adaptive value (e.g. you may share the resource), this influence should be kept in mind when generalizing results to different social contexts (as we discuss further later).

To summarise, a heuristic graph indicating the relationships between preference, avoidance and affective valence, plus potential false leads and false nulls, is shown in Figure 3.6.

Generalisability and external validity

The previous sections focus on examples where preference and avoidance tests provide valid or misleading information about the affective valence of the subject being tested. However, there are also situations in which the tests do accurately reflect the subject's valence, but the subject's state or testing context mean that results may not generalise to other states, situations or populations (i.e. have low external validity). This is an issue that has been greatly discussed with respect to preference tests (although it actually applies to every chapter of this book).

To illustrate, the underlying motivation for approaching or avoiding a resource may differ between sexes, or be affected by the state of the individual (reproductive state, health, current motivations), season and the extent to which different motivational systems are likely to influence choice behaviour (Fraser and Matthews 1997; Kirkden and Pajor 2006). Just as for some other test-based indicators (e.g. see Chapter 7 on judgement biases), measured preferences can also be shaped by aspects of the test design. Where choices depend on these factors, generalisability of findings may be limited. We consider some examples here.

Previous experiences can be influential, as illustrated by chicks preferring the type of flooring they were reared on, as opposed to a potentially more comfortable alternative (Petherick et al. 1990). Animals may thus often be 'primed' to prefer a resource they know well (Shields et al. 2004); such priming effects are well described, e.g. sexual motivation (Pfaus et al. 2013). Additionally, neophobia will cause animals to initially prefer environments they are used to. For example, Dawkins (1977) showed that hens raised in battery cages were more likely to choose a battery cage over an open hen-run than hens with previous experience with the larger enclosure. A wildlife management effort to reduce human–bear conflicts illustrates another effect of prior experience. Black bears received aversive conditioning via thrown rocks, pepper spray or non-lethal rubber slugs shot from a 12-gauge shotgun when consuming human food or entering human-occupied areas. This avoidance conditioning was effective for bears without prior experience of human food, but was less successful for bears who had already eaten human food (Mazur 2010; Figure 3.7). This failure of aversive conditioning was suggested to be linked to the animals' previous feeding experience not being associated with negative experiences (see also the phenomenon of latent inhibition: Lubow 1973). Hence, individual experiences can influence test outcomes (see e.g. Petherick et al. 1990). Checking if subjects have had prior contact with the stimulus and/or choice can help produce more generalisable outcomes, and choosing subjects that resemble the external population of interest is key (as it is for all welfare research).

Figure 3.7 Bear enjoying a soda. Aversive conditioning that works with some individuals often fails if bears have previously experienced human food. Source: Photo by Chad Linquist / Nevada Appeal.

Another form of priming can reflect the role of eliciting stimuli in the environment: animals exposed to cues from resources may be more motivated to reach them than animals without such cues. For example, mink worked less hard for a toy when they could not see it (Warburton and Mason 2003). Thus, toy-exposed mink may not be good models of toy-naive mink.

Likewise, preference and aversion responses can also change with current affective states. For example, mink in an under-stimulated state (e.g. housed in a non-enriched environment) will approach faster and spend more time in contact with aversive stimuli (glove, air puff, predator urine) in comparison to mink housed in an enriched environment (Meagher et al. 2017). This suggests that the animal's motivational balance between avoiding a negative stimulus and seeking any type of stimulation shifts in accordance with their current affective state. To put it another way, 'bored' test subjects may have motivations for stimulation (even potentially aversive stimulation) that more stimulated subjects do not (something that could also help explain the differences between the preferences of 'two-chamber' versus 'seven-chamber' ferrets outlined earlier, and a topic explored in detail in Chapter 9).

Memory biases are another process through which current affect influences aversion responses. In humans, 'mood congruent memory' theory proposes that recall of information is facilitated if mood matches information valence: it is easier to remember negative events when feeling bad, and positive events when feeling good (Lewis et al. 2005). Although still seldom explored (Burman and Mendl 2018), this bias has also been observed in animals: rats that had been exposed to social isolation (which induced depressive-like behaviours) reinstated an aversion response to a bright light and loud noise (Takatsu-Coleman et al. 2013), implicating that previous experiences can have consequences for current preference and avoidance results.

This all means that the context in which an animal is tested can also influence the generalisability of test results, as it contains several elements that can influence motivation for resources (including timing within the day, how animals were transported there and whether animals were tested solitarily or with company). It is therefore preferable to use a long-term closed economy set-up (e.g. Reijgwart et al. 2016), where the animal can only gain access to the resource in the test situation (as opposed to an open economy where the animal will have 'free' access to the resource outside of the test situation). Because animals live in the apparatuses for days or even weeks, the closed economy set-up also rules out any confounding influences of handling, transport, access time and test duration. Additionally, in an open economy, the animal's motivation for a resource(s) may be influenced by its availability, or the availability of motivationally-related 'substitutes', outside the test, which could decrease differences in motivation between resources (Hursh 1984; Mason et al. 1998; Ladewig et al. 2002).

Use in practice

Strong inferences can be made from preference and aversion tests if the aforementioned issues with responsiveness, selectivity and generalisability are considered. With these caveats in mind, in this section, we will mention some great examples of studies that used preference/avoidance paradigms.

Preference and motivation tests have been very useful in identifying cage designs that greatly improve animal welfare in win–win situations. For example, in one of the first simple choice tests published, Hughes and Black (1973) introduced four floor types and assessed which type of flooring hens prefer: they preferred a fine hexagonal mesh over other types of

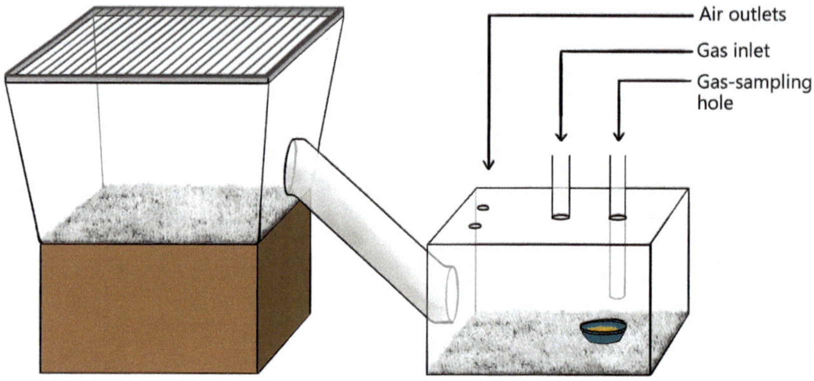

Figure 3.8 Approach-avoidance apparatus for the study of CO_2 in rats. The top cage was the rat's home cage, and the bottom cage contained a sizeable treat (20 Cheerios®). The bottom cage was gradually filled with CO_2 as the rat started consuming the reward, until they climbed back to the top cage. Source: Redrawn from Améndola and Weary (2019) / PLOS / CC BY 4.0.

flooring, with the authors suggesting hen comfort to be driven by number of points of support for the bird's feet rather than width of the wire material. Using different set-ups, researchers have used motivation tests to identify resources that are likely to be essential for an animals' well-being, such as a perch and a nest box for hens (Cooper and Appleby 1996; Olsson and Keeling 2002), nest-building material for mice (van de Weerd et al. 1998), rooting material for pigs (Pedersen et al. 2002, 2005) and water for swimming for mink (Mason et al. 2001).

Similarly, clever aversion test designs have helped evaluate the impact of negative procedures. For example, carbon dioxide (CO_2) is a common euthanasia agent for rodents. Rats were put in a conflicting situation of a food reward being placed in a chamber filled by different concentrations of CO_2 (Figure 3.8). At 5% and 10% CO_2, rats were willing to enter and spend time in the chamber to eat the reward. At 15% and 20% CO_2, rats were reluctant to enter the chamber, with most subjects giving up the reward entirely to avoid gas exposure. Argon, a supposed humane alternative to CO_2, caused similar avoidance responses from rats (Niel and Weary 2007), indicating that both gases induced negative affect. When offering a valued food reward in a chamber gradually filled with CO_2, rats stopped eating and left the chamber prior to losing consciousness, confirming the aversiveness of the gas for the animals (Niel et al., 2008).

Conclusion

Preference and avoidance paradigms allow us to answer a fundamental question when it comes to animal welfare: does the animal like or dislike something? The theory is simple and intuitive: if something aids survival and reproduction, it will be perceived as positive and sought out; if it is harmful, it will be perceived as negative and avoided; this is the foundation for validation Test 2 (Chapter 2). Instead of relying on the natural surroundings and behaviour of wild ancestors to infer what we should provide for domesticated animals, we can thus ask the animals themselves what they value in life, which is not always what we would expect: female mink appear to prefer lower ceilings instead of higher ceilings in their cage (Díez-León et al. 2017), which was not expected in a species that is a good climber.

Tests to measure preference and avoidance can be as straightforward as the underlying theory, as simple as presenting two feeds and seeing which one is consumed, or observing whether cattle stay away from a 'mean' handler. The tests can be made more sophisticated by asking animals how much energy or time they will invest in approaching or acquiring a resource, or how much pain they will tolerate to obtain something. By imposing costs such as weighted gates or lever presses, the value of an event or resource can even be quantified, allowing researchers to compare different resources using a common currency.

Not only are preference and avoidance tests valuable in assessing how important or aversive particular stimuli or resources are to subject animals, but they also provide information that allows validation of other welfare indicators (Test 2 in Chapter 2). Resources that are approached or worked for are assumed to generate positive affective states whilst avoided ones generate negative states. Therefore, if an indicator changes reliably in the presence of these preferred or avoided resources, this is evidence that it is a valid marker of, respectively, positive and negative affective valence. Furthermore, preference and avoidance tests can be used to assess changes in underlying moods that influence responsiveness to all rewards, such as anhedonia and anxiety (see Chapter 9).

References

Adámková, J., Svoboda, J., Benediktová, K., Martini, S., Nováková, P., Tůma, D., Kučerová, M., Divišová, M., Begall, S., Hart, V. and Hynek, B. (2017) *Directional preference in dogs: laterality and 'pull of the north'. Plos One*, 12(9), e0185243. https://doi.org/10.1371/journal.pone.0185243

Améndola, L. and Weary, D.M. (2019) *Evidence for consistent individual differences in rat sensitivity to carbon dioxide. Plos One*, 14(4), e0215808. https://doi.org/10.1371/journal.pone.0215808

Anderson, E.E. (1938) *The interrelationship of drives in the male albino rat. Comparative Psychology Monographs*, 14(6,72), 1–75.

Ballantyne, J.C., Carr, D.B., Chalmers, T.C., Dear, K.B., Angelillo, I.F. and Mosteller, F. (1993) *Postoperative patient-controlled analgesia: meta-analyses of initial randomized control trials. Journal of Clinical Anesthesia*, 5(3), 182–193. https://doi.org/10.1016/0952-8180(93)90013-5

Batsell, W.R. and Brown, A.S. (1998) *Human flavor-aversion learning: A comparison of traditional aversions and cognitive aversions. Learning and Motivation*, 29, 383–396. https://doi.org/10.1006/lmot.1998.1016

Berridge, K.C. (2004) *Motivation concepts in behavioral neuroscience. Physiology & Behavior*, 81(2), 179–209. https://doi.org/10.1016/j.physbeh.2004.02.004

Berridge, K.C. and Dayan, R. (2021) *Liking. Current Biology Magazine*, 31(24), R1555–R1557. https://doi.org/10.1016/j.cub.2021.09.069

Black, T.E., Fofah, O., Giray, T., Wells, H., Le Conte, Y. and Abramson, C.I. (2018) *Influence of environmental experience on aversive conditioning in honey bees (Apis mellifera L.). Apidologie*, 49, 647–659. https://doi.org/10.1007/s13592-018-0592-0

Bradshaw, R.H. (1992) *Individual attributes as predictors of social status in small groups of laying hens. Applied Animal Behaviour Science*, 34(4), 359–363. https://doi.org/10.1016/S0168-1591(05)80095-4

Burman, O.H.P. and Mendl, M.T. (2018) *A novel task to assess mood congruent memory bias in non-human animals. Journal of Neuroscience Methods*, 308, 269–275. https://doi.org/10.1016/j.jneumeth.2018.07.003

Cabanac, M. and Lafrance, L. (1990) *Postingestive alliesthesia: the rat tells the same story. Physiology & Behavior*, 47(3), 539–543. https://doi.org/10.1016/0031-9384(90)90123-L

Cabanac, M. (1990) Taste: The maximization of multidimensional pleasure. In Capaldi, E. D., Powley, T. L. (eds) *Taste, Experience, and Feeding*. American Psychological Association, pp. 28–42. https://doi.org/10.1037/10075-019

Cabanac, M. (1992) *Pleasure: the common currency. Journal of Theoretical Biology*, 155(2), 173–200. https://doi.org/10.1016/S0022-5193(05)80594-6

Colpaert, F.C., De Witte, P., Maroli, A.N., Awouters, F., Niemegeers, C.J.E. and Janssen, P.A.J. (1980) *Self-administration of the analgesic suprofen in arthritic rats: Evidence of Mycobacterium butyricum-induced arthritis as an experimental model of chronic pain. Life Sciences*, 27(11), 921–928. https://doi.org/10.1016/0024-3205(80)90101-0

Colpaert, F.C., Tarayre, J.P., Alliaga, M., Bruins Slot, L.A., Attal, N. and Koek, W. (2001) *Opiate self-administration as a measure of chronic nociceptive pain in arthritic rats. Pain*, 91(1–2), 33–45. https://doi.org/10.1016/S0304-3959(00)00413-9

Cooper, J.J. and Appleby, M.C. (1996) *Demand for nest boxes in laying hens. Behavioural Processes*, 36(2), 171–182. https://doi.org/10.1016/0376-6357(95)00027-5

Cooper, J.J., Appleby, M.C. (2003) *The value of environmental resources to domestic hens: a comparison of the work-rate for food and for nests as a function of time. Animal Welfare* 12(1):39–52. 10.1017/S0962728600025367

Cooper, J.J. and Mason, G.J. (2001) *The use of operant technology to measure behavioral priorities in captive animals. Behavior Research Methods, Instruments, & Computers*, 33, 427–434. https://doi.org/10.3758/BF03195397

Danbury, T.C., Weeks, C.A., Waterman-Pearson, A.E., Kestin, S.C. and Chambers, J.P. (2000) *Self-selection of the analgesic drug carprofen by lame broiler chickens. Veterinary Record*, 146(11), 307–311. https://doi.org/10.1136/vr.146.11.307

Dawkins, M.S. (1977) *Do hens suffer in battery cages? Environmental preferences and welfare. Animal Behaviour*, 25, 1034–1046. https://doi.org/10.1016/0003-3472(77)90054-9

Dawkins, M.S. (1980) *Animal Suffering: The Science of Animal Welfare*. Springer Science & Business Media.

Dawkins, M.S. (1983) *Battery hens name their price: consumer demand theory and the measurement of ethological 'needs'. Animal Behaviour*, 31, 351–364. https://doi.org/10.1016/S0003-3472(83)80026-8

Dawkins, M.S. (1988) *Behavioural deprivation: a central problem in animal welfare. Applied Animal Behaviour Science*, 20(3–4), 209–225. https://doi.org/10.1016/0168-1591(88)90047-0

Dawkins, M.S. (1990) *From an animal's point of view: motivation, fitness, and animal welfare. Behavioral and Brain Sciences*, 13(1), 1–9. https://doi.org/10.1017/S0140525X00077104

de Passillé, A.M., Rushen, J., Ladewig, J. and Petherick, C. (1996) *Dairy calves' discrimination of people based on previous handling. Journal of Animal Science*, 74, 969–974. https://doi.org/10.2527/1996.745969x

Díez-León, M., Quinton, M. and Mason, G. (2017) *How tall should a mink cage be? Using animals' preferences for different ceiling heights to improve cage design. Applied Animal Behaviour Science*, SI: Ethology and sustainability, 192(July 1, 2017), 24–34. https://doi.org/10.1016/j.applanim.2017.03.002

Domjan, M. (1983) Biological constraints on instrumental and classical conditioning: Implications for general process theory. In Bower, G. H. (ed.) *Psychology of Learning and Motivation*. Academic Press, pp. 215–277. https://doi.org/10.1016/S0079-7421(08)60100-0

Dumont, B. and Boissy, A. (2000) *Grazing behaviour of sheep in a situation of conflict between feeding and social motivations. Behavioural Processes*, 49(3), 131–138. https://doi.org/10.1016/S0376-6357(00)00082-6

Ede, T., Lecorps, B., Keyserlingk, M.A.G. and Weary, D.M. (2019) *Calf aversion to hot-iron disbudding. Scientific Reports*, 9, 5344. https://doi.org/10.1038/s41598-019-41798-7

Ede, T., Weary, D.M. and von Keyserlingk, M.A. (2022) *Calves are socially motivated. JDS communications*, 3(1), 44–48. https://doi.org/10.3168/jdsc.2021-0132

Elwood, R.W. and Appel, M. (2009) *Pain experience in hermit crabs? Animal Behavior*, 77, 1243–1246. https://doi.org/10.1016/j.anbehav.2009.01.028

Faure, J.M. and Lagadic, H. (1994) *Elasticity of demand for food and sand in laying hens subjected to variable wind speed. Applied Animal Behaviour Science*, 42, 49–59. https://doi.org/10.1016/0168-1591(94)90006-X

Forgas, J.P. (1995) *Mood and judgement: The affect infusion model. Psychological Bulletin Journal*, 117, 39–66. https://doi.org/10.1.37/0033-2909.117.1.39

Fraser, D. and Matthews, L.R. (1997) Preference and motivation testing. In Appleby, M. C., Hughes, B. O. (eds) *Animal Welfare*. CAB International, Wallingford, UK, pp. 159–173.

Fraser, D. (2009) *Animal behaviour, animal welfare and the scientific study of affect. Applied Animal Behaviour Science*, 118(Special Issue: Animal Suffering and Welfare), 108–117. https://doi.org/10.1016/j.applanim.2009.02.020

Galhardo, L., Almeida, O. and Oliveira, R.F. (2011) *Measuring motivation in a cichlid fish: an adaptation of the push-door paradigm. Applied Animal Behaviour Science*, 130(1–2), 60–70. https://doi.org/10.1016/j.applanim.2010.12.008

Garcia, J., Kimeldorf, D.J., Koelling, R.A. (1955). *Conditioned Aversion to Saccharin Resulting from Exposure to Gamma Radiation. Science* 122, no. 3160 (July 22, 1955): 157–58. https://doi.org/10.1126/science.122.3160.157

Grigg, E.K., Pick, L. and Nibblett, B. (2012) *Litter box preference in domestic cats: covered versus uncovered. Journal of Feline Medicine and Surgery*, 15(4), 280–284. https://doi.org/10.1177/1098612X12465606

Holm, L., Jensen, M.B. and Jeppesen, L.L. (2002) *Calves' motivation for access to two different types of social contact measured by operant conditioning. Applied Animal Behaviour Science*, 79(3), 175–194. https://doi.org/10.1016/S0168-1591(02)00137-5

Houston, A.I. (1997) *Demand curves and welfare. Animal Behaviour*, 53, 983–990. https://doi.org/10.1006/anbe.1996.0397

Hughes, B.O. and Black, A.J. (1973) *The preference of domestic hens for different types of battery cage floor. British Poultry Science*, 14(6), 615–619. https://doi.org/10.1080/00071667308416071

Hursh, S.R. (1984) *Behavioral economics. Journal of the Experimental Analysis of Behavior*, 42(3), 435–452. https://doi.org/10.1901/jeab.1984.42-435

Jensen, R.A., Martinez, J.L., Vasquez, B.J. and McGaugh, J.L. (1979) *Benzodiazepines alter acquisition and retention of an inhibitory avoidance response in mice. Psychopharmacology*, 64, 125–126. https://doi.org/10.1007/BF00427358

Kirkden, R.D., Edwards, J.S.S. and Broom, D.M. (2003) *A theoretical comparison of the consumer surplus and the elasticities of demand as measures of motivational strength. Animal Behaviour*, 65, 157–178. https://doi.org/10.1006/anbe.2002.2035

Kirkden, R.D. and Pajor, E.A. (2006) *Motivation for group housing in gestating sows. Animal Welfare*, 15, 119–130. https://doi.org/10.1017/S0962728600030207

Ladewig, J., Sørensen, D.B., Nielsen, P.P. and Matthews, L.R. (2002) *The quantitative measurement of motivation: generation of demand functions under open versus closed economies. Applied Animal Behaviour Science*, 79(4), 325–331. https://doi.org/10.1016/S0168-1591(02)00156-9

Lamon, S., Wilson, G.T. and Leaf, R.C. (1977) *Human classical aversion conditioning: Nausea versus electric shock in the reduction of target beverage consumption. Behaviour Research and Therapy*, 15, 313–320. https://doi.org/10.1016/0005-7967(77)90062-6

Lang, P.J., Bradley, M.M. and Cuthbert, B.N. (1998) *Emotion and motivation: measuring affective perception. Journal of Clinical Neurophysiology*, 15(5), 397–408. https://doi.org/10.1097/00004691-199809000-00004

Lerner, J.S. and Keltner, D. (2000) *Beyond valence: Toward a model of emotion-specific influence on judgement and choice. Cognition and Emotion*, 14, 473–493. https://doi.org/10.1080/026999300402763

Lewis, P.A., Critchley, H.D., Smith, A.P. and Dolan, R.J. (2005) *Brain mechanisms for mood congruent memory facilitation. NeuroImage*, 25, 1214–1223. https://doi.org/10.1016/j.neuroimage.2004.11.053

Lubow, R.E. (1973) *Latent Inhibition. Psychological Bulletin*, 79(6), 398–407. https://doi.org/10.1037/h0034425

Mason, G., Cooper, J. and Clarebrough, C. (2001) *Frustrations of fur-farmed mink. Nature*, 410, 35–36. https://doi.org/10.1038/35065157

Mason, G., Garner, J. and McFarland, D. (1998) *Assessing animal priorities: Future directions. Animal Behaviour*, 55, 1082–1083. https://doi.org/10.1006/anbe.1998.0730

Masterson, F.A. and Crawford, M. (1982) *The defense motivation system: A theory of avoidance behavior. Behavioral and Brain Sciences*, 5(4), 661–675. https://doi.org/10.1017/S0140525X00014114

Mauss, I.B. and Robinson, M.D. (2009) *Measures of emotion: A review. Cognition and Emotion*, 23(2), 209–237. https://doi.org/10.1080/02699930802204677

Mazur, R.L. (2010) *Does Aversive Conditioning Reduce Human—Black Bear Conflict? Journal of Wildlife Management*, 74(1), 48–54. https://doi.org/10.2193/2008-163

McConnachie, E., Smid, A.M.C., Thompson, A.J., Weary, D.M., Gaworski, M.A. and von Keyserlingk, M.A.G. (2018) *Cows are highly motivated to access a grooming substrate. Biology Letters*, 14(8), 20180303. https://doi.org/10.1098/rsbl.2018.0303

Meagher, R.K., Campbell, D.L.M. and Mason, G.J. (2017) *Boredom-like states in mink and their behavioural correlates: A replicate study. Applied Animal Behaviour Science*, 197, 112–119. https://doi.org/10.1016/j.applanim.2017.08.001

Mendl, M. and Paul, E.S. (2020) *Animal affect and decision-making. Neuroscience & Biobehavioral Reviews*, 112, 144–163. https://doi.org/10.1016/j.neubiorev.2020.01.025

Meyerson, B.J. and Lindström, L.H. (1973) *Sexual motivation in the female rat. A methodological study applied to the investigation of the effect of estradiol benzoate. Acta Physiologica Scandinavica*, 389(Suppl), 1–80.

Møe, R.O., Bakken, M., Kittilsen, S., Kingsley-Smith, H. and Spruijt, B.M. (2006) *A note on reward-related behaviour and emotional expressions in farmed silver foxes (Vulpes vulpes) – basis for a novel tool to study animal welfare. Applied Animal Behaviour Science*, 101(3–4), 362–368. https://doi.org/10.1016/j.applanim.2006.02.004

Mucha, R.F. (1987) *Is the motivational effect of opiate withdrawal reflected by common somatic indices of precipitated withdrawal? A place conditioning study in the rat. Brain Research*, 418, 214–220. https://doi.org/10.1016/0006-8993(87)90088-6

Ng, Y.K. (1983) *Welfare Economics*. Macmillan, Basingstoke.

Ng, Y.K. (1990) *The case for and difficulties in using demand area to measure challenges in well-being. Behavioral and Brain Sciences*, 13, 30–31. https://doi.org/10.1017/S0140525X00077359

Niel, L. and Weary, D.M. (2007) *Rats avoid exposure to carbon dioxide and argon. Applied Animal Behaviour Science*, 107, 100–109. https://doi.org/10.1016/j.applanim.2006.08.002

Niel, L., Stewart, S.A. and Weary, D.M. (2008) *Effect of flow rate on aversion to gradual-fill carbon dioxide exposure in rats. Applied Animal Behaviour Science*, 109(1), 77–84. https://doi.org/10.1016/j.applanim.2007.02.004

O'Donnell, S., Webb, J.K. and Shine, R. (2010) *Conditioned taste aversion enhances the survival of an endangered predator imperilled by a toxic invader. Journal of Applied Ecology*, 47, 558–565. https://doi.org/10.1111/j.1365-2664.2010.01802.x

Ogi, A., Licitra, R., Naef, V., Marchese, M., Fronte, B., Gazzano, A. and Santorelli, F.M. (2021) *Social preference tests in zebrafish: A systematic review. Frontiers in Veterinary Science*, 722, 590057. https://doi.org/10.3389/fvets.2020.590057

Olsson, I.A.S., Keeling, L.J. (2002) *The push-door for measuring motivation in hens: Laying hens are motivated to perch at night. Animal Welfare* 11:11–19. 10.1017/S0962728600024283

Öst, L.G. (1985) *Ways of acquiring phobias and outcome of behavioral treatments. Behaviour Research and Therapy*, 23, 683–689. https://doi.org/10.1016/0005-7967(85)90066-X

Pajor, E.A., Rushen, J. and de Passillé, A.M.B. (2000) *Aversion learning techniques to evaluate dairy cattle handling practices. Applied Animal Behaviour Science*, 69, 89–102. https://doi.org/10.1016/S0168-1591(00)00119-2

Pedersen, L.J., Jensen, M.B., Hansen, S.W., Munksgaard, L., Ladewig, J. and Matthews, L. (2002) *Social isolation affects the motivation to work for food and straw in pigs as measured by operant conditioning techniques. Applied Animal Behaviour Science*, 77, 295–230. https://doi.org/10.1016/S0168-1591(02)00066-7

Pedersen, L.J., Holm, L., Jensen, M.B. and Jorgensen, E. (2005) *The strength of pigs' preferences for different rooting material measured using concurrent schedules of reinforcement. Applied Animal Behaviour Science*, 94, 31–48. https://doi.org/10.1016/j.applanim.2004.11.023

Petherick, J.C., Waddington, D. and Duncan, I.J.H. (1990) *Learning to gain access to a foraging and dustbathing substrate by domestic fowl: is "out of sight out of mind"? Behavioural Processes*, 22, 213–226. https://doi.org/10.1016/0376-6357(91)90095-H

Pfaus, J.G., Erickson, K.A. and Talianakis, S. (2013) *Somatosensory conditioning of sexual arousal and copulatory behavior in the male rat: A model of fetish development. Physiology & Behavior*, 120, 1–7. https://doi.org/10.1016/j.physbeh.2013.08.005

Pham, M.T. (1998) *Representativeness, relevance and the use of feelings in decision making. Journal of Consumer Research*, 25, 144–159. https://doi.org/10.1086/209532

Piazza, C.C., Fisher, W.W., Hagopian, L.P., Bowman, L.G. and Toole, L. (1996) *Using a choice assessment to predict reinforcer effectiveness. Journal of Applied Behavior Analysis*, 29, 1–9. https://doi.org/10.1901/jaba.1996.29-1

Podturkin, A.A., Krebs, B.L. and Watters, J.V. (2022) *A quantitative approach for using anticipatory behaviour as a graded welfare assessment. Journal of Applied Animal Welfare Science*, 9, 1–15. https://doi.org/10.1080/10888705.2021.2012783

Reijgwart, M.L., Vinke, C.M., Hendriksen, C.F.M., van der Meer, M., Schoemaker, N.J. and van Zeeland, Y.R.A. (2015) *Workaholic ferrets: Does a two-chamber consumer demand study give insight in the preferences of laboratory ferrets (Mustela putorius furo)? Applied Animal Behaviour Science*, 171, 161–169. https://doi.org/10.1016/j.applanim.2015.08.032

Reijgwart, M.L., Vinke, C.M., Hendriksen, C.F.M., van der Meer, M., Schoemaker, N.J. and van Zeeland, Y.R.A. (2016) *Ferrets (Mustela Putorius furo) enrichment priorities and preferences as determined in a seven-chamber consumer demand study. Applied Animal Behaviour Science*, 180, 114–121. https://doi.org/10.1016/j.applanim.2016.04.022

Reijgwart, M.L., Vinke, C.M., Hendriksen, C.F., van der Meer, M., Schoemaker, N.J. and van Zeeland, Y.R. (2018) *Are all motivation tests the same? The effect of two adaptations to a three-chamber consumer demand study in ferrets. Animal Behaviour*, 135, 1–11. https://doi.org/10.1016/j.anbehav.2017.10.026

Rolls, E.T. (2005) *Emotion Explained*. Oxford University Press.
Rolls, E.T. (2014) *Emotion and decision-making explained: A précis. Cortex*, 59, 185–193. https://doi.org/10.1016/j.cortex.2014.01.020
Rushen, J. (1986) *Aversion of sheep to electro-immobilization and physical restraint. Applied Animal Behaviour Science*, 15, 315–324. https://doi.org/10.1016/0168-1591(86)90124-3
Saker, P., Farrell, M.J., Adib, F.R., Egan, G.F., McKinley, M.J. and Denton, D.A. (2014) *Regional brain responses associated with drinking water during thirst and after its satiation. Proceedings of the National Academy of Sciences*, 111(14), 5379–5384. https://doi.org/10.1073/pnas.1403382111
Schroeder, P., Jones, S., Young, I.S. and Sneddon, L.U. (2014) *What do zebrafish want? Impact of social grouping, dominance and gender on preference for enrichment. Laboratory Animals*, 48(4), 328–337. https://doi.org/10.1177/0023677214538239
Schulteis, G., Ahmed, S.H., Morse, A.C., Koob, G.F. and Everitt, B.J. (2000) *Conditioning and opiate withdrawal. Nature*, 405, 1013–1014. https://doi.org/10.1038/35016630
Seaman, S.C., Waran, N.K., Mason, G. and D'Eath, R.B. (2008) *Animal economics: assessing the motivation of female laboratory rabbits to reach a platform, social contact and food. Animal Behaviour*, 75, 31–42. https://doi.org/10.1016/j.anbehav.2006.09.031
Sherwin, C.M. (2003) *Social context affects the motivation of laboratory mice, Mus musculus, to gain access to resources. Animal Behaviour*, 66, 649–655. https://doi.org/10.1006/anbe.2003.2239
Shields, S.J., Garner, J.P. and Mench, J.A. (2004) *Dustbathing by broiler chickens: A comparison of preference for four different substrates. Applied Animal Behaviour Science*, 87, 69–82. https://doi.org/10.1016/j.applanim.2004.01.003
Small, D., Zatorre, R.J., Dagher, A., Evans, A.C. and Jones-Gotman, M. (2001) *Changes in brain activity related to eating chocolate: From pleasure to aversion. Brain*, 124(9), 1720–1733. https://doi.org/10.1093/brain/124.9.1720
Smith, S.F., Appleby, M.C. and Hughes, B.O. (1990) *Problem solving by domestic hens: opening doors to reach nest sites. Applied Animal Behaviour Science*, 28(3), 287–292. https://doi.org/10.1016/0168-1591(90)90108-P
Smith, M.E., Linnell, J.D.C., Odden, J. and Swenson, J.E. (2000) *Review of methods to reduce livestock depredation II. Aversive conditioning, deterrents and repellents. Acta Agriculturae Scandinavica Section A Animal Science*, 50, 304–315. https://doi.org/10.1080/090647000750069502
Spruijt, B.M., Van den Bos, R. and Pijlman, F.T. (2001) *A concept of welfare based on reward evaluating mechanisms in the brain: anticipatory behaviour as an indicator for the state of reward systems. Applied Animal Behaviour Science*, 72(2), 145–171. https://doi.org/10.1016/S0168-1591(00)00204-5
Stinus, L., Caille, S. and Koob, G.F. (2000) *Opiate withdrawal-induced place aversion lasts for up to 16 weeks. Psychopharmacology*, 149, 115–120. https://doi.org/10.1007/s002139900358
Takatsu-Coleman, A.L., Patti, C.L., Zanin, K.A., Zager, A., Carvalho, R.C., Borçoi, A.R., Ceccon, L.M.B., Berro, L.F., Tufik, S., Andersen, M.L., and Frussa-Filho, R. (2013) *Short-term social isolation induces depressive-like behaviour and reinstates the retrieval of an aversive task: Mood-congruent memory in male mice? Journal of Psychiatry and Neuroscience*, 38:259–268. https://doi.org/10.1503/jpn.120050
Tapp, P.D., Siwak, C.T., Estrada, J., Head, E., Muggenburg, B.A., Cotman, C.W. and Milgram, N.W. (2003) *Size and reversal learning in the beagle dog as a measure of executive function and inhibitory control in aging. Learning & Memory*, 10(1), 64–73. https://doi.org/10.1101/lm.54403
Thorndike, E.L. (1911) *Animal Intelligence: Experimental Studies*. The Macmillan company, New York.

Tucker, C.B., Munksgaard, L., Mintline, E.M. and Jensen, M.B. (2018) *Use of a pneumatic push gate to measure dairy cattle motivation to lie down in a deep-bedded area. Applied Animal Behaviour Science*, 201, 15–24. https://doi.org/10.1016/j.applanim.2017.12.018

Tucker, C.B., Weary, D.M. and Fraser, D. (2003) *Effects of three types of free-stall surfaces on preferences and stall usage by dairy cows. Journal of Dairy Science*, 86, 521–529. https://doi.org/10.3168/jds.S0022-0302(03)73630-3

Tzschentke, T.M. (1998) *Measuring reward with the conditioned place preference paradigm: a comprehensive review of drug effects, recent progress and new issues. Progress in Neurobiology*, 56, 613–672. https://doi.org/10.1016/S0301-0082(98)00060-4

van de Weerd, H.A., van Loo, P.L.P, van Zutphen, L.F.M., Koolhaas, J.M., Baumans, V. (1998) *Strength of preference for nesting material as environmental enrichment for laboratory mice. Applied Animal Behaviour Science* 55: 369–382. https://doi.org/10.1016/S0168-1591(97)00043-9

Varian, H.R. (1996) *Intermediate Microeconomics: A Modern Approach*, 4th ed. W.W. Norton, New York.

Venable, C.P., Adams, T.S. and Langkilde, T. (2019) *Aversion learning in response to an invasive venomous prey depends on stimulus strength. Biological Invasions*, 21, 1973–1980. https://doi.org/10.1007/s10530-019-01949-3

Vinke, C.M., Houx, B.B., van den Bos, R. and Spruijt, B.M. (2006) *Anticipatory behaviour and stereotypical behaviour in farmed mink (Mustela vison) in the presence, absence and after removal of swimming water. Applied Animal Behaviour Science*, 96, 129–142. https://doi.org/10.1016/j.applanim.2005.04.022

von Keyserlingk, M.A.G., Amorim Cestari, A., Franks, B., Fregonesi, J.A. and Weary, D.M. (2017) *Dairy cows value access to pasture as highly as fresh feed. Scientific Reports*, 7, 44953. https://doi.org/10.1038/srep44953

Warburton, H. and Mason, G. (2003) *Is out of sight out of mind? The effects of resource cues on motivation in mink, Mustela vison. Animal Behaviour*, 65(4), 755–762. https://doi.org/10.1006/anbe.2003.2097

Warner, L.H. (1927) *A study of sex drive in the white rat by means of the obstruction method. Comparative Psychology Monographs*, 4, 1–67.

Wenker, M.L., Bokkers, E.A., Lecorps, B., von Keyserlingk, M.A., van Reenen, C.G., Verwer, C.M. and Weary, D.M. (2020) *Effect of cow-calf contact on cow motivation to reunite with their calf. Scientific Reports*, 10(1), 14233. https://doi.org/10.1038/s41598-020-70927-w

Winkielman, P., Berridge, K.C. and Wilbarger, J. (2005) *Unconscious affective reactions to masked happy versus unhappy faces influence consumption behaviour and judgements of value. Personality and Social Psychology Bulletin*, 31, 121–135. https://doi.org/10.1177/0146167204271309

Wong, D., von Keyserlingk, M.A.G., Richards, J.G. and Weary, D.M. (2014) *Conditioned place avoidance of Zebrafish (Danio rerio) to three chemicals used for euthanasia and anaesthesia. Plos One*, 9(2), e88030. https://doi.org/10.1371/journal.pone.0088030

Wystrach, A., Buehlmann, C., Schwarz, S., Cheng, K., and Graham, P. (2020) *Rapid aversive and memory trace learning during route navigation in desert ants. Current Biology*, 30(10):1927–1933.e2. https://doi.org/10.1016/j.cub.2020.02.082

Vocal Signals

4

Avelyne S. Villain[1,2] and Elodie F. Briefer[1]

[1]Behavioural Ecology Group, Section for Ecology and Evolution, Department of Biology, University of Copenhagen, Copenhagen, Denmark
[2]GenPhySE, Université de Toulouse, INRAE, ENVT, Castanet-Tolosan, France

Introduction

According to the componential view, emotions are characterised by coordinated changes in five components: neurophysiological activity, motivation (or action), cognition, conscious awareness (or feeling), and expressions (Paul et al. 2005; Scherer and Ekman 2014; see also Chapter 1). Expressions of emotion (facial, vocal, and postural appearance) thus constitute a component on their own. The emotional aspects of these expressions likely first appeared as unintentional cues to a producer's inner state, emitted as mere consequences of emotion-related changes in underlying physiology, before evolving to have a communicative function through a ritualisation process and becoming signals aimed at conspecifics (Shariff and Tracy 2011). The perception of these signals by conspecifics can also lead to emotional contagion (e.g. Briefer 2018; Herrando and Constantinides 2021).

The most studied modes of expression are facial (Waller and Micheletta 2013) and vocal (e.g. Briefer 2012, 2020). Whether these expressions are cues (i.e. unintentional or involuntary, e.g. resulting from changes in the tension and action of muscles used for voice production), or signals evolved to communicate emotions or behavioural intentions, has been highly debated (Heesen et al. 2022; Waller et al. 2017). Nevertheless, there is ample evidence, particularly for vocal expressions, of a link to the emotion of the producer, in both non-human animals and humans (Scheiner and Fischer 2011; Zimmermann et al. 2013), and vocal indicators of emotions have been observed across species (Briefer 2012, 2020). Because they can be recorded non-invasively and can provide a real-time measure of the producer's affective state, vocalisations are therefore promising indicators of animal welfare (Manteuffel et al. 2004).

Vocalisations can be classified into different *types* (e.g. dog bark vs growl) based on their function in an animal's life (e.g. contact call, alarm call, food calls) and/or their acoustic structure (e.g. whinny, grunt, bark; Bradbury and Vehrencamp 2011). The type of vocalisation emitted may vary according to the animal's emotional state (i.e. certain emotions are associated with specific call types), as may the acoustic structure of a specific type of vocalisation (e.g. emotion-related variation in amplitude, frequency or duration; Briefer 2012). To capture all possible emotion-related changes to vocalisations, we will provide an overview of the effect of emotional valence on (i) vocalisation types, (ii) their occurrence over time (rate) and (iii) the main acoustic parameters that characterise their structure (Table 4.1; Figure 4.1).

In this chapter, we first investigate if, based on current evidence, vocalisations have construct validity as indicators of emotional valence across species. We then explore the responsiveness and selectivity of vocalisation types, rates and acoustic parameters. Last, we give guidelines and recommendations for best practice when assessing emotions using vocalisations, based on known indicators already described *and* when investigating a new species. Since the literature on vocal expression of emotions has focused heavily on mammals (Briefer 2012, 2020; Scheiner and Fischer 2011; Zimmermann et al. 2013), most examples will be mammalian, with a few from bird species. In addition, since the aim of this book is to highlight indicators of animal welfare, we mainly give examples of farm and laboratory animals as well as some species kept in zoos.

Because most vocalisations are produced under a certain level of bodily activation or arousal (see e.g. Kiley 1972) and associated with action (e.g. active defence or appetitive behaviour), they are particularly useful indicators of valenced states that are of medium to high arousal. For these types of affective state, it has been argued that arousal/activation corresponds to 'motive intensity', or degree of motivation to 'do something', which in turn may reflect the intensity of emotional valence (Bradley et al. 2001; referred to as the 'arousal as intensity' model or 'symmetric V-shaped relation' between arousal and valence, Kuppens et al. 2013). If so, and if vocalisations reliably indicate affective arousal for these emotional states (many studies have investigated this relationship), they may also be useful markers of the intensity of valence, similarly to the glucocorticoid and heart-rate indicators described in Chapters 5 and 6, as long as the direction of valence is already known. However, it should be noted that this ignores those affective states in the lower half of the valence/arousal space, where the intensity of the valenced state may not correlate positively with arousal at all (e.g. boredom, intense depression, calmness, intense feelings of relaxation/serenity). For these states, vocalisations may not be useful indicators of valence and, as mentioned earlier, may rarely be produced.

The construct validation of vocal expressions as indicators of emotional valence

In this section, we review evidence suggesting that vocal expressions of emotional valence have construct validity, according to the five key validation methods (Tests 1–5) described in Chapter 2.

Do vocalisations reliably change in animals exposed to stimuli that they prefer or avoid?

Exposure to preferred or avoided stimuli (which is Test 2 in Chapter 2) has been used in several species to study vocal expressions of emotions. In pigs (*Sus scrofa domesticus*), Reimert et al. (2013) demonstrated that when pairs of pigs were conditioned to expect access to a preferred environment (an enriched room associated with tail movements and play behaviours; see Chapter 8) versus an aversive environment (social isolation and unpredictable human intrusions, associated with the expression of negative behaviours, such as more

Table 4.1 Definitions of measurable parameters that can be extracted in vocalisations.

	Abbreviation	Name	Unit[a]	Definition
Descriptors of vocal behaviour:	**Vocalisation/ call type**			Category of vocalisation or call defined by a specific acoustic structure and/or function (e.g. *contact call*, *alarm call*, *bark*, *growl*).
	Rate	Rate of vocalisations	s^{-1}	Number of vocalisations per time unit.
Descriptors of each vocalisation (acoustic parameters)				
In the time domain:	**Duration**		s^{-1}	Duration of a single vocalisation.
In the frequency domain:	**Energy distribution**		Hz	Distribution of energy in the spectrum (i.e. frequency vs. time), representing how low or high in frequency the vocalisation is overall.
	Q50	Median	Hz	Parameter characterising the energy distribution; frequency below which 50% of the energy is contained.
	F_{peak}	Frequency peak	Hz	Frequency of highest amplitude (i.e. loudest frequency in the vocalisation).
	F1, F2, F3, etc.	Formants	Hz	Concentration of acoustic energy (band) around particular frequencies in the vocalisation, corresponding to the resonances of the vocal tract.
	f_o	Fundamental frequency	Hz	Fundamental frequency, i.e. lowest frequency of the vocalisation, usually accompanied by a set of 'harmonics' (multiples of f_o). Related to perceived pitch.
	FM	Frequency modulation		Variation in the main frequency parameters relative to the duration of a vocalisation (includes down sweeps and up sweeps).
In the amplitude domain:	**Amplitude**		dB	Level of energy in the vocalisation (intensity or energy). Related to perceived loudness.
	AM	Amplitude modulation		Variation in amplitude relative to the duration of the vocalisation.

[a] dB, decibel; Hz, Hertz; s, seconds.

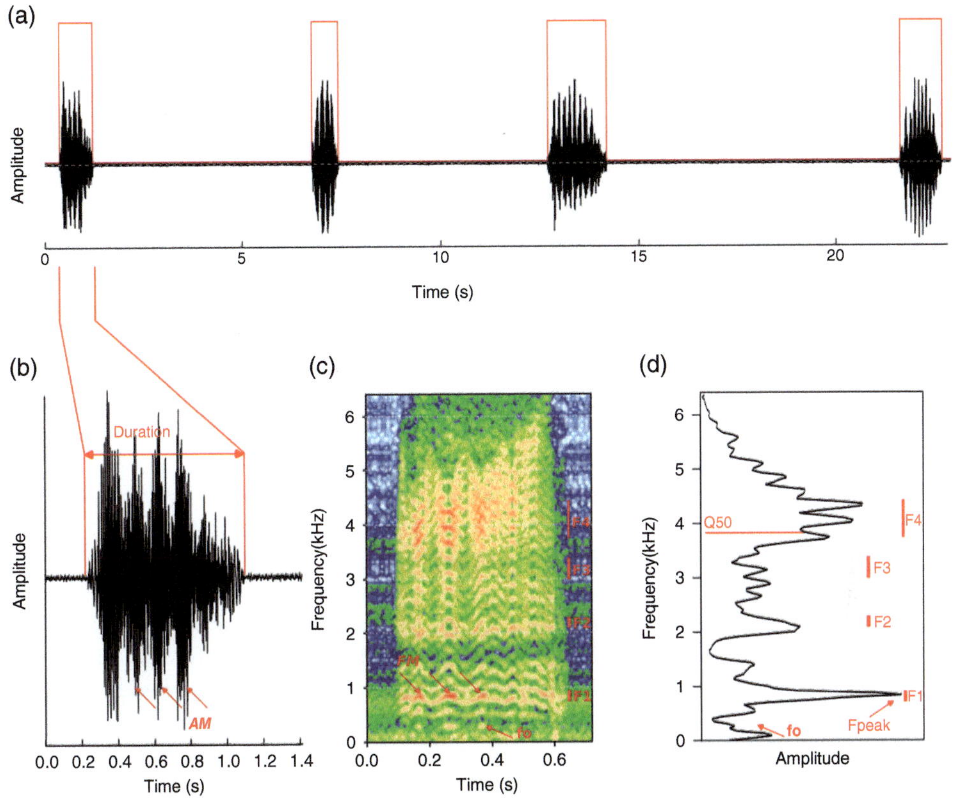

Figure 4.1 Descriptors of vocal behaviour (a) and vocalisation structure (b–d). (a) Oscillogram (amplitude vs. time) of a series of four goat *bleats*, from which the rate can be extracted. (b) Oscillogram of the first *bleat*, indicating the *duration* and *amplitude modulations* (*AM*). (c) Spectrogram (frequency vs. time; relative amplitude is indicated with a colour code) of the first *bleat*, indicating the fundamental frequency (f_o), the *formants* (*F*1–4) and *frequency modulations* (*FMs*). (d) Normalised spectrum (frequency vs. relative amplitude, Fast Fourier Transformation window length = 0.02 s, overlap = 50%) of the first *bleat*, indicating the f_o, *F*1–4, the *frequency peak* (F_{peak}) and the *Median* (*Q50*, i.e. frequency below which 50% of the energy is contained).

defaecation and 'low tail' postures), the pigs produced different call types; *barks* were only heard during the positive treatment, whereas *screams* and *squeals* were only heard in the negative treatment. *Grunts* were produced in both, but in a larger quantity in the negative treatment (see also Briefer et al. 2019 for a similar result). These results suggest that pig *play barks* may be a valid indicator of positive emotional states (but see Chan 2011 for a description of *alarm barks* produced during negative situations), whereas *screams* and *squeals* are more likely to indicate negative emotional states. Other studies have focused on analysing the acoustic structure of vocalisations rather than the type of call given. For example, pigs conditioned to expect preferred stimuli (applesauce, toys, straw) produced *grunts* with a higher *peak frequency* and *energy distribution* than those produced by pigs anticipating aversive stimuli (e.g. sudden noises; Leliveld et al. 2016). Similarly, Villain et al. (2020a) conditioned pigs to anticipate a reunion with familiar conspecifics (a preferred event). This resulted in subjects spending a lot of time next to or watching the door through which the conspecifics were about to enter. The anticipated reunion was then delayed to generate an aversive frustration phase. *Grunt duration* significantly increased from the non-delayed anticipation phase (positive) to the frustration phase (negative), providing evidence that this parameter is a valid indicator of valence. In another pig study using a spatial task, Friel et al. (2019) studied valence-related changes in the structure of pig *grunts* when either approaching a reward, versus withholding such responses, or showing avoidance during negative trials. Subjects showing approach in positive trials produced shorter grunts than those of subjects in negative trials, further suggesting that *grunt duration* is a good indicator of emotional valence (see also Briefer et al. 2022).

In chickens (*Gallus gallus domesticus*), similar paradigms were used to study vocal expression during reward anticipation (preferred; positive) and frustration (avoided; negative). Chickens produced more *gakel calls* during frustration compared to pre-frustration trials, supporting the idea that this call type could indicate negative affect (Zimmerman et al. 2000). Likewise, hens anticipating a conditioned reward (food; dustbathing substrates) were more likely to produce *food calls* and *fast clucks* than *gakel calls*, supporting the idea that *food calls* and *fast clucks* could indicate positive states (McGrath et al. 2017).

The effect of preferred or aversive stimuli on vocalisations has also been studied through rewarding electrical stimulation of the brain in rats (*Rattus norvegicus*, Burgdorf et al. 2000) and squirrel monkeys (*Saimiri sciureus*, Fichtel et al. 2001). Anticipation of a conditioned experimenter-delivered brain simulation of the reward system led to an increase in the *rate* of *50 kHz* ultrasonic vocalisations (USVs) in rats (Burgdorf et al. 2000). Furthermore, rat *50 kHz* vocalisations are also emitted during social play (Burke et al. 2018). Since play is generally a rewarding activity (Ahloy-Dallaire et al. 2018; see Chapter 8), this further supports these vocalisations as indicating positive affect. Indeed, in several other species, the occurrence of certain vocalisations, such as kea parrot (*Nestor notabilis*) play calls (Schwing et al. 2017), also co-varies with play (Winkler and Bryant 2021).

Do vocalisations reliably change in animals exposed to ancestrally fitness-boosting or fitness-harming situations?

The effect of exposure to threats and/or promoters of ancestral fitness (Test 3 in Chapter 2) on vocal production has been studied in several farm and zoo animals. For example, in domestic horses (*Equus caballus*), exposure to contexts in which their companion or whole group was removed (likely fitness harming for this prey species, and hence negative; van Dierendonck 2006), induced *whinnies* with longer *durations* and a higher f_0 than exposure to contexts in which the companion or whole group returned (likely fitness boosting, and hence positive; Briefer et al. 2015a). In goats (*Capra hircus*), *bleats* produced during food anticipation (watching the experimenter approaching with a food bucket, considered as a fitness-boosting situation) had a more stable f_0 (i.e. which showed less variation) than *bleats* produced during short-term social isolation or food frustration (considered as fitness-harming situations; Briefer et al. 2015b). Using a similar approach, Briefer et al. (2019) compared domestic pig *grunts* produced during short-term isolation (fitness harming) versus while having access to food, toys and water together with a pen mate (fitness boosting). This test showed that positive *grunts* are notably shorter in *duration*, with higher *formants* but a lower f_0, when compared to negative *grunts*. In cattle, milk deprivation of the calf and mother–calf separation (both assumed to be fitness-harming situations) resulted in higher call *rates*, and calls with a higher f_0 and *amplitude* (see Green et al. 2018 for a review). However, the lack of comparison with recordings of calls in positive situations precludes us from making clear conclusions regarding the extent to which these parameters reflect emotional valence (an issue revisited in the selectivity section later). These findings nevertheless reveal that vocal expression in cows changes according to the intensity of negative valence.

In zoo animals (who cannot be easily manipulated), vocal indicators of emotions have been assessed in naturally occurring social contexts that are likely fitness boosting/harming. In wild boars (*Sus Scrofa*), more *screams* and *squeals* were produced during agonistic interactions (aversive and potentially injurious, hence fitness harming) while more *grunts* were produced during anticipation of food or affiliative interactions (fitness boosting). In addition, *formants* and *energy distribution* were lower, and call *durations* were shorter, across all call types during positive compared to negative contexts (Maigrot et al. 2018). In Przewalski's horses (*Equus przewalskii*), more *squeals* and *whinnies* were produced during agonistic interactions and social isolation (fitness harming), while *nickers* were more specific to affiliative interactions and food anticipation (fitness boosting). In addition, the *energy distribution* in the calls decreased from negative to positive contexts across all call types (Maigrot et al. 2017).

Do vocalisations reliably change in animals given affect-modulating drugs?

Effects of affect-modulating drugs (Test 4 in Chapter 2) have been investigated not only in rats and mice, but also other species such as pigs. In rats, injecting amphetamine (a drug that rats are willing to work for) increased the *rate* of *50 kHz* USVs (Burgdorf et al. 2001a), while injections of naloxone or lithium chloride (LiCl) (drugs both generating sickness behaviour) decreased their *rate*, and instead elicited the production of *22 kHz* vocalisations (Burgdorf et al. 2001b). In mice, amphetamine induced frequency-modulated USVs in adult males, suggesting that these vocalisations indicate positive affect (Wang et al. 2008). In zebra finches (*Taeniopygia guttata*), some effects of social isolation on the structure of the *distance call* (a contact call) were mimicked after oral administration of corticosterone (stress-mediating hormone) in socially housed subjects, demonstrating stress-related changes (negative valence intensity) in vocalisation structure; longer *duration* and higher *frequencies* (Perez et al. 2012). Finally, in pigs, stimulation of the amygdala with acetylcholine provoked vocalisations with a similar structure as *scream* calls. As the amygdala mediates fear and stress, this further suggests a link between fear, stress and *screams*, and suggests that pig *screams* indicate a state of negative valence of high intensity (Manteuffel et al. 2007). We will also present further examples later, from pigs castrated while treated with different types and degrees of pharmacological pain control.

Do vocalisations reliably co-vary with pre-validated affect indicators?

Another method to assess the construct validity of vocal indicators of emotions is to test if they co-vary with validated indicators of affect (Test 5 in Chapter 2). This has often been tested in the aforementioned studies, in order to obtain further validation of the underlying emotional states of vocalising animals. For instance, when studying vocal expression of emotions in goats, horses, pigs, wild boars and Przewalski's horses, we used Test 2 or 3 (see previous sections) from Chapter 2 to infer valence *a priori* based on characteristics of the recording contexts, while data on other variables were also collected for both *a posteriori* validation and to obtain information on emotional intensity (Briefer et al. 2015a, 2015b; Maigrot et al. 2017, 2018). In all cases, changes in acoustic parameters in valenced contexts were associated with behavioural changes previously validated as indicators of negative and positive emotions (e.g. ear and tail postures, head movements and position).

Further, in much work on domestic species, animals' heart rates were used to assess the intensity of the emotional state from the degree of physiological arousal induced (see also Chapter 6). Thus, several studies showed correlations between physiological parameters likely to increase with emotional intensity (see Chapters 5 and 6 on cortisol and heart rate, respectively), and acoustic parameters. Generally, increased heart rate, cortisol or adrenaline plasma levels in situations of known negative valence correlate with higher vocalisation *rates* (long *grunts* and *squeal-grunt* in pigs, Marchant et al. 2001; Schrader and Todt 1998), higher f_0 (rhesus macaque, *Macaca mulatta*, Schwartz et al. 2022; prairie voles, *Microtus ochrogaster*, Stewart et al. 2015), and also shorter *durations* (prairie voles, Stewart et al. 2015). Of course, these aspects may potentially increase in positive states as well, which needs to be checked, since this would then cause selectivity problems if the direction of valence is not known (as we revisit later).

Do vocalisations accurately track self-reported affective states in humans?

Assuming that changes in vocal production are homologous across species, we finally investigate if vocalisations accurately track self-reported affective states in humans (Test 1 in Chapter 2), and whether vocal indicators of affect in humans may be transferred to other species. Based on the theories of vocal production, Scherer (1986) made detailed predictions about how speech acoustic features should vary according to the emotional state of the producer. Effects related to emotional valence are harder to predict than those related to intensity/arousal (e.g. increased call *rates* linked to increased respiration rate, and increased *amplitude* and *frequencies* linked to increased sub-glottal pressure and muscle tension at level of the larynx and vocal tract walls). However, effects related to emotional valence could result from valence-specific body postures and facial expressions (e.g. mouth opening, smiling; Scherer 1986). For example, positive outcomes seem to lead to a *wider* voice, characterised by a lower *energy distribution* and a lower frequency of the first *formant*. By contrast, negative outcomes are associated with a *narrow* voice, characterised by a higher *energy distribution* and frequency of the first *formant*, but lower values of the second and possibly also third *formants* (Scherer 1986). Higher *frequencies* are also commonly observed in contexts of negative compared to positive valence in non-human animals, at least when focusing on the frequency of the fundamental (f_0; Briefer 2020).

Expression of emotions in human speech is, however, also impacted by sociocultural effects. Indeed, vocal production learning affects both speech ontogeny (long-term vocal plasticity) and social communication strategies (short-term controlled vocal flexibility through social representation). To counteract such biases, some studies have focused either on spontaneous non-verbal vocalisations in adults (e.g. Anikin et al. 2018) or on vocalisations emitted before the development of speech (in babies), which are supposedly less subjected to sociocultural effects and imitation processes. Such studies have shown, for example, that baby cries emitted during heel-pricks for neonatal screening under different analgesic procedures differ according to the degree of pain (assessed based on facial expression and behaviour; Bellieni et al. 2004). Overall, across ages, higher levels of pain lead notably to cries of higher *amplitudes* and higher f_0 (Helmer et al. 2020; Lautenbacher et al. 2017). These findings are similar to those quantified in non-human animals when negative valence intensity increases (Briefer 2020).

Earlier, we focused on homologies of vocal *structure* between humans and other species. But can vocalisation *types* ever show homology? This is probably far less likely, as species-specific selection pressures drove the evolution of vocalisations. Nevertheless, some have argued that rat *50 kHz* vocalisations might be evolutionarily similar to human laughter (Burgdorf and Moskal 2010), and even that *22 kHz* vocalisations could be an evolutionary homologue of human crying (negative affect; Brudzynski 2019). Furthermore, *screams* appear to be conserved in structure, function and affective significance across many diverse species (Schwartz et al. 2020).

Responsiveness and selectivity

In this section, we review the responsiveness (i.e. risks of false nulls) and selectivity (risks of false leads) of vocal indicators of emotional valence. We focus here on biological effects, rather than additional methodological issues like recorder proximity or the problems of comparing animals recorded indoors versus outdoors.

Responsiveness issues – When are changes in emotional valence not accompanied by vocal production/change in vocalisation structure?

Changes in valence might not always be reflected by a change in vocal production. This can occur in several specific cases, as outlined here. First, due to their communicative function, vocalisations generally require an audience (e.g. conspecifics, and/or humans in the case of domestic animals). There might thus be situations (e.g. total isolation) where animals will call less despite experiencing negative emotions of high intensity, while they will be more vocal if conspecifics are around (e.g. Siebert et al. 2011). Such 'audience effects' have been shown in species as varied as chickens (Karakashian et al. 1988) and chimpanzees (*Pan troglodytes*; e.g. Soldati et al. 2022). In a classic example, Marler et al. (1986) showed that roosters produce more *food calls* when females (but not males) are around. Karakashian et al. (1988) also showed that roosters only produce *alarm calls* when conspecifics are around, and refrain from calling in the presence of heterospecifics. As food and alarm contexts are both expected to elicit rather intense emotions (positive and negative, respectively), the absence of vocal production in the absence of a certain audience suggests that animals refrain from calling despite experiencing intense emotions (although of course such effects could also be true negatives, the absence of appropriate audience decreasing the emitter's emotional intensity). The effect of the audience should thus be accounted for when using vocalisations to assess emotions, and potential audience effects should be tested (see Townsend et al. 2017).

In addition to audience effects, threshold effects might occur. As mentioned in the Introduction, most vocalisations are produced during states of medium to high bodily activation or arousal. Low arousal states, no matter how intensely valenced (e.g. calm, bored, depressed) will thus, in most cases, not be accompanied by vocalisations. Vocal expressions may therefore show low or no responsiveness at all to changes of valence in low arousal states (although in these cases, a reduction in vocal *rate* might be relevant). Furthermore, large variation between species exists in the activation threshold for eliciting a vocalisation, determining how vocal individuals are. For instance, pigs or chickens, who naturally live in more closed habitats (i.e. forests), vocalise when foraging for food to maintain group cohesion, and will therefore produce many vocalisations even in low to medium arousal states (Zimmerman et al. 2000). By contrast, other species such as cattle, horses and sheep that live in more open habitats are less vocal since they can rely on other modalities to communicate (e.g. vision) that are less prone to predator detection than acoustic signals. These species will thus only start vocalising under medium to high arousal states (Kiley 1972). Similarly, since call *types* usually have a specific function (e.g. indicating the presence of a predator or food, attracting mates or male competition), their production will often be bound to a specific context or a limited range of contexts, and to a specific sex (e.g. fallow deer groans, bird territorial songs) or age (e.g. mouse pup isolation calls). Some call *types* will thus reflect specific affective subtypes, rather than emotional valence, and only in specific individuals (e.g. very young or old). In these contexts, vocalisations also have low responsiveness to changes in valence.

Finally, many vocal indicators might be inclined to ceiling effects. Since their features are constrained by the dimensions and functioning of the vocal apparatus, there might be instances where valence intensity increases without a corresponding change in *amplitude*, which is limited by lung capacity; f_o, which depends on the characteristics of the vocal folds in mammals; or *formant* frequencies, which depend on the length and shape of the vocal tract (Titze and Martin 1998).

Selectivity issues – When does variation in vocal production/structure not reflect changes in underlying emotional valence?

Vocal indicators of emotions are also prone to selectivity issues, such as changes that are unrelated to emotional valence. For example, as communicative signals, vocalisations function to transmit many pieces of information that are not limited to affective states. The main types of information encoded in vocal signals, which can be more or less static, are producer identity (species, population, sex and individual) and characteristics (body size, weight, age and dominance rank), as well as more affectively-relevant external events (predators, presence of food; *functionally referential signals*, Townsend and Manser 2013), and internal states (motivation, emotions) (Aubin and Mathevon 2020). As a result, acoustic parameters of vocalisations (*amplitude*, *duration*, *frequencies*; Table 4.1) will often be influenced by factors unrelated to affect. For instance, in mammals, older individuals, males and more dominant individuals are often larger than younger, female and less dominant individuals, respectively. This will result in larger lungs, longer and thicker vocal folds and longer and larger vocal tracts, meaning an ability to produce longer vocalisations of higher *amplitude*, lower f_o and lower *formants* and a lower *energy distribution* (Taylor and Reby 2010). Not controlling for age, sex or individuality might thus, for instance, lead to a wrong assessment of vocalisations produced by females and younger individuals as indicating high emotional arousal, as revealed in the case of zebra finch *distance calls*, for which the identity of the caller must be known to interpret these calls accurately in terms of stress and negative affective valence (Soula et al. 2018).

Call *types* can also have very different features, ranging from low-frequency (LF) *groans* to high-pitched *squeals*. Such features are often adapted to the call function (e.g. short- vs long-range communication, aggressive vs. appeasing function; Morton 1977). Due to these constraints acting on the acoustic structure of vocalisations, the way emotional valence, and even sometimes valence intensity, is reflected in acoustic parameters can vary between vocalisation types and species (Linhart et al. 2015; Maigrot et al. 2017, 2018). This presence of different vocalisation *types* also means that acoustic parameters will be much more different between than within call types. A jump in acoustic parameter, such as f_o for instance, could thus reflect a change in vocalisation *type* (e.g. from *grunts* to *barks* in pigs), which might not necessarily be accompanied by the same level of change in emotional state.

Some audience effects (as described earlier) can also decrease the selectivity of vocal indicators of emotions. For example, attacked chimpanzees were shown to emit *screams* that reflected a higher level of received aggression than the assessed level (i.e. *screams* similar to those given in cases of very severe aggression although the actual assessed level was less severe), when a conspecific in the audience was of equal or higher rank to the aggressor (Slocombe and Zuberbühler 2007). Such *exaggeration* of acoustic parameters in the presence of a specific audience can be interpreted as evidence for intentionality in vocal production (Townsend et al. 2017) and might hinder emotion assessment, as there is a risk that vocal expressions do not closely match the internal state of the producer, particularly in species that can control the features of their vocalisations (e.g. play-acted vocal emotion expressions in humans; Jürgens et al. 2011).

Finally, as a result of the strong impact of arousal-related physiological changes to the vocal production apparatus, arousal modifies many acoustic parameters. For instance, according to Scherer (1986), increased arousal (which may reflect either positive or negative valence intensity for some types of affective state as discussed in the Introduction) should lead to higher respiration rates, higher subglottal pressure, higher muscle tension in the larynx, inhibition of salivation at the level of the vocal tract and pharyngeal constrictions. Consequently, these anatomical and physiological changes should lead to an increased *rate* of vocal production, the production of vocalisations of higher *amplitude*, with a higher f_0, higher *energy distribution* and higher *formants* (Scherer 1986). These predictions made for humans should be valid for most vertebrates, at least those sharing similar vocal production mechanisms, due to the highly conserved physiological arousal mechanism and neurophysiological processes of voice production (Andics and Faragó 2018). These highly predictable changes have been evidenced across species and should be kept in mind when investigating vocal expression of emotional valence (reviewed in Briefer 2012, 2020), as this might result in a similar acoustic structure in both positive and negative states of comparable intensity. For example, in African elephants (*Loxodonta africana*), rumbles produced during both dominance interactions (avoided/potentially injurious; negative) and affiliative interactions (preferred; positive) had a more variable f_0, higher *formants*, and were longer in *duration* compared to those produced in neutral social contexts (minimal social activity), suggesting that changes in these parameters reflect intensity, but not whether the state was positively or negatively valenced (Soltis et al. 2011). In general, contact calls, such as elephant rumbles, are produced during both isolation (negative) and reunion (positive). Their production will hence not be valence specific. However, their acoustic parameters (i.e. structure) will often differ between negative and positive contexts and should thus be useful to assess emotional valence (e.g. Briefer et al. 2015a, 2015b).

Heuristic relationships between vocalisations and emotional valence: a case study of pig calls

Based on the evidence reviewed in the construct validity section, and the sensitivity and selectivity issues explained earlier, heuristic graphs can be drawn that illustrate how vocalisations may change with different intensities of emotional valence. Since differences exist between species in how vocalisation parameters relate to emotional valence (Briefer 2020), we will here focus on pigs: an intensively studied species, providing a decent corpus of data to inform the discussion on the validity of several acoustic parameters for tracking emotional states. We will, again, assume that arousal reflects the intensity of emotional valence, and hence that more negative or more positive emotional states are also higher in arousal (apart from low-arousal emotional states, as discussed in the Responsiveness and selectivity section). Since emotion-related changes to vocal indicators in pigs and other species are often call-type specific, we will also take into account differences between pig low-frequency calls (*LF*; mainly *grunts* and *barks*) and high-frequency (*HF*) calls (mainly *screams* and *squeals*).

Call types

The number of different call types produced by pigs has been estimated to be up to 12 (Kiley 1972), the most common being *grunts*, *screams* and *squeals* (Borell et al. 2009). More recently, Tallet et al. (2013) redefined pig vocal repertoire using a clustering approach and investigated their corresponding contexts of production and associated emotional valence. For this chapter, we will map the results on piglet vocalisations of Tallet et al. (2013) based on Test 2 (preference/avoidance) and Test 3 (affecting ancestral fitness) from Chapter 2. Accordingly, situations that should be preferred by piglets and/or can be labelled as fitness boosting are assumed to reflect positive valence (e.g. before nursing, huddling, social reunion and after nursing), while those that should be avoided by piglets and/or labelled as fitness harming are estimated to reflect negative valence (fighting during nursing, missed nursing, restraint in human arms, social isolation, simulated crushing and castration; see Tallet et al. (2013) for description).

The approach by Tallet et al. (2013) led to two possibilities: a two call-type clustering (*HF* and *LF* calls), or a five call-type clustering (see legend of Figure 4.2: *HFs* – high-frequency stable calls (*squeals*), *HFm* – high-frequency-modulated calls (mainly *screams*), *LFs* – low-frequency stable calls (*low grunts*), *LFm* – low-frequency-modulated calls (*high grunts/barks*) and *LFt* – tonal calls (*croaks/barks*)). *Squeals* described in previous studies correspond to

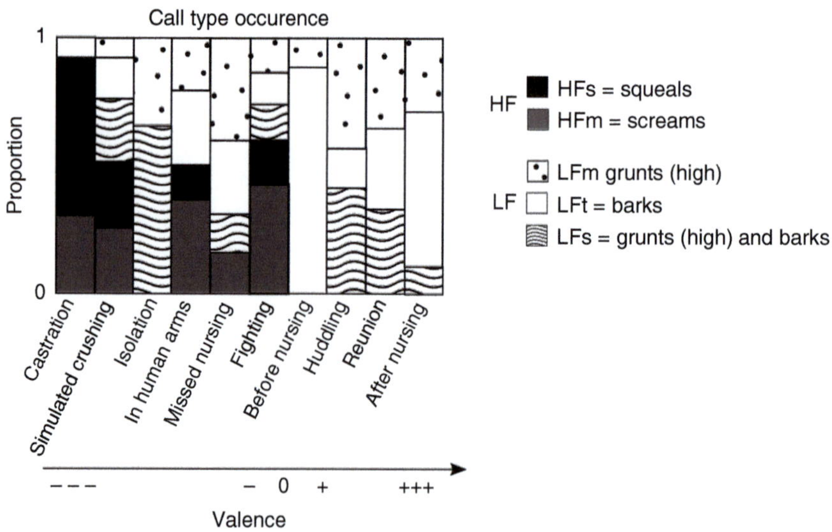

Figure 4.2 Occurrence of call types as a function of emotional valence in piglet vocalisations, according to recorded contexts, from most negative to most positive. High-frequency calls (*HF*) are divided into *HFs* (stable frequency) and *HFm* (modulated frequency). Low-frequency calls (*LF*) are divided into *LFm* (modulated frequency), *LFs* (stable frequency) and *LFt* (tonal). The correspondence with other classifications is also indicated. *HF* calls are produced during negative emotional valence mostly, while *LF* calls are produced during both negative and positive emotional valences. Source: Adapted from Tallet et al. (2013).

either *HFs* or *HFm*, while *barks* may be classified as *LFt* or *LFm*. *HF* calls were produced in higher proportions during castration, crushing, fighting and human handling, with more *HFm* during fighting and human handling than *HFs*. *LF* calls were produced in higher proportions during social reunion, huddling and isolation, with more *LFt* calls before and after nursing, more *LFs* calls during isolation and intermediate *LFs* and *LFt* during social reunion. In addition, there was a significant association between the occurrence of *HFs* calls and negatively-ranked contexts on one hand, and between the occurrence of *LFt* calls and less negatively-ranked contexts on the other hand (while the proportion of *LFm* was unaffected). Therefore, overall, *HF* calls specifically reflect negative valence, while *LF* calls can be produced in both negative and positive states and are thus less valence specific (Figure 4.2).

Vocalisation rate

Temporal aspects of vocal activity, such as series of elements within a vocalisation or series of vocalisations, may also give information about emotional state. In pigs, as mentioned previously, the *rates* (number of calls per time unit; Table 4.1) of the main pig call types (*grunts* [*LF*] and *screams* and *squeals* [*HF*]) vary as a function of valence intensity. As evidence of a graded responses, Marx et al. (2003) found that the *rate* of all piglet call types increased in the following order: restraint with anaesthesia, restraint without anaesthesia, castration with anaesthesia and castration without anaesthesia. This suggests that call *rate* is sensitive to changes in negative valence intensity (see Puppe et al. (2005) for similar results). *LF rate*, although rather low, increased from the pre-castration presurgical handling period to the surgical period during castration, and further during the postsurgical handling period, while *HF rate* remained high throughout, suggesting that the *LF rate* is more sensitive to small changes in negative valence intensity compared to *HF rate* (Puppe et al. 2005). This could, however, also be explained by a ceiling effect in the production of *HF*, as Hillmann et al. (2004) found that *HF rate* increased rather steadily with decreasing temperature (i.e. increasing discomfort; Test 2 in Chapter 2). Weary and colleagues also found higher call *rates* (mainly *HF* calls) in situations causing more intense negative valence (Weary et al. 1997, 1998; Weary and Fraser 1995). Therefore, most studies suggest that the *rates* of both *HF* and *LF* increase with negative valence intensity (but see Schrader and Todt 1998). Call *rate* seems also sensitive to valence intensity within positive contexts: Villain et al. (2020a) found that pigs produced grunts (*LF*) at a higher *rate* when anticipating a reunion with familiar conspecifics, a context suggested to be of higher intensity, and hence to be more positive, based on behavioural indicators (higher mobility indicating higher motivation to reunite) than a reunion with a familiar human. When comparing positive and negative valence of the same intensity (as shown by a similar heart rate), Briefer et al. (2019) found a higher *rate* of *grunts* in the negative (social isolation) compared to the positive situation (in a pair, with food and toys). Overall, call *rate* for both *HF* and *LF* seems to indicate valence intensity and could be useful to assess emotion intensity when the valence is known (Figure 4.3).

Call duration

Call *duration* (*duration* of a single vocalisation; Table 4.1) is a temporal parameter that is often measured. In pigs, *grunt* (*LF*) *duration* was repeatedly found to indicate the valence of the producer, as described in the construct validity section (Test 2 on preference (Friel et al. 2019; Villain et al. 2020a); Test 3 on ancestral fitness (Briefer et al. 2019)), with shorter *grunts* indicating more positive states. Concerning *screams* and *squeals* (*HF*), Weary et al. (1997, 1998) and Weary and Fraser (1995) found that hungrier piglets in isolation, as well as

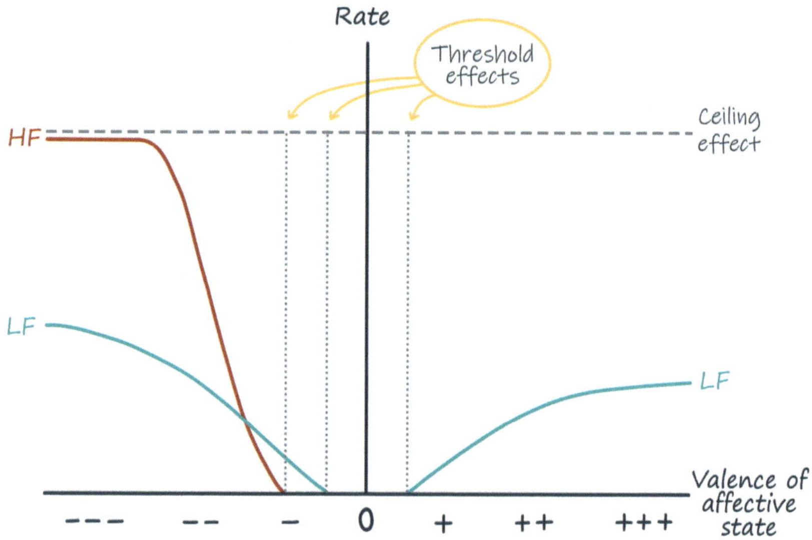

Figure 4.3 Rate of call production as a function of the valence in high-frequency (*HF, red line*) and low-frequency (*LF, blue lines*) pig vocalisations. *HF* calls are produced during negative emotional valence mostly, while *LF* calls are produced during both negative and positive emotional valences. Threshold effects (respectively for *HF* and *LF* calls) are indicated as low-intensity emotional valence usually results in no vocal activity. Ceiling effects are indicated as acoustic parameters are constrained by the anatomy of the animal. The occurrence of *HF* in positive contexts is negligible and hence not shown here.

piglets isolated in a colder enclosure, produced *HF* calls with *longer durations* than isolated but fed piglets or isolated piglets in a warmer enclosure, suggesting that longer *HF* calls reflect negative valence intensity. Concerning *LF* calls, *duration* increased during castration in the study by Puppe et al. (2005) mentioned earlier. Across call types, Tallet et al. (2013) found a significant negative correlation between piglet call *duration* and the ranking of contexts according to their valence (from more negative to more positive; see Figure 4.2). Similarly, Briefer et al. (2022) investigated vocal indicators of valence across 19 contexts encountered by pigs, from birth to slaughter. These were assigned a positive or negative valence based on whether they were preferred/fitness boosting or avoided/fitness harming. Call *duration* (all call types combined) was higher in negative than positive contexts. Therefore, substantial evidence suggests that pig call *duration* indicates the emotional valence of the producer, across call types, age classes and contexts (Figure 4.4).

Call frequency (Hz)

Multiple parameters related to the call spectrum can be measured (i.e. *frequencies* vs. their relative *amplitude*; measured in Hertz; see Figure 4.1 and Table 4.1). Here, for simplicity, we combine all measures in the frequency domain (f_o, *Peak Frequency*, descriptors of the *frequency distribution*, *formants*) to investigate if pig call *frequencies*, in general, are sensitive to valence intensity. As mentioned earlier, *frequencies* of *LF* calls were repeatedly found to be higher in positive compared to negative contexts (Briefer et al. 2019, 2022; Leliveld et al. 2016). Within negative contexts specifically, the *frequencies* of both *LF* calls (*grunts*) and *HF* calls (*screams*) increased with struggling during a standard back test in piglets

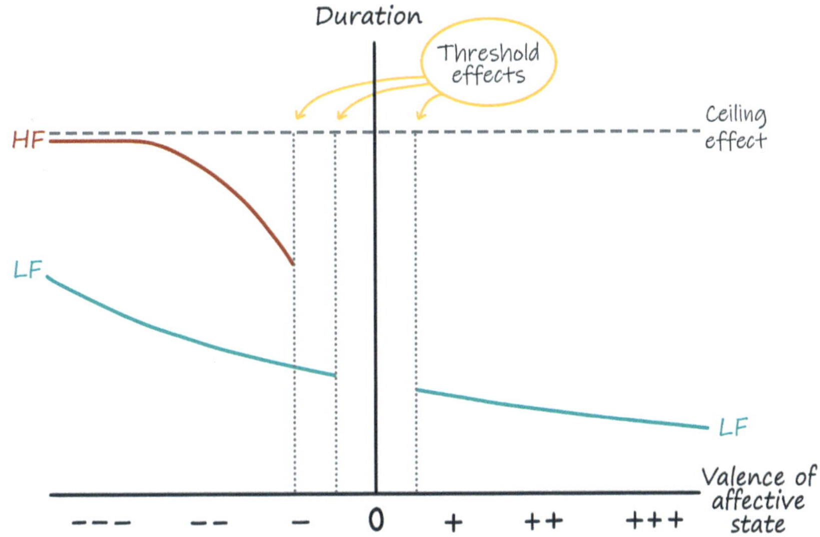

Figure 4.4 Call duration as a function of valence in high-frequency (*HF*, red line) and low-frequency (*LF*, blue line) pig vocalisations (see Figure 4.3 for definitions of other aspects).

(Linhart et al. 2015), and in colder or more hungry piglets in isolation (Weary and Fraser 1995; Weary et al 1997, 1998), suggesting that *frequencies* reflect negative valence intensity (especially for *HF* calls, which are produced in higher proportions). Similarly, piglets castrated without anaesthesia produced calls with higher *frequencies* than those given anaesthesia (Marx et al. 2003). Within positive contexts, in contrast, Villain et al. (2020a) did not find any change in *frequency* with valence intensity, suggesting that this parameter is not always sensitive in these contexts. To summarise, both *LF* and *HF* call *frequencies* seem to indicate negative valence intensity, while *LF* call *frequency* also indicates valence (being higher in positive than negative calls; Figure 4.5).

Summary of vocal indicators of emotions in pigs

Based on results from the literature, we have attempted to draw heuristic graphs for several parameters: the *occurrence* of call types and their *rate*, *duration* and *frequencies* (Figures 4.3–4.5). It should be noted that knowledge of the effect of positive valence intensity on these parameters is currently scarce; our heuristic graphs for positive valence are thus approximations, which would need validation. According to these heuristic graphs, emotional valence can be assessed by the occurrence of *LF* calls (*grunts* and *barks*) versus *HF* calls (*screams* and *squeals*): *HF* calls indicate negative valence, while *LF* calls are not valence specific. If *LF* calls are recorded, one should then look at the *frequency* (e.g. *energy distribution*, *peak frequency*, *formants*) and *duration* of these calls (Table 4.1). Higher *frequencies* and shorter *LF duration* indicate more positive valence. Positive valence intensity could be indicated by an increase in *LF rate* and a decrease in their *duration*. Negative valence intensity can be indicated by an increase in *LF* and *HF rate*, *frequency* and *duration*. Therefore, faster call *rates*, higher *frequencies* and longer *durations* will (particularly in the case of *HF* calls) usually indicate more negative emotions.

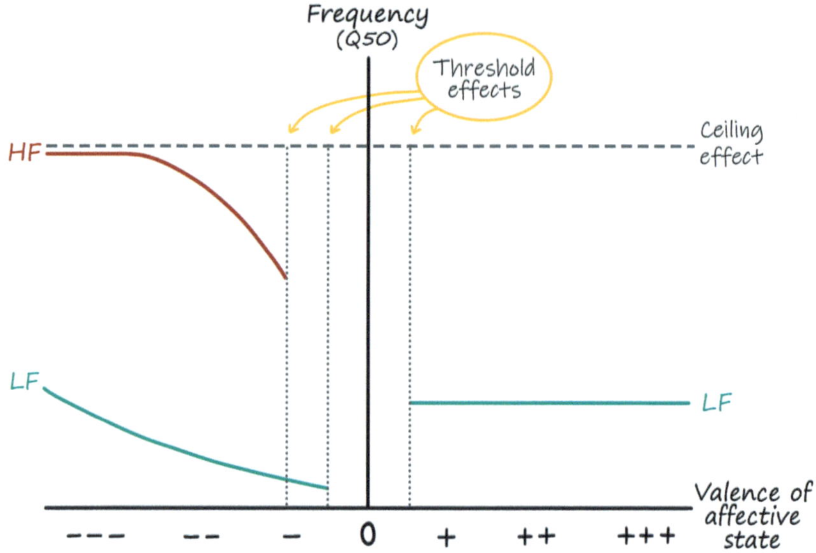

Figure 4.5 Call frequency (Hz) (specifically Q50) as a function of valence in the high-frequency (*HF*) and low-frequency (*LF*) calls of pigs (see Figure 4.3. for definitions of other aspects).

Conclusions and take-home messages for valid use

Opportunities for and limitations of vocal indicators of affect

The evidence suggests that vocal indicators of emotional valence can have strong construct validity, based on the five tests described in Chapter 2 and considered in the construct validation section earlier. The most studied species so far have been pigs and rats. The evidence reviewed earlier thus shows that some species produce valence-specific call *types*; for example, rat *50 kHz* vocalisations, horse *nickers*, pig short-duration *grunts*, chicken *food calls* and *fast clucks* and kea parrot *play calls* indicate positive affect, while rat *22 kHz* vocalisations, horse *squeals*, pig *high-frequency calls* (*screams* and *squeals*) and chicken *gakel calls* indicate negative affect. In addition, the acoustic structure of calls produced in both positive and negative situations (e.g. contact calls) can be affected by emotional valence, with shorter *durations* and lower and less variable f_o often indicating more positive affect (Briefer 2020). These changes are, however, often species specific, or even call-type specific, and would thus have to be established (validated) on a species-by-species basis.

Many parameters also change with the intensity of emotions, independently of the valence (Figures 4.3–4.5). Unlike vocal indicators of the positivity or negativity of affect (valence), these parameter changes are highly conserved across species, at least within negative emotions; faster *rates* of production, higher *frequencies* (f_o, *energy distribution* and/or *formants*) and higher *amplitude* reflect physiological changes linked to arousal/intensity that impact vocal production mechanisms. As discussed in the Introduction, these parameter changes can be useful indicators of the intensity of emotions, as long as the direction of valence is already known. Although studies are still lacking to allow general conclusions to be drawn about vocal parameters linked to the intensity of positive valence, the current literature seems to suggest that higher vocalisation rates may also occur in connection with more positive emotions.

As a consequence, such well-validated vocalisations can be used in practice in welfare assessment, and in the validation of other indicators of affect. For example, one study (as per Test 5 in Chapter 2) used rat *50 kHz* vocalisations as well-validated markers of positive emotions to validate a novel judgement bias task (Rygula et al. 2012). In this study, the rats that produced *50 kHz* vocalisations during tickling, which was used to induce positive affect, also displayed a positive/optimistic response bias to the ambiguous, intermediate tone, in comparison to the rats not emitting *50 kHz* vocalisations during tickling. More recently, studies on pigs compared their perception of an enrichment object and the presence of a human associated with positive contacts (Villain et al. 2020b), showing that shorter grunts (associated with more positive affect) were produced in the presence of the human but not the object. Another study tested the use of vocalisation structure to assess the quality of human–animal relationship/interactions (Villain et al. 2023). Shorter *grunts* were produced in piglets that had received repeated positive contacts from the human. In addition, shorter *grunts* with higher *frequencies*, indicating more positive emotions, were produced when the human was providing positive contacts.

Vocal indicators of emotional valence are, like most indicators, prone to responsiveness and selectivity issues (see earlier section on this topic). The responsiveness issues (risks of false nulls) mostly involve specific situations when animals might not vocalise, or not produce a certain call *type* despite experiencing a change in emotions. These could occur when the audience to whom the call is addressed (e.g. conspecifics, females, familiar individuals) is absent (audience effects); when bodily activation (arousal) is too low and when the context does not match the call *type* (for context-specific *types*). Finally, some ceiling effects exist for all acoustic parameters, as they are tightly constraint by the limitations of the vocal production apparatus. Selectivity issues (risks of false leads) mostly concern the acoustic structure of the vocalisations, as these can be affected by many factors other than the emotional valence. These range from more static factors (e.g. individuality, sex, population) to more dynamic ones (e.g. audience effects leading to exaggeration of call-type structure). Among dynamic factors, often, intensity/arousal can have a stronger effect than valence, resulting in parameter changes that are intensity and not valence specific (see *LF* calls in Figure 4.3). However, when the valence is known, these changes are useful to reveal the intensity of affect for certain types of affective state (see Introduction). In addition, the call-*type* function (e.g. close or long-range communication) might constrain the emotion-related changes, resulting in different patterns depending on the *type*. Finally, when comparing different call *types* (e.g. pig *LF* and *HF*), one might also find very large differences ('jumps') in parameters that do not closely match the change in affective states. To avoid false alarms and not make any wrong assumptions regarding the internal state of the producer of the vocalisation, it is advised to focus on within-call *type* and within-individual changes (repeated measures) and to acquire a good knowledge of how acoustic parameters are affected by valence for a specific call *type* and species beforehand (e.g. for pigs, see Figures 4.3–4.5).

The increased knowledge of welfare that vocal indicators could lead to may be relevant not only for understanding the states of producers, but also the states of *receivers* of these sounds. In some cases, vocalisations have been shown to induce emotional contagion, which plays an important role in group coordination and cooperation (and may even explain the selection of these signals through evolution; Briefer 2018; Herrando and Constantinides 2021). For example, in kea parrots (*Nestor notabilis*), playbacks of *play calls* induced play in conspecifics (Schwing et al. 2017), while in rats, playbacks of *50 kHz* calls induced an optimistic bias – and *22 kHz* vocalisations a pessimistic bias – in a validated judgement bias task (Saito et al. 2016; see also Hammond et al. 2024). In zebra finches, females showed a state matching to that of their partner after hearing their partner's stressed calls (Perez et al. 2015). Thus, emotion-related vocalisations may not only indicate welfare states, but also influence these states when played back; a fascinating topic for future applied research.

How to study a new species

When studying a species for which there is little or no evidence for vocal expression of emotional valence, there are several steps to take to reach solid conclusions about validity for use in welfare assessment. First, one should get familiar with the species' vocal repertoire. How many vocalisation *types* are produced, in which contexts are they used, and what are their functions? If this information is not available, one will have to either investigate the call repertoire (e.g. cluster analyses; Tallet et al. 2013) or focus on a commonly produced call *type*, ideally of known function. Many vocal repertoires are unfortunately not 'discrete' (composed of very distinct call *types*) but 'graded' (composed of many structural intermediates between defined call *types*). Such gradation in call *types* may blur their link to underlying emotions, by making them less specific to contexts and thus emotional states. However, in this case, one could focus on the acoustic parameters rather than the call types.

Next, one should choose which of the five validation tests listed in Chapter 2 to use and also carefully select recording conditions in which the animals are likely to vocalise. These situations can differ between species, depending on how vocal they are, and which level of arousal triggers a call; as mentioned before, contexts inducing low-arousal affective states are unlikely to be accompanied by vocalisations (see the threshold effects in the section on Responsiveness and selectivity issues). Audience effects should also be considered, as animals likely will not produce a certain call *type* when the relevant audience is absent. Since intensity of emotion has a strong effect on most acoustic parameters, it is also important to measure it, using for example heart rate monitors (Briefer et al. 2015a, 2015b), or infrared thermography (Travain and Valsecchi 2021 for a review; see also Chapter 6), and also to pick situations of *a priori* known valence. In addition, acoustic data should not be recorded under conditions with relatively high or various background noise that may impact the quality of the measures (for example, vocalisations produced indoors and outdoors may not be comparable).

Finally, one should choose which parameters to extract and analyse. This will first depend on the vocalisation structure; for instance, *formant* frequencies cannot be detected or measured in high f_0 and tonal calls that are not modulated. Therefore, in such calls, it is not advised to extract formants. Similarly, f_0 cannot be measured in noisy, chaotic calls, and can be hard to find in very low-pitch calls (e.g. below 50 Hz). However, in these cases, one can always extract parameters characterising the *energy distribution* (see Table 4.1). In addition, some parameters, such as overall *amplitude* (loudness) and *duration*, can be controlled by most species and are hence more prone to audience effects than others (see section on Responsiveness and selectivity issues). By contrast, only few species can control vocalisation *frequencies* (Janik and Slater 1997). Some parameters should nevertheless directly reflect the affective state of the producer. This could be through, as mentioned, major anatomical and physiological changes constraining several of the systems involved in vocal production (respiratory and resonance systems; for more details about the mechanisms of vocal expression of emotions, see Briefer 2020). With the aim of drawing conclusions (either at the species level or across species), we would also encourage measuring and reporting robust parameters, which can be computed using different programs with similar output, on all kinds of calls, and which are not too dependent on the recording distance (at least when recording over short distances), such as the call *rate*, *duration*, *energy spectrum* or *peak frequency* and potentially also parameters characterising *frequency* or *amplitude modulations* (for more details about recording advice, equipment and sound analyses, see e.g. Fischer et al. 2013; Gerhardt 1998; Sueur 2018).

Concluding remarks

Vocal signals can encode the emotion of the signaller and serve a communication function, which makes these signals likely to be detectable by conspecifics and also us humans, especially via detailed analysis of vocalisation structure. As a result, vocalisations are promising indicators of non-human animal emotions and may be useful for improving welfare through contagion of positive emotions. The field of bioacoustics has also made great progress in the past years, from the type of equipment available for recording and playing sounds of good quality (e.g. high-quality microphones and speakers), the analysis techniques and accessibility of sound analyses software (e.g. open-source software with possibility for automated extraction of acoustic parameters), to the related computer processing (e.g. machine learning algorithms), allowing researchers to process large amounts of data and extract data in an objective and fast manner. We hope that this chapter will prompt further study of vocalisations as indicators of emotions, allowing advances in the valid assessment of emotions and welfare of captive animals.

References

Ahloy-Dallaire, J., Espinosa, J. and Mason, G. (2018) *Play and optimal welfare: Does play indicate the presence of positive affective states? Behavioural Processes*, 156, 3–15. https://doi.org/10.1016/j.beproc.2017.11.011

Andics, A. and Faragó, T. (2018) Voice perception across species. In Frühholz, S., Belin, P. (eds) *The Oxford Handbook of Voice Perception*. Oxford University Press, pp. 363–392. https://doi.org/10.1093/oxfordhb/9780198743187.013.16

Anikin, A., Bååth, R. and Persson, T. (2018) *Human non-linguistic vocal repertoire: Call types and their meaning. Journal of Nonverbal Behavior*, 42(1), 53–80. https://doi.org/10.1007/s 10.19-017-0267-y

Aubin, T. and Mathevon, N. (2020) *Coding Strategies in Vertebrate Acoustic Communication.* Springer Cham. https://link.springer.com/book/10.1007/978-3-030-39200-0

Bellieni, C.V., Sisto, R., Cordelli, D.M. and Buonocore, G. (2004) *Cry features reflect pain intensity in term newborns: An alarm threshold. Pediatric Research*, 55(1), 142–146. https://doi.org/10.1203/01.pdr.0000099793.99608.cb

Borell, E., von Bünger, B., Schmidt, T. and Horn, T. (2009) *Vocal-type classification as a tool to identify stress in piglets under on-farm conditions*. Proceedings of the 4th International Workshop on the Assessment of Animal Welfare at Farm and Group Level. *Animal Welfare*, 18(4), 407–416. doi: 10.1017/S0962728600000816

Bradbury, J.W. and Vehrencamp, S.L. (2011). Principles of Animal Communication (Second). http://www.sinauer.com/principles-of-animal-communication.html

Bradley, M.M., Codispoti, M., Cuthbert, B.N. and Lang, P.J. (2001). Emotion and motivation I: Defensive and appetitive reactions in picture processing. Emotion, 1(3), 276–298. https://doi.org/10.1.37/1528-3542.1.3.276

Briefer, E.F. (2012) *Vocal expression of emotions in mammals: Mechanisms of production and evidence. Journal of Zoology*, 288(1), 1–20. https://doi.org/10.1111/j.1469-7998.2012.00920.x

Briefer, E.F. (2018) *Vocal contagion of emotions in non-human animals. Proceedings of the Royal Society B: Biological Sciences*, 285(1873), 20172783. https://doi.org/10.1098/rspb.2017.2783

Briefer, E.F. (2020) Coding for 'Dynamic' information: Vocal expression of emotional arousal and valence in non-human animals. In Aubin, T., Mathevon, N. (eds) *Coding Strategies in Vertebrate Acoustic Communication*. Springer International Publishing, pp. 137–162. https://doi.org/10.1007/978-3-030-39200-0_6

Briefer, E.F., Maigrot, A.-L., Mandel, R., Freymond, S.B., Bachmann, I. and Hillmann, E. (2015a) *Segregation of information about emotional arousal and valence in horse whinnies. Scientific Reports*, 5(1), 9989. https://doi.org/10.1038/srep09989

Briefer, E.F., Sypherd, C.C.-R., Linhart, P., Leliveld, L.M.C., Padilla de la Torre, M., Read, E.R., Guérin, C., Deiss, V., Monestier, C., Rasmussen, J.H., Špinka, M., Düpjan, S., Boissy, A., Janczak, A.M., Hillmann, E. and Tallet, C. (2022) *Classification of pig calls produced from birth to slaughter according to their emotional valence and context of production. Scientific Reports*, 12(1), 3409. https://doi.org/10.1038/s41598-022-07174-8

Briefer, E.F., Tettamanti, F. and McElligott, A.G. (2015b) *Emotions in goats: Mapping physiological, behavioural and vocal profiles. Animal Behaviour*, 99, 131–143. https://doi.org/10.1016/j.anbehav.2014.11.002

Briefer, E.F., Vizier, E., Gygax, L. and Hillmann, E. (2019) *Expression of emotional valence in pig closed-mouth grunts: Involvement of both source- and filter-related parameters. The Journal of the Acoustical Society of America*, 145(5), 2895–2908. https://doi.org/10.1121/1.5100612

Brudzynski, S.M. (2019) *Emission of 22 kHz vocalizations in rats as an evolutionary equivalent of human crying: Relationship to depression. Behavioural Brain Research*, 363, 1–12. https://doi.org/10.1016/j.bbr.2019.01.033

Burgdorf, J., Knutson, B. and Panksepp, J. (2000) *Anticipation of rewarding electrical brain stimulation evokes ultrasonic vocalization in rats. Behavioral Neuroscience*, 114(2), 320–327. https://doi.org/10.1037/0735-7044.114.2.320

Burgdorf, J., Knutson, B., Panksepp, J. and Ikemoto, S. (2001a) *Nucleus accumbens amphetamine microinjections unconditionally elicit 50-kHz ultrasonic vocalizations in rats. Behavioral Neuroscience*, 115, 940–944. https://doi.org/10.1037/0735-7044.115.4.940

Burgdorf, J., Knutson, B., Panksepp, J. and Shippenberg, T.S. (2001b) *Evaluation of rat ultrasonic vocalizations as predictors of the conditioned aversive effects of drugs. Psychopharmacology*, 155(1), 35–42. https://doi.org/10.1007/s002130100685

Burgdorf, J. and Moskal, J.R. (2010) Chapter 6.2 - Frequency modulated 50 kHz ultrasonic vocalizations reflect a positive emotional state in the rat: Neural substrates and therapeutic implications. In Brudzynski, S. M. (ed.) *Handbook of Behavioral Neuroscience*, vol 19. Elsevier, pp. 209–214. https://doi.org/10.1016/B978-0-12-374593-4.0021-8

Burke, C.J., Kisko, T.M., Euston, D.R. and Pellis, S.M. (2018) *Do juvenile rats use specific ultrasonic calls to coordinate their social play? Animal Behaviour*, 140, 81–92. https://doi.org/10.1016/j.anbehav.2018.03.019

Chan, W.Y. (2011). The meaning of barks: Vocal communication of fearful and playful affective states in pigs. Ph.D. Thesis, Washington State University. https://www.proquest.com/docview/924232657/abstract/67B435F6A43D44D2PQ/1

Fichtel, C., Hammerschmidt, K. and Jürgens, U. (2001) *On the vocal expression of emotion. A multi-parametric analysis of different states of aversion in the squirrel monkey. Behaviour*, 138(1), 97–116. https://doi.org/10.1163/15685390151067094

Fischer, J., Noser, R. and Hammerschmidt, K. (2013) *Bioacoustic field research: A primer to acoustic analyses and playback experiments with primates. American Journal of Primatology*, 75(7), 643–663. https://doi.org/10.1002/ajp.22153

Friel, M., Kunc, H.P., Griffin, K., Asher, L. and Collins, L.M. (2019) *Positive and negative contexts predict duration of pig vocalisations. Scientific Reports*, 9(1), Article 1. https://doi.org/10.1038/s41598-019-38514-w

Gerhardt, H.C. (1998) Acoustic signals of animals: Recording, field measurements, analysis and description. In Hopp, S. L., Owren, M. J., Evans, C. S. (eds) *Animal Acoustic Communication: Sound Analysis and Research Methods*. Springer, pp. 1–25. https://doi.org/10.1007/978-3-642-76220-8_1

Green, A.C., Johnston, I.N. and Clark, C.E.F. (2018) *Invited review: The evolution of cattle bioacoustics and application for advanced dairy systems. Animal*, 12(6), 1250–1259. https://doi.org/10.1017/S1751731117002646

Hammond, T.J., Brown, S.M., Meddle, S.L., Nielsen, B.L., Lawrence, A.B. and Bombail, V. (2024) *Feel-good songs: Application of a novel playback paradigm to induce a positive affective state in juvenile male Wistar rats. Applied Animal Behaviour Science*, 275, 106296. https://doi.org/10.1016/j.applanim.2024.106296

Heesen, R., Sievers, C., Gruber, T. and Clay, Z. (2022). Primate communication: Affective, intentional, or both? (In Press). In: Schwartz, Bennett L.; Beran, Michael J. (eds.) *Primate Cognitive Studies*. Cambridge University Press, Cambridge. https://www.cambridge.org/core/books/primate-cognitive-studies/43e0134a306b44f570457f391e86b752#fndtn-information

Helmer, L.M.L., Weijenberg, R.A.F., de Vries, R., Achterberg, W.P., Lautenbacher, S., Sampson, E.L. and Lobbezoo, F. (2020) *Crying out in pain—A systematic review into the validity of vocalization as an indicator for pain. European Journal of Pain*, 24(9), 1703–1715. https://doi.org/10.1002/ejp.1623

Herrando, C. and Constantinides, E. (2021) *Emotional Contagion: A Brief Overview and Future Directions. Frontiers in Psychology*, 12. https://www.frontiersin.org/articles/10.3389/fpsyg.2021.712606

Hillmann, E., Mayer, C., Schön, P.-C., Puppe, B. and Schrader, L. (2004) *Vocalisation of domestic pigs (Sus scrofa domestica) as an indicator for their adaptation towards ambient temperatures. Applied Animal Behaviour Science*, 89(3), 195–206. https://doi.org/10.1016/j.applanim.2004.06.008

Janik, V. M. and Slater, P. J. B. (eds). (1997) Vocal learning in mammals. In *Advances in the Study of Behavior*, vol Volume 26. Academic Press, pp. 59–99. http://www.sciencedirect.com/science/article/pii/S0065345408603770

Jürgens, R., Hammerschmidt, K. and Fischer, J. (2011) *Authentic and Play-Acted Vocal Emotion Expressions Reveal Acoustic Differences. Frontiers in Psychology*, 2. https://www.frontiersin.org/articles/10.3389/fpsyg.2011.00180

Karakashian, S.J., Gyger, M. and Marler, P. (1988) *Audience effects on alarm calling in chickens (Gallus gallus). Journal of Comparative Psychology*, 102, 129–135. https://doi.org/10.1037/0735-7036.102.2.129

Kiley, M. (1972) *The Vocalizations of Ungulates, their Causation and Function. Zeitschrift Für Tierpsychologie*, 31(2), 171–222. https://doi.org/10.1111/j.1439-0310.1972.tb01764.x

Kuppens, P., Tuerlinckx, F., Russell, J.A. and Barrett, L.F. (2013) *The relation between valence and arousal in subjective experience. Psychological Bulletin*, 139(4), 917–940. https://doi.org/10.1037/a0030811

Lautenbacher, S., Salinas-Ranneberg, M., Niebuhr, O. and Kunz, M. (2017) *Phonetic characteristics of vocalizations during pain. Pain Reports*, 2(3), e597. https://doi.org/10.1097/pr9.0000000000000597

Leliveld, L.M.C., Düpjan, S., Tuchscherer, A. and Puppe, B. (2016) *Behavioural and physiological measures indicate subtle variations in the emotional valence of young pigs. Physiology & Behavior*, 157, 116–124. https://doi.org/10.1016/j.physbeh.2016.02.002

Linhart, P., Ratcliffe, V.F., Reby, D. and Špinka, M. (2015) *Expression of Emotional Arousal in Two Different Piglet Call Types. PLoS One*, 10(8), e0135414. https://doi.org/10.1371/journal.pone.0135414

Maigrot, A.-L., Hillmann, E., Anne, C. and Briefer, E.F. (2017) *Vocal expression of emotional valence in Przewalski's horses (Equus przewalskii). Scientific Reports*, 7(1), 8779. https://doi.org/10.1038/s41598-017-09437-1

Maigrot, A.-L., Hillmann, E. and Briefer, E.F. (2018) *Encoding of Emotional Valence in Wild Boar (Sus scrofa) Calls. Animals*, 8(6), 85. https://doi.org/10.3390/ani8060085

Manteuffel, G., Puppe, B. and Schön, P.C. (2004) *Vocalization of farm animals as a measure of welfare. Applied Animal Behaviour Science*, 88(1), 163–182. https://doi.org/10.1016/j.applanim.2004.02.012

Manteuffel, G., Schön, P.C., Düpjan, S., Tuchscherer, A. and Bellmann, O. (2007) *Acetylcholine injection into the amygdala elicits vocalization in domestic pigs (Sus scrofa). Behavioural Brain Research*, 178(1), 177–180. https://doi.org/10.1016/j.bbr.2006.12.007

Marchant, J.N., Whittaker, X. and Broom, D.M. (2001) *Vocalisations of the adult female domestic pig during a standard human approach test and their relationships with behavioural and heart rate measures. Applied Animal Behaviour Science*, 72(1), 23–39. https://doi.org/10.1016/s0168-1591(00)00190-8

Marler, P., Dufty, A. and Pickert, R. (1986) *Vocal communication in the domestic chicken: II. Is a sender sensitive to the presence and nature of a receiver? Animal Behaviour*, 34, 194–198. https://doi.org/10.1016/0003-3472(86)90023-0

Marx, G., Horn, T., Thielebein, J., Knubel, B. and von Borell, E. (2003) *Analysis of pain-related vocalization in young pigs. Journal of Sound and Vibration*, 266(3), 687–698. https://doi.org/10.1016/s0022-460x(03)00594-7

McGrath, N., Dunlop, R., Dwyer, C., Burman, O. and Phillips, C.J.C. (2017) *Hens vary their vocal repertoire and structure when anticipating different types of reward. Animal Behaviour*, 130, 79–96. https://doi.org/10.1016/j.anbehav.2017.05.025

Morton, E.S. (1977) *On the occurrence and significance of motivation-structural rules in some bird and mammal sounds. The American Naturalist*, 111(981), 855–869.

Paul, E.S., Harding, E.J. and Mendl, M. (2005) *Measuring emotional processes in animals: The utility of a cognitive approach. Neuroscience and Biobehavioral Reviews*, 29(3), 469–491. https://doi.org/10.1016/j.neubiorev.2005.01.002

Perez, E.C., Elie, J.E., Boucaud, I.C.A., Crouchet, T., Soulage, C.O., Soula, H.A., Theunissen, F.E. and Vignal, C. (2015) *Physiological resonance between mates through calls as possible evidence of empathic processes in songbirds. Hormones and Behavior*, 75, 130–141. https://doi.org/10.1016/j.yhbeh.2015.09.002

Perez, E.C., Elie, J.E., Soulage, C.O., Soula, H.A., Mathevon, N. and Vignal, C. (2012) *The acoustic expression of stress in a songbird: Does corticosterone drive isolation-induced modifications of zebra finch calls? Hormones and Behavior*, 61(4), 573–581. https://doi.org/10.1016/j.yhbeh.2012.02.004

Puppe, B., Schön, P.C., Tuchscherer, A. and Manteuffel, G. (2005) *Castration-induced vocalisation in domestic piglets, Sus scrofa: Complex and specific alterations of the vocal quality. Applied Animal Behaviour Science*, 95(1), 67–78. https://doi.org/10.1016/j.applanim.2005.05.001

Reimert, I., Bolhuis, J.E., Kemp, B. and Rodenburg, T.B. (2013) *Indicators of positive and negative emotions and emotional contagion in pigs. Physiology & Behavior*, 109, 42–50. https://doi.org/10.1016/j.physbeh.2012.11.002

Rygula, R., Pluta, H. and Popik, P. (2012) *Laughing Rats Are Optimistic. PLoS One*, 7(12), e51959. https://doi.org/10.1371/journal.pone.0051959

Saito, Y., Yuki, S., Seki, Y., Kagawa, H. and Okanoya, K. (2016) *Cognitive bias in rats evoked by ultrasonic vocalizations suggests emotional contagion. Behavioural Processes*, 132, 5–11. https://doi.org/10.1016/j.beproc.2016.08.005

Scheiner, E. and Fischer, J. (2011) Emotion Expression: The Evolutionary Heritage in the Human Voice. In Welsch, W., Singer, W. J., Wunder, A. (eds) *Interdisciplinary Anthropology: Continuing Evolution of Man*. Springer, pp. 105–129. https://doi.org/10.1007/978-3-642-11668-1_5

Scherer, K.R. (1986) *Vocal affect expression: A review and a model for future research. Psychological Bulletin*, 99(2), 143–165.

Scherer, K.R. and Ekman, P. (2014) *Approaches To Emotion*. Psychology Press, New York, p. 448.

Schrader, L. and Todt, D. (1998) *Vocal quality is correlated with levels of stress hormones in domestic pigs. Ethology*, 104(10), 859–876. https://doi.org/10.1111/j.1439-0310.1998.tb00036.x

Schwartz, J.W., Engelberg, J.W.M. and Gouzoules, H. (2020) *Was that a scream? Listener agreement and major distinguishing acoustic features. Journal of Nonverbal Behavior*, 44(2), 233–252. https://doi.org/10.1007/s10.19-019-00325-y

Schwartz, J.W., Sanchez, M.M. and Gouzoules, H. (2022) *Vocal expression of emotional arousal across two call types in young rhesus macaques. Animal Behaviour*, 190, 125–138. https://doi.org/10.1016/j.anbehav.2022.05.017

Schwing, R., Nelson, X.J., Wein, A. and Parsons, S. (2017) *Positive emotional contagion in a New Zealand parrot. Current Biology: CB*, 27(6), R213–R214. https://doi.org/10.1016/j.cub.2017.02.020

Shariff, A.F. and Tracy, J.L. (2011) *What Are Emotion Expressions For? Current Directions in Psychological Science*, 20(6), 395–399. https://doi.org/10.1177/0963721411424739

Siebert, K., Langbein, J., Schön, P.-C., Tuchscherer, A. and Puppe, B. (2011) *Degree of social isolation affects behavioural and vocal response patterns in dwarf goats (Capra hircus). Applied Animal Behaviour Science*, 131(1), 53–62. https://doi.org/10.1016/j.applanim.2011.01.003

Slocombe, K.E. and Zuberbühler, K. (2007) *Chimpanzees modify recruitment screams as a function of audience composition. Proceedings of the National Academy of Sciences*, 104(43), 17228–17233. https://doi.org/10.1073/pnas.0706741104

Soldati, A., Fedurek, P., Dezecache, G., Call, J. and Zuberbühler, K. (2022) *Audience sensitivity in chimpanzee display pant hoots. Animal Behaviour*, 190, 23–40. https://doi.org/10.1016/j.anbehav.2022.05.010

Soltis, J., Blowers, T.E. and Savage, A. (2011) *Measuring positive and negative affect in the voiced sounds of African elephants (Loxodonta africana). The Journal of the Acoustical Society of America*, 129(2), 1059–1066. https://doi.org/10.1121/1.3531798

Soula, H.A., Carnino, D., Fernandez, M.S.A., Perez, E.C., Villain, A.S. and Vignal, C. (2018) *Stress-induced flexibility and individuality in female and male zebra finch distance calls. Behavioural Processes*. https://doi.org/10.1016/j.beproc.2018.03.018

Stewart, A.M., Lewis, G.F., Yee, J.R., Kenkel, W.M., Davila, M.I., Sue Carter, C. and Porges, S.W. (2015) *Acoustic features of prairie vole (Microtus ochrogaster) ultrasonic vocalizations covary with heart rate. Physiology & Behavior*, 138, 94–100. https://doi.org/10.1016/j.physbeh.2014.10.011

Sueur, J. (2018) *Sound Analysis and Synthesis with R*. Springer International Publishing. https://doi.org/10.1007/978-3-319-77647-7

Tallet, C., Linhart, P., Policht, R., Hammerschmidt, K., Šimeček, P., Kratinova, P. and Špinka, M. (2013) *Encoding of situations in the vocal repertoire of piglets (Sus scrofa): A comparison of discrete and graded classifications. PLoS One*, 8(8), e71841. https://doi.org/10.1371/journal.pone.0071841

Taylor, A.M. and Reby, D. (2010) *The contribution of source–filter theory to mammal vocal communication research. Journal of Zoology*, 280(3), 221–236. https://doi.org/10.1111/j.1469-7998.2009.00661.x

Titze, I.R. and Martin, D.W. (1998) *Principles of Voice Production. The Journal of the Acoustical Society of America*, 104(3), 1148–1148. https://doi.org/10.1121/1.424266

Townsend, S.W., Koski, S.E., Byrne, R.W., Slocombe, K.E., Bickel, B., Boeckle, M., Braga Goncalves, I., Burkart, J.M., Flower, T., Gaunet, F., Glock, H.J., Gruber, T., Jansen, D.A.W.A.M., Liebal, K., Linke, A., Miklósi, Á., Moore, R., van Schaik, C.P., Stoll, S., et al. (2017) *Exorcising Grice's ghost: An empirical approach to studying intentional communication in animals. Biological Reviews*, 92(3), 1427–1433. https://doi.org/10.1111/brv.12289

Townsend, S.W. and Manser, M.B. (2013) *Functionally referential communication in mammals: the past, present and the future. Ethology*, 119(1), 1–11. https://doi.org/10.1111/eth.12015

Travain, T. and Valsecchi, P. (2021) *Infrared thermography in the study of animals' emotional responses: a critical review. Animals*, 11(9), 2510. https://doi.org/10.3390/ani11092510

Van Dierendonck, M.C. (2006). The Importance of Social Relationships in Horses. Ph.D. Thesis, Utrecht University; 181 pp.

Villain, A.S., Guérin, C. and Tallet, C. (2023) *The use of pigs vocalisation structure to assess the quality of human-pig relationship. Peer Community Journal*, 3. https://doi.org/10.24072/pcjournal.260

Villain, A.S., Hazard, A., Danglot, M., Guérin, C., Boissy, A. and Tallet, C. (2020a) *Piglets vocally express the anticipation of pseudo-social contexts in their grunts. Scientific Reports*, 10(1), 18496. https://doi.org/10.1038/s41598-020-75378-x

Villain, A.S., Lanthony, M., Guérin, C. and Tallet, C. (2020b) *Manipulable object and human contact: Preference and modulation of emotional states in weaned pigs. Frontiers in Veterinary Science*, 7, 577433. https://doi.org/10.3389/fvets.2020.577433

Waller, B.M. and Micheletta, J. (2013) *Facial expression in nonhuman animals. Emotion Review*, 5(1), 54–59. https://doi.org/10.1177/1754073912451503

Waller, B.M., Whitehouse, J. and Micheletta, J. (2017) *Rethinking primate facial expression: A predictive framework. Neuroscience and Biobehavioral Reviews*, 82, 13–21. https://doi.org/10.1016/j.neubiorev.2016.09.005

Wang, H., Liang, S., Burgdorf, J., Wess, J. and Yeomans, J. (2008) *Ultrasonic vocalizations induced by sex and amphetamine in M2, M4, M5 Muscarinic and D2 Dopamine receptor knockout mice. PLoS One*, 3(4), e1893. https://doi.org/10.1371/journal.pone.0001893

Weary, D.M., Braithwaite, L.A. and Fraser, D. (1998) *Vocal response to pain in piglets. Applied Animal Behaviour Science*, 56(2), 161–172. https://doi.org/10.1016/s0168-1591(97)00092-0

Weary, D.M. and Fraser, D. (1995) *Calling by domestic piglets: Reliable signals of need? Animal Behaviour*, 50(4), 1047–1055. https://doi.org/10.1016/0003-3472(95)80105-7

Weary, D.M., Ross, S. and Fraser, D. (1997) *Vocalizations by isolated piglets: A reliable indicator of piglet need directed towards the sow. Applied Animal Behaviour Science*, 53(4), 249–257. https://doi.org/10.1016/s0168-1591(96)01173-2

Winkler, S.L. and Bryant, G.A. (2021) *Play vocalisations and human laughter: A comparative review. Bioacoustics*, 30(5), 499–526. https://doi.org/10.1080/09524622.2021.1905065

Zimmerman, P.H., Koene, P. and van Hooff, J.A.R.A.M. (2000) *The vocal expression of feeding motivation and frustration in the domestic laying hen, Gallus domesticus. Applied Animal Behaviour Science*, 69(4), 265–273. https://doi.org/10.1016/s0168-1591(00)00136-2

Zimmermann, E., Leliveld, L. and Schehka, S. (2013) Toward the evolutionary roots of affective prosody in human acoustic communication: A comparative approach to mammalian voices. In *Evolution of Emotional Communication*. Oxford University Press. https://doi.org/10.1093/acprof:oso/9780199583560.003.0008

Acute Glucocorticoid Responses

5

Georgia J. Mason[1] and Michael T. Mendl[2]

[1]Department of Integrative Biology & Campbell Centre for the Study of Animal Welfare, University of Guelph, Guelph, Ontario, Canada
[2]Bristol Veterinary School, University of Bristol, Bristol, UK

Introduction

This chapter reviews how to use glucocorticoid (GC) hormones to assess negative emotions. We first summarise relevant aspects of basic biology, before reviewing evidence that acute GC responses reflect acute negative affect. Next, we cover the types of error that can occur if these GC responses are naively assumed to always indicate negative affect: false nulls, where negative emotions are assumed absent despite actually being present, and false leads, where negative emotions are assumed present despite actually being absent. We then provide simple principles for avoiding such errors, and end with examples illustrating how useful GC measures can be.

Our focus here is short-term emotions only, because GC responses accompanying longer-lasting affective states are diverse and, at least with current states of knowledge, unpredictable: as mentioned in Chapter 2, GCs can essentially increase, stay at or return to baseline, or even decrease during sustained negative affective states or following recurrent adversity over the lifespan (e.g. Dickens and Romero 2013; Guilliams and Edwards 2010; Miller et al. 2007; see also Conner and Barrett 2012). Consequently, any given pattern or difference between groups can then be interpreted in contradictory ways for welfare, perhaps using *post hoc* reasoning or HARKing: clearly problematic for making accurate inferences about long-term animal well-being. Consequently, we currently do not recommend using GC levels to assess negative moods or cumulative stress (with Chapters 7–13 reviewing preferable ways to assess these long-term states).

Some basic biology: the function and regulation of glucocorticoids

GCs are steroid hormones released into the blood by the adrenal glands. The two main forms, cortisol and corticosterone, are typically co-released (to different extents in different species, e.g. Koren et al. 2012; Sheriff et al. 2011), and in undisturbed conditions they show diurnal rhythms that anticipate periods of physical activity, as well as moment-to-moment ultradian pulsatility (e.g. Lightman and Conway-Campbell 2010). In the blood, GCs are in either in 'free' form or 'bound' to protein molecules, particularly corticosteroid binding globulins (CBGs; e.g. Mormède et al. 2007; Porterfield et al. 2001). Only free GC molecules are small enough to leave the capillaries and reach target tissues, where their effects include promoting catabolism (e.g. Porterfield et al. 2001; Sapolsky et al. 2000; Wingfield and Romero 2015) by stimulating the breakdown of fat to free fatty acids and glycogen to glucose. At the high concentrations occurring after elicitation by challenge, GCs' functions also include promoting the metabolism of proteins to amino acids and then glucose. They inhibit glucose uptake by fat and the brain too, while enhancing its uptake by muscles; and (amongst other effects) they stimulate appetite after a challenge so that energy reserves can be restored (e.g. Wingfield and Romero 2015; Sacta et al. 2016; Sapolsky et al. 2000).

Acute super-normal releases of GCs are triggered by challenges that initiate the classic hypothalamic–pituitary–adrenal (HPA) 'stress' response. Briefly, brain regions that respond to affectively significant events (e.g. the amygdalae) signal the hypothalamus to release corticosteroid releasing hormone (CRH). This stimulates the pituitary to secrete adrenocorticotrophic hormone (ACTH) into the bloodstream, which then promotes GC release from the adrenal glands. Blood GC levels typically rise 2–5 minutes after a challenge, peaking 15–60 minutes later (e.g. Mormède et al. 2007 and many others, e.g. Figure 5.1), and this is accompanied by an increase in the proportion of GC that is free, partly because CBGs saturate as GC levels rise. GC responses end when levels return to baseline. This can take between 30 minutes and many hours: a time course influenced by the nature and affective relevance of the eliciting stimulus (see Figure 5.1). Free GCs are then cleared from the blood by diffusion into saliva and sweat, and by active excretion by the liver and kidney into urine and faeces.

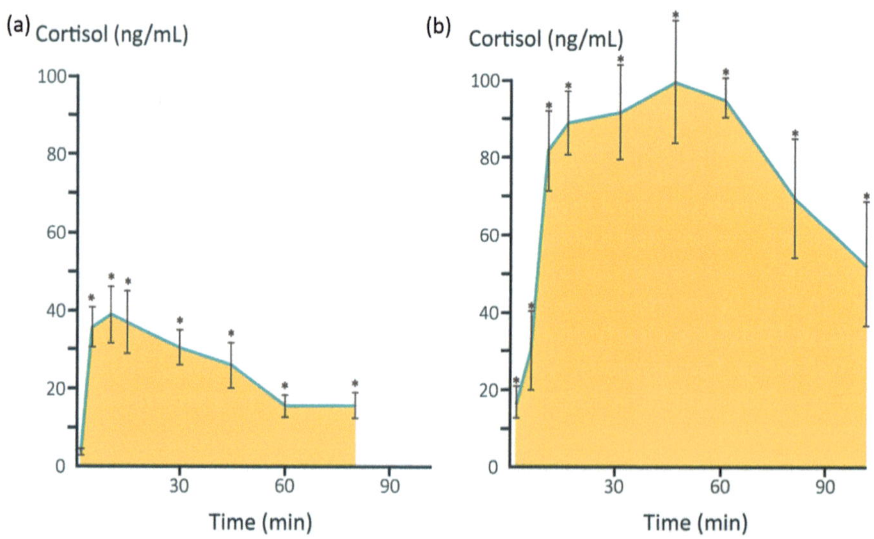

Figure 5.1 Plasma (thus total, bound plus free) cortisol levels in domestic sheep exposed at *Time 0* to a 2- to 5-minute long aversive event: a person shouting and banging their pen (a) or a sheepdog (b). Source: Harlow et al. (1987) / with permission of Canadian Science Publishing.

Sampling glucocorticoids

Blood and saliva sampling provides the most sensitive way of detecting even small changes in circulating levels of GCs. Saliva GCs are essentially at equilibrium with free GCs in the blood, while plasma assays typically measure total GCs, not distinguishing free from bound components (e.g. Sheriff et al. 2010). For assessing acute responses to challenge, measurement of free GCs is often favoured because they are the biologically active fraction, and generally show larger response magnitudes than do total GCs (e.g. Breuner et al. 2013; Schoech et al. 2013). The pulsatility of GC release will affect blood and saliva samples, however, because they are sensitive to moment-to-moment changes in GC levels. Repeated sampling or large sample sizes may therefore be required to obtain representative data on circulating levels. Importantly, such sampling must also be sufficiently calm and quick that it does not itself cause GC levels to rise in ways that influence sampled values (e.g. under 2 minutes from start of animal capture to acquisition of blood sample; Mormède et al. 2007; Novak et al. 2013 and others).

The resulting data can provide information about both the *peak* and *entire* GC response to a challenge. The peak is the maximum level reached or the maximum increase above a previously measured baseline. Its timing can be estimated from previous studies, allowing targeting of sampling times, or pinpointed by repeated sampling following a challenge. The latter approach, if sufficiently prolonged, can also allow an across-time response curve to be plotted, with the area under the curve (AUC; Figure 5.1) providing information on the *entire magnitude* of an integrated GC response. The AUC is less prone to ceiling effects than peak values: at high levels of ACTH, the adrenal gland's rate of GC output can reach a maximum, but response duration can vary, hence yielding differences in entire response magnitude (e.g. Anisman et al. 2001; Kent et al. 1998; Rushen et al. 2008). For example, pigs held around the snout by a snare for 5 minutes or moved from a pen to a crate in another building showed similarly large peak GC responses (Figure 5.2), but the latter had larger AUC values indicating that they recovered more slowly. However, the AUC may be insensitive to small differences in GC responses as these are 'diluted out' across time (something covered further below).

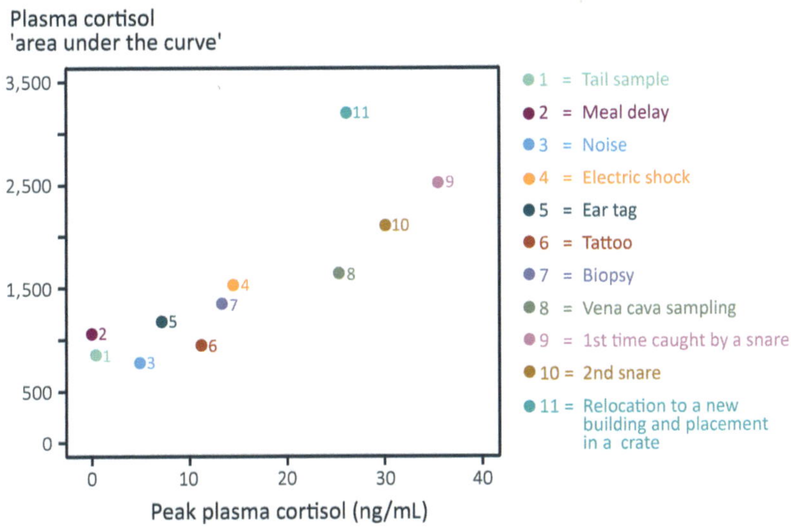

Figure 5.2 Plasma GC responses of pigs to a range of challenges, with peak values plotted against the area under the entire GC response curve (AUC). Source: Adapted from Merlot et al. (2011).

Minimally invasive or non-invasive measures, such as saliva sampling, can offer a specific advantage over blood sampling in that the sampling procedure itself is unlikely to alter GC levels. Other non-invasive ways of assessing free GC levels include assaying them or their metabolites in urine, faeces, milk of ungulates and even, for aquatic species, aquarium water. Unlike for blood and saliva, for such samples, the assayed levels reflect the integrated accumulation of free GCs over several hours, and so can be used to estimate AUCs if sampled at a time when GCs are likely to have returned to baseline. They also smooth out the ultradian pulses seen in blood and saliva samples. Because free GCs diffuse rapidly into them, milk and aquarium water are best sampled soon after the acute plasma GC response to a challenge is estimated to be over, to capture the area under the whole response curve. For urine and faeces, acute elevations in GC take a while to be processed and excreted (e.g. for faecal GC metabolites, the key factor is how long material takes to move from duodenum to rectum). For these, the best time to sample may therefore be some hours after an acute GC release into the blood (e.g. Cook 2012; Sheriff et al. 2011).

Using glucocorticoid responses to assess negative emotions: construct validity

Hans Selye, the pioneer of GC research (e.g. 1956), was perhaps the first to catalogue the pronounced GC responses of people faced with threatening stimuli. He also explored this acute GC response via a range of animal experiments that today seem appalling (with rabbits suspended by their heads, rats with their paws taped together injured by being tumbled in rotating drums, and more). Given this start, it is perhaps ironic how widely these hormones have been used since by researchers trying to improve animal welfare (e.g. Tiemann et al. 2023). This section reviews further evidence for the links between GC responses and negative emotions.

Evidence from humans experiencing negative emotions (Test 1)

Human participants experimentally exposed to acute aversive challenges in the laboratory (e.g. mental arithmetic, public speaking and other challenges we either *infer* are aversive from empathy or *know* are aversive due to self-report data from subjects) typically show elevations in the most commonly used GC measure in such studies, salivary cortisol (e.g. Table 5.1). In contrast, subjects exposed to *positive* experiences (e.g. watching an enjoyable movie or dancing the tango) can sometimes show falls in cortisol levels (e.g. Budde et al. 2010; Murcia et al. 2009).

In non-laboratory 'ambulatory' studies, where participants living their daily lives record how they feel several times a day while simultaneously collecting saliva samples, negative states covary consistently with elevated GC levels (Table 5.1). For example, the 'cortisol awakening response', the difference between salivary cortisol levels immediately at waking and the higher levels 30 minutes later, is often larger in people anticipating a stressful, aversive day.

Non-laboratory studies may also involve more dramatic challenges, as well as more clearly demonstrating graded relationships with the severity of the challenge. In one, ballroom dancers showed cortisol responses when competing that were not present on mere training days; furthermore, those who reported the greatest psychological stress and the least satisfaction with their performances had the highest GC levels over several hours. In another, first-time parachute jumpers made three jumps over two days. Each caused large increases in GC levels, though the

Table 5.1 Acute salivary cortisol responses in humans exposed to a range of experiences (arranged by the size of the mean GC increase; *N. B. values are typically approximate since read from graphs*).

Experience *(sex of subjects)*	Mean baseline salivary cortisol (nmol/L)	Mean peak salivary cortisol (nmol/L)	Increase (peak-baseline) (nmol/L)	References and notes
Three daily stressors compared to zero in an 'ambulatory' study *(M & F)*	6.9	8.6	1.7	Smyth et al. (1998) (who also report that the most stressful days elevated salivary GCs by c. 2.5 nmol/L)
Lab-based social stress test, e.g. public speaking; mock interview *(M & F)*	4.0	6.0	2.0	Yim et al. (2010)
As above (F)	5.0	9.0	4.0	Kirschbaum et al. (1992)
As above (M & F)	7.0	11.0	4.0	Rohleder et al. (2007)
Describing a past rape *(F)*	10.0	17.0	7.0	Gola et al. (2012)
Making the third parachute jump of one's life *(M & F)*	2.5	24.0	21.5	Deinzer et al. (1997)
Ballroom dancing competition *(M & F)*	6	35	29	Rohleder et al. (2007)
Making the first parachute jump of one's life *(M & F)*	5	44.5	c. 39.5	Deinzer et al. (1997)

third jump less so than the first (Table 5.1). Other human research has targeted individuals who have had terrible experiences like serious car accidents, finding very elevated cortisol levels for hours following the incident, with values being higher in people with more severe injuries and reduced by sedatives or analgesia (e.g. Delahanty et al. 2003; Brorsson et al. 2014). Similar patterns hold for rape victims: in the hours following an attack, cortisol levels are typically elevated, being especially high if the rape was severe (e.g. involving injury; Resnick et al. 1995). Overall, there is thus considerable evidence that this indicator passes Chapter 2's Test 1.

Evidence from animals exposed to stimuli that are aversive and/or harmful for ancestral fitness

Acute GC responses are reliably elicited in animals exposed to stimuli that they would avoid if given the chance (cf Chapter 2's Test 2) and/or by events that would be bad for ancestral fitness in the wild (Test 3). The magnitude of these responses may also reflect the strength of the threatening stimulus in a graded way.

For example, for rats exposed to 0, 2 mA or 5 mA shocks, GC responses were greatest for those receiving the strongest (Pitman et al. 1990). Predator cues also reliably induce acute GC responses, as seen in primates exposed to model hawks and live snakes, wild birds to stuffed owls and weasels, and laboratory mice to cat or fox urine. Thus, in one mouse study, plasma corticosterone levels approximately tripled after 5 minutes in a brightly lit, aversive 'open field', but increased over *fivefold* if the arena was also sprayed with fox urine (Anisman et al. 2001). Acute GC elevations also occur when animals are frustrated by denial of expected food or other rewards. For example, urine samples revealed significant elevations in GC output when American mink were prevented from reaching food or swimming water, two highly preferred resources, but not when prevented from reaching a less-valued extra cage or nest box (Mason et al. 2001). Fighting with conspecifics, especially being attacked and losing, elevates GCs too (e.g. Buwalda et al. 2012; Walker et al. 2009). Being social isolated (an aversive experience for social species) also raise GC levels in many species, and again this can be in a graded way depending on how complete the isolation is. For example, acute GC responses in isolated sheep are smaller if they can see a mirror or even just a picture of another sheep (Da Costa et al. 2004; Parrott et al. 1988).

Acute GC responses likewise occur in young animals prematurely separated from their mothers (e.g. Kim et al. 2011; Mears and Brown 1997; Rhind et al. 1998). And again, these may be graded according to the severity of this challenge. Thus in a study of mice, these were greatest in those individuals least ready for weaning as judged by their age or weight (Würbel and Stauffacher 1997); while in young rhesus monkeys, these are greater if they are then housed alone rather than with young conspecifics (Novak et al. 2013).

Further evidence that the strength of animal GC responses tracks the aversiveness of a scenario (so reflecting the assumed intensity of the associated negative emotion) is provided by studies of individual differences in temperament (inferred from animals' relative willingness to approach various stimuli, or from variation in other affect indicators cf. Test 5): bolder individuals generally show smaller GC responses than do more timid, cautious individuals. For example, rats and mice spending most time in cautious 'stretched attend' postures in an Elevated Plus Maze correspondingly showed the highest post-test plasma GC levels (e.g. Rodgers 1997), while calves that took longest to touch a novel object (a tambourine attached to a yellow container) again showed the greatest acute GC responses (van Reenen et al. 2005).

Responsiveness issues causing false nulls: when negative emotions are not mirrored by GC responses

So far, it would seem that acute GC responses reliably occur during negative emotions and, to a degree, that they are responsive to how negative these experiences are, sometimes even showing graded increases with increasing intensity of the experience. But do these endocrine markers accurately track the occurrence of, and changes in, *all* negative emotions? Unfortunately not. To give some examples from humans, experimental lab stressors like making subjects do mental arithmetic increase sympathetic responses like heart rate much more reliably than they induce GC responses (e.g. Kajantie and Phillips 2006). Other minor challenges also often fail to impact GCs (e.g. sitting through an uninteresting lecture or listening to music that reduces self-reported positive affect: Kreutz et al. 2004; West et al. 2004). But even *highly* aversive experiences can generate false nulls: in the rape study outlined above, not all women responded to the assault with GC increases (Resnick et al. 1995). In all such cases, relying on GCs alone could thus lead to false null conclusions: mistaken inferences that negative emotions are not impacted by situations or treatments when in fact they are.

Animal and human data reveal that three main issues seem to underlie this lack of responsiveness: methodological factors; certain attributes of individual subjects; and the possibility that some negative affective responses, especially those not involving physical activity, are not accompanied by acute GC responses. We review these next.

Methodological factors reducing responsiveness

The time of day when stressors are experienced and sampling occurs can influence responsiveness. This is due to circadian changes in adrenal reactivity to ACTH: the adrenal glands are typically less responsive to ACTH towards the end of a species' circadian active period (evening for most humans; morning for nocturnal animals) (e.g. Dallman et al. 1978; Ulrich-Lai et al. 2006; Sage et al. 2002). Stressors applied at this time may therefore elicit smaller GC responses than when applied during the rising phase of the circadian rhythm. To illustrate, laboratory stressors (e.g. public speaking or experiencing an icy-cold compress) induce smaller GC responses if applied to people in the afternoon or evening rather than the morning (Kudielka et al. 2004; Van Eekelen et al. 2003), with similar reduced evening responsiveness shown by other naturally diurnal species (e.g. pigs: Ruis et al. 1997). Similarly, in naturally nocturnal rats, aversive events like restraint cause less prolonged GC responses if experienced in the light rather than the dark phase of the day (e.g. Atkinson et al. 2006; Bradbury et al. 1991).

Sample and measurement type affects responsiveness as well, in part, by determining whether free or total hormone is assayed and whether the response's peak or AUC is measured. Total GCs (in blood samples) may be less sensitive to challenging situations than free GCs, because during acute responses the proportion of unbound GC in blood increases, thus yielding greater response sizes than do total GCs. For example, zebra finches in one study showed a 20-fold increase in *free* GCs 30 minutes after handling, but only a sixfold increase in *total* GC levels: a much smaller effect (Breuner et al. 2006). Likewise, total cortisol concentrations did not increase in horses introduced to a group of agonistic conspecifics, even though free levels did (Alexander and Irvine 1998).

When GC responses are large, measures that estimate AUCs (e.g. urine, faeces, cisternal milk) are also likely to be more sensitive than those estimating peak GC levels, since peak values are prone to ceiling effects. For example, in one study mice exhibited maximum peak GC levels at a certain shock intensity, beyond which their levels plateaued, and yet their AUC values continued to increase beyond this shock intensity (e.g. Anisman et al. 2001; Armario et al. 2012). On the other hand, when GC responses are small, AUC measures often show reduced sensitivity, because (as mentioned earlier) minor, transient elevations become too diluted or masked by background values to be detectable. Cook (2012) thus states '*fecal corticoid metabolites measurements are insensitive to all but the largest increases in adrenocortical output*', providing similar evidence for urinary samples too.

Blunted glucocorticoid reactivity in individuals

GC reactivity often varies with reproductive state or season, and this can be one source of false nulls. In women, for example, acute GC responses are lower during the follicular phase of the oestrous cycle, even though stressors are not perceived as any less aversive (e.g. Baron and Brush 1979; Kirschbaum et al. 1999; Kajantie and Phillips 2006). Similarly, in the moulting season, many wild birds show smaller, sometimes even negligible, acute GC responses to being captured compared to at other times, even though '*they all struggle, show elevated respiration rates, defecate profusely, and may give distress vocalizations*' (Wingfield and Romero 2015).

Reductions in GC reactivity can also occur after sustained adversity. Thus the women showing relatively small GC reactions to being raped turned out to be those with a long history of being abused (Resnick et al. 1995). Likewise, one large-scale study found that people who had experienced more than one 'adverse life event' (e.g. assault) showed minimal GC responses to stressors like public speaking, despite no reductions in how psychologically stressful they found these (Lovallo et al. 2012). Depressed people also often show relatively small 'cortisol awakening responses', even if anticipating a challenging, worrying day (e.g. Chida and Steptoe 2009; Fries et al. 2009). In all these cases, acute GC responses are insensitive tools for assessing negative emotions, showing low responsiveness to affective changes.

Type of affective state: negative emotions typically unaccompanied by acute GC responses

Certain acute negative states seem rather reliably not associated with GC responses. In humans, the emotion of disgust is one (e.g. Rohrmann et al. 2008). Losing a computer or sports game is another (e.g. Donzella et al. 2000; Salvador et al. 1987; Wirth et al. 2006). Passively enduring slightly negative, monotonous experiences (known or suspected to be boring) also seems to fail to elevate GCs (e.g. Kreutz et al. 2004; Yamakoshi et al. 2009), and the same may hold for sadness (or at least the mild levels induced in laboratory studies: Ottowitz et al. 2004). Perhaps these are negative emotions associated with low arousal or low physical activity, in which the catabolic effects of GC rises would not be functionally useful.

Functional arguments have been used to explain why, in animals too, some types of aversive situation do not reliably induce acute GC responses. Young quail, hens and nestling puffins, for instance, show no elevation in total plasma GCs after 24 hours without food or 24 hours without water (Freeman et al. 1983; Kitaysky et al. 2005; Wall and Cockrem, 2009). A study on rat also found that total GCs were low '*when there is a very high or **very low probability** that food is coming*' (our emphasis) (Coover et al. 1984). Such null responses may reflect that quietly 'sitting it out' to conserve resources can be adaptive when faced with temporary shortages (Toates 1986; Wingfield and Romero 2015).

Selectivity issues causing false leads: when GC responses occur in positive or non-affective states

Thinking about the catabolic function of acute GC responses can also help us understand another type of counter-example: cases where elevated GCs do not reflect negative emotions, but instead positively valanced ones or ones with no clear affective valence at all. Thus there is reliable evidence of eating and sexual behaviour being linked with elevated GCs from a wide range of species, including people and calves, pigs and other farm animals during and after eating meals (e.g. reviewed by reviewed by Mormède et al. 2007); from young men chatting to young women (e.g. van der Meij et al. 2010); and from studies of copulation in male bulls, goats, boars, rhesus monkeys and rats (e.g. Borg et al. 1991; Thun and Eggenberger 1996; Howland et al. 1985; Elias and Weil 1989; Figure 5.3). Increases in plasma corticosterone are also associated with nursing behaviour in female goats, rats and cattle (Hoffman et al. 1996; Walker et al. 1992; von Walter et al. 2010); with voluntary wheel-running, a potent positive reinforcer for mice (e.g. Girard and Garland 2002, reviewed Malisch et al. 2008); and with playing with toys in dogs (Flint et al. 2024).

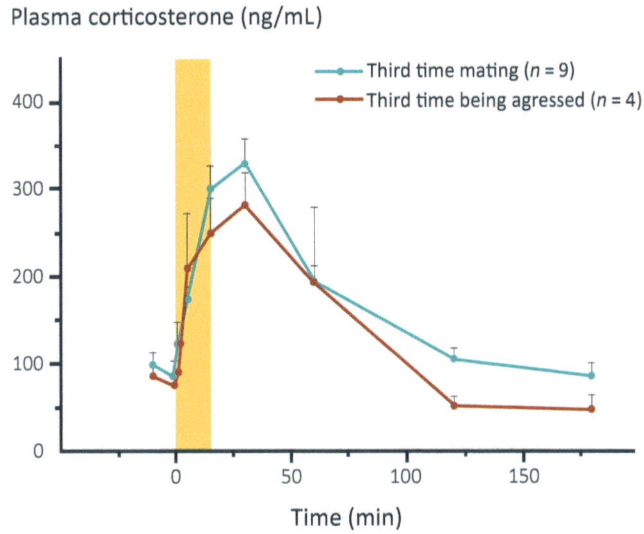

Figure 5.3 GC levels in plasma repeatedly sampled from catheterised male rats after an encounter with another aggressive and unfamiliar male (which resulted in them being threatened, and sometimes being physically attacked), or after interacting with a receptive female (which often led to mating). Source: Buwalda et al. (2012) / with permission of Elsevier.

In such cases, GC elevations may even be graded in ways that reflect how *positive* the experience is! GCs thus increase in humans and rhesus monkeys self-administering recreational drugs like cocaine, with monkeys exhibiting larger rises to stronger, more preferred doses (Broadbear et al. 1999). GCs also elevated in people taking a dance class, and the more fun they had, the higher their cortisol levels became (West et al. 2004). And playing games also increases cortisol, with 'Type A' people often showing marked GC elevations when winning: something one study called the 'thrill of victory' (Jones et al. 1986; Salvador et al. 1987).

Another type of false positive occurs when individuals rendered unconscious, and hence 'emotion-free', are exposed to health or fitness threats. For example, GCs elevate in fully anaesthetised people undergoing surgery, and in rats exposed to ether, even though this chemical very rapidly renders them unconscious (Desborough 2000; Gibbs 1970; Hagen et al. 1980).

Thus, diverse situations that involve alertness, activity, autonomic responses and/or catabolism can cause acute GC responses, even when affectively neutral (or unfelt) or affectively positive. In such cases, GC responses have poor selectivity as indicators of negative affective states.

Conclusions and take home messages for valid use

The research in this chapter reveals some key principles for using acute GC responses as indicators of emotion and welfare. First, humans reporting negative emotions and animals exposed to aversive and/or fitness-threatening events do often show acute GC elevations. Furthermore, in such cases the magnitude of these often seems to reflect how negative the experiences are. Measuring the peak or AUC of a GC response can therefore be a very useful tool in the welfare researcher's arsenal (with the added benefit that these hormones can readily be sampled non-invasively).

However, like acute increases in heart rate – an indicator of another classic physiological 'stress' system (see Chapter 6) – elevations in GCs should be seen as reflecting the arousal or activity that often accompanies emotions, both positive and negative, rather than being unique to negative valence. As shown in Figure 5.4, at least for high arousal or metabolically demanding states, GCs have a V-shaped relationship with emotional intensity. Thus as Koolhaas and colleagues summarise for rats (2011), '*the physiological 'stress' response to appetitive, rewarding stimuli. . . can be as large as the response to negative stimuli*"; *and* as Pollard argues for humans (1995), '*Cortisol levels may be as much influenced by positive emotions as by negative emotions*'. Therefore, a second important principle is that it is inappropriate (despite being rather common) to use elevated GCs alone as evidence that an animal is in a negative state: naively assuming that high GCs indicate negative affect can generate false leads about animal welfare.

This leads to a third principle. GCs can help us infer the intensity of negative emotions only when there are good *independent* reasons for assuming such states. Thus, if we have other evidence that an animal's state is to the left of the *y*-axis in Figure 5.4 (e.g. from avoidance responses [Test 2, see Chapter 3]; alarm or distress calls [Chapter 4]; negative judgement biases [Chapter 7];, or inferences about a context's effects on ancestral fitness

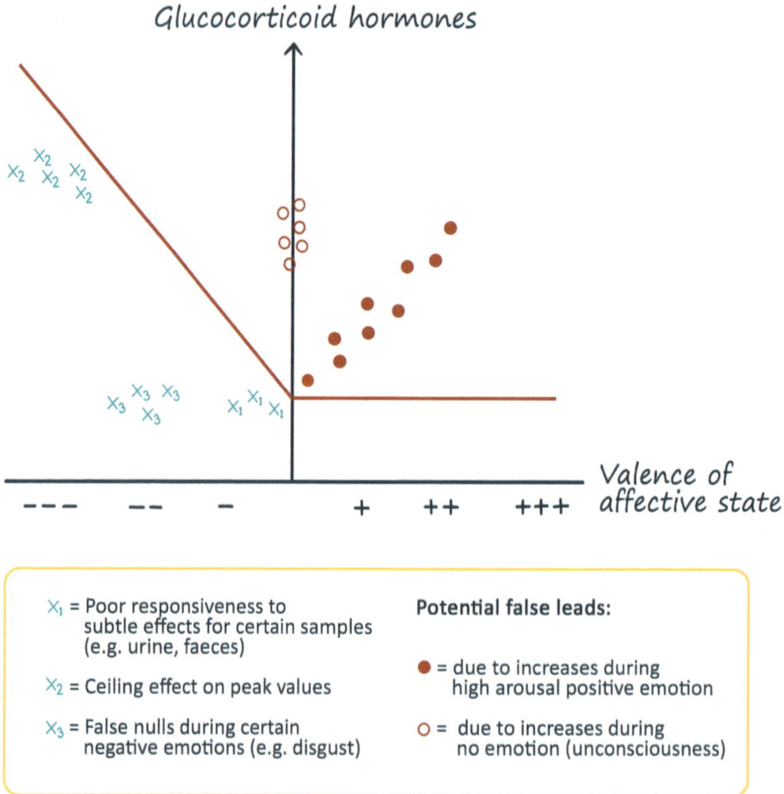

Figure 5.4 GC responses during positive and negative emotions. *When states are high arousal or energy-demanding, GCs may show an essentially V-shaped relationship with affective intensity. If we are using GCs to estimate negative affect, such patterns can potentially lead to false leads if GC levels rise during arousing positive (or absent) emotions, or false nulls if GCs fail to rise during negative emotions involving low arousal, inactivity or a need to conserve resources.*

[Test 3]), then (and only then) GC levels can provide very useful information about the intensity of that state. More speculatively, if instead we have independent evidence that an animal's state is positive (i.e. to the right of the *y*-axis in Figure 5.4), it may even be the case that GC levels tell us just how positive (though this is far less well validated).

A fourth principle for using GCs to assess negative emotions is that animals must be awake or conscious throughout the situation being investigated. As we have seen, GC responses can occur in the absence of subjective experiences (e.g. during anaesthesia). Not appreciating this could lead to potential false leads suggesting that animals are distressed when in fact they are not feeling anything at all. This makes GCs inappropriate for assessing the welfare implications of treatments that render animals insensible (e.g. certain killing methods).

Finally, acute GC responses can also lead to false null findings: during some types of aversive experience, in some types of subject, and when certain methods are used, GC responses may be minimal or even absent, despite the presence of negative affective states. Thus, a fifth principle is that one should know *a priori* that the situation being studied elicits detectable GC responses. Without this knowledge, we could draw false null inferences that the lack of a response indicates the absence of negative affect, when in fact the animal *is* experiencing poor welfare.

Using GCs in practice

Examples of the valid use of GC measures therefore often come from studies assessing the mitigation of (awake) animals' responses to stimuli that are already known to be aversive (e.g. evaluating the effectiveness of various forms of pain control). In this context, GC responses have been successfully, validly used to demonstrate the benefits of local anaesthetics and/or NSAID analgesics during the 'disbudding' of calves (removal of horn-producing tissue using caustic chemicals or cauterisation), castration of lambs and cattle, and vasectomy of laboratory mice (e.g. Morisse et al. 1995; Heinrich et al. 2009; Nogues et al. 2025). GC responses have sometimes even revealed effects that were hard to detect by behavioural observations alone (e.g. Milligan et al. 2004); while in the mouse study, GC and behavioural data combined to reveal dosing effects: the lowest NSAID dose affected pain-related behaviours but only the highest dose decreased GC levels (Wright-Williams et al. 2007).

GC levels have also been used validly to assess the relative impact of variations in the duration of events likely to be highly aversive, such as the capture of wild animals. For example, a study investigating how to minimise capture stress in little brown bats found that GCs rose linearly as post-capture sampling and handling times increased from baseline (<3 minutes) to 15 to 30 minutes (Edwards et al. 2022; Figure 5.5), the authors concluded that, whenever possible, captured bats should be released before 30 minutes.

Non-invasive GC measures have also been used to good effect by researchers harvesting faecal droppings and 'yellow snow' (snow that has been urinated on) from free-living wild animals, to assess the impact of a range of aversive human activities ranging from snowmobiling to hunting (e.g. Creel et al. 2002).

Such studies reveal that, used well, GCs can potentially be more sensitive than other measures for assessing the emotional impacts of aversive events, and that the use of samples which can be collected without disturbing subjects can also be a superb research tool. These are just a few of many examples of the valid use of GC responses to infer animal welfare or affective state, illustrating how very valuable GC measurements can be. GCs are very clearly not hormones that track psychological stress in any simple way, but for known negative emotions, especially high arousal ones, they can be excellent tools for ranking and reducing the aversiveness of short-term events or practices that may compromise welfare.

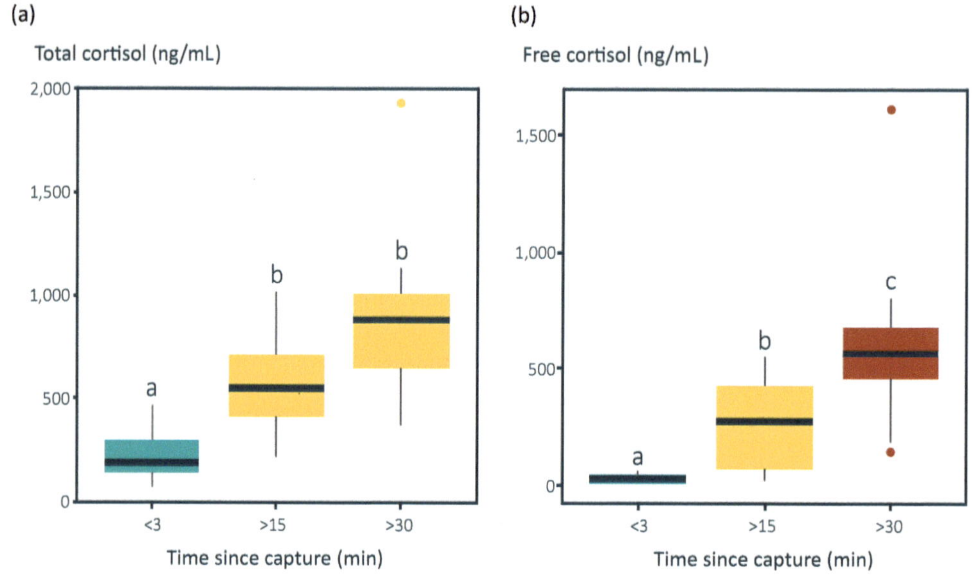

Figure 5.5 (a) Free blood cortisol levels and (b) total blood cortisol levels in little brown bats captured and handled for different lengths of time. Source: Edwards et al. (2022) / with permission of Elsevier.

References

Alexander, S.L. and Irvine, C.H. (1998) *The effect of social stress on adrenal axis activity in horses: The importance of monitoring corticosteroid-binding globulin capacity. Journal of Endocrinology*, 157(3), 425–432. https://doi.org/10.1677/joe.0.1570425

Anisman, H., Hayley, S., Kelly, O., Borowski, T. and Merali, Z. (2001) *Psychogenic, neurogenic, and systemic stressor effects on plasma corticosterone and behavior: Mouse strain-dependent outcomes. Behavioral Neuroscience*, 115(2), 443–454.

Atkinson, H.C., Wood, S.A., Kershaw, Y.M., Bate, E. and Lightman, S.L. (2006) *Diurnal variation in the responsiveness of the hypothalamic-pituitary-adrenal axis of the male rat to noise stress. Journal of Neuroendocrinology*, 18(7), 526–533.

Armario, A., Daviu, N., Muñoz-Abellán, C., Rabasa, C., Fuentes, S., Belda, X., Gagliano, H. and Nadal, R. (2012) *What can we know from pituitary–adrenal hormones about the nature and consequences of exposure to emotional stressors? Cellular and Molecular Neurobiology*, 32(5), 749–758. https://doi.org/10.1007/s10571-012-9814-6

Baron, S. and Brush, F.R. (1979) *Effects of acute and chronic restraint and estrus cycle on pituitary-adrenal function in the rat. Hormones and Behavior*, 12(3), 218–224. https://doi.org/10.1016/0018-506X(79)90004-7

Borg, K.E., Esbenshade, K.L. and Johnson, B.H. (1991) *Cortisol, growth hormone, and testosterone concentrations during mating behavior in the bull and boar. Journal of Animal Science*, 69(8), 3230–3240. https://doi.org/10.2527/1991.6983230x

Bradbury, M.J., Cascio, C.S., Scribner, K.A. and Dallman, M.F. (1991) *Stress-induced adrenocorticotropin secretion: Diurnal responses and decreases during stress in the evening are not dependent on corticosterone. Endocrinology*, 128(2), 680–688. https://doi.org/10.1210/endo-128-2-680

Breuner, C.W., Delehanty, B. and Boonstra, R. (2013) *Evaluating stress in natural populations of vertebrates: Total CORT is not good enough. Functional Ecology*, 27(1), 24–36. https://doi.org/10.1111/1365-2435.12016

Breuner, C.W., Lynn, S.E., Julian, G.E., Cornelius, J.M., Heidinger, B.J., Love, O.P., Sprague, R.S., Wada, H. and Whitman, B.A. (2006) *Plasma-binding globulins and acute stress response. Hormone and Metabolic Research*, 38, 260–268. https://doi.org/10.1055/s-2006-925347

Broadbear, J.H., Winger, G., Cicero, T.J. and Woods, J.H. (1999) *Effects of self-administered cocaine on plasma adrenocorticotropic hormone and cortisol in male rhesus monkeys. Journal of Pharmacology and Experimental Therapeutics*, 289(3), 1641–1647.

Brorsson, C., Dahlqvist, P., Nilsson, L., Thunberg, J., Sylvan, A. and Naredi, S. (2014) *Adrenal response after trauma is affected by time after trauma and sedative/analgesic drugs. Injury*, 45(8), 1149–1155. https://doi.org/10.1016/j.injury.2014.02.001

Budde, H., Windisch, C., Kudielka, B.M. and Voelcker-Rehage, C. (2010) *Saliva cortisol in school children after acute physical exercise. Neuroscience Letters*, 483(1), 16–19. https://doi.org/10.1016/j.neulet.2010.07.036

Buwalda, B., Scholte, J., de Boer, S.F., Coppens, C.M. and Koolhaas, J.M. (2012) *The acute glucocorticoid stress response does not differentiate between rewarding and aversive social stimuli in rats. Hormones and Behavior*, 61(2), 218–226. https://doi.org/10.1016/j.yhbeh.2011.12.012

Chida, Y. and Steptoe, A. (2009) *Cortisol awakening response and psychosocial factors: A systematic review and meta-analysis. Biological Psychology*, 80(3), 265–278. https://doi.org/10.1016/j.biopsycho.2008.10.004

Conner, T.S. and Barrett, L.F. (2012) *Trends in ambulatory self-report: The role of momentary experience in psychosomatic medicine. Psychosomatic Medicine*, 74(4), 327–337. https://doi.org/10.1097/PSY.0b013e3182546f18

Cook, N.J. (2012) *Review: Minimally invasive sampling media and the measurement of corticosteroids as biomarkers of stress in animals. Canadian Journal of Animal Science*, 92(3), 227–259.

Coover, G.D., Murison, R., Sundberg, H., Jellestad, F. and Ursin, H. (1984) *Plasma corticosterone and meal expectancy in rats: Effects of low probability cues. Physiology & Behavior*, 33(2), 179–184. https://doi.org/10.1016/0031-9384(84)90097-0

Creel, S., Fox, J.E., Hardy, A., Sands, J., Garrott, B. and Peterson, R.O. (2002) *Snowmobile activity and glucocorticoid stress responses in wolves and elk. Conservation Biology*, 16(3), 809–814. https://doi.org/10.1046/j.1523-1739.2002.00554.x

da Costa, A.P., Leigh, A.E., Man, M.-S. and Kendrick, K.M. (2004) *Face pictures reduce behavioural, autonomic, endocrine and neural indices of stress and fear in sheep. Proceedings of the Royal Society of London. Series B: Biological Sciences*, 271(1552), 2077–2084. https://doi.org/10.1098/rspb.2004.2831

Dallman, M.F., Engeland, W.C., Rose, J.C., Wilkinson, C.W., Shinsako, J. and Siedenburg, F. (1978) *Nycthemeral rhythm in adrenal responsiveness to ACTH. American Journal of Physiology - Regulatory, Integrative and Comparative Physiology*, 235(5), R210–R218. https://doi.org/10.1152/ajpregu.1978.235.5.R210

Deinzer, R., Kirschbaum, C., Gresele, C. and Hellhammer, D.H. (1997) *Adrenocortical responses to repeated parachute jumping and subsequent h-CRH challenge in inexperienced healthy subjects. Physiology & Behavior*, 61(4), 507–511. https://doi.org/10.1016/S0031-9384(96)00465-9

Delahanty, D.L., Raimonde, A.J., Spoonster, E. and Cullado, M. (2003) *Injury severity, prior trauma history, urinary cortisol levels, and acute PTSD in motor vehicle accident victims. Journal of Anxiety Disorders*, 17(2), 149–164. https://doi.org/10.1016/S0887-6185(02)00185-8

Desborough, J.P. (2000) *The stress response to trauma and surgery. British Journal of Anaesthesia*, 85(1), 109–117. https://doi.org/10.1093/bja/85.1.109

Dickens, M.J. and Romero, L.M. (2013) *A consensus endocrine profile for chronically stressed wild animals does not exist. General and Comparative Endocrinology*, 191, 177–189.

Donzella, B., Gunnar, M.R., Krueger, W.K. and Alwin, J. (2000) *Cortisol and vagal tone responses to competitive challenge in preschoolers: Associations with temperament. Developmental Psychobiology*, 37(4), 209–220. https://doi.org/10.1002/1098-2302(2000)37:4<209::aid-dev1>3.0.co;2-s

Edwards, P.D., Boonstra, R., Bosson, C.O., Jane Harms, N., Kukka, P.M., Willis, C.K. and Jung, T.S. (2022) *Effects of capture on stress-axis measures in endangered little brown bats (Myotis lucifugus). Journal of Mammalogy*, 103(1), 91–99.

Elias, E. and Weil, S. (1989) *Serum cortisol levels in camels (Camelus dromedarius) during the reproductive cycle. Comparative Biochemistry and Physiology. A, Comparative Physiology*, 94(4), 787–790. https://doi.org/10.1016/0300-9629(89)90634-8

Flint, H.E., Weller, J.E., Parry-Howells, N., Ellerby, Z.W., McKay, S.L. and King, T. (2024) *Evaluation of indicators of acute emotional states in dogs. Scientific Reports*, 14(1), 6406.

Freeman, B.M., Manning, A.C.C. and Flack, I.H. (1983) *Adrenal cortical activity in the domestic fowl, Gallus domesticus, following withdrawal of water or food. Comparative Biochemistry and Physiology Part A: Physiology*, 74, 639–641.

Fries, E., Dettenborn, L. and Kirschbaum, C. (2009) *The cortisol awakening response (CAR): Facts and future directions. International Journal of Psychophysiology*, 72(1), 67–73. https://doi.org/10.1016/j.ijpsycho.2008.03.014

Gibbs, F.P. (1970) *Circadian variation of ether-induced corticosterone secretion in the rat. The American Journal of Physiology*, 219(2), 288–292. https://doi.org/10.1152/ajplegacy.1970.219.2.288

Girard, I. and Garland, T. (2002) *Plasma corticosterone response to acute and chronic voluntary exercise in female house mice. Journal of Applied Physiology (Bethesda, MD: 1985)*, 92(4), 1553–1561. https://doi.org/10.1152/japplphysiol.00465.2001

Gola, H., Engler, H., Schauer, M., Adenauer, H., Riether, C., Kolassa, S., Elbert, T. and Kolassa, I.-T. (2012) *Victims of rape show increased cortisol responses to trauma reminders: A study in individuals with war- and torture-related PTSD. Psychoneuroendocrinology*, 37(2), 213–220. https://doi.org/10.1016/j.psyneuen.2011.06.005

Guilliams, T.G. and Edwards, L. (2010) *Chronic stress and the HPA axis. The Standard*, 9(2), 1–12.

Hagen, C., Brandt, M.R. and Kehlet, H. (1980) Prolactin, LH, FSH, GH and cortisol response to surgery and the effect of epidural analgesia. *European Journal of Endocrinology*, 94(2), 151–154. https://doi.org/10.1530/acta.0.0940151

Harlow, H.J., Thorne, E.T., Williams, E.S., Belden, E.L. and Gern, W.A. (1987) *Adrenal responsiveness in domestic sheep (Ovis aries) to acute and chronic stressors as predicted by remote monitoring of cardiac frequency. Canadian Journal of Zoology*, 65(8), 2021–2027. https://doi.org/10.1139/z87-307

Heinrich, A., Duffield, T.F., Lissemore, K.D., Squires, E.J. and Millman, S.T. (2009) *The impact of meloxicam on postsurgical stress associated with cautery dehorning. Journal of Dairy Science*, 92(2), 540–547. https://doi.org/10.3168/jds.2008-1424

Hoffman, D.P., Stevenson, J.S. and Minton, J.E. (1996) *Restricting calf presence without suckling compared with weaning prolongs postpartum anovulation in beef cattle. Journal of Animal Science*, 74(1), 190–198. https://doi.org/10.2527/1996.741190x

Howland, B.E., Sanford, L.M. and Palmer, W.M. (1985) *Changes in serum levels of LH, FSH, prolactin, testosterone, and cortisol associated with season and mating in male pygmy goats. Journal of Andrology*, 6(2), 89–96. https://doi.org/10.1002/j.1939-4640.1985.tb00822.x

Jones, K.V., Copolov, D.L. and Outch, K.H. (1986) *Type A, test performance and salivary cortisol. Journal of Psychosomatic Research*, 30(6), 699–707. https://doi.org/10.1016/0022-3999(86)90104-2

Kajantie, E. and Phillips, D.I.W. (2006) *The effects of sex and hormonal status on the physiological response to acute psychosocial stress. Psychoneuroendocrinology*, 31(2), 151–178. https://doi.org/10.1016/j.psyneuen.2005.07.002

Kent, J.E., Molony, V. and Graham, M.J. (1998) *Comparison of methods for the reduction of acute painproduced by rubber ring castration or tail docking of week-old lambs. The Veterinary Journal*, 155(1), 39–51. https://doi.org/10.1016/S1090-0233(98)80033-6

Kim, M.-H., Yang, J.-Y., Upadhaya, S.D., Lee, H.-J., Yun, C.-H. and Ha, J.K. (2011) *The stress of weaning influences serum levels of acute-phase proteins, iron-binding proteins, inflammatory cytokines, cortisol, and leukocyte subsets in Holstein calves. Journal of Veterinary Science*, 12(2), 151–157.

Kirschbaum, C., Bartussek, D. and Strasburger, C.J. (1992) *Cortisol responses to psychological stress and correlations with personality traits. Personality and Individual Differences*, 13(12), 1353–1357. https://doi.org/10.1016/0191-8869(92)90181-N

Kirschbaum, C., Kudielka, B.M., Gaab, J., Schommer, N.C. and Hellhammer, D.H. (1999) *Impact of gender, menstrual cycle phase, and oral contraceptives on the activity of the hypothalamic-pituitary-adrenal axis. Psychosomatic Medicine*, 61(2), 154–162. https://doi.org/10.1097/00006842-199903000-00006

Kitaysky, A.S., Romano, M.D., Piatt, J.F., Wingfield, J.C. and Kikuchi, M. (2005) *The adrenocortical response of tufted puffin chicks to nutritional deficits. Hormones and Behavior*, 47(5), 606–619. https://doi.org/10.1016/j.yhbeh.2005.01.005

Koren, L., Whiteside, D., Fahlman, S., Ruckstuhl, K., Kutz, S., Checkley, S., Dumond, M. and Wynne-Edwards, K. (2012) *Cortisol and corticosterone independence in cortisol-dominant wildlife. General and Comparative Endocrinology*, 177(1), 113–119. https://doi.org/10.1016/j.ygcen.2012.02.020

Koolhaas, J.M., Bartolomucci, A., Buwalda, B., de Boer, S.F., Flügge, G., Korte, S.M., Meerlo, P., Murison, R., Olivier, B., Palanza, P., Richter-Levin, G., Sgoifo, A., Steimer, T., Stiedl, O., van Dijk, G., Wöhr, M. and Fuchs, E. (2011) *Stress revisited: A critical evaluation of the stress concept. Neuroscience & Biobehavioral Reviews*, 35(5), 1291–1301. https://doi.org/10.1016/j.neubiorev.2011.02.003

Kreutz, G., Bongard, S., Rohrmann, S., Hodapp, V. and Grebe, D. (2004) *Effects of choir singing or listening on secretory immunoglobulin A, cortisol, and emotional state. Journal of Behavioral Medicine*, 27(6), 623–635.

Kudielka, B.M., Schommer, N.C., Hellhammer, D.H. and Kirschbaum, C. (2004) *Acute HPA axis responses, heart rate, and mood changes to psychosocial stress (TSST) in humans at different times of day. Psychoneuroendocrinology*, 29(8), 983–992. https://doi.org/10.1016/j.psyneuen.2003.08.009

Lightman, S.L. and Conway-Campbell, B.L. (2010) *The crucial role of pulsatile activity of the HPA axis for continuous dynamic equilibration. Nature Reviews. Neuroscience*, 11(10), 710–718. https://doi.org/10.1038/nrn2914

Lovallo, W.R., Farag, N.H., Sorocco, K.H., Cohoon, A.J. and Vincent, A.S. (2012) *Lifetime adversity leads to blunted stress axis reactivity: Studies from the Oklahoma Family Health Patterns Project. Biological Psychiatry*, 71(4), 344–349. https://doi.org/10.1016/j.biopsych.2011.10.018

Malisch, J.L., Breuner, C.W., Gomes, F.R., Chappell, M.A. and Garland, T. (2008) *Circadian pattern of total and free corticosterone concentrations, corticosteroid-binding globulin, and physical activity in mice selectively bred for high voluntary wheel-running behavior. General and Comparative Endocrinology*, 156(2), 210–217. https://doi.org/10.1016/j.ygcen.2008.01.020

Mason, G.J., Cooper, J. and Clarebrough, C. (2001) *Frustrations of fur-farmed mink. Nature*, 410(6824), 35–36. https://doi.org/10.1038/35065157

Mears, G.J. and Brown, F.A. (1997) *Cortisol and β-endorphin responses to physical and psychological stressors in lambs. Canadian Journal of Animal Science*, 77(4), 689–694. https://doi.org/10.4141/A97-051

Merlot, E., Mounier, A.M. and Prunier, A. (2011) *Endocrine response of gilts to various common stressors: A comparison of indicators and methods of analysis. Physiology & Behavior*, 102(3), 259–265. https://doi.org/10.1016/j.physbeh.2010.11.009

Miller, G.E., Chen, E. and Zhou, E.S. (2007) *If it goes up, must it come down? Chronic stress and the hypothalamic-pituitary-adrenocortical axis in humans. Psychological Bulletin*, 133(1), 25–45.

Milligan, B.N., Duffield, T. and Lissemore, K. (2004) *The utility of ketoprofen for alleviating pain following dehorning in young dairy calves. The Canadian Veterinary Journal = La Revue Veterinaire Canadienne*, 45(2), 140–143.

Morisse, J.P., Cotte, J.P. and Huonnic, D. (1995) *Effect of dehorning on behaviour and plasma cortisol responses in young calves. Applied Animal Behaviour Science*, 43(4), 239–247. https://doi.org/10.1016/0168-1591(95)00569-E

Mormède, P., Andanson, S., Aupérin, B., Beerda, B., Guémené, D., Malmkvist, J., Manteca, X., Manteuffel, G., Prunet, P., van Reenen, C.G., Richard, S. and Veissier, I. (2007) *Exploration of the hypothalamic-pituitary-adrenal function as a tool to evaluate animal welfare. Physiology & Behavior*, 92(3), 317–339. https://doi.org/10.1016/j.physbeh.2006.12.003

Murcia, C., Bongard, S. and Kreutz, G. (2009) *Emotional and neurohumoral responses to dancing Tango Argentino: The effects of music and partner. Music and Medicine*, 1, 14–21. https://doi.org/10.1177/1943862109335064

Nogues, E., Stojkov, J., Stojkova, B.J., von Keyserlingk, M.A. and Weary, D.M. (2025) *Efficacy of pain management for cattle castration: A systematic review and meta-analysis. Animal Welfare*, 34, e43.

Novak, M.A., Hamel, A.F., Kelly, B.J., Dettmer, A.M. and Meyer, J.S. (2013) *Stress, the HPA axis, and nonhuman primate well-being: A review. Applied Animal Behaviour Science*, 143(2), 135–149. https://doi.org/10.1016/j.applanim.2012.10.012

Ottowitz, W.E., Dougherty, D.D., Sirota, A., Niaura, R., Rauch, S.L. and Brown, W.A. (2004) *Neural and endocrine correlates of sadness in women: Implications for neural network regulation of HPA activity. The Journal of Neuropsychiatry and Clinical Neurosciences*, 16(4), 446–455. https://doi.org/10.1176/jnp.16.4.446

Parrott, R.F., Houpt, K.A. and Misson, B.H. (1988) *Modification of the responses of sheep to isolation stress by the use of mirror panels. Applied Animal Behaviour Science*, 19(3), 331–338. https://doi.org/10.1016/0168-1591(88)90015-9

Pitman, D.L., Ottenweller, J.E. and Natelson, B.H. (1990) *Effect of stressor intensity on habituation and sensitization of glucocorticoid responses in rats. Behavioral Neuroscience*, 104(1), 28–36. https://doi.org/10.1037/0735-7044.104.1.28

Pollard, T.M. (1995) *Use of cortisol as a stress marker: Practical and theoretical problems. American Journal of Human Biology*, 7(2), 265–274. https://doi.org/10.1002/ajhb.1310070217

Porterfield, D., Laskin, D., Jung, S.-K., Malchow, R., Billack, B., Smith, P. and Heck, D. (2001) *Proteins and lipids define the diffusional field of nitric oxide. American Journal of Physiology. Lung Cellular and Molecular Physiology*, 281, L904–L912. https://doi.org/10.1152/ajplung.2001.281.4.L904

Resnick, H.S., Yehuda, R., Pitman, R.K. and Foy, D.W. (1995) *Effect of previous trauma on acute plasma cortisol level following rape. The American Journal of Psychiatry*, 152(11), 1675–1677. https://doi.org/10.1176/ajp.152.11.1675

Rhind, S.M., Reid, H.W., McMillen, S.R. and Palmarini, G. (1998) *The rôle of cortisol and β-endorphin in the response of the immune system to weaning in lambs. Animal Science*, 66(2), 397–402. https://doi.org/10.1017/S135772980000953X

Rodgers, R.J. (1997) *Animal models of "anxiety": Where next? Behavioural Pharmacology*, 8(6), 477–496.

Rohleder, N., Beulen, S.E., Chen, E., Wolf, J.M. and Kirschbaum, C. (2007) *Stress on the dance floor: The cortisol stress response to social-evaluative threat in competitive ballroom dancers. Personality and Social Psychology Bulletin* https://doi.org/10.1177/0146167206293986, 33, 69–84.

Rohrmann, S., Hopp, H. and Quirin, M. (2008) *Gender differences in psychophysiological responses to disgust. Journal of Psychophysiology*, 22(2), 65–75.

Ruis, M.A.W., Te Brake, J.H.A., Engel, B., Ekkel, E.D., Buist, W.G., Blokhuis, H.J. and Koolhaas, J.M. (1997) *The circadian rhythm of salivary cortisol in growing pigs: Effects of age, gender, and stress. Physiology & Behavior*, 62(3), 623–630. https://doi.org/10.1016/S0031-9384(97)00177-7

Rushen, J., de Passillé, A. M., von Keyserlingk, M. A. G. and Weary, D. M. (eds) (2008) *The Welfare of Cattle*. Springer Netherlands, pp. 229–253. https://doi.org/10.1007/978-1-4020-6558-3_9

Sacta, M.A., Chinenov, Y. and Rogatsky, I. (2016) *Glucocorticoid signaling: An update from a genomic perspective. Annual Review of Physiology*, 78, 155–180. https://doi.org/10.1146/annurev-physiol-021115-105323

Sage, D., Maurel, D. and Bosler, O. (2002) *Corticosterone-dependent driving influence of the suprachiasmatic nucleus on adrenal sensitivity to ACTH. American Journal of Physiology - Endocrinology and Metabolism*, 282(2), E458–E465. https://doi.org/10.1152/ajpendo.00287.2001

Salvador, A., Simón, V., Suay, F. and Llorens, L. (1987) *Testosterone and cortisol responses to competitive fighting in human males: A pilot study. Aggressive Behavior*, 13(1), 9–13. https://doi.org/10.1002/1098-2337(1987)13:1<9::AID-AB2480130103>3.0.CO;2-4

Sapolsky, R.M., Romero, L.M. and Munck, A.U. (2000) *How do glucocorticoids influence stress responses? Integrating permissive, suppressive, stimulatory, and preparative actions. Endocrine Reviews*, 21(1), 55–89. https://doi.org/10.1210/edrv.21.1.0389

Schoech, S.J., Romero, L.M., Moore, I.T. and Bonier, F. (2013) *Constraints, concerns and considerations about the necessity of estimating free glucocorticoid concentrations for field endocrine studies. Functional Ecology*, 27(5), 1100–1106. https://doi.org/10.1111/1365-2435.12142

Selye, H. (1956) *The Stress of Life*. McGraw-Hill, New York.

Sheriff, M.J., Krebs, C.J. and Boonstra, R. (2010) *Assessing stress in animal populations: Do fecal and plasma glucocorticoids tell the same story? General and Comparative Endocrinology*, 166(3), 614–619.

Sheriff, M.J., Dantzer, B., Delehanty, B., Palme, R. and Boonstra, R. (2011) *Measuring stress in wildlife: Techniques for quantifying glucocorticoids. Oecologia*, 166(4), 869–887. https://doi.org/10.1007/s00442-011-1943-y

Smyth, J., Ockenfels, M.C., Porter, L., Kirschbaum, C., Hellhammer, D.H. and Stone, A.A. (1998) *Stressors and mood measured on a momentary basis are associated with salivary cortisol secretion. Psychoneuroendocrinology*, 23(4), 353–370. https://doi.org/10.1016/s0306-4530(98)00008-0

Thun, R. and Eggenberger, E. (1996) *Relationship between cortisol and testosterone during resting conditions, after acute stress and hormone stimulation in steers. Schweizer Archiv fur Tierheilkunde*, 138(5), 225–233.

Tiemann, I., Fijn, L.B., Bagaria, M., Langen, E.M., Van Der Staay, F.J., Arndt, S.S., et al. (2023) *Glucocorticoids in relation to behavior, morphology, and physiology as proxy indicators for the assessment of animal welfare. A systematic mapping review. Frontiers in Veterinary Science*, 9, 954607.

Toates, F.M. (1986) *Motivational Systems*. CUP Archive.

Ulrich-Lai, Y.M., Arnhold, M.M. and Engeland, W.C. (2006) *Adrenal splanchnic innervation contributes to the diurnal rhythm of plasma corticosterone in rats by modulating adrenal sensitivity to ACTH. American Journal of Physiology - Regulatory, Integrative and Comparative Physiology*, 290(4), R1128–R1135. https://doi.org/10.1152/ajpregu.00042.2003

van der Meij, L., Buunk, A.P. and Salvador, A. (2010) *Contact with attractive women affects the release of cortisol in men. Hormones and Behavior*, 58(3), 501–505. https://doi.org/10.1016/j.yhbeh.2010.04.009

van Eekelen, A.P.J., Kerkhof, G.A. and van Amsterdam, J.G.C. (2003) *Circadian variation in cortisol reactivity to an acute stressor. Chronobiology International*, 20(5), 863–878. https://doi.org/10.1081/cbi-120024212

Van Reenen, C.G., OConnell, N.E., Van der Werf, J.T.N., Korte, S.M., Hopster, H., Jones, R.B. and Blokhuis, H.J. (2005) *Responses of calves to acute stress: Individual consistency and relations between behavioral and physiological measures. Physiology & Behavior*, 85(5), 557–570. https://doi.org/10.1016/j.physbeh.2005.06.015

von Walter, L.W., Lidfors, L., Madej, A., Dahlborn, K. and Hydbring-Sandberg, E. (2010) *Cardiovascular, endocrine and behavioural responses to suckling and permanent separation in goats. Acta Veterinaria Scandinavica*, 52(1), 51. https://doi.org/10.1186/1751-0147-52-51

Walker, C.D., Lightman, S.L., Steele, M.K. and Dallman, M.F. (1992) *Suckling is a persistent stimulus to the adrenocortical system of the rat. Endocrinology*, 130(1), 115–125. https://doi.org/10.1210/endo.130.1.1309321

Walker, F.R., Masters, L.M., Dielenberg, R.A. and Day, T.A. (2009) *Coping with defeat: Acute glucocorticoid and forebrain responses to social defeat vary with defeat episode behaviour. Neuroscience*, 162(2), 244–253. https://doi.org/10.1016/j.neuroscience.2009.04.041

Wall, J.P. and Cockrem, J.F. (2009) *Effects of corticosterone treatment on responses to fasting in Japanese quail. Comparative Biochemistry and Physiology Part A: Molecular & Integrative Physiology*, 154(2), 211–215. https://doi.org/10.1016/j.cbpa.2009.06.006

West, J., Otte, C., Geher, K., Johnson, J. and Mohr, D.C. (2004) *Effects of Hatha yoga and African dance on perceived stress, affect, and salivary cortisol. Annals of Behavioral Medicine: A Publication of the Society of Behavioral Medicine*, 28(2), 114–118. https://doi.org/10.1207/s15324796abm2802_6

Wingfield, J.C. and Romero, L. M. (2015) *Tempests, Poxes, Predators, and People: Stress in Wild Animals and How They Cope*. Oxford University Press.

Wirth, M.M., Welsh, K.M. and Schultheiss, O.C. (2006) *Salivary cortisol changes in humans after winning or losing a dominance contest depend on implicit power motivation. Hormones and Behavior*, 49(3), 346–352. https://doi.org/10.1016/j.yhbeh.2005.08.013

Wright-Williams, S.L., Courade, J.-P., Richardson, C.A., Roughan, J.V. and Flecknell, P.A. (2007) *Effects of vasectomy surgery and meloxicam treatment on faecal corticosterone levels and behaviour in two strains of laboratory mouse. Pain*, 130(1), 108–118. https://doi.org/10.1016/j.pain.2006.11.003

Würbel, H. and Stauffacher, M. (1997) *Age and weight at weaning affect corticosterone level and development of stereotypies in ICR-mice. Animal Behaviour*, 53(5), 891–900. https://doi.org/10.1006/anbe.1996.0424

Yamakoshi, T., Park, S.-B., Jang, W.-C., Kim, K., Yamakoshi, Y. and Hirose, H. (2009) *Relationship between salivary Chromogranin-A and stress induced by simulated monotonous driving. Medical & Biological Engineering & Computing*, 47(4), 449–456. https://doi.org/10.1007/s11517-009-0447-y

Yim, I.S., Quas, J.A., Cahill, L. and Hayakawa, C.M. (2010) *Childrens and adults salivary cortisol responses to an identical psychosocial laboratory stressor. Psychoneuroendocrinology*, 35(2), 241–248. https://doi.org/10.1016/j.psyneuen.2009.06.014

Body Temperature and Heart Rate

6

Dorothy E.F. McKeegan and Neil P. Evans

School of Biodiversity, One Health and Veterinary Medicine, University of Glasgow, Glasgow, UK

Introduction

The sympathetic nervous system plays a primary role in regulating homeostasis in a wide range of body systems. It responds to external challenge by rapidly triggering neural, endocrine, behavioural and muscular activity, known collectively as the sympatho-adrenal response. When relevant stimuli (positive and negative) are perceived, alongside direct nervous control of various organ systems, a neuronal pathway is activated that results in release of catecholamines (adrenaline and noradrenaline) from the medulla of the adrenal gland. For example, within the cardiopulmonary system, sympathetic nervous system activation triggers increased cardiac output, respiration rate, bronchodilation and redistribution of blood flow (Gordan et al. 2015). Hence, the sympatho-adrenal response is commonly termed the '*fight or flight response*'. Conditions that induce increased activity within the sympathetic nervous system overlap with those activating the hypothalamo-pituitary–adrenal axis (Smith and Vale 2006) – the other system associated with classical 'stress physiology' (see Chapter 5). In this chapter, we review the value of two sympathetically mediated system responses that are relevant to the assessment of emotional states in animals: body temperature and heart rate (HR). These metrics are attractive since they may be readily and objectively measured, including by non-invasive means. In general, they have been applied as indicators of the response to acute stimuli, so we will not discuss them in relation to chronic challenges. Furthermore, discussion of body temperature will consider only endotherms (mammals and birds) that meet important assumptions about underlying thermal physiology.

Body temperature

Homeotherms utilise physiological and behavioural thermoregulatory mechanisms to closely regulate a body temperature (in the range of 36–42 °C) that is generally higher than ambient. Although each species has its own 'set point', body temperature is not completely uniform and variation is observed across body areas and in relation to a range of factors including season, time of day, hormonal status and physical activity. In homeotherms, core body temperature is 'protected' and varies least, whereas the temperature of the periphery exhibits greater fluctuations facilitating heat loss or gain in response to environmental challenges. Indeed, some extremities (such as rabbit ears, chicken combs and rat tails) are adapted for peripheral heat dissipation/conservation due to their high densities of arteriovenous anastomoses. The 'thermostat' which monitors the set point of body temperature resides in the hypothalamus of the brain and controls the nervous and endocrine systems involved in thermoregulation. When environmental temperatures fall, a homeotherm can increase metabolic rate to boost heat production using isometric muscle contractions (shivering), and/or employ strategies to conserve heat, such as peripheral vasoconstriction, piloerection and/or behavioural responses (e.g. huddling, postural changes to reduce surface area). Some homeotherms may also exhibit non-shivering thermogenesis, whereby brown fat is metabolised specifically for heat production. Conversely, in warm ambient conditions, or following the production of heat from vigorous exercise, vasodilation and sweating or panting (depending on the species) can be triggered, alongside a series of behavioural responses (e.g. open stance, seeking shade) to promote evaporative heat loss.

Sympatho-adrenal activation induces rapid redistribution of blood from the periphery to the core via cutaneous vasoconstriction, resulting in a consequent increase in core body temperature. This response is commonly termed '*stress induced hyperthermia*' (SIH), and for convenience and compatibility with the literature this abbreviation will be used here, though we note that the term 'stress' is problematic and risks circular reasoning. For the purposes of this chapter, SIH can be assumed to mean '*stimulus induced hyperthermia*' which makes no assumptions about the accompanying emotional response or the stimulus source (external or internal). Bidirectional changes in body temperature may be simultaneously observed during SIH, depending on location: warming of the core, cooling of the periphery and no change in some tissues. Alongside other sympathetically mediated reactions, SIH serves the adaptive function of diverting blood flow to organs necessary for responding to the stimulus (primarily muscles and pulmonary organs) as well as minimising risk of blood loss from extremities, in the case of injury. Simultaneously, the sympathetic nervous system induces thermogenesis, by activation of adipose tissue which can also contribute to core warming. Thus, SIH can be detected by measurement of core temperature increase and/or peripheral cooling. The changes are transient and not extreme (e.g. in mice, core temperature increases 1.2–1.9 °C in 8 minutes following repeated gentle touching in their home cage, and returns to baseline in 60 minutes (Van der Heyden et al. 1997). At the body surface, SIH is detected by a rapid temperature reduction (e.g. a 1.3–2.2 °C decline in the comb/wattles of hens starting within one minute following handling; Herborn et al. 2015). Figure 6.1 shows successive thermal images of a hen following exposure to an aversive stimulus (white noise; Jones et al. 2012). Interestingly, rapid surface cooling was followed by subsequent warming of the comb which exceeded the baseline, followed by a slower decline. This later 'overshoot' probably reflects dissipation of heat generated in the core. Figure 6.2 shows an idealised surface temperature SIH timeline, noting individual components that may be extracted to characterise the response.

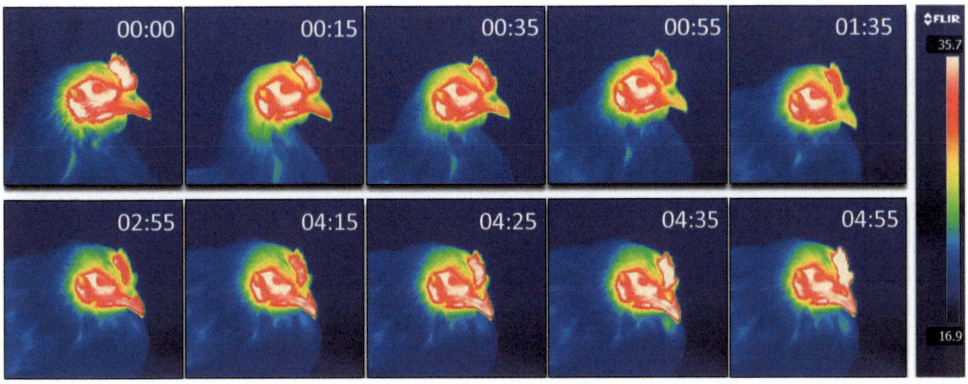

Figure 6.1 A series of thermal images illustrating stimulus-induced hyperthermia (SIH) in a hen exposed to an aversive audio stimulus (10 s white noise; Jones et al. 2012), observable on the comb (unfeathered region; images courtesy of Dorothy McKeegan et al. unpublished). Time in seconds is indicated in the top right of each image. Zero is the start of the post stimulus period, and within 15 s there is a visible reduction in comb temperature (the white colour represents the highest temperatures in the image), progressing up to around three minutes, when the start of recovery is evident. By around five minutes after imposition of the aversive stimulus, the comb surface is warmer than at baseline, reflecting heat dissipation following core thermogenesis.

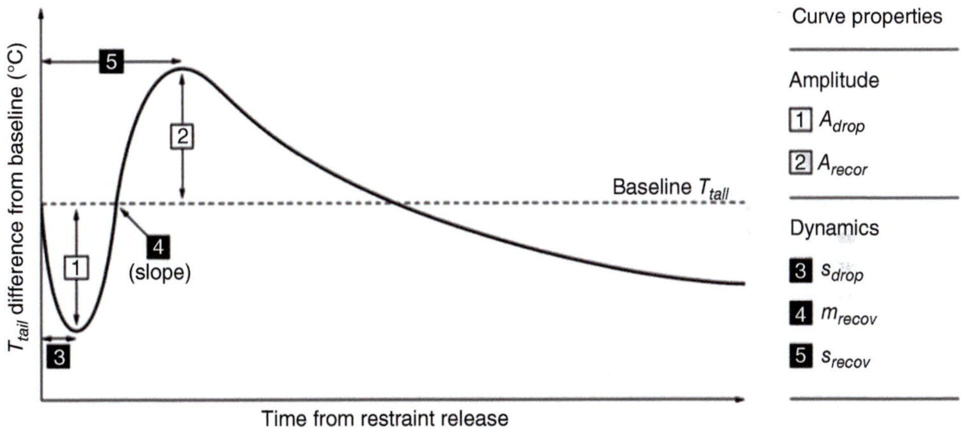

Figure 6.2 Idealised graph of body surface stimulus-induced hyperthermia (SIH), showing that five distinct components may be extracted. The amplitude ([1] A_{drop}) and timing ([3] S_{drop}) of the initial decrease in temperature from baseline captures the maximum temperature drop in response to the stimulus. This is followed by a temperature rise back towards and above the baseline, which will have a maximum amplitude ([2] A_{recov}) and timing ([5] S_{recov}). The slope of the recovery trajectory may also be determined ([4] m_{Recov}). Source: Jerem et al. (2019) / Elsevier / CC BY 4.0.

Heart rate

Heart rate (HR) is the number of contractions (beats) of the heart per minute (bpm). Cardiac activity occurs as a result of inherent phasic electrical activity that, in the normal animal, is initiated in the sinoatrial node, spreads through the cardiac muscle and results in contraction of the atria and ventricles. This inherent electrical activity is regulated by the sympathetic and parasympathetic branches of the autonomic nervous system (ANS). Increased sympathetic nervous system activity results in the release of adrenaline and noradrenaline from efferent nerve fibres and the adrenal gland, leading to an increase in HR, while increased parasympathetic activity induces acetylcholine release at parasympathetic nerve terminals, which lowers HR (e.g. during sleep). As the two branches of the ANS are always active, HR is influenced by the interaction between both the sympathetic and parasympathetic nervous systems. Their typically opposing actions, superimposed on basal tone within the ANS, results in a system wherein HR is exquisitely controlled and can be modified with great speed. Activity within the ANS is influenced by a wide variety of factors, including physiological variables such as energy balance, physical activity, hormones and psychological factors. Despite the range of inputs, in all instances, the outputs will be changes in the release of noradrenaline/adrenaline and acetylcholine which in turn influence HR (and the force of cardiac output).

Sampling body temperature and heart rate

Body temperature

Core body temperature (and often other physiological variables) is typically tracked and recorded over long periods of time using implanted telemetric or logging devices. If placed appropriately, this approach can provide accurate (±0.1 °C), remotely collected data, but it necessitates invasive surgical implantation which is associated with welfare harm and limits the method for use only in captive contexts. Orally introduced gastrointestinal temperature loggers can also be used to assess core temperature; however, it should be noted that their data can be affected by ingestion of food and water. Body temperatures considered equivalent to core can also be obtained by use of non-injurious probes inserted in body openings (most often rectal but also vaginal and aural).

Body surface temperature can be measured with contact sensors (thermistors), affixed directly to the skin surface. While these are accurate (±0.01 °C), and low cost, they may require plucking or shaving to expose bare skin, can be difficult to reliably attach and may modify the immediate environment of the underlying skin. Their presence may also interfere with normal behaviour, and these factors can influence subsequent results. These drawbacks for assessing peripheral temperature are overcome by infrared technologies which allow remote surface temperature detection. Non-contact infrared thermography (IRT) cameras convert the infrared light emitted by an object (or animal) to images in which temperature variations are represented by a grey or colour scale (McCafferty 2007; see also Figure 6.1). Since the body surface of birds and mammals is mostly covered with an insulative layer of fur or feathers, to assess peripheral temperature, IRT studies usually focus on so called 'thermal windows' – uninsulated areas, such as the nose, ear canals and extremities such as tails, combs/wattles and feet (areas important for thermoregulation, see before). Furthermore, due to its ease of visualisation, many SIH studies utilise the thermal window in and around the eye (e.g. Stewart et al. 2008). While the extraction of IRT data (average, minimum and maximum temperature) from individual thermal camera images can be very time consuming with multiple defined regions of interest (ROIs) and large numbers of images collected over time, automated systems have been/are being developed (e.g. Franco et al. 2019). The availability of cheaper and more portable thermal cameras has led to an explosion of animal-based IRT studies in the last ten years, but the apparent ease of data capture with IRT does not reflect the wide range of methodological considerations required for data validity.

Heart rate

As with body temperature, HR can be assessed in a variety of ways and different approaches have strengths and weaknesses. The simplest way, extensively used in human and veterinary clinical medicine, is palpation of an artery. This requires compliance by the test subject and direct interaction between the subject and investigator, which for most animal studies is problematic. The 'gold standard' method for the measurement of HR is from a full electrocardiogram (ECG; Figure 6.3), which requires the placement of multiple wired electrodes directly onto or under the skin. This can be difficult to achieve for animal studies, for the reasons outlined earlier in relation to the use of thermistors for the assessment of skin temperature and as it limits animal movement. While wearable ECG equipment, such as Holter monitors and miniature ECG-radio transmitters, have been developed, they still require close electrode-skin contact, which in some cases has required invasive surgical subcutaneous implantation, which limits their general applicability. Furthermore, the use of wearable ECG equipment is limited by animal size and the ability of an animal to be fitted with the equipment such that it does not affect normal activity. Photoplethysmography (PPG) is a non-invasive technology that can be used to estimate HR. It is based on volumetric variations of blood circulation at the surface of skin, measured using a light source and a photodetector (one sensor). PPG is used widely in clinical medicine and has been embraced by the public through wearable PPG devices (e.g. Apple Watch, Fitbit Blaze). However, PPG requires good, sustained, optical skin contact to ensure data accuracy, which may be impaired by pigmentation and hairiness for many animals. To alleviate some of these difficulties, injectable subcutaneous wireless PPG sensors are being developed for animal use (Reynolds et al. 2019), but they do require an initial invasive procedure.

Wearable electrode-based strap monitors, initially developed for human use as fitness trackers (two sensors, wireless transmission of data to a logger), have also been embraced by many researchers to non-invasively monitor cardiac electrical activity in animals. Studies have indicated that in humans, across a range of conditions, results obtained using wearable electrode-based systems compare well to data obtained by ECG and may be more accurate than wrist-worn PPG devices, which can be affected by skin tone (Koerber et al. 2022). Straps/monitors marketed for human use have been used in dogs, goats, sheep, pigs and calves (Bowman et al. 2015; Kitajima et al. 2021; Marchant-Forde et al. 2004; Mohr et al. 2002; Wojniusz et al. 2011) though

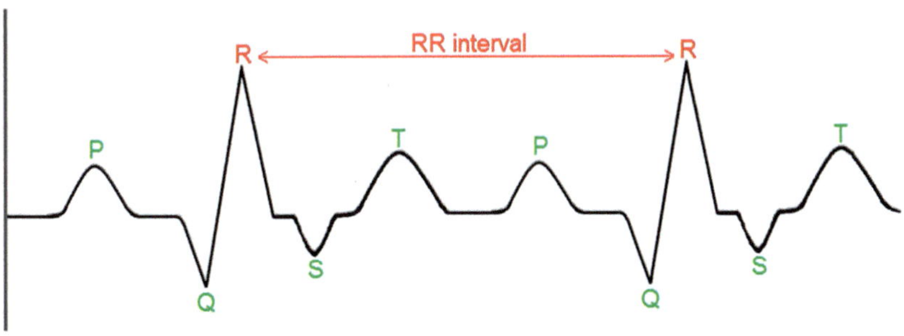

Figure 6.3 The electrocardiogram (ECG) is a record of the average electrical potential generated in the heart muscle graphed in terms of voltage (V) against time (ms). Each wave corresponds to electrical activity of the heart in a specific area. Recognisable deflections are atrial depolarisation (P wave), ventricular depolarisation (QRS complex) and ventricular repolarisation (T wave). The R–R interval (shown in red) is the time between ventricular depolarisations and therefore represents the length of time between each heartbeat, which is the basis for HR measurement.

they may not be suitable for small animals/individuals due to the fixed spacing of the electrodes precluding accurate data collection. An alternate system that uses moveable electrode pads, designed for use in equine performance training has also been used with larger animals such as cattle (Frei et al. 2022; Mott et al. 2021). Figures 6.4 and 6.5 show some examples of animals wearing HR recording equipment. The nature of the electrode-based strap monitors means that it is necessary to handle the animal both to fit and to recover the data collection apparatus and electrode/skin contact is required.

HR can also be assessed non-invasively using computer vision or video analysis, where different techniques assess subtle changes (colour or motion) caused by the pulsatile activity of the beating heart in an ROI which ideally has large capillaries near the skin surface and minimal hair covering. These technologies have been trialled in cattle (Jorquera-Chavez et al. 2019), primates (Unakafov et al. 2018), rodents (Kunczik et al. 2019), sheep (Fuentes et al. 2020) and pigs (Wang et al. 2021). Based on work reported to date, this approach can provide a measure that represents HR, but animal movement and lighting conditions may contribute to lower accuracy and greater variability relative to other HR assessment methods such as ECG.

Cardiac implantable electronic devices and subcutaneous cardiac monitors are used in human clinical practice as tools for monitoring HR, often when arrhythmias are suspected. They have been used in a series of wildlife studies in brown bears (*Ursus arctos*), grey wolves

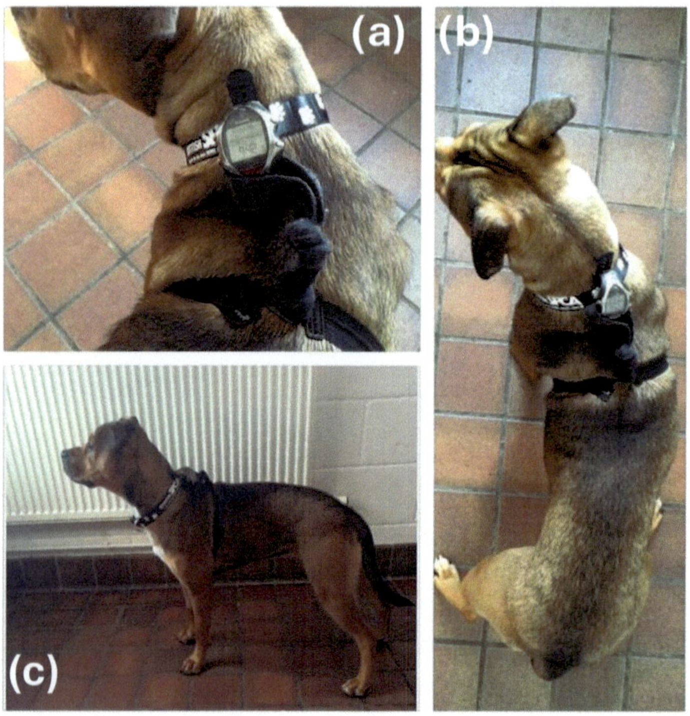

Figure 6.4 Photographs demonstrating how the Polar® RS800CX equipment can be used with canine participants: (a) and (b) the way in which the watch-computer was positioned at the back of the head by fastening through the collar and excess strap material. (c) The positioning of the strap immediately behind the axilla and around the cranio-ventral thorax. Source: Photo by Author.

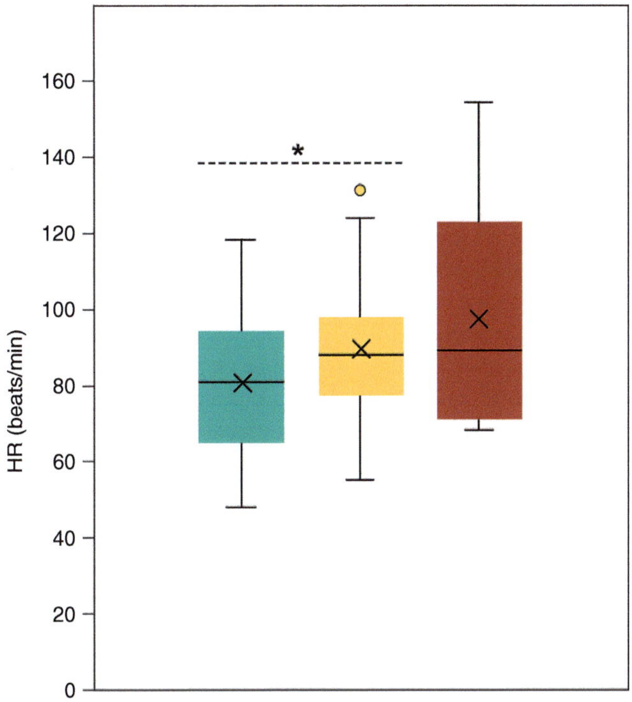

Figure 6.5 Mean (x) heart rate (beats per minute) for dogs (n = 53) in a school classroom situation as part of animal-assisted education. Dogs were categorised into three types: low stress (n = 22; blue), higher stress/non-threatening (n = 22; yellow) and higher stress/threatening dogs (n = 6; red) based on their behaviour during interactions with pupils and their owner. Source: Adapted from Bidoli et al. (2022).

(*Canis lupus*), moose (*Alces alces*), maned wolves (*Chrysocyon brachyurus*) and southern elephant seals (*Mirounga leonine*; Laske et al. 2021). While the use of these devices requires a surgical intervention, once implanted they can generate long-term data sets, and advances in wireless telemetry (external and wearable) mean that with appropriate telemetry stations it is possible to have real-time streaming of data.

The construct validation of body temperature and heart rate as indicators of emotional state

Body temperature

Evidence from humans

There is ample evidence of body temperature change in humans experiencing negative emotional states. Also called 'psychogenic fever', SIH has been primarily studied in university students taking part in written exams (e.g. Briese 1995; Marazziti et al. 1992), but data are also available relative to other challenges such as sporting contests (e.g. Renbourn 1960). The valence of some of these challenges is not always clear and may vary between individuals, and we will therefore focus on examples where there is less ambiguity. The focus of initial research utilising body temperature measures was core warming, but later work suggests a complex picture, where the amplitude and direction of stimulus-induced temperature changes depends on the site of temperature measurement. For example, when volunteers were exposed to the Trier Social Stress Test, which requires participants to make an interview-style presentation, followed by a surprise mental arithmetic test, in front of an interview panel who do not provide feedback or encouragement (Vinkers et al. 2013), a drop in skin temperature was seen. This happened in a variety of locations (fingertips, noses) but at some sites temperature changes were sexually dimorphic and even showed an increase (male cheeks). Alterations in body temperature have also been reported to be associated with negatively-valenced psychiatric disorders such as schizophrenia (corneal temperature elevation; Shiloh et al. 2005) and depression (disruption to circadian rhythm; Daimon et al. 1992). Finally, pharmacological studies have shown that acute and chronic elevations in core temperature associated with negative emotional states that are resistant to the effects of antipyretic drugs do decrease in response to psychotropic drugs that display anxiolytic properties (Oka 2015). Based on these varied findings, a complex pattern is emerging for the relationships between affective valence and temperature change in humans. There is a lack of data for positive states, but decreases in facial skin temperature were observed in infants with joyful expressions (Nakanishi and Imai-Matsumura 2008), while skin temperature increases accompanied listening to calming music in adults (McFarland 1985). This impacts on construct validity for this indicator, and this will be discussed further in the responsiveness and selectivity sections.

Evidence from animals

The SIH response appears to be highly conserved among homeotherms, having been observed in all mammal and bird species that have been investigated. SIH has been observed in response to a range of stimuli that are aversive or harmful, and where multiple strands of evidence support the presence of a negative experience (reviewed by Travain and Valsecchi 2021). These include pain during disbudding without anaesthesia in cattle (Stewart et al. 2008), manual restraint in rats (Wongsaengchan et al. 2023), foot-shock-conditioned fear in rats (Vianna and Carrive 2005) and agonistic encounters in pigs (Boileau et al. 2019). The SIH response to negative stimuli has been shown to covary with indicators of arousal, including HR, plasma adrenocorticotropic hormone (ACTH) and glucocorticoids (e.g. Groenink et al. 1994; van Bogaert et al. 2006). Animals exposed to open field and elevated plus mazes (Lecorps et al. 2016), and hence in an assumed negative state (e.g. if exhibiting defensive, avoidance, or escape behaviour), show reliable SIH changes suggesting that SIH may also reflect valence (although arousal is also likely to be elevated in these subjects).

SIH can be activated by mild stimuli at levels that do not trigger other 'classical physiological stress' responses (for example, the surface temperature response to a 10 s burst of white noise shown in Figure 6.1 was not accompanied by a measurable release of plasma corticosterone [McKeegan et al. unpublished]). Further, there is evidence that the strength of the stimulus is matched by the extent of body temperature alteration. For example, mouse core body temperature showed a proportional increase by up to 0.8, 2.0 and 2.4 °C following exposure to progressively more negative stimuli – tapping the cage with a finger, tail handling and tail handling plus placement in a novel cage, respectively (van Bogaert et al. 2006). There has been limited systematic work on the effect of changing stimulus magnitude on surface temperature during SIH, though handling severity (cradling vs side-pinning) predicted the extent of initial skin cooling, plus delayed skin warming in hens (Herborn et al. 2015) and rats (Wongsaengchan et al. 2023). It is important to note that if these manipulations generate elevations in arousal as well as negative states, SIH may reflect the former and/or the latter.

As much of the early work on SIH was conducted in rodents, the association between SIH and aversive situations is robustly evidenced in these species, and SIH (primarily in mice) is used as a model to screen potential anxiolytic drugs. This work has demonstrated that the SIH paradigm, in rodents, is sensitive to clinically effective anxiolytics such as 5-HT_{1A} receptor agonists and benzodiazepines, but that it is not acutely sensitive to antidepressants (Olivier et al. 2003; Vinkers et al. 2009). As with humans, administration of drugs that target inflammatory systems also do not affect SIH in rodents, indicating that the increase in temperature is not a 'fever' response mediated by the immune system (induced by interleukins and lipopolysaccharide, for example).

Although SIH is an element of a generalised arousal response, there has been almost total focus on how it manifests during assumed negative affective states. There is not enough research on positive states to draw robust conclusions, but the available findings broadly support the notion that SIH either captures arousal independently of valence or that the changes associated with arousal mask the direction of affective valence. Examples of this will be covered in the section on selectivity problems.

Heart rate

Evidence from humans

Heart rate is exquisitely responsive to a variety of stimuli over very short timescales (the timescale between beats being 0.5–1.2 seconds). However, the relationship between emotional state and HR appears to be complex. An extensive review reported that HR could be increased, decreased or unchanged in response to anger, happiness and amusement; increased or decreased in response to anxiety, fear, sadness, disgust and pride; increased in response to embarrassment and joy and decreased in response to affection and contentment (Kreibig 2010). Due to its metabolic effects, physical activity can also affect HR. While this can be controlled for in human trials, effects of challenging life events on resting HR have been reported, and the changes can be either an increase or a decrease in resting HR (Lee et al. 2014). This diversity in the HR response measured is potentially a consequence of vagal rebound (a sudden and transient increase in vagal variability; Mezzacappa et al. 2001). The link between HR and negative states is supported by the pharmacological depression of HR possible with drugs that antagonise sympathetic action such as Beta blockers and are recognised treatments for psychological distress (Viola et al. 2022), anxiety and stress (Noyes 1985), although these likely reduce arousal too.

Evidence from animals

There is a wealth of data available related to changes in HR in animals where it has been reported that HR or changes in HR can be associated with disease status, physiological functioning, and individual characteristics, e.g. temperament and coping strategies (reviewed by Wascher 2021). As outlined in the section on human studies, the interpretation of HR as a measure of emotional state in animals also requires caution, to ensure that observed changes are due to a physiological or psychological response and are associated with affective valence rather than other factors such as physical activity (which may or may not also be a consequence of the stimulus). Changes in HR in response to exposure to events likely to induce negative emotional states have been reported. For example, HR is increased in dogs exposed to unfamiliar environments, introduced to strangers or separated from their owners (Palestrini et al. 2005), in response to loud noises (Hydbring-Sandberg et al. 2004), all of which generate behavioural changes that are valid markers of negative affect (see Test 5 in Chapter 2). HR in greater white-fronted geese, *Anser albifrons*, is elevated during agonistic encounters to levels associated with the physical activity of flight (Ely et al. 1999) and is also raised in horses subject to transportation, which is thought to be an aversive situation in which physical activity is controlled or limited (Schmidt et al. 2010). The effect of transport on HR in horses is supported by data from sheep which also indicate that HR can be increased by aversive situations such as social isolation and introduction to a novel group of animals where rank order must be established (Baldock and Sibly 1990). Data from sheep also suggest that changes in HR may be related to emotional valence, as increased HR is associated with social isolation (negative valence), but is decreased when animals are voluntarily groomed by a familiar human (positive valence; Reefmann et al. 2009a, 2009b). Even in this controlled study, the aversive stimulus (social isolation) may be associated with greater arousal relative to the positive stimulus (grooming), and as such the HR data may not be isolated from effects of arousal (Reefmann et al. 2009b). A relationship between HR and positive stimuli has also been reported in cattle, where HR was decreased in association with additional positive handling and gentle human interactions during ovarian palpation per rectum (Waiblinger et al. 2004).

Few studies have reported the direct comparison of stimulus aversiveness on changes in HR in animals, however, Baldock and Sibly (1990) reported that exposure to an unfamiliar human resulted in an increase in sheep HR of 45 bpm, but exposure to both a strange human and a dog, i.e. an increase in aversiveness, increased HR by 79 bpm.

Pharmacological treatments that block the sympathetic nervous system have been shown to reduce or eliminate the increase in HR associated with a variety of challenges, including cold stress in sheep (Webster and Hays 1968), capture in roe deer (Montané et al. 2007) and transport in Southern chamois (López-Olvera et al. 2006). The involvement of the sympathetic nervous system to drive an increase in HR has been demonstrated in rats, in response to psychological stress (an open field test), by the loss of the tachycardic response where rats are treated with a ganglion blocker (pentolinium), a b1-adrenoceptor antagonist (atenolol) and the classical anxiolytic drug, diazepam (van den Buuse et al. 2001). In addition, HR and anxiety-related behaviours are lowered in dogs with a diagnosis of anxiety disorder by use of a 'thunder shirt' that applies deep pressure stimulation (King et al. 2014), which has been reported to reduce anxiety in humans and animals (Grandin 1992; Reynolds et al. 2015). More recently, it has been shown in mice that an elevation of HR independent of changes in activity (through optogenetics) is associated with the expression of anxiety-like behaviour (Hsueh et al. 2023). With respect to both body temperature and HR, there is therefore evidence linking these variables with affective state, including pharmacological support that they can be used as a valid measure, along with evidence that the size of the stimulus is reflected in the degree of change noted in the variable. Figure 6.5 shows graded HR responses related to overt stress-related behaviour in dogs. However, in each case, it is important to consider any results in the context of the situation in which the measurement is being collected to ensure that it is not unduly influenced by peripheral factors and any accompanying change in arousal.

Responsiveness – do elevations in HR and core temperature reliably accompany negatively-valenced states?

Measurement and experimental design issues

Some methodological considerations are relevant to both responsiveness *and* selectivity, and will briefly be dealt with here, before focusing on issues that specifically increase risks of false-negative results (false nulls). When considering body temperature, the body ROI/implant location/sensor type should be standardised within a study, and thermographic cameras calibrated and corrected with respect to environmental temperatures and humidities. When using thermographic cameras, variations in measurement distance and angle should be minimised or, where they cannot be controlled for, included in analysis (e.g. Herborn et al. 2015). While there can be variation across a population, each individual has a relatively consistent baseline surface temperature (e.g. hens; Herborn et al. 2015), so it can be advantageous to express SIH data in terms of the within-subject deviation from a baseline. There is also evidence of sex differences in body temperature, with females tending to have slightly higher core body temperatures than males (e.g. in mice; Sanchez-Alavez et al. 2011), so studies have to be balanced by sex or sex should be included as a variable in the analysis. Baseline body temperature (De Castro 1978) and HR (e.g. Black et al. 2019) are both subject to circadian variation, so time of day must also be controlled for.

Evidence for limits to responsiveness and potential biological causes of false nulls

Given their homeostatic control, both body temperature and HR are subject to strong ceiling and floor effects, with physiological limits to the range of measurements. These have been captured in the heuristic graphs shown in Figure 6.6. For body temperature, the response range is limited by the need to keep core temperature within a narrow range (± 1–3 °C), although peripheral temperatures are less constrained. For HR, elevation is limited by the electrical properties of cardiac tissue, which do vary between species, individuals and sex and age. Thus, beyond a certain level of challenge, a flattening of the body temperature and HR response will be seen, with a further change not being possible, even though the situation may result in a more intense experience for the animal. This explains the strong ceiling effects illustrated in Figure 6.6.

As with glucocorticoids, HR and body temperature are also likely to have poor responsiveness for detecting low-arousal negative emotional states, though there is some evidence for increased HR during boredom in humans (Merrifield and Danckert 2014). Effects on HR depend on the stimulus – for example, it has been reported in humans that psychological stressors such as preoperative anxiety and some emotional experiences do not always result in changes in HR. This is despite changes in ANS activity detected using other methodologies such as HR variability (Sleigh and Henderson 1995), indicating that the activity of the sympathetic and parasympathetic branch of the ANS may change without observable changes in HR (Boissy et al. 2007). To our knowledge, nothing is known about SIH in animals in low-arousal states, though decreases in HR during freezing or tonic immobility are well recognised (Carli and Farabollini 2021).

In cold ambient conditions, the ROIs used for body temperature measurement (usually highly vascularised peripheral tissues) are likely to already be rather cool as attempts to reduce heat loss result in vasoconstriction, so a further reduction may not be possible, even though negative states may be experienced. For example, in ptarmigans (a bird of the grouse

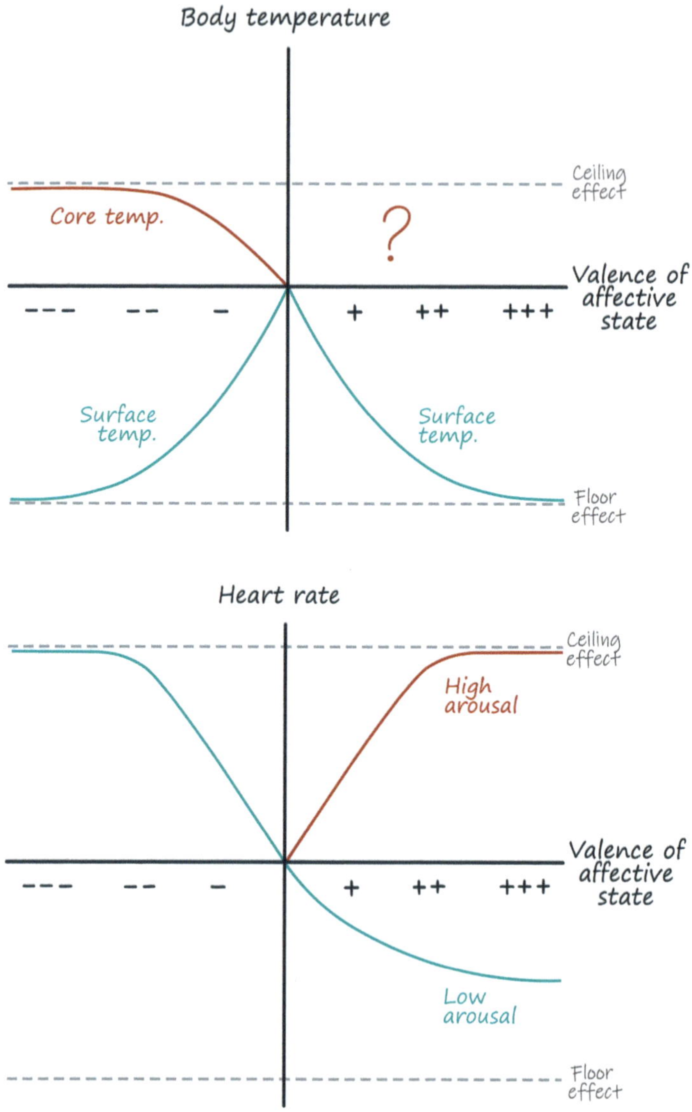

Figure 6.6 Body temperature (BT) and heart rate (HR) plotted against positive and negative emotional valence. In both cases, changes are expressed relative to individual baseline: resting HR and BT in a thermoneutral environment. For body temperature, only initial responses are plotted (since the recovery phase adds problematic complexity and variation across time and individuals). Core temperature rises during negatively-valenced emotional states, but almost no data are available on positive valence. Surface temperature falls initially in response to both positive and negative emotional states. The control of temperature in the core is tighter than at the periphery, hence the relatively lower ceiling. Heart rate increases to a ceiling during negative and positive emotional states with no threshold that must be exceeded before a response is seen. Low-arousal positive states may be accompanied by mild reductions in heart rate.

family) subjected to an aversive situation (handling stress), at −20 °C compared to 0 °C, cloacal temperature increased less and back skin temperature decreased more during SIH (Nord and Folkow 2019). These effects of homeostatic thermoregulation can be avoided in housed animals but can be challenging for outdoor-kept or wild animal studies. Inflammation in certain body regions can also affect blood flow and the temperature of local tissues, thus preventing measurement of a drop in peripheral temperature in response to an aversive event; indeed, IRT has been used to detect/assess inflammation (e.g. digital dermatitis in dairy cattle; Harris-Bridge et al. 2018), so the health status of the subjects should also be considered before drawing conclusions about SIH.

Selectivity problems – when core body temperature and/or heart rate increases do not reflect negative emotional states (false leads)

This section aims to explain situations where rises in core body temperature and HR occur in the absence of negative affective states and hence generate false leads and low selectivity. Arguably the biggest issue in this regard is that sympathetically mediated body temperature and HR responses are related to general physiological arousal and so may be altered when both negative and positive emotional states are experienced. As previously stated, insufficient evidence exists on the relationship between body temperature and positive states, but SIH may capture arousal independently of valence. For example, in hens, comb temperature decreased following exposure to a cue signalling reward as well as consumption of the associated food item, reflecting emotional arousal in both cases (Moe et al. 2012). In dogs, eye temperature increase (indicative of core warming) was shown using IRT during delivery of palatable treats (Travain et al. 2016), and also separation anxiety leading to the conclusion that while arousal was detected, temperature assessment failed to discriminate emotional valence (Taylor et al. 2020).

Whilst there is less known about positive states in relation to SIH, there is ample evidence of this lack of valence specificity for HR elevations. In humans, elevated HR has been associated with pleasant experiences such as exposure to a favoured food (Horio 2000), monetary incentives during a motor task (Fowles et al. 1982), sexual activity (Xue-Rui et al. 2008) and dancing (Blanksby and Reidy 1988), although in some of these situations physical activity may underlie any increase in HR. Similarly in animals, food reward (e.g. Zupan et al. 2016), sexual behaviour (e.g. Terada et al. 2003) and partner reunion (Ellenberg et al. 2013) are accompanied by increased HR. HR can also be progressively elevated, in apparently affectively neutral situations, by increasing levels of physical activity (Karvonen and Vuorimaa 1988) and exercise can lead to heat dissipation that could also mimic the later phase of SIH. As an element of a highly generalised arousal response, this physiological activation is a major source of false leads. Indeed, some causal stimuli are not experienced at all, since we know that HR and body temperature responses may occur in unconscious animals. Core body temperature reduction is observed in anaesthetised subjects (Redondo et al. 2012) and HR may increase during surgery in unconscious individuals as a result of sympathetic reflex activation (for example, in brain-dead organ donors; Fitzgerald et al. 1995). Further, concurrent health challenges such as moderate blood loss, dehydration, low and high blood pressure and nonpathologic cardiac dysrhythmia in the absence of a negative affective state can also be associated with elevated HR (Adams et al. 2014; Gupta and Fahim 2005; Zaza et al. 2018), potentially leading to false leads. Underlying physiological state such as cardiovascular fitness level, age and pregnancy can also influence HR (Loerup et al. 2019; Reimers et al. 2018; Santos et al. 2013) and therefore should be considered.

Despite these issues with selectivity, however, it may be possible to draw conclusions about the intensity of experience, whether pleasant or unpleasant, as long as the direction of valence is known (e.g. animals are showing strong preference or avoidance responses, for example). In this regard, sympathetically mediated physiological responses have similar limitations to glucocorticoids (see Chapter 5).

For body temperature, as with the normal physiological regulation of HR, SIH responses are also overlaid on ongoing thermoregulatory effort, since maintaining a constant deep body temperature is crucial for survival in homeotherms. A combination of afferent sensing, central control and efferent activity underpins this system, and there are both heat and cold receptors throughout body tissues. In the thermoneutral zone (where the rate of heat production is essentially the same as heat loss), minimal action is required, and core temperature can be maintained by minor alterations to skin blood flow. This is the thermal status in which SIH excursions are most readily observed. During thermal challenge, animals may exhibit adaptive behavioural changes that indicate thermoregulatory efforts (e.g. panting), but thermal discomfort may also be present without their manifestation, since certain behavioural strategies might be prevented or constrained by the situation (e.g. broiler chickens unable to stand up and open their wings to dissipate heat due to lack of headroom in transport crates). Although body temperature can be temporarily altered by exercise or SIH, these excursions are less extreme and more transient compared to those causing concern during hyper- or hypothermia (which usually refer to core temperature alterations of at least 3–4 °C). In certain circumstances, the hypothalamic set point can be increased (e.g. during fever in response to infectious pathogens) or decreased (e.g. during anaesthesia), which will affect individual baselines. As with HR, it is also possible that changes in body temperature may occur as a general effect of arousal but not valence (Kosonogov et al. 2017, Salazar-Lopez et al. 2015, Nakanishi and Imai-Matsumura 2008, Ermatinger et al. 2019).

Another major source of potential false leads is related to measurement approaches for body temperature and HR. In both cases, methods that require handling or invasive actions will trigger a rapid sympathetic response, which, while potentially associated with a genuinely negative experience (e.g. fear) in the test subject, tells you nothing about the population or intervention that is the subject of the study. In addition, where HR is assessed by palpation, human error (mitigated by the assessor experience) can result in overestimation compared to more objective methods.

Conclusions and take-home messages for valid use

Body temperature and HR are both output measures of ANS activity, and this commonality is reflected in some aspects of the heuristic graphs (Figure 6.6). As highly responsive indicators that can be accessed non-invasively, they have considerable value in welfare assessment. However, their poor selectivity (especially well understood for HR) means that particularly strong controls and careful experimental conditions need to be in place to ensure that conclusions are valid.

Although both measures are not able to independently indicate the valence of emotional responses, this can often be overcome by inference from behavioural context or additional data collection. Their graded responses are very useful, but ceiling and floor effects prevent accurate assessment at the extremes of their ranges, as both are subject to robust homeostatic control. Both are also subject to modulation in unconscious animals, so they lack validity as welfare indicators in any circumstances where consciousness is not assured. Further, since they reflect activation of the same system, their use in parallel has limited value and they cannot be treated as independent variables.

Perhaps more than other potential indicators, there are a great many methods available to capture body temperature and HR data. Various factors will dictate what is appropriate for a particular investigation, and in line with the 3Rs we would always recommend that non-invasive sensors are fully considered. It is increasingly apparent that IRT has great advantages for body temperature assessment, but it is highly methodologically complex, with many controls to consider (despite the apparent ease of actually obtaining images).

Body temperature and HR are well-studied indicators, but knowledge gaps remain. More well-designed and controlled experiments are needed to determine the relationship between SIH and positive states. These should aim to determine whether there are (possibly species- and tissue-specific) thermal signatures which differ between negative and positive arousal, and whether proportionality is present (i.e. whether the magnitude of a temperature change relates to the assumed intensity of the valenced state, ideally controlling for arousal). Thermal lateralisation, where temperature differences are observed between symmetrical body parts (such as eyes and ears), may have the potential to inform us of emotional valence, but few data have been collected. Further work is also needed to determine if temperature measurements can detect emotional states that do not involve marked changes in arousal. In summary, body temperature and HR provide readily measurable, sensitive and proportional welfare measures with great utility, as long as care is taken to determine valence from context or other measures.

References

Adams, W.M., Ferraro, E.M., Huggins, R.A. and Casa, D.J. (2014) *Influence of body mass loss on changes in heart rate during exercise in the heat: A systematic review. Journal of Strength and Conditioning Research*, 28(8), 2380–2389. https://doi.org/10.1519/jsc.0000000000000501

Baldock, N.M. and Sibly, R.M. (1990) *Effects of handling and transportation on the heart rate and behaviour of sheep. Applied Animal Behaviour Science*, 28(1–2), 15–39. https://doi.org/10.1016/0168-1591(90)90044-e

Bidoli, E.M.Y., Erhard, M.H. and Döring, D. (2022) *Heart rate and heart rate variability in school dogs. Applied Animal Behaviour Science*, 248, 105574. https://doi.org/10.1016/j.applanim.2022.105574

Black, N., D'Souza, A., Wang, Y., Piggins, H., Dobrzynski, H., Morris, G. and Boyett, M.R. (2019) *Circadian rhythm of cardiac electrophysiology, arrhythmogenesis, and the underlying mechanisms. Heart Rhythm*, 16, 298–307. https://doi.org/10.1016/j.hrthm.2018.08.026

Blanksby, B.A. and Reidy, P.W. (1988) *Heart rate and estimated energy expenditure during ballroom dancing. British Journal of Sports Medicine*, 22, 57–60. https://doi.org/10.1136/bjsm.22.2.57

Boileau, A., Farish, M., Turner, S.P. and Camerlink, I. (2019) *Infrared thermography of agonistic behaviour in pigs. Physiology and Behavior*, 210, 112637. https://doi.org/10.1016/j.physbeh.2019.112637

Boissy, A., Manteuffel, G., Jensen, M.B., Moe, R.O., Spruijt, B., Keeling, L.J., Winkler, C., Forkman, B., Dimitrov, I., Langbein, J., Bakken, M., Veissier, I. and Aubert, A. (2007)

Assessment of positive emotions in animals to improve their welfare. Physiology and Behavior, 92, 375–397. https://doi.org/10.1016/j.physbeh.2007.02.003

Bowman, A., Scottish, S.P.C.A., Dowell, F.J. and Evans, N.P. (2015) *'Four Seasons' in an animal rescue centre; classical music reduces environmental stress in kennelled dogs. Physiology and Behavior*, 143, 70–82. https://doi.org/10.1016/j.physbeh.2015.02.035

Briese, E. (1995) *Emotional hyperthermia and performance in humans. Physiology & Behavior*, 58, 615–618. https://doi.org/10.1016/0031-9384(95)00091-v

Carli, G. and Farabollini, F. (2021) *Cardiovascular correlates of human emotional vasovagal syncope differ from those of animal freezing and tonic immobility. Physiology and Behavior*, 238, 113463. https://doi.org/10.1016/j.physbeh.2021.113463

Daimon, K., Yamada, N., Tsujimoto, T. and Takahashi, S. (1992) *Circadian rhythm abnormalities of deep body temperature in depressive disorders. Journal of Affective Disorders*, 26, 191–198. https://doi.org/10.1016/0165-0327(92)90015-x

De Castro, J.M. (1978) *Diurnal rhythms of behavioral effects on core temperature. Physiology & Behavior*, 21, 883–886. https://doi.org/10.1016/0031-9384(78)90160-9

Ellenberg, U., Mattern, T. and Seddon, P.J. (2013) *Heart rate responses provide an objective evaluation of human disturbance stimuli in breeding birds. Conservation Physiology*, 1, cot013. https://doi.org/10.1093/conphys/cot013

Ely, C.R., Ward, D.H. and Bollinger, K.S. (1999) *Behavioral correlates of heart rates of free-living greater white-fronted geese. The Condor: Ornithological Applications*, 101, 390–395.

Ermatinger, F.A., Brügger, R.K. and Burkart, J.M. (2019) *The use of infrared thermography to investigate emotions in common marmosets. Physiology & Behavior*, 211, 112672. https://doi.org/10.1016/j.physbeh.2019.112672

Fitzgerald, R.D., Templ, D.E. and Lackner, F.F.X. (1995) *Cardiovascular and catecholamine response to surgery in brain-dead organ donors. Anaesthesia*, 50, 388–392. https://doi.org/10.1111/j.1365-2044.1995.tb05989.x

Fowles, D.C., Fisher, A.E. and Tranel, D.T. (1982) *The heart beats to reward: The effect of monetary incentive on heart rate. Psychophysiology*, 19, 506–513. https://doi.org/10.1111/j.1469-8986.1982.tb02577.x

Franco, N.H., Geros, A., Oliveira, L., Olsson, A.S. and Aguiar, P. (2019) *ThermoLabAnimal - a high-throughput analysis software for non-invasive thermal assessment of laboratory mice. Physiology & Behavior*, 207, 113–121. https://doi.org/10.1016/j.physbeh.2019.05.004

Frei, A., Evans, N.P., King, G., McAloon, C.G. and Viora, L. (2022) *Associations between cow-level parameters and heart rate variability as a marker of the physiological stress response in dairy cows. Journal of Dairy Research*, 1-6. https://doi.org/10.1017/s0022029922000565, 265–270.

Fuentes, S., Viejo, C.G., Chauhan, S.S., Joy, A., Tongson, E. and Dunshea, F.R. (2020) *Non-invasive sheep biometrics obtained by computer vision algorithms and machine learning modeling using integrated visible/infrared thermal cameras. Sensors*, 20, 6334. https://doi.org/10.3390/s20216334

Gordan, R., Gwathmey, J.K. and Xie, L. (2015) *Autonomic and endocrine control of cardiovascular function. World Journal of Cardiology*, 7, 204–214. https://doi.org/10.4330/wjc.v7.i4.204

Grandin, T. (1992) *Calming effects of deep touch pressure in patients with autistic disorder, college students and animals. Journal of Child and Adolescent Psychopharmacology*, 2, 63–72. https://doi.org/10.1089/cap.1992.2.63

Groenink, L., van der Gugten, J., Zethof, T.Z., van der Heyden, J. and Olivier, B. (1994) *Stress-induced hyperthermia in mice: Hormonal correlates. Physiology & Behavior*, 56, 747–749. https://doi.org/10.1016/0031-9384(94)90237-2

Gupta, R.K. and Fahim, M. (2005) *Regulation of cardiovascular functions during acute blood loss. Indian Journal of Physiology and Pharmacology*, 49, 213–219.

Harris-Bridge, G., Young, L., Handel, I., Farish, M., Mason, C., Mitchell, M.A. and Haskell, M.J. (2018) *The use of infrared thermography for detecting digital dermatitis in dairy cattle: What is the best measure of temperature and foot location to use? Veterinary Journal*, 237, 26–33. https://doi.org/10.1016/j.tvjl.2018.05.008

Herborn, K.A., Graves, J.L., Jerem, P., Evans, N.P., Nager, R., McCafferty, D.J. and McKeegan, D.E.F. (2015) *Skin temperature reveals the intensity of acute stress. Physiology & Behavior*, 152, 225–230. https://doi.org/10.1016/j.physbeh.2015.09.032

Horio, T. (2000) *Effects of various taste stimuli on heart rate in humans. Chemical Senses*, 25, 149–153. https://doi.org/10.1093/chemse/25.2.149

Hsueh, B., Chen, R., Jo, Y., Tang, D., Raffiee, M., Kim, Y.S., Inoue, M., Randles, S., Ramakrishnan, C., Patel, S., Kim, D.K., Liu, T.X., Kim, S.H., Longzhi, T., Mortazavi, L., Cordero, A., Shi, J., Zhao, M., Ho, T.T., Crow, A., Yoo, A.W., Raja, C., Evans, K., Bernstein, D., Zeineh, M., Goubran, M. and Deisseroth, K. (2023) *Cardiogenic control of affective behavioural state. Nature*, 615, 292–299. https://doi.org/10.1038/s41586-023-05748-8

Hydbring-Sandberg, E., von Walter, L.W., Hoglund, K., Svartberg, K., Swenson, L. and Forkman, B. (2004) *Physiological reactions to fear provocation in dogs. Journal of Endocrinology*, 180, 439–448. https://doi.org/10.1677/joe.0.1800439

Jerem, P., Jenni-Eiermann, S., McKeegan, D., McCafferty, D.J. and Nager, R.G. (2019) *Eye region surface temperature dynamics during acute stress relate to baseline glucocorticoids independently of environmental conditions. Physiology & Behavior*, 210, 112627. https://doi.org/10.1016/j.physbeh.2019.112627

Jones, A.R., Bizo, L.A. and Foster, T.M. (2012) *Domestic hen chicks' conditioned place preferences for sound. Behavioural Processes*, 89, 30–35. https://doi.org/10.1016/j.beproc.2011.10.007

Jorquera-Chavez, M., Fuentes, S., Dunshea, F.R., Warner, R.D., Poblete, T. and Jongman, E.C. (2019) *Modelling and validation of computer vision techniques to assess heart rate, eye temperature, ear-base temperature and respiration rate in cattle. Animals*, 9(12), 1089. https://doi.org/10.3390/ani9121089

Karvonen, J. and Vuorimaa, T. (1988) *Heart rate and exercise intensity during sports activities. Practical application. Sports Medicine*, 5, 303–311. https://doi.org/10.2165/00007256-198805050-00002

King, C., Buffington, L., Smith, T.J. and Grandin, T. (2014) *The effect of a pressure wrap (ThunderShirt®) on heart rate and behavior in canines diagnosed with anxiety disorder. Journal of Veterinary Behavior*, 9, 215–221. https://doi.org/10.1016/j.jveb.2014.06.007

Kitajima, K., Oishi, K., Miwa, M., Anzai, H., Setoguchi, A., Yasunaka, Y., Himeno, Y., Kumagai, H. and Hirooka, H. (2021) *Effects of heat stress on heart rate variability in free-moving sheep and goats assessed with correction for physical activity. Frontiers in Veterinary Science*, 8, 658763. https://doi.org/10.3389/fvets.2021.658763

Koerber, D., Khan, S., Shamsheri, T., Kirubarajan, A. and Mehta, S. (2023) *Accuracy of heart rate measurement with wrist-worn wearable devices in various skin tones: A systematic review. Journal of Racial and Ethnic Health Disparities*, 10, 2676–2684. https://doi.org/10.1007/s40615-022-01446-9

Kosonogov, V., De Zorzi, L., Honore, J., Martinez-Velazquez, E.S., Nandrino, J., Martinez-Selva, J.M. and Sequeira, H. (2017) *Facial thermal variations: A new marker of emotional arousal. PLoS One*, 12(2017), 1–15. https://doi.org/10.1371/journal.pone.0183592

Kreibig, S.D. (2010) *Autonomic nervous system activity in emotion: A review. Biological Psychology*, 84, 394–421. https://doi.org/10.1016/j.biopsycho.2010.03.010

Kunczik, J., Pereira, C.B., Zieglowski, L., Tolba, R., Wasserman, L., Hager, C., Bleich, A., Janssen, H., Thum, T. and Czaplik, M. (2019) *Remote vitals monitoring in rodents using video recordings. Biomedical Optics Express*, 10, 4422–4436. https://doi.org/10.1364/boe.10.004422

Laske, T.G., Garshelis, D.L., Iles, T. and Iaizzo, P.A. (2021) *An engineering perspective on the development and evolution of implantable cardiac monitors in free-living animals. Philosophical Transactions of the Royal Society, B: Biological Sciences*, 376, 20200217. https://doi.org/10.1098/rstb.2020.0217

Lecorps, B., Rodel, H.G. and Feron, C. (2016) *Assessment of anxiety in open field and elevated plus maze using infrared thermography. Physiology & Behavior*, 157, 209–216. https://doi.org/10.1016/j.physbeh.2016.02.014

Lee, J.M., Kim, C.H., Kange, J.I. and Suh, I. (2014) *Association between stressful life events and resting heart rate. BMC Psychology*, 2, 29. https://doi.org/10.1186/s40359-014-0029-0

Loerup, L., Pullon, R.M., Birks, J., Fleming, S., Mackillop, L.H., Gerry, S. and Watkinson, P.J. (2019) *Trends of blood pressure and heart rate in normal pregnancies: A systematic review and meta-analysis. BMC Medicine*, 17, 167. https://doi.org/10.1186/s12916-019-1399-1

López-Olvera, J.R., Marco, I., Montane, J. and Lavin, S. (2006) *Transport stress in Southern chamois (Rupicapra pyrenaica) and its modulation by acepromazine. The Veterinary Journal*, 172, 347–355. https://doi.org/10.1016/j.tvjl.2005.06.007

Marazziti, D., Di Muro, A. and Castrogiovanni, P. (1992) *Psychological stress and body temperature changes in humans. Physiology & Behavior*, 52, 393–395. https://doi.org/10.1016/0031-9384(92)90290-i

Marchant-Forde, R.M., Marlin, D.J. and Marchant, J.N. (2004) *Validation of a cardiac monitor for measuring heart rate variability in adult female pigs: Accuracy, artefacts and editing. Physiology & Behavior*, 80, 449–458. https://doi.org/10.1016/j.physbeh.2003.09.007

McCafferty, D.J. (2007) *The value of infrared thermography for research on mammals: Previous applications and future directions. Mammal Review*, 37, 207–223. https://doi.org/10.1111/j.1365-2907.2007.00111.x

McFarland, R.A. (1985) *Relationship of skin temperature changes to the emotions accompanying music. Biofeedback and Self Regulation*, 10(3), 255267. https://doi.org/10.1007/bf00999346

Merrifield, C. and Danckert, J. (2014) *Characterizing the psychophysiological signature of boredom. Experimental Brain Research*, 232, 481–491. https://doi.org/10.1007/s00221-013-3755-2

Mezzacappa, E.S., Kelsey, R.M., Katkin, E.S. and Sloan, R.P. (2001) *Vagal rebound and recovery from psychological stress. Psychosomatic Medicine*, 63, 650–657. https://doi.org/10.1097/00006842-200107000-00018

Moe, R.O., Stubsjoen, S.M., Bohlin, J., Flo, A. and Bakken, M. (2012) *Peripheral temperature drop in response to anticipation and consumption of a signaled palatable reward in laying hens (Gallus domesticus). Physiology & Behavior*, 106, 527–533. https://doi.org/10.1016/j.physbeh.2012.03.032

Mohr, E., Langbein, J. and Nurnberg, G. (2002) *Heart rate variability: A noninvasive approach to measure stress in calves and cows. Physiology & Behavior*, 75, 251–259. https://doi.org/10.1016/s0031-9384(01)00651-5

Montané, J., Marco, I., Lopez-Olvera, J.R., Rossi, L., Manteca, X. and Lavin, S. (2007) *Effect of acepromazine on the signs of capture stress in captive and free-ranging roe deer (Capreolus capreolus). Veterinary Record*, 160, 730–738. https://doi.org/10.1136/vr.160.21.730

Mott, R., Dowell, F. and Evans, N. (2021) *Use of the Polar V800 and Actiheart 5 heart rate monitors for the assessment of heart rate variability (HRV) in horses. Applied Animal Behaviour Science*, 241, 105401. https://doi.org/10.1016/j.applanim.2021.105401

Nakanishi, R. and Imai-Matsumura, K. (2008) *Facial skin temperature decreases in infants with joyful expression. Infant Behavior and Development*, 31(1), 137–144. https://doi.org/10.1016/j.infbeh.2007.09.001

Nord, A. and Folkow, L.P. (2019) *Ambient temperature effects on stress-induced hyperthermia in Svalbard ptarmigan. Biology Open*, 8(6), bio 043497. https://doi.org/10.1242/bio.043497

Noyes, R. (1985) *Beta-adrenergic blocking drugs in anxiety and stress. The Psychiatric Clinics of North America*, 8, 119–132.

Oka, T. (2015) *Psychogenic fever: How psychological stress affects body temperature in the clinical population. Temperature (Austin)*, 2, 368–378. https://doi.org/10.1080/23328940.2015.1056907

Olivier, B., Zethof, T., Pattij, T., van Boogaert, M., van Oorschot, R., Leahy, C., Oosting, R., Bouwknecht, A., Veening, J., van der Gugten, J. and Groenink, L. (2003) *Stress-induced hyperthermia and anxiety: Pharmacological validation. European Journal of Pharmacology*, 463, 117–132. https://doi.org/10.1016/s0014-2999(03)01326-8

Palestrini, C., Previde, E.P., Spiezio, C. and Verga, M. (2005) *Heart rate and behavioural responses of dogs in the Ainsworth's Strange Situation: A pilot study. Applied Animal Behaviour Science*, 94, 75–88. https://doi.org/10.1016/j.applanim.2005.02.005

Redondo, J.I., Suesta, P., Serra, I., Soler, C., Soler, G., Gil, L. and Gomez-Villamandos, R.J. (2012) *Retrospective study of the prevalence of postanaesthetic hypothermia in dogs. Veterinary Record*, 171, 374. https://doi.org/10.1136/vr.100476

Reefmann, N., Kaszas, F.B., Wechsler, B. and Gygax, L. (2009a) *Physiological expression of emotional reactions in sheep. Physiology & Behavior*, 98, 235–241. https://doi.org/10.1016/j.physbeh.2009.05.017

Reefmann, N., Wechsler, B. and Gygax, L. (2009b) *Behavioural and physiological assessment of positive and negative emotion in sheep. Animal Behaviour*, 78, 651–659. https://doi.org/10.1016/j.anbehav.2009.06.015

Reimers, A.K., Knapp, G. and Reimers, C. (2018) *Effects of exercise on the resting heart rate: A systematic review and meta-analysis of interventional studies. Journal of Clinical Medicine*, 7(12), 503. https://doi.org/10.3390/jcm7120503

Renbourn, E.T. (1960) *Body temperature and pulse rate in boys and young men prior to sporting contests. A study of emotional hyperthermia: With a review of the literature. Journal of Psychosomatic Research*, 4, 149–175. https://doi.org/10.1016/0022-3999(60)90008-8

Reynolds, J., Ahmmed, P. and Bozkurt, A. (2019) *An injectable system for subcutaneous photoplethysmography, accelerometry, and thermometry in animals. IEEE Transactions on Biomedical Circuits and Systems*, 13, 825–834. https://doi.org/10.1109/tbcas.2019.2923153

Reynolds, S., Lane, S.J. and Mullen, B. (2015) *Effects of deep pressure stimulation on physiological arousal. The American Journal of Occupational Therapy*, 69(3), 6903350010. https://doi.org/10.5014/ajot.2015.015560

Salazar-López, E., Dominguez, E., Ramos, V.J., de la Fuente, J., Meins, A., Ilboraa, O., Galvez, G., Rodriguez-Artacho, M.A. and Gomez-Milan, E. (2015) *The mental and subjective skin: Emotion, empathy, feelings and thermography. Consciousness and Cognition*, 34, 149–162. https://doi.org/10.1016/j.concog.2015.04.003

Sanchez-Alavez, M., Alboni, S. and Conti, B. (2011) *Sex- and age-specific differences in core body temperature of C57Bl/6 mice. Age (Dordrecht, Netherlands)*, 33, 89–99. https://doi.org/10.1007/s11357-010-9164-6

Santos, M.A., Sousa, A.C.S., Reis, F.P., Santos, T.R., Lima, S.O. and Barreto-Filho, J.A. (2013) *Does the aging process significantly modify the Mean Heart Rate? Arquivos Brasileiros de Cardiologica*, 101, 388–398. https://doi.org/10.5935/abc.2013018

Schmidt, A., Mostl, E., Wehnert, C., Aurich, J., Muller, J. and Aurich, C. (2010) *Cortisol release and heart rate variability in horses during road transport. Hormones and Behaviour*, 57, 209–215. https://doi.org/10.1016/j.yhbeh.2009.11.003

Shiloh, R., Munitz, H., Portuguese, S., Gross-Isseroff, R., Sigler, M., Bodinger, L., Katz, N., Stryjer, R., Hermesh, H. and Weizman, A. (2005) *Corneal temperature in schizophrenia patients. International Journal of Neuropsychopharmacology*, 8, 537–547. https://doi.org/10.1017/S1461145705005377

Sleigh, J.W. and Henderson, J.D. (1995) *Heart rate variability and preoperative anxiety. Acta Anaesthesiologica Scandinavica*, 39, 1059–1061. https://doi.org/10.1111/j.1399-6576.1995.tb04229.x

Smith, S.M. and Vale, W.W. (2006) *The role of the hypothalamic-pituitary-adrenal axis in neuroendocrine responses to stress. Dialogues in Clinical Neuroscience*, 8, 383–395. https://doi.org/10.31887/dcns.2006.8.4/ssmith

Stewart, M., Stafford, K.J., Dowling, S.K., Schaefer, A.L. and Webster, J.R. (2008) *Eye temperature and heart rate variability of calves disbudded with or without local anaesthetic. Physiology & Behavior*, 93, 789–797. https://doi.org/10.1016/j.physbeh.2007.11.044

Taylor, S., Webb, L., Montrose, V.T. and Williams, J. (2020) *The behavioral and physiological effects of dog appeasing pheromone on canine behavior during separation from the owner. Journal of Veterinary Behavior*, 40, 36–42. https://doi.org/10.1016/j.jveb.2020.08.001

Terada, M., Wato, S. and Saito, T.R. (2003) *Changes of heart rate during sexual behavior in the female rat. Experimental Animals*, 52, 85–88. https://doi.org/10.1538/expanim.52.85

Travain, T., Colombo, E.S., Grandi, L.C., Heinzl, E., Pelosi, A., Previde, E.P. and Valsecchi, P. (2016) *How good is this food? A study on dogs' emotional responses to a potentially pleasant event using infrared thermography. Physiology & Behavior*, 159, 80–87. https://doi.org/10.1016/j.physbeh.2016.03.019

Travain, T. and Valsecchi, P. (2021) *Infrared thermography in the study of animals' emotional responses: A critical review. Animals (Basel)*, 11. https://doi.org/10.3390/ani11092510

Unakafov, A.M., Moller, S., Kagan, I., Gail, A., Treue, S. and Wolf, F. (2018) *Using imaging photoplethysmography for heart rate estimation in non-human primates. PLoS One*, 13, e0202581. https://doi.org/10.1371/journal.pone.0202581

van Bogaert, M.J., Groenink, L., Oosting, R.S., Westphal, K.G.C., van der Gugten, J. and Olivier, B. (2006) *Mouse strain differences in autonomic responses to stress. Genes, Brain, and Behavior*, 5, 139–149. https://doi.org/10.1111/j.1601-183X.2005.00143.x

Van den Buuse, M., van Acker, S.A.B.E., Fluttert, M. and de Kloet, E.R. (2001) *Blood pressure, heart rate, and behavioral responses to psychological "novelty" stress in freely moving rats. Psychophysiology*, 38, 490–499. https://doi.org/10.1111/1469-8986.3830490

Van der Heyden, J.A., Zethof, T.J. and Olivier, B. (1997) *Stress-induced hyperthermia in singly housed mice. Physiology & Behavior*, 62, 463–470. https://doi.org/10.1016/s0031-9384(97)00157-1

Vianna, D.M. and Carrive, P. (2005) *Changes in cutaneous and body temperature during and after conditioned fear to context in the rat. European Journal of Neuroscience*, 21, 2505–2512. https://doi.org/10.1111/j.1460-9568.2005.04073.x

Vinkers, C.H., de Jong, N.M., Kalkman, C.J., Westphal, K.G.C., van Oorschot, R., Olivier, B., Korte, S.M. and Groenink, L. (2009) *Stress-induced hyperthermia is reduced by rapid-acting anxiolytic drugs independent of injection stress in rats. Pharmacology, Biochemistry and Behaviour*, 93, 413–418. https://doi.org/10.1016/j.pbb.2009.05.017

Vinkers, C.H., Penning, R., Hellhammer, J., Verster, J.C., Klaessens, J.H.G.M., Olivier, B. and Kalkman, C.J. (2013) *The effect of stress on core and peripheral body temperature in humans. Stress*, 16, 520–530. https://doi.org/10.3109/10253890.2013.807243

Viola, M., Ouyang, D., Xu, J., Maciejewski, P.K., Prigerson, H.G. and Derry, H.M. (2022) *Associations between beta-blocker use and psychological distress in bereaved adults with cardiovascular conditions. Stress and Health*, 38, 147–153. https://doi.org/10.1002/smi.3061

Waiblinger, S., Menke, C., Korff, J. and Bucher, A. (2004) *Previous handling and gentle interactions affect behaviour and heart rate of dairy cows during a veterinary procedure. Applied Animal Behaviour Science*, 85, 31–42. https://doi.org/10.1016/j.applanim.2003.07.002

Wang, M., Youssef, A., Larsen, M., Rault, J., Berckmans, D., Marchant-Forde, J., Hartung, J., Bleich, A., Lu, M. and Norton, T. (2021) *Contactless video-based heart rate monitoring of a resting and an anesthetized pig. Animals*, 11(2), 442. https://doi.org/10.3390/ani11020442

Wascher, C.A.F. (2021) *Heart rate as a measure of emotional arousal in evolutionary biology. Philosophical Transactions of the Royal Society, B: Biological Sciences*, 376, 20200479. https://doi.org/10.1098/rstb.2020.0479

Webster, A.J.F. and Hays, F.L. (1968) *Effects of beta-adrenergic blockade on the heart rate and energy expenditure of sheep during feeding and during acute cold exposure. Canadian Journal of Physiology and Pharmacology*, 46, 577–583. https://doi.org/10.1139/y68-084

Wojniusz, S., Vogele, C., Ropstad, E., Evans, N., Robinson, J., Sutterlin, S., Erhard, H.W., Solbakk, A., Endestad, T., Olberg, D.E. and Haraldsen, I.R.H. (2011) *Prepubertal gonadotropin-releasing hormone analog leads to exaggerated behavioral and emotional sex differences in sheep. Hormones and Behavior*, 59, 22–27. https://doi.org/10.1016/j.yhbeh.2010.09.010

Wongsaengchan, C., McCafferty, D.J., Evans, N.P., McKeegan, D.E.F. and Nager, R.G. (2023) *Body surface temperature of rats reveals both magnitude and sex differences in the acute stress response. Physiology & Behavior*, 264, 114138. https://doi.org/10.1016/j.physbeh.2023.114138

Xue-Rui, T., Ying, L., Da-Zhong, Y. and Xiao-Jun, C. (2008) *Changes of blood pressure and heart rate during sexual activity in healthy adults. Blood Pressure Monitoring*, 13, 211–217. https://doi.org/10.1097/mbp.0b013e3283057a71

Zaza, A., Ronchi, C. and Malfatto, G. (2018) *Arrhythmias and heart rate: Mechanisms and significance of a relationship. Arrhythmia & Electrophysiology Review*, 7, 232–237. https://doi.org/10.15420/aer.2018.12.3

Zupan, M., Buskas, J., Altimiras, J. and Keeling, L.J. (2016) *Assessing positive emotional states in dogs using heart rate and heart rate variability. Physiology & Behaviour*, 155, 102–111. https://doi.org/10.1016/j.physbeh.2015.11.027

Part 2

Indicators of Mood

7 Judgement Biases

Elizabeth S. Paul and Vikki Neville

Bristol Veterinary School, University of Bristol, Bristol, UK

Introduction: judgement bias task design

Judgement bias tasks, also referred to as 'cognitive bias tasks' and 'ambiguous cue interpretation tasks' (Enkel et al. 2010; Hales et al. 2014), are discrimination-based tests that tap animals' decision-making processes with the aim of inferring affective states (Harding et al. 2004; Mendl et al. 2009). They form part of a wider group of cognitive bias measures (e.g. memory and attention bias tasks) designed to assess cognitive components of animal affective state. At the core of judgement bias tasks is the requirement of tested animals to make decisions about affectively ambiguous stimuli; specifically, to decide whether these are likely to predict reward or punishment (or a greater or lesser reward).

Several methodologies for judgement bias testing in animals have been developed, most relying on instrumental conditioning. The main division is between 'Go/No-go' designs and 'Go/Go' tasks. In a typical reward–punishment 'Go/No-go' variant, the animal is first trained to show a conditional discrimination: to perform an instrumental response (the 'Go' response–lever press, screen peck, food-bowl lid-flip, etc.) to a reward-predicting stimulus (visual, auditory, tactile or olfactory) and to refrain from performing that response (i.e. 'No-go') to a punishment-predicting stimulus in the same sensory domain. Correct Go responding gives rise to delivery of a reward, while correct No-go responding enables the animal to avoid a punishment. Incorrect Go responses, on the other hand, following a punishment-predicting stimulus, give rise to the delivery of a punishment. Incorrect No-go responses, following a reward-predicting stimulus, lead to the non-delivery of reward. In other words, a Go response is risky – it leads to either reward or punishment, while a No-go response is safe – it leads to neither punishment nor reward. When this discrimination has been learned to a pre-assigned criterion (and to the same degree in groups being compared), the affect-testing element of the task begins: occasional, ambiguous and probe stimuli are included amongst the already-trained, reward-predicting and punishment-predicting 'anchor' stimuli (also known as 'reference' stimuli). These are cues that are sensorily intermediate between the anchors (e.g. intermediate tones [auditory] or coloured stimuli [visual]). Because they are ambiguous, the animal needs to judge whether to respond (a) as if witnessing a reward predictor and make a Go response, or (b) as if witnessing a punishment predictor and make a No-go response. This is a probabilistic decision – to risk punishment (if an incorrect Go is performed) or risk missing a reward (if an incorrect No-go is performed) – that often leads researchers to describe animals' responses in validated judgement bias tests as more or less 'optimistic-like' (i.e. if more responses are made as if the ambiguous stimulus was the reward-predicting anchor, and/or such responses are made with short latencies), or 'pessimistic-like' (i.e. if more responses are withheld, as if the ambiguous stimulus was the punishment-predicting anchor; see Figure 7.1). The terms 'optimistic' and 'pessimistic' must be used cautiously, however, as the task only reveals *relative* differences between groups in their willingness to make Go or No-go responses, and cannot identify the absolute affective processes or states that guide these decisions. Nevertheless, they are convenient short-hand terms and, in animal welfare research, responding to ambiguous cues as if reward-predicting is taken as reflecting a relatively positively-valenced state, while responding to them as if punishment-predicting is taken as reflecting a relatively negatively-valenced state. Mendl, Paul and colleagues (Mendl et al. 2010a; see also Mendl and Paul 2020 and Chapter 1) hypothesised that such effects may be functional adaptations to enable optimal decision-making under uncertainty; their proposal is that animals' background affective states represent accumulated information about the rewarding and punishing nature of their environments, thereby predicting the likelihood of future rewards and punishments (see also Bateson et al. 2011). For this reason, these tasks are thought to be especially sensitive to long-lasting affective states (e.g. moods).

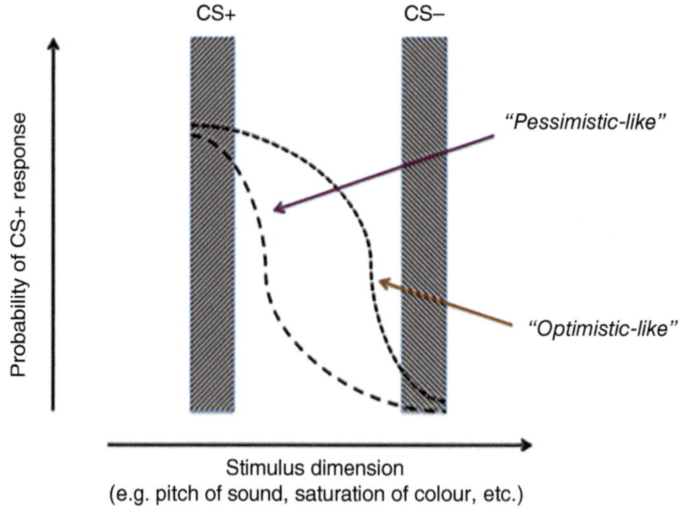

Figure 7.1 Optimistic-like and pessimistic-like responding in a typical judgement bias task (Go/No-go variant). Animals are trained to respond to positive, conditioned anchor stimuli (CS+ stimuli) and not respond to negative, conditioned anchor stimuli (CS– stimuli). When probe-tested with intermediate, ambiguous stimuli, optimistic-like responses involve making a greater proportion of CS+ responses, while pessimistic-like responses involve making a smaller proportion of CS+ responses.

For an illustration of a Go/No-go judgement bias task, and another common variant, the Go/Go task, see Figures 7.2 and 7.3. In Go/Go tasks (also known as 'active response' or 'active choice' tasks), two instrumental responses are learned – one to obtain a reward and the other to avoid a punisher, or to obtain a lesser (or greater) reward. Another type of task, essentially a variant of the Go/No-go task, is the 'Spatial' task, in which animals learn to move quickly to a spatial location which regularly provides reward (e.g. a food bowl containing food rewards) and slowly, if at all, to a location which does not (e.g. a food bowl containing inaccessible food rewards). During testing, latencies to move to intermediate (ambiguous) locations are recorded. A fourth type of task is a non-trained, 'Unconditioned response' design (e.g. Brilot et al. 2009; Salmeto et al. 2011; see Figure 7.4), in which animals automatically respond differentially to affectively meaningful stimuli (e.g. approaching and avoiding images of predators or conspecifics). Their attraction to, or repulsion from, ambiguous stimuli is then tested (Figure 7.5). These differences between task designs are important to highlight, as the primary measurements made can differ and, as we will see in the section on construct validation, some methods may have greater validity than others.

In every case, a task must go through several checks before it can be used in welfare assessment. First, a check of a fundamental form of technical (or 'face') validity is needed, to establish that anchor cues have been learned appropriately, and that intermediate probe cues

Figure 7.2 A Go/No-go task for chickens, designed by Deakin et al. (2016). Birds were trained to peck a coloured disc on a screen to receive a food reward (mealworm) and to refrain from pecking a disc of differing colour saturation to avoid a punisher (air puff); these were the 'reference' or 'anchor' stimuli. Ambiguous probe cues were intermediately coloured discs. Birds' Go (i.e. peck) responses, when plotted as a proportion of the total number of each disc colour shown, produced a sigmoid curve (see Figure 7.1): all/almost all unambiguous food-predicting discs were pecked, while no/few unambiguous air puff-predicting discs were pecked. Deakin et al. (2016) then found that chickens show more optimistic-like biases when kept at a preferred ambient temperature, and more pessimistic-like ones at a less preferred (cooler) temperature. Source: Adapted from Deakin et al. (2016).

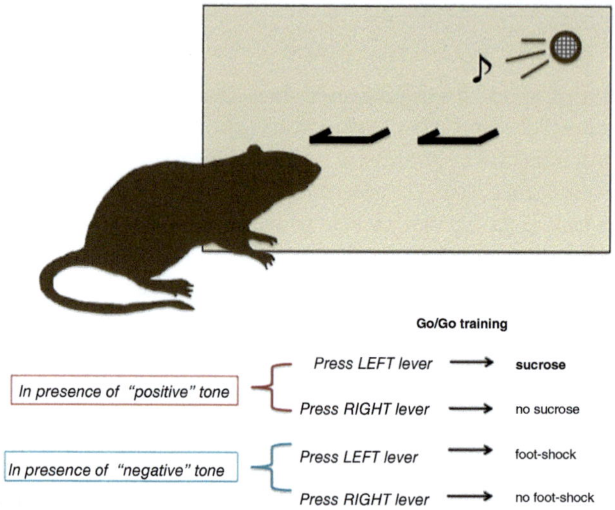

Figure 7.3 A Go/Go task for rats (e.g. Enkel et al. 2010; Papciak et al. 2013). Rats are trained on a conditional discrimination: to press one lever (in this example, the left) to receive a reward (sucrose solution) in the presence of one tone cue (e.g. 2 kHz) and to press the other lever (right) to avoid a punishment (0.5 mA foot-shock) in the presence of another (e.g. 9 kHz). Probe cues are intermediate tones. Enkel et al. (2010) validated this task as an indicator of affect by comparing congenitally helpless rats (a genetic model of depression) with non-helpless rats. Source: Enkel et al. (2010), Papciak et al. (2013).

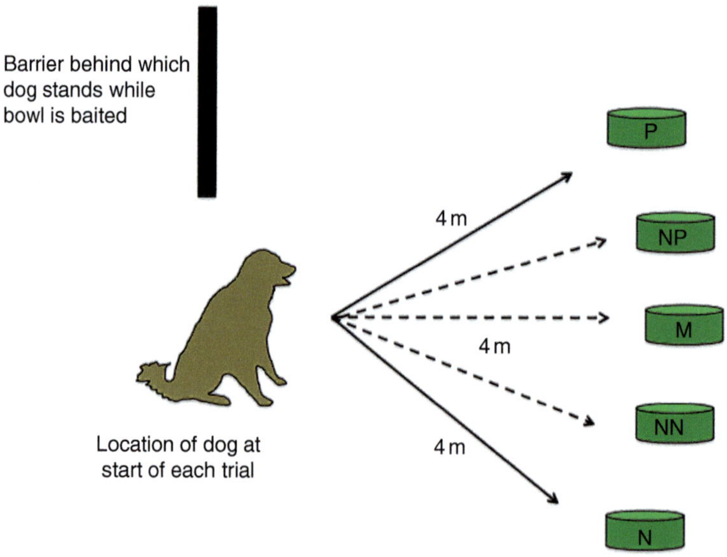

Figure 7.4 A spatial (Go/No-go) task for dogs (Mendl et al. 2010b); this task was based on a task validated in rats by Burman et al. (2008) living in unenriched cages versus cages enriched with preferred features such as sawdust substrate, wooden block and nest box (see also Neville et al. 2023). In each trial of the dog task, one bowl was presented. Dogs were trained that a bowl in location P (positive) contained a food reward and that a bowl in location N (negative) contained none. During testing, occasional probe bowls were placed in intermediate positions between the rewarded and unrewarded locations (NP = near positive; M = middle; NN = near negative). Dogs showing higher levels of anxious 'separation behaviour' (in a previously validated separation-related behaviour test conducted approx. 1–2 days before the judgement bias test) showed more pessimistic-like responding in the judgement bias test (M and NN locations). Source : Mendl et al. (2010b), Burman et al. (2008), Neville et al. (2023).

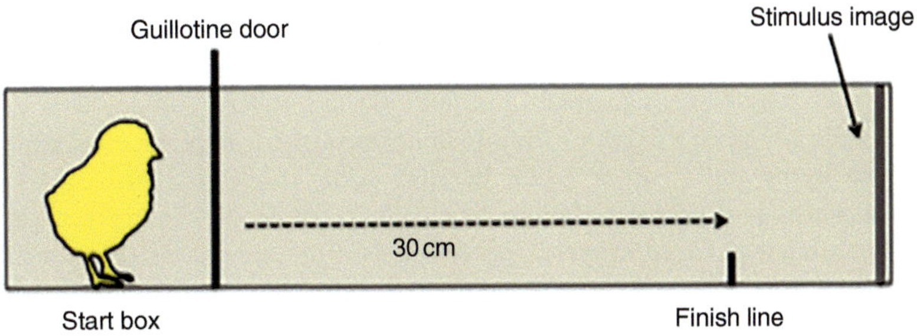

Figure 7.5 An unconditioned response judgement bias task conducted with chicks. A stimulus image of a chick silhouette (or a mirror) was used as the positive anchor cue, and an image of an owl silhouette was used as the negative anchor cue. Because these cues are innately appetitive and aversive to chicks (respectively), this methodology obviated the need for training. Morphed chick/owl silhouettes were then used as ambiguous probe cues for testing. Source: Adapted from Salmeto et al. (2011).

are indeed responded to as intermediate (i.e. that generalisation has occurred). In Go/No-go tasks, for example, this means that most reward- and punishment-predicting anchor cues are responded to appropriately, and that the animals' responses to ambiguous probe cues are made to an intermediate degree (i.e. that a smaller proportion are responded to than to the reward-predicting anchor cue). It also means that animals will show longer latencies to respond to probe cues than trained reward cues, and that probe cues more dissimilar to this anchor produce longer latencies still. Second, once an animal judgement bias task has passed such checks, it then must pass at least one of the five tests listed in Chapter 2, with the exception of Test 1 (i.e. whether the indicator accurately tracks self-reported affective states in humans). This is because the conditioned tasks used in animals are typically different from the unconditioned tasks shown to be sensitive to affect in humans, although in recent years some back-translations have been made (i.e. from the animal task designs to human ones – e.g. Aylward et al. 2020; Iigaya et al. 2016). Furthermore, while relatively stronger (more positive or negative) judgement biases might reflect relatively better or poorer welfare states, they could also reflect differences in factors that may be unrelated to affect or welfare in a predictable way, such as hunger levels, perception of the cues and/or lapses in attention, thus giving rise to false leads or false nulls as we will see in the following sections.

Basic construct validity of judgement bias as an indicator of affect

Evidence of construct validity from humans

As mentioned, judgement biases were first shown to be modulated by affect in humans, in work conducted several decades ago. Beck's schema theory of depression proposed that people learn to carry expectations (schemas) about the rewarding and punishing nature of the world around them (Beck et al. 1961, 1979; Beck 1967). When faced with new and/or ambiguous stimuli, Beck suggested that those with depression-prone schemas would tend to 'expect the worst', interpreting otherwise neutral stimuli as negative or threatening. Evidence for these ideas came from a variety of tests, although none of these early tasks used conditioning like the animal tasks described above. Many focused on ambiguous linguistic stimuli. For example, Mathews et al. (1989) found that clinically anxious patients were more likely to interpret homophones such as die/dye, pane/pain and guilt/gilt, negatively, compared to non-anxious individuals, while Mogg et al. (2006) found similar results for depressed vs. control participants (see also Byrne and Eysenck 1993; Eysenck et al. 1987; Hadwin et al. 1997; Mogg et al. 1994; Richards et al. 1993). Other studies made use of people's interpretations of images of affectively ambiguous facial expressions: either neutral expressions or ones produced by morphing two or more expressions together. In one such study, socially anxious participants showed enhanced identification of fear expressions when compared with low anxious participants (Richards et al. 2002), while other studies have found that people with social anxiety and social phobia disorders have biased tendencies to judge expressions as angry (Bell et al. 2011; Maoz et al. 2016; Mohlman et al. 2007). Many of these human studies thus focused on clinical mood disorders. However, some studies found similar effects in non-clinical settings in which short-term emotion manipulations were made (e.g. Nygren et al. 1996).

Recently, researchers have focused on developing trained judgement bias studies for humans that directly mirror the conditioned, non-human animal tasks reviewed in the introduction of this chapter. Although this research is still in its infancy, progress has been made, particularly in the development of novel judgement bias tasks for use with humans that employ arbitrary, non-linguistic stimuli. In such tasks, largely consistent relationships have been found between judgement biases being more negative and self-reports of ongoing non-clinical low mood and mood-related disorders (e.g. Aylward et al. 2020; Iigaya et al. 2016; Paul et al. 2011; Schick et al. 2013).

Evidence of construct validity from animals

The first animal-focused judgement bias task to be conducted used a Go/No-go method: rats were required to press a lever to obtain a food reward when they heard a particular tone (2 kHz), and to refrain from pressing the lever when they heard another tone (4 kHz; Harding et al. 2004). Rats living in unpredictable housing conditions (i.e. similar to a 'chronic mild stress' regime – see Chapter 2), designed to generate mild depression-like states, were slower to respond to, and tended to show fewer responses to, ambiguous (intermediate) probe tones. This study offered initial evidence of the potential validity of conditioning-based judgement bias tests for animals.

Since then, many experiments have compared animals in preferred/worked-for situations with animals in less-preferred/aversive situations (Test 2 in Chapter 2), like Deakin et al. (2016) did for their warmer/cooler hens (Figure 7.2). In one study of chickens, birds kept in housing full of preferred resources appeared more optimistic in a judgement bias task than birds kept in smaller, more barren conditions. However, this only held for more exploratory individuals, not for less exploratory birds (Ross et al. 2019). This might reflect a requirement for an individual-level assessment of preference, since not all members of a species may agree on what kinds of environments or living conditions are preferable. Thus, Paul et al. (2022) found that laying hens' judgement biases were not more optimistic-like in generally preferred and non-preferred living conditions, but when birds' *individual* preferences were considered (i.e. taking into account that some hens preferred the *generally* non-preferred conditions), there was a significant association between living condition preference and judgement bias in the expected direction. Judgement bias tests have also been used to investigate the affective impact of treatments that are putatively positive, such as tickling rats (Rygula et al. 2012) or negative, such as punishment-based training methods for dogs (Casey et al. 2021). Tickling has been found to be to be reinforcing and to generate conditioned place preferences in rats

(Burgdorf and Panksepp 2001), while punishing training treatments are aversive to dogs, at least in the short term (Casey et al. 2021).

Another class of validation approaches is to compare judgement biases in animals experiencing or not experiencing threats to fitness (Test 3 in Chapter 2). Examples include Neave et al.'s (2013) comparison of calves' judgement biases before and after hot-iron disbudding (showing more pessimistic-like responses following the procedure) and Cockburn et al.'s (2018) investigation of Cavalier King Charles dogs suffering from the pain-inducing disorder Syringomyelia (showing more pessimistic-like responding than non-affected dogs). Social stress is also arguably a fitness threat and has been investigated in a number of studies. For example, as we saw in Figure 7.5, Salmeto et al. (2011) found that isolation stress was associated with more negative judgement biases in chicks (a naturally sociable infant bird), while Papciak et al. (2013) found that chronic stress induced by three weeks of daily social defeat was also associated with increased pessimistic-like responding in rats. In recent meta-analyses of judgement bias studies, in which a wide range of species, judgement bias methodologies and short-term/long-term affect manipulations were considered (broadly those of Tests 2 and 3 in Chapter 2), overall support for construct validity was found. However, effect sizes varied considerably (from small to large; Lagisz et al. 2020; Neville et al. 2020b), and there were also exceptions to this pattern, as we will discuss in the next section.

Turning to pharmacological validation, several judgement bias validation studies have made use of drugs with known psycho-affective influences in humans. Most of these have involved rodents (rats) as subjects. Neville et al. (2020b) meta-analysed these psychopharmacological studies and found that overall, drug-based manipulations using a variety of antidepressants, anxiolytics and anxiogenics altered judgement bias at the probe cues as predicted (although again there were exceptions: see the following section). Finally, a few studies have investigated whether individual animals' judgement bias scores co-vary with other known welfare measures, such as signs of separation anxiety in dogs (see Figure 7.4). In sum, various approaches to validating of animal judgement bias tasks, in all their varieties, have offered largely supportive– though not universal– evidence for its value as an affective state, and thus welfare, indicator.

Responsiveness and selectivity

False nulls indicating problems with responsiveness

One potential concern for any measure of affective state is its responsiveness (its sensitivity), or lack thereof, to changes in valence, and thus the possibility of false null results. Such results may sometimes occur for judgement bias tasks. For example, Neville et al. (2021b) developed a human task using primary reinforcers and punishers (instead of the usual monetary rewards and losses): apple juice was given as a reward, and salty tea as a punisher. To their surprise, moment-by-moment self-reports of affective state during judgement bias testing did not reveal convincing associations with the individual judgements made (Neville et al. 2021a, 2021b). Whether this reveals lack of responsiveness on the part of the judgement bias task itself, or the self-report measure of affect used, is not clear. Such studies of the relationship between participants' self-reports of affective state and optimistic/pessimistic performance on such 'new generation' human judgement bias tasks are still in their infancy and need further investigation (e.g. Iigaya et al. 2016; Aylward et al. 2020).

Likewise, for animals, several studies have found null results despite using strong and apparently well-validated affect manipulations. For example, Crump et al. (2021) found that dairy cows did not show differential responses to ambiguous stimuli when living in preferred or non-preferred environments (indoors vs. access to pasture). Walker et al. (2014) found that dogs separated from a bonded conspecific showed no change in judgement bias. Other studies similarly failed to find significant differences in the judgement biases of animals exposed to differently valued environments, stimuli or resources across a variety of species (e.g. Gott et al. 2019; Hernandez et al. 2014; Wichman et al. 2012). And when relationships between judgement biases and telomere developmental attrition have been assessed, patterns appeared uncertain (Bateson et al. 2015; Gott et al. 2019). Further examples of false null results for mice can be found in Chapter 2.

These findings might help to explain why the judgement bias meta-analysis by Lagisz et al. (2020) showed that some judgement bias studies generate effect sizes close to zero. In this meta-analysis, a number of moderating factors were found that might help understand why (Lagisz et al. 2020); for example, the effects of affect manipulations on judgement bias were significant for males but not for females, and task type (Go/Go vs. Go/No-go) was also influential. Further research is now needed to establish the implications of these findings – e.g. whether males and females do actually differ in associations between affect and judgement bias (since the sex of subject is often confounded by species and experimental protocols; farm animal research tends to focus on females, for example, while experimental rodent research tends to focus on males). That Lagisz et al. (2020) found that effect sizes ranged from small to large (average Hedges' g from 0.2 to 0.6) could also mean that some studies with small sample sizes were prone to Type II errors (when one fails to reject a null hypothesis that is actually false): another potential reason for false nulls.

Other situations, in which extremes of affect may not be detected, relate to ceiling and floor effects within the judgement bias task. For example, if two groups of animals are both experiencing negative affective states, their pessimistic-like responding to ambiguous probe cues during testing may both be so strong as to fail to discriminate between them. Although we are not aware of any specific examples of this, it is also clear that in some cases of extreme affect (e.g. fear, excitement), or excessive pharmacological manipulations, animals may simply be unable to engage with the task at all (e.g. Rygula et al. 2014a). As a possible illustration, Paul et al. (2023) mentioned a judgement-bias trained hen who suddenly became a 'pariah' (i.e. became fearful and was ostracised by fellow pen-mates) and refused to take part in judgement bias tests completely. On some occasions, then, false nulls could arise from animals suffering from extremely fearful or depressive-like states and losing the capacity to take part in any kind of test (e.g. due to behavioural suppression and/or lack of food motivation). A related issue is where there is an imbalance in the positive and negative decision outcomes of a judgement bias task. For example, if a negative outcome is much more aversive than a positive outcome is rewarding, all subjects may opt to make a pessimistic-like response to ambiguity, irrespective of their affective states.

Another potential source of false null results concerns the observation that the strongest effects of affective manipulations are not necessarily observed at what experimenters perceive to be the midpoint between anchor cues (Lagisz et al. 2020; Neville et al. 2020b). This means that there are likely to be more false nulls if experimenters use fewer graded probes, because a human observer's notional midpoint may not be the same as the animal's experienced midpoint. Finally, some null results may reflect the possibility that judgement bias tasks may be more sensitive to long-term moods and mood disorders than to transient emotions (Lagisz et al. 2020). The problem here seems to be that affective state can change significantly across short time scales, with the simple act of taking part in the judgement bias test potentially inverting a pre-existing state (e.g. from distress to relief).

Several results thus illustrate sources of failures of responsiveness, and further research is now needed to develop a more comprehensive picture of the situations, tasks and species involved when such problems occur.

False leads indicating problems with selectivity

Another concern that has been voiced is the possibility that the data obtained from judgement bias studies may sometimes lack selectivity (e.g. Mendl et al. 2009; Ross et al. 2019), such that, for example, apparent optimistic-like biases do not reflect relatively positive affect, but instead reflect neutral or even negative affect. For example, Verbeek et al. (2014) found that sheep who had been food restricted for one week (a presumed negative event) showed a *positive* bias in a spatial judgement bias task: they approached the ambiguous probe locations on a greater proportion of tests than did sheep who had been given a high level of feed (150% of maintenance levels). Some other studies comparing animals in preferred/non-preferred situations have likewise found results that are opposite to those that were originally predicted (e.g. Bateson et al. 2015; Brydges et al. 2012; Burman et al. 2011; Gott et al. 2019; Doyle et al. 2010). And release from restraint and shearing stressors in sheep – arguably similar in some ways to predator attacks – have been found to be associated with positive judgement biases (Sanger et al. 2011). Similarly, Neville et al. (2020a) found that rats exposed to rewarding stimuli immediately prior to testing were more pessimistic-like in their responses than rats exposed to punishing stimuli immediately prior to testing.

Likewise, pharmacological manipulations do not always alter judgement bias in predicted ways. When Neville et al. (2020b) meta-analysed studies using psychopharmacological affect manipulations, they found unexpected dosage effects: greater effects were seen when there were relatively smaller differences between treatments. Furthermore, when the neurobiological target of a drug presumed to *reduce* negative states was adrenergic (via acute pharmacological manipulations of the adrenergic system with reboxetine or clonidine), pessimistic-like judgement biases were unexpectedly generated.

A number of factors might help explain such false positive results (i.e. false leads). One could be the timing of the task relative to the timing of the affect manipulation. A task conducted on animals who have just been through a negative experience may not detect their negative state because of ensuing positive, relief-like responses during the test itself (e.g. Doyle et al. 2010; Sanger et al. 2011). If so, the results would be true reflections of affective state at the time of the test, and not false leads. Activity may also potentially be a confound, at least in Go/No-go tasks. For example, in basic Go/No-go reward-punishment tasks, a less active animal (e.g. one that is partially sedated as a result of an experimental drug treatment) may be prone towards showing more No-go (i.e. pessimistic-like) responses, and may also show fewer or slower Go responses to the rewarding anchor cues, while a more active animal might show more or faster Go responses simply due to their higher physical activity levels.

Another potential issue is that groups of animals may differ in their *valuation* of the rewards and punishers on offer in the task. That is, biased responses in the task may reveal *value* judgements and not *subjective probability* judgements. The most common example concerns feeding motivation: hungry animals, or those experiencing unreliable food supplies in the past, may well show more apparently positive (optimistic-like) biases in a scenario involving food, simply because their relative valuation of a food reward is greater (i.e. their positive valuation of food outweighs their negative valuation of the punisher, as this latter is unlikely to be influenced by hunger). Computational modelling of judgement bias data offers a way of estimating how an individual's judgement of the *likelihood*, and/or *value* of a decision outcome influences their choices (Iigaya et al. 2016; Neville et al. 2020a; 2021a).

Finally, how animals are trained may have an effect, perhaps by changing how the ambiguous or midpoint cues of the judgement bias test are perceived. For example, Jones et al. (2018) observed that counterbalancing anchor stimuli had significant effects on the proportion of optimistic-like responses made to ambiguous cues, which differed between groups depending on which cue had been presented as their positive anchor.

Turning to the drug studies showing unexpected pessimistic-like responses, what causes these is unknown. One possibility is that these could perhaps reflect *true* results rather than false leads (i.e. the expectation was wrong), the drugs rendering animals more negative rather than more positive in their affective state. Thus for the adrenergic drugs, one possible explanation is the acute method of administration used: in humans, this worsens clinical anxiety and depression in the early stages of treatment (see Neville et al. 2020b for full discussion) and these drugs can also induce anxiety-like states in rodents (Inoue et al. 2006; Tanaka et al. 2000). Furthermore, the unexpected dosage effects could perhaps be the result of a U-shaped distribution of drug-response magnitude. However, these are hypotheses generated after the results were known (HARKing), and so they must now be empirically tested, if circular reasoning is to be avoided (indeed, the same argument holds for hypothesised relief-like states following removal of an acute negative affect manipulation).

Use of judgement bias tasks in practice

Animal welfare scientists have used judgement bias tests to investigate a broad range of potential welfare issues and/or improvements. For example, validated judgement bias tasks have been used to investigate the welfare impact of long-term health or husbandry conditions, including environmental complexity in trout and chickens (Anderson et al. 2021; Anderson et al. 2022), and disease in dogs and mice (Hobbs et al. 2020; Resasco et al. 2021), as well as of more acute husbandry procedures, such as disbudding in calves (Neave et al. 2013). This work, and the knowledge gained from both development studies and validation studies, gives rise to several suggestions for best practise. Methodology across judgement bias tasks can vary considerably, so animal welfare scientists need to make a number of decisions when designing their research. As mentioned previously, we first need to ensure that – in a new animal judgement bias task – the ambiguous cues really are treated as being intermediate between the positive and negative training cues. Pilot work is thus important to assess what subject animals regard as the 'midpoint' (or at least, a 'mid-area') between two trained anchor cues, i.e. to identify truly ambiguous probes from the perspective of the animal (see Jones et al. 2018). These checks are best made using at least two ambiguous cues (one closer to the positive training cue and one closer to the negative training cue) to produce the predicted generalisation curve. Intermediate responding to a single ambiguous cue provides weaker evidence of generalisation, but single cues offer an efficient form of testing in tasks where generalisation has already been demonstrated. In reality, many studies use three ambiguous cues (e.g. Deakin et al. 2016; Harding et al. 2004).

Activity levels may sometimes be a confound, as also described previously. Using active choice tasks (i.e. Go/Go tasks), in which both 'Go' responses require similar activity levels, may help to avoid this problem. Furthermore, while latency data (e.g. latency to press the Go lever in a rodent Go/Go paradigm) can offer a fine-grained insight into judgement biases, they are also potentially complicated by confounding variables such as activity or arousal levels (which make animals move faster), and/or by animals' uncertainty about the outcome (which may make them decide on an action more slowly). Decision data (i.e. the proportion of positive responses) may therefore offer a more reliable means of assessing affective state (see also Barker et al. 2018).

The possibility that judgement bias results can reflect feeding motivation rather than affective state was also briefly discussed in the previous section. It can therefore be helpful to test food motivation alongside judgement bias to check whether reward valuation and reward probability could be conflated (see e.g. Paul et al. 2022). Further methods for reducing this risk are to avoid using food as a reward or to employ 'treat' food rewards whose valuations may be less affected by hunger (i.e. highly preferred items, such as sucrose for rats and mealworms for chickens – e.g. Rygula et al. 2014b; Deakin et al. 2016). Another basic precaution when food rewards are used is to check that levels of responding to the (unambiguously) rewarding anchor cue have not changed as a result of any affect manipulations. Indeed, much relevant information can come from assessing animals' responses to anchor cues. The key information obtained from judgement bias studies obviously relates to how animals respond to ambiguous cues, as ultimately we anticipate affective state (and hence welfare) to provide information to the animal about how best to act when there is uncertainty about the outcome of an action. However, responses to the anchor cues may provide additional useful information. For example, if a particular treatment does alter responding to these cues, it may indicate that it has affected how animals perceive reinforcers (not just via hunger, but also potentially, via anhedonia or stress sensitisation; see Chapter 9 on responses to rewards) or instead that it has impacted some other aspect of cognition or behaviour unrelated to welfare. Checking whether this has occurred is thus another useful aspect of testing the technical properties (and thus face validity) of a judgement bias task before moving on to construct validation.

Animal welfare scientists must also decide whether and how to reinforce ambiguous probe cues. Many take the approach of not reinforcing any; in most judgement bias studies, ambiguous probe trials are neither rewarded nor punished. But over extended periods of time, this may lead to subject animals learning about the null value of these cues (see review by Roelofs et al. 2016). To minimise this possibility, in most studies, probe cues are interspersed infrequently amongst numerous anchor trials. However, additional measures are available. For example, partial reinforcement of the anchor cues can decrease the possibility that the unreinforced nature of the probe cues is learned (Jones et al. 2018; see also Hales et al. 2016 and Lecorps et al. 2019 for other suggestions). Each approach has advantages and disadvantages, but with such precautions in place, it should be possible to avoid the deleterious effects of animals learning about the null value of probe cues.

Practically, the methodology selected may also be dictated by the time and resources available. Training periods can vary considerably in judgement bias studies, from an hour or so (e.g. as in the spatial Go/No-go task used with dogs – e.g. Mendl et al. 2010b) to several weeks (e.g. as in a probabilistic version of the task used with calves, where 22 training sessions were required; Lecorps et al. 2019). One way to reduce experimental durations is to make use of existing, unconditioned responses to anchor cues (e.g. approach/avoid), as shown in Figure 7.5, and such tasks have been used on a small number of occasions (e.g. Salmeto et al. 2011). Another approach is to make use of automated methods, allowing for multiple animals to be tested simultaneously and therefore expand sample sizes with little or no increase in training time. Importantly, automated methods such as this will also reduce the risk of observer bias, as there is no possibility of experimenters miscalculating latencies or proportions of responses made by subject animals. However, some development is still needed.

For example, an automatic task for rats designed by Jones et al. (2018) showed face validity, but the authors concluded that further validation, using long-term and pharmacological affect manipulations, was still required.

Finally, as mentioned in the *Responsiveness and selectivity* section, interpretation of the welfare impact of treatments, particularly shorter-term treatments, can sometimes be muddied by the timing and context of judgement bias testing. Typically, judgement bias testing and administration of a treatment do not occur simultaneously: animals are moved to a testing arena, away from conspecifics, where they can focus on and attend to the stimuli presented to them. Consequently, judgement bias data on the impact of short-term treatments are often collected immediately after administration of this treatment. The task may then sometimes measure the affective impact of *release from these treatments* as opposed to the *treatment itself* (e.g. a disappointment-like state following the end of a positive treatment, or a relief-like state at the end of a negative one). Although this can be a significant problem, it is possible that repeated experiences with such affect manipulations may be sufficient to generate longer-term (on-going) positive and negative states that can be detected using judgement bias testing, while testing in the home cage may also help to ameliorate some such problems. Relatedly, judgement bias tasks may even have their own effects on the affective states of animals exposed to them. For example, because of the rewards associated with most well-learned judgement bias tasks, animals may find that the entire process of taking part in such tests is rewarding. Such effects may then overshadow any affect manipulations being investigated, especially if they are mild.

Conclusions and future research

When first developed for use in non-human animals, the judgement bias task offered the novel possibility to investigate affective states at a cognitive level. The task was characterised by tightly controlled (mostly) instrumental behavioural decisions; it focused explicitly on affective valence (rather than physiological/affective arousal) and it offered experimenters the opportunity to compare the cognitive-affective processes of humans and animals (i.e. in 'translational' studies, e.g. see Anderson et al. 2012; Neville et al. 2021b). After nearly two decades of research, the role of judgement bias tasks as investigative tools in the study of animal affect continues to develop (e.g. Raoult et al. 2017). As potential indicators of welfare across a wide range of species, judgement bias studies have taken a range of validatory approaches (looking to see whether judgement biases differ between animals with presumed welfare or affective differences; e.g. those with known and tested differences in health, affective states or satisfied preferences). Much work has also been more exploratory: often valuable when research into judgement biases was in its infancy, enabling the potential development of novel methods for investigating decision biases in a wide range of species. However, there have also been problems in the interpretation of judgement bias studies, with some researchers failing to distinguish between experiments designed to validate a task or an affect manipulation (i.e. via Tests 2–5 in Chapter 2) and those seeking to use judgement biases in tasks simply *assumed* to be valid, to investigate questions about previously unknown affective or welfare states. Indeed, making inferences about affect and welfare on the bases of judgement bias results is not always a simple matter. The species most extensively and successfully studied in the field of judgement biases is the rat (*Rattus novegicus*, e.g. Burman et al. 2008; Enkel et al. 2010; Hales et al. 2016; Nguyen et al. 2020; Papciak et al. 2013; Rygula et al. 2014b). For this animal, well-established tasks and validated designs are available, and findings made when using such methods are likely to be both useful and informative. This is not the case with all species and tasks, however, and further task development and validatory research is still needed (e.g. Burani et al. 2022; Resasco et al. 2021). Before use in welfare assessment, we need to be certain that responses to a particular type of judgement bias task are sensitive and specific to affective variation in the species concerned, by seeing if it passes one or more construct validation tests. Such checks may also give an idea of the task's responsiveness (or, indeed, the relative sensitivities of different task methods), since the likely effect sizes will determine the sample sizes then needed to avoid Type II errors. And even then, careful consideration must be given to how best to analyse and interpret judgement bias data. In all cases, even with well-validated tasks, it is important to note that animals' judgement bias scores can only provide information about welfare in *relative* terms. It cannot be stated that an animal is experiencing absolutely 'good' or 'bad' welfare on the basis of their judgement bias scores, only that their welfare state may be better or poorer than a conspecific completing an identical task and showing a more optimistic-like or pessimistic-like bias, respectively (see Figure 7.6).

In conclusion, there is still much to be understood about the links between affective states and judgements about ambiguity. For example, it has been suggested that judgement bias tasks may be especially sensitive to relatively long-lasting affective states (e.g. moods), but whether this is really the case, and why that might be so, remain to be fully investigated. Moreover, we still need to understand the extent to which judgement biases arise from heightened or reduced expectations of the *probability* of rewards (following the presentation of ambiguous cues), and/or from variation in the *valuation* of rewards or punishers. To do this, computational modelling may be a helpful way of disentangling the various cognitive-affective processes underlying judgement biases, as computational approaches allow identification of the parameters that are most related to welfare to be extracted from data for analysis (e.g. Hales et al. 2017; Neville et al. 2020a). Further experimental studies will also be needed to

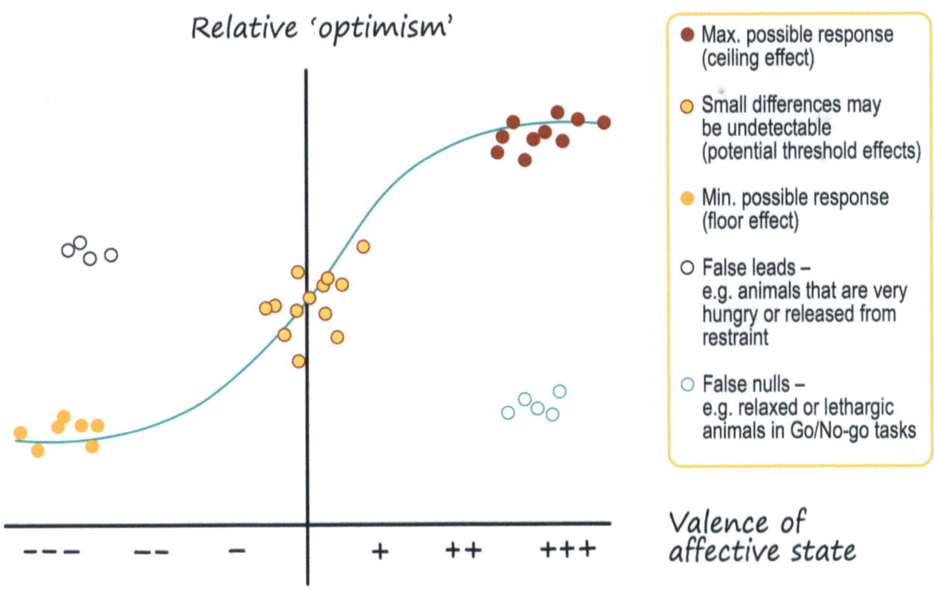

Figure 7.6 A schematic illustration of the results expected from judgement bias tests, showing floor and ceiling effects and potential threshold effects. Potential false leads and false nulls are also illustrated. Note that only *relatively* optimistic-like and pessimistic-like responses can be obtained with this test (i.e. relative to other animals being tested or relative to the same animal at a different time point), as e.g. proportion of ambiguous probe cues responded to.

discover whether certain judgement bias task methods simply fail to reflect affect (while others succeed) or fail to reflect affect in certain species or phyla. For example, it is possible that some animals do not possess affective systems that are cognitively penetrable. However, with such investigations of the complexities of links between affect and cognition, and with a rigorous approach to construct validation, the future of the judgement bias task looks likely to be bright.

References

Anderson, M.G., Campbell, A.M., Crump, A., Arnott, G. and Jacobs, L. (2021) *Environmental complexity positively impacts affective states of broiler chickens. Scientific Reports*, 11, 16966. https://doi.org/10.1038/s41598-021-95280-4

Anderson, M.G., Campbell, A.M., Kuhn, D.D., Smith, S.A. and Jacobs, L. (2022) *Impact of environmental complexity and stocking density on affective states of rainbow trout (Oncorhynchus mykiss). Animal Cognition*, 25(5), 1331–1343. https://doi.org/10.1007/s10071-022-01616-6

Anderson, M.H., Hardcastle, C., Munafo, M.R. and Robinson, E.S.J. (2012) *Evaluation of a novel translational task for assessing emotional biases in different species. Cognitive, Affective, and Behavioural Neuroscience*, 12, 373–381. https://doi.org/10.3758/s13415-011-0076-4

Aylward, J., Hales, C., Robinson, E. and Robinson, O.J. (2020) *Translating a rodent measure of negative bias into humans: the impact of induced anxiety and unmedicated mood and anxiety disorders. Psychological Medicine*, 50, 237–246. https://doi.org/10.1017/s0033291718004117

Barker, T.H., Howarth, G.S. and Whittaker, A.L. (2018) *Increased latencies to respond in a judgment bias test are not associated with pessimistic biases in rats. Behavioural Processes*, 146, 64–66. https://doi.org/10.1016/j.beproc.2017.11.016

Bateson, M., Brilot, B. and Nettle, D. (2011) *Anxiety: An evolutionary approach. Canadian Journal of Psychiatry / Revue Canadienne de Psychiatrie*, 56(12), 707–715. https://doi.org/10.1177/070674371105601202

Bateson, M., Emmerson, M., Ergun, G., Monaghan, P. and Nettle, D. (2015) *Opposite effects of early life competition and developmental telomere attrition on cognitive biases in Juvenile European starlings. PLoS One*, 10, e0132602. https://doi.org/10.1371/journal.pone.0132602

Beck, A.T. (1967) *Depression: Clinical, Experimental and Theoretical Aspects*. Harper Row, New York, p. 370.

Beck, A.T., Rush, A.J., Shaw, B.F. and Emery, G. (1979) *Cognitive Therapy of Depression*. The Guilford Press, New York, p. 406. ISBN: 978-0898629194

Beck, A.T., Ward, C.H., Mendelson, M., Mock, J. and Erbaugh, J. (1961) *An inventory for measuring depression. Archives of General Psychiatry*, 4, 561–571. https://doi.org/10.1001/archpsyc.1961.01710120031004

Bell, C., Bourke, C., Colhoun, H., Carter, F., Frampton, C. and Porter, R. (2011) *The misclassification of facial expressions in generalised social phobia. Journal of Anxiety Disorders*, 25(2), 278–283. https://doi.org/10.1016/j.janxdis.2010.10.001

Brilot, B.O., Normandales, C.L., Parkin, A. and Bateson, M. (2009) *Can we use starlings' aversion to eyespots as the basis for a novel 'cognitive bias' task? Applied Animal Behaviour Science*, 118(3–4, SI), 182–190. https://doi.org/10.1016/j.applanim.2009.02.015

Brydges, N.M., Hall, L., Nicholson, R., Holmes, M.C. and Hall, J. (2012) *The effects of juvenile stress on anxiety, cognitive bias and decision making in adulthood: a rat model. PLoS One*, 7, e48143. https://doi.org/10.1371/journal.pone.0048143

Burani, C., Pelosi, A. and Valsecchi, P. (2022) *A promising novel judgement bias test to evaluate affective states in dogs (Canis familiaris). Animal Cognition*, 25, 837–852. https://doi.org/10.1007/s10071-021-01596-z

Burgdorf, J. and Panksepp, J. (2001) *Tickling induces reward in adolescent rats. Physiology & Behaviour*, 72(1–2), 167–173. https://doi.org/10.1016/s0031-9384(00)00411-x

Burman, O.H.P., McGowan, R., Mendl, M., Norling, Y., Paul, E., Rehn, T. and Keeling, L. (2011) *Using judgement bias to measure positive affective state in dogs. Applied Animal Behaviour Science*, 132(3–4), 160–168. https://doi.org/10.1016/j.applanim.2011.04.001

Burman, O.H.P., Parker, R., Paul, E.S. and Mendl, M.T. (2008) *A spatial judgement bias task to determine background emotional state in laboratory rats, Rattus novegicus. Animal Behaviour*, 76(3), 801–809. https://psycnet.apa.org/doi/10.1016/j.anbehav.2008.02.014

Byrne, A. and Eysenck, M.W. (1993) *Individual differences in positive and negative interpretative biases. Personality and Individual Differences*, 14, 849–851. https://doi.org/10.1016/0191-8869(93)90100-h

Casey, R.A., Naj-Oleari, M., Campbell, S., Mendl, M. and Blackwell, E.J. (2021) *Dogs are more pessimistic if their owners use two or more aversive training methods. Scientific Reports*, 11(1), 1–8. https://doi.org/10.1038/s41598-021-97743-0

Cockburn, A., Smith, M., Rusbridge, C., Fowler, C., Paul, E.S., Murrell, J.C., Blackwell, E.J., Casey, R.A., Whay, H.R. and Mendl, M. (2018) *Evidence of negative affective state in Cavalier King Charles Spaniels with syringomyelia. Applied Animal Behaviour Science*, 201, 77–84. https://psycnet.apa.org/doi/10.1016/j.applanim.2017.12.008

Crump, A., Jenkins, K., Bethell, E.J., Ferris, C.P., Kabboush, H., Weller, J. and Arnott, G. (2021) *Optimism and pasture access in dairy cows. Scientific Reports*, 11, 4882. https://doi.org/10.1038/s41598-021-84371-x

Deakin, A., Browne, W.J., Hodge, J.J.L., Paul, E.S. and Mendl, M. (2016) *A screen-peck task for investigating cognitive bias in laying hens. PLoS One*, 11(7), e0158222. https://doi.org/10.1371/journal.pone.0158222

Doyle, R.E., Fisher, A.D., Hinch, G.N., Boissy, A. and Lee, C. (2010) *Release from restraint generates a positive judgement bias in sheep. Applied Animal Behaviour Science*, 122(1), 28–34. https://doi.org/10.1016/j.applanim.2009.11.003

Enkel, T., Gholizadeh, D., von Bohlen Und Halbach, O., Sanchis-Segura, C., Hurlemann, R., Spanagel, R., Gass, P. and Vollmayr, B. (2010) *Ambiguous-cue interpretation is biased under stress- and depression-like states in rats. Neuropsychopharmacology*, 35, 1008–1015. https://doi.org/10.1038/npp.2009.204

Eysenck, M.W., MacLeod, C. and Matthews, A. (1987) *Cognitive functioning and anxiety. Psychological Research*, 49, 189–195. https://doi.org/10.1007/bf00308686

Gott, A., Andrews, C., Bedford, T., Nettle, D. and Bateson, M. (2019) *Developmental history and stress responsiveness are related to response inhibition, but not judgement bias, in a cohort of European starlings (Sturnus vulgaris). Animal Cognition*, 22, 99–111. https://doi.org/10.1007/s10071-018-1226-7

Hadwin, J., Frost, S., French, C.C. and Richards, A. (1997) *Cognitive processing and trait anxiety in typically developing children: Evidence for interpretation bias. Journal of Abnormal Psychology*, 106, 486–490. https://doi.org/10.1037//0021-843x.106.3.486

Hales, C.A., Houghton, C.J. and Robinson, E.S.J. (2017) *Behavioural and computational methods reveal differential effects for how delayed and rapid onset antidepressants effect decision-making in rats. European Neuropsychopharmacology*, 27, 1268–1280. https://doi.org/10.1016/j.euroneuro.2017.09.008

Hales, C.A., Robinson, E.S. and Houghton, C.J. (2016) *Diffusion modelling reveals the decision making processes underlying negative judgement bias in rats. PLoS One*, 11(3), e0152592. https://doi.org/10.1371/journal.pone.0152592

Hales, C.A., Stuart, S.A., Anderson, M.H. and Robinson, E.S.J. (2014) *Modelling cognitive affective biases in major depressive disorder using rodents. British Journal of Pharmacology*, 171, 4524–4538. https://doi.org/10.1111/bph.12603

Harding, E.J., Paul, E.S. and Mendl, M.T. (2004) *Animal behaviour: Cognitive bias and affective state. Nature*, 427, 312–312. https://doi.org/10.1038/427312a

Hernandez, C.E., Hinch, G., Lea, J., Ferguson, D. and Lee, C. (2014) *Acute stress enhances sensitivity to a highly attractive food reward without affecting judgement bias in laying hens. Applied Animal Behaviour Science*, 163, 135–143. https://doi.org/10.1016/j.applanim.2014.12.002

Hobbs, S.L., Law, T.H., Volk, H.A., Younis, C., Casey, R.A. and Packer, R.M.A. (2020) *Impact of canine epilepsy on judgement and attention biases. Scientific Reports*, 10(1), 17719. https://doi.org/10.1038/s41598-020-74777-4

Iigaya, K., Jolivald, A., Jitkrittum, W., Gilchrist, I.D., Dayan, P., Paul, E. and Mendl, M. (2016) *Cognitive bias in ambiguity judgements: Using computational models to dissect the effects of mild mood manipulation in humans. PLoS One*, 11(11), e0165840. https://doi.org/10.1371/journal.pone.0165840

Inoue, T., Nakagawa, S., Izumi, T., Kitaichi, Y. and Koyama, T. (2006) *Effect of combined treatment with noradrenaline and serotonin reuptake inhibitors on conditioned freezing. European Journal of Pharmacology*, 540(1–3), 91–95. https://doi.org/10.1016/j.ejphar.2006.04.013

Jones, S., Neville, V., Higgs, L., Paul, E.S., Dayan, P., Robinson, E.S. and Mendl, M. (2018) *Assessing animal affect: an automated and self-initiated judgement bias task based on natural investigative behaviour. Scientific Reports*, 8(1), 1–12. https://doi.org/10.1038/s41598-018-30571-x

Lagisz, M., Zidar, J., Nagakawa, S., Neville, V., Sorato, E., Paul, E.S., Bateson, M., Mendl, M. and Lovlie, H. (2020) *Optimism, pessimism and judgement bias in animals: A systematic review and meta-analysis. Neuroscience and Biobehavioural Reviews*, 118, 3–17. https://doi.org/10.1016/j.neubiorev.2020.07.012

Lecorps, B., Ludwig, B.R., von Keyserlingk, M.A. and Weary, D.M. (2019) *Pain-induced pessimism and anhedonia: Evidence from a novel probability-based judgment bias test. Frontiers in Behavioral Neuroscience*, 13, 54. https://doi.org/10.3389/fnbeh.2019.00054

Maoz, K., Elder, S., Stoddard, J., Pine, D.S., Leibenluft, E. and Bar-Haim, Y. (2016) *Angry-happy interpretations of ambiguous faces in social anxiety disorder. Psychiatry Research*, 241, 122–127. https://doi.org/10.1016/j.psychres.2016.04.100

Mathews, A., Richards, A. and Eysenck, M. (1989) *Interpretation of homophones related to threat in anxiety states. Journal of Abnormal Psychology*, 98(1), 31–34. https://doi.org/10.1.37//0021-843x.98.1.31

Mendl, M., Brooks, J., Basse, C., Burman, O., Paul, E., Blackwell, E. and Casey, R. (2010b) *Dogs showing separation-related behaviour exhibit a 'pessimistic' cognitive bias. Current Biology*, 20(19), R839–R840. https://doi.org/10.1016/j.cub.2010.08.030

Mendl, M., Burman, O.H.P., Parker, R.M.A. and Paul, E.S. (2009) *Cognitive bias as an indicator of animal emotion and welfare: Emerging evidence and underlying mechanisms. Applied Animal Behaviour Science*, 188, 161–181. https://psycnet.apa.org/doi/10.1016/j.applanim.2009.02.023

Mendl, M., Burman, O.H.P. and Paul, E.S. (2010a) *An integrative and functional framework for the study of animal emotion and mood. Proceedings of the Royal Society B*, 277, 2895–2904. https://doi.org/10.1098/rspb.2010.0303

Mendl, M.T. and Paul, E.S. (2020) *Animal affect and decision-making. Neuroscience and Biobehavioural Reviews*, 112, 144–163. https://doi.org/10.1016/j.neubiorev.2020.01.025

Mogg, K., Bradbury, K.E. and Bradley, B.P. (2006) *Interpretation of ambiguous information in clinical depression. Behaviour Research and Therapy*, 44(10), 1411–1419. https://doi.org/10.1016/j.brat.2005.10.008

Mogg, K., Bradley, B.P., Miller, T., Potts, H., Genwright, J. and Kentish, J. (1994) *Interpretation of homophones related to threat – anxiety or response bias effects. Cognitive Therapy and Research*, 18(5), 461–477. https://doi.org/10.1007/bf02357754

Mohlman, J., Carmin, C.N. and Price, R.B. (2007) *Jumping to interpretations: Social anxiety disorder and the identification of emotional facial expressions. Behaviour Research and Therapy*, 45(3), 591–599. https://doi.org/10.1016/j.brat.2006.03.007

Neave, H.W., Daros, R.R., Costa, J.H., von Keyserlingk, M.A. and Weary, D.M. (2013) *Pain and pessimism: Dairy calves exhibit negative judgement bias following hot-iron disbudding. PLoS One*, 8(12), e80556. https://doi.org/10.1371/journal.pone.0080556

Neville, V., Dayan, P., Gilchrist, I.D., Paul, E.S. and Mendl, M. (2021a) *Dissecting the links between reward and loss, decision-making, and self-reported affect using a computational approach. PLoS Computational Biology*, 17(1), e1008555. https://doi.org/10.1371/journal.pcbi.1008555

Neville, V., Dayan, P., Gilchrist, I.D., Paul, E.S. and Mendl, M. (2021b) *Using primary reinforcement to enhance translatability of a human affect and decision-making judgment bias task. Journal of Cognitive Neuroscience*, 33(12), 2523–2535. https://doi.org/10.1162/jocn_a_01776

Neville, V., King, J., Gilchrist, I.D., Dayan, P., Paul, E.S. and Mendl, M. (2020a) *Reward and punisher experience alter rodent decision-making in a judgement bias task. Scientific Reports*, 10(1), 1–14. https://doi.org/10.1038/s41598-020-68737-1

Neville, V., Lind, J., Mendl, E., Cozma, N.E., Paul, E.S. and Mendl, M. (2023) *A mapping review of refinements to laboratory rat housing and husbandry. Lab Animal*, 52, 63–74. https://doi.org/10.1038/s41684-023-01124-1

Neville, V., Nakagawa, S., Zidar, J., Paul, E.S., Lagisz, M., Bateson, M., Løvlie, H. and Mendl, M. (2020b) *Pharmacological manipulations of judgement bias: A systematic review and meta-analysis. Neuroscience and Biobehavioral Reviews*, 108, 269–286. https://doi.org/10.1016/j.neubiorev.2019.11.008

Nguyen, H.A.T., Guo, C. and Homberg, J.R. (2020) *Cognitive bias under adverse and rewarding conditions: A systematic review of rodent studies. Frontiers in Behavioral Neuroscience*, 14, 14. https://doi.org/10.3389/fnbeh.2020.00014

Nygren, T.E., Isen, A.M., Taylor, P.J. and Dulin, J. (1996) *The influence of positive affect on the decision rule in risk situations: focus on outcome (and especially avoidance of loss) rather than probability. Organizational Behavior and Human Decision Processes*, 66, 59–72. https://doi.org/10.1006/obhd.1996.0038

Papciak, J., Popik, P., Fuchs, E. and Rygula, R. (2013) *Chronic psychosocial stress makes rats more 'pessimistic' in the ambiguous-cue interpretation paradigm. Behavioural Brain Research*, 256, 305–310. https://doi.org/10.1016/j.bbr.2013.08.036

Paul, E.S., Browne, W., Mendl, M.T., Caplen, G., Held, S., Trevarthen, A. and Nicol, C.J. (2023) *Affective trajectories: Are hens influenced by positive and negative changes in their living conditions? Applied Animal Behaviour Science*, 261, 105883. https://doi.org/10.1016/j.applanim.2023.105883

Paul, E.S., Browne, W., Mendl, M.T., Caplen, G., Trevarthen, A., Held, S. and Nicol, C.J. (2022) *Assessing animal welfare: a triangulation of preference, judgement bias and other candidate welfare indicators. Animal Behaviour*, 186, 151–177. https://doi.org/10.1016/j.anbehav.2022.02.003

Paul, E.S., Cuthill, I., Kuroso, G., Norton, V., Woodgate, J. and Mendl, M. (2011) *Mood and the speed of decisions about anticipated resources and hazards. Evolution and Human Behaviour*, 32, 21–28. https://doi.org/10.1016/j.evolhumbehav.2010.07.005

Raoult, C.M.C., Moser, J. and Gygax, L. (2017) *Mood as cumulative expectation mismatch: A test of theory based on data from non-verbal cognitive bias tests. Frontiers in Psychology*, 8, 2197. https://doi.org/10.3389/fpsyg.2017.02197

Resasco, A., MacLellan, A., Ayala, M.A., Kitchenham, L., Edwards, A.M., Lam, S., Dejardin, S. and Mason, G. (2021) *Cancer blues? A promising judgment bias task indicates pessimism in nude mice with tumors. Physiology & Behavior*, 238, 113465. https://doi.org/10.1016/j.physbeh.2021.113465

Richards, A., French, C.C., Calder, A.J., Webb, B., Fox, R. and Young, A.W. (2002) *Anxiety-related bias in the classification of emotionally ambiguous facial expressions. Emotion*, 2, 273–287. https://doi.org/10.1037/1528-3542.2.3.2.3

Richards, A., Reynolds, A. and French, C.C. (1993) *Anxiety and the spelling and use in sentences of threat/neutral homophones. Current Psychology: Developmental, Learning, Personality, Social*, 12, 18–25. https://doi.org/10.1007/bf02737089

Roelofs, S., Boleij, H., Nordquist, R.E. and van der Staay, F.J. (2016) *Making decisions under ambiguity: Judgement bias tasks for assessing emotional state under ambiguity. Frontiers in Behavioural Neuroscience*, 10, 119. https://doi.org/10.3389/fnbeh.2016.00119

Ross, M., Garland, A., Harlander-Matauschek, A., Kitchenham, L. and Mason, G. (2019) *Welfare-improving enrichments greatly reduce hens' startle responses, despite little change in judgement bias. Scientific Reports*, 9, 11881. https://doi.org/10.1038/s41598-019-48351-6

Rygula, R., Golebiowska, J., Kregiel, J., Holuj, M. and Popik, P. (2014a) *Acute administration of lithium, but not valproate, modulates cognitive judgement bias in rats. Psychopharmacology*, 232, 2149–2156. https://doi.org/10.1007/s00213-014-3847-0

Rygula, R., Papciak, J. and Popik, P. (2014b) *The effects of acute pharmacological stimulation of the 5-HT, NA and DA systems on the cognitive judgement bias of rats in the ambiguous-cue interpretation paradigm. European Neuropsychopharmacology*, 24, 1103–1111. https://doi.org/10.1016/j.euroneuro.2014.01.012

Rygula, R., Pluta, H. and Popik, P. (2012) *Laughing rats are optimistic. PLoS One*, 7(12), e51959. https://doi.org/10.1371/journal.pone.0051959

Salmeto, A.L., Hymel, K.A., Carpenter, E.C., Brilot, B.O., Bateson, M. and Sufka, K.J. (2011) *Cognitive bias in the chick anxiety–depression model. Brain Research*, 1373, 124–130. https://doi.org/10.1016/j.brainres.2010.12.007

Sanger, M.E., Doyle, R.E., Hinch, G.N. and Lee, C. (2011) *Sheep exhibit a positive judgement bias and stress-induced hyperthermia following shearing. Applied Animal Behaviour Science*, 131(3–4), 94–103. https://doi.org/10.1016/j.applanim.2011.02.001

Schick, A., Wessa, M., Vollmayr, B., Kuehner, C. and Kanske, P. (2013) *Indirect assessment of an interpretation bias in humans: Neurophysiological and behavioural correlates. Frontiers in Human Neuroscience*, 7, 272. https://doi.org/10.3389/fnhum.2013.00272

Tanaka, M., Yoshida, M., Emoto, H. and Ishii, H. (2000) *Noradrenaline systems in the hypothalamus, amygdala and locus coeruleus are involved in the provocation of anxiety: basic studies. European Journal of Pharmacology*, 405(1–3), 397–406. https://doi.org/10.1016/s0014-2999(00)00569-0

Verbeek, E., Ferguson, D. and Lee, C. (2014) *Are hungry sheep more pessimistic? The effects of food restriction on cognitive bias and the involvement of ghrelin in its regulation. Physiology & Behaviour*, 123, 67–75. https://doi.org/10.1016/j.physbeh.2013.09.017

Walker, J.K., Waran, N.K. and Phillips, C.J.C. (2014) *The effect of conspecific removal on the behaviour and physiology of pair-housed shelter dogs. Applied Animal Behaviour Science*, 158, 46–56. https://doi.org/10.1016/j.applanim.2014.06.010

Wichman, A., Keeling, L.J. and Forkman, B. (2012) *Cognitive bias and anticipatory behavior of hens housed in basic and enriched pens. Applied Animal Behaviour Science*, 140, 62–69. https://doi.org/10.1016/j.applanim.2012.05.006

Play Behaviour

8

Suzanne D.E. Held

Bristol Veterinary School, University of Bristol, Bristol, UK

Introduction

Play behaviour is central to conceptual frameworks of what positive animal welfare is – including how to measure and achieve it (e.g. Yeates and Main 2008; Mellor and Beausoleil 2015; Lawrence et al. 2019; Fife-Cook and Franks 2019; Webb et al. 2019; Williams 2021) – and has long been linked with animal welfare (e.g. Fagen 1981; Lawrence 1987; Fraser and Duncan 1998). However, play is biologically so complex that it still has not fully realised its potential as a welfare indicator. Play takes many forms and differs between species. Within species, forms and frequencies of play vary with factors such as age and sex that can reasonably be assumed to be independent of affective state. Frequencies can also vary between breeds, genetic lines or litters- and with individual predispositions, that is with factors that may not be affect-neutral. Furthermore, biological factors that are fundamental to the way in which play behaviour reflects affective states are not yet fully understood and probably vary between play types. Such complexity and uncertainty have implications for how we can and should use play as a welfare indicator based on affective state: the focus of this chapter.

At our current state of knowledge, play in humans and animals is best thought of as an umbrella term for a heterogenous range of behaviours with shared features (e.g. Burghardt 1998; Nijhof et al. 2018; Spinka et al. 2001). Features include that play peaks before adulthood and appears to have no obvious function; that it is performed in a repeated pattern; and, in animals, it is typically marked by small body movements such as head shakes and body twists (Figure 8.1) or by specific play signals (Figure 8.2; e.g. Fagen 1981; Martin and Caro 1985; Newberry et al. 1988; Pellis and Pellis 1996; Palagi et al. 2016). Burghardt (2005, 2010) provided the most used list of defining features (for another example, see Bateson and Martin 2013).

Early scientific arguments for play as a welfare indicator in animals were based on yet another feature. In young animals especially, play is most frequent when they are well and

Figure 8.1 Lambs displaying jumps and body twists indicative of locomotor play behaviour. Source: Photo by Suzanne Held.

Figure 8.2 Dog in a body position (front legs and head low, rear end held high) signalling motivation to play, Source: drawing after photo by Julie Bedford.

their immediate biological needs are met; and it often decreases or disappears when conditions become more challenging, such as during illness, increased predation risk or malnutrition (Fagen 1981; Lawrence 1987; Fraser and Duncan 1998; Spinka et al. 2001). In these instances, play frequencies thus pass Test 3 from Chapter 2 (i.e. reflects the fitness situation of the animal). However, the expression of play behaviour is highly variable between individuals and can be affected by factors *not* dependent on affective states (Held and Ŝpinka 2011; Ahloy-Dallaire et al. 2018; Lawrence et al. 2019). A recent resurgence in research on animal play biology has improved the understanding of such factors. It has also clarified instances where play behaviour does not respond in expected ways to changed background conditions as considered later in the chapter.

Different play types

Animal play is typically categorised into three types: locomotor-rotational, object and social play including play-fighting (also called 'rough-and-tumble play'). What form these take depends on species; for example, see the overview tables in Held and Ŝpinka (2011) or Ahloy-Dallaire et al. (2018). Whether or not the different play types share a single biological function and are thus 'motivationally unitary' (Ahloy Dallaire and Mason 2016) is unresolved, but relevant to how playfulness involving the different types may reflect affective valence. Hypotheses for a single shared function focus on the developmental benefits of training neural, musculoskeletal, psychological and emotional systems in infancy for success and competence in adult life (humans e.g. Fredrickson 2001; animals, e.g. Spinka et al. 2001). Supporting evidence includes findings of different play types working as motivational substitutes for each other, for example, in some carnivores who perform more object play when no social partners are available (Ahloy Dallaire and Mason 2016). Stable individual differences in children and dogs across several play types similarly suggest a single motivational basis (Barnett 1991; Svartberg and Forkman 2002; Trevlas et al. 2003).

Other evidence instead supports the idea of at least some motivational aspects *not* being shared across play types (see Pellis et al. 2019; Kellman and Radwan 2022, a full review being beyond this chapter). In mink, for example, play-fighting and object play are not full motivational substitutes since experimentally increased object play does not decrease play-fighting (Ahloy Dallaire and Mason 2016). In an arena test, morphine administration enhances social play but not locomotor play in calves (Sutherland et al. 2019), and only locomotor play but not social play showed a motivational rebound after four days of no access to a play arena (Bertelsen and Jensen 2019). In summary, studies in pigs, rats, dogs, cats and calves show that different play types may not always work as motivational substitutes; may be differently affected by factors influencing, or independent of, affective state and may react differently to changed conditions. The strongest evidence for diverging motivational systems is for object play and play-fighting (the latter typically subsumed under social play; pigs: e.g. Brown et al. 2018, Franchi et al. 2023; rats: e.g. Lampe et al. 2017, Pellis and Pellis 2009; cats: review by Delgado and Hecht, 2019; dogs: e.g. Rooney et al. 2000; calves: e.g. Bertelsen and Jensen 2019; Bailly-Caumette et al. 2023). Different play types may therefore reflect different motivational states which may be associated with different affective states in different ways (a heterogeneity problem echoed in the chapters on vocalisation and stereotypic behaviour; Chapters 4 and 11).

Playfulness and how it is measured

The term playfulness is used in two ways in the psychology and welfare science literature. In the first sense, playfulness simply refers to play levels, that is how much animals play in a given situation. I will refer to this as playfulness or play levels. It can be spontaneous play observed in the wild, in a field, cage or pen over a defined period; or it can be play that is experimentally elicited by providing animals with stimuli known to trigger play, either in the home environment or in a test arena, with the play responses of test subjects then being quantified. When we look to associations with background affective state (e.g. prior moods), spontaneous and elicited play are important variables to consider. In the other sense, playfulness refers to an underlying, intrinsic personality trait, already known in some species to consistently influence an individual's propensity to play across diverse play situations. I will here refer to this as 'individual predisposition to play', '*a priori* playfulness' or 'playfulness trait', and in the section on construct validity below, I will ignore any potential associations it may have with background affective state but return to it later.

Construct validity

Evidence from humans

Evidence in humans for playfulness reflecting background affective state (mood)

The evidence for a predictable relationship between playfulness and mood (Test 1 in Chapter 2) is particularly strong in children, where many harmful conditions that result in negative affective states also lead to reductions in play and/or changes in play types (without always reducing overall activity levels). This effect is so robust that decreased overall play levels, changes in the relative proportions of different types of play, or both, are used to assess mental and physical illness, pain and trauma in children (e.g. severity of Rhinovirus infections in school-age children: Monto et al. 1987; pain in children with cancer: McGrath et al. 1990), with recent reviews by Ahloy-Dallaire et al. (2018) and Nijhof et al. (2018). To summarise and illustrate here, children suffering from hunger, pain or physical illness typically experience low mood, fear, anxiety and depression (e.g. Graves 1976; Sigman et al. 1989; Patenaude and Kupst 2005; Pinquart and Shen 2011). They also, crucially, show changes in *how much* and *how* they play. An example is from children in hospital with leukaemia playing significantly less overall and differently compared to healthy controls. The play of the sick children was less varied, more repetitive in play themes and activities; they did not play as much fully socially, playing less next to others and less interactively (Gariépy and Howe 2003). Playing is also changed in children with traumatic life experiences that are closely linked to poor mental health, such as abuse or homelessness. Again, the evidence is for social withdrawal, including reduced social play, sometimes accompanied by increases in solitary play such that overall play levels can be unaffected (e.g. Fantuzzo et al. 1996). More switching between play types, or in and out of play, is also found (e.g. Mol Lous et al. 2000, 2002). And finally, play is affected in children and adolescents suffering from affective disorders such as clinical depression or anxiety. An example is from the studies by Mol Lous and colleagues. They showed that clinically depressed 3–6-year olds were as physically active as non-depressed peers but, again, were socially withdrawn and played less; specifically showing lower levels of fantasy play (Mol Lous et al. 2000, 2002). Less is known about whether playfulness further increases as affective states improve in children who are already well, not hungry, nor depressed, anxious and so on (Ahloy-Dallaire et al. 2018).

In human adults, the likelihood that they will play, and how much, is related to generalised positive mood and the discrete, high-arousal emotion of 'joy' (Fredrickson 2001; Proyer 2012). Adults also self-report lower levels of playfulness with depression or anxiety (Leung 2014, cited in Berger et al. 2018; Versluys, 2017).

Taken together, the evidence from humans for playfulness reflecting affective states is strong and highlights two aspects to consider for animal play. The first is that the nature of play may change as well as, or even instead of, overall quantity (Pellis and Pellis 2010; Held and Špinka 2011, Ahloy-Dallaire et al. 2018). Secondly, reductions in social play can be caused by social withdrawal and social anxiety, rather than primarily by lower motivation to play itself. Still, less social play caused by social anxiety or withdrawal in itself may reflect a more negative affective background state.

Evidence from animals

Experimental studies on animal play as a welfare indicator can suffer from problems of construct validity for several reasons. A reason specific to play is that the understanding of its biology has lagged behind recognising its potential as a welfare indicator. Recent progress in the understanding of fundamental aspects of play biology has helped here. It has clarified aspects that are or are not important for indicating affective states in animals. These are mentioned briefly later and expanded on in subsequent sections. A further reason is another of Burghardt's five defining features of play. It is the one about play being expressed in a 'relaxed field' when the animal is free of threats to its biological fitness and 'intense competing systems', by which he means motivational systems (Burghardt 2005, 2010). This creates a *circular trap for using play as a welfare indicator*: if animals play, they must by definition be free of fitness challenges or intense motivational conflict, and their welfare therefore okay to good, or the behaviour cannot be said, by definition, to have been play. To be clear, Burghardt's features were not intended as absolute criteria but rather as operational aids to help recognise play (Burghardt 2005). In practice, however, they are often seen as absolutely defining, thereby creating the trap. This section thus focuses on examples that have avoided circularity: where animals had equal opportunity to express play (e.g. the same space available); where factors independent of affective state such as age and sex are controlled for and in which conditions can justifiably be assumed to modulate affect based on the arguments laid out in Chapter 2. It highlights studies that are not covered in previous reviews (Boissy et al. 2007; Held and Špinka 2011; Ahloy-Dallaire et al. 2018) and/or are particularly good contributions to construct validity.

Evidence for changes in play when animals are exposed to stimuli that they avoid or prefer

Studies across several species have shown decreases in play levels in response to known aversive stimuli (Test 2 in Chapter 2). Induction of pain provides strong evidence, with young animals exposed to tissue damage predictably decreasing or stopping playing. Good examples are castration in lambs (Thornton and Waterman-Pearson 2002) and hot-iron disbudding in calves (burning of horn buds to prevent horns developing; e.g. Mintline et al. 2012; Rushen and de Passillé 2012; Bučková et al. 2022). Note that less locomotory and/or social play was reported after disbudding for calves observed in their home pens (Winder et al. 2017) and also when tested in a play arena (Mintline et al. 2013), suggesting this as a robust finding, independent of how play was measured.

Aversive environmental conditions can also reduce playfulness. For example, O'Connor et al. (2010) tested the effects on weaner pigs of being housed in low or high levels of ambient noise, light or ammonia. Pigs are known to actively avoid ammoniated environments

(Jones et al. 1996; Smith et al. 1996), and high ammonia levels had the strongest negative effect on locomotor play (no other play type recorded; O'Connor et al. 2010). This was independent of overall physical activity levels (Parker et al. 2010).

Conversely, access to preferred stimuli increases solo and social play; examples are water baths for mink or food treats hidden in straw for pigs. In pigs, rough handling reduces all play types and social isolation reduces solitary play frequencies (Vinke et al. 2005; Dudink et al. 2006; Reimert et al. 2013). The findings by Vinke and colleagues built on earlier consumer demand studies showing water baths to be a strongly preferred resource in farmed mink (Mason et al. 2001). In the study by Vinke et al. (2005), mink in the treatment group had access to a separate water bath. Control mink did not but were otherwise housed in equivalent conditions. In all five weeks of the study, young mink in the water-bath group played significantly more (all three play types) in their home cages than controls (Vinke et al. 2005). Young pigs prefer straw as a bedding to bare flooring (Beattie et al. 1998) and showed more locomotory and object play in their home pens when provided with straw daily, compared to no-straw controls in same-sized pens (e.g. Bolhuis et al. 2005). Similarly, Pempek et al. (2017) provided pre-weaned individually housed calves in their home enclosures with objects known to be preferred (review by Mandel et al. 2016). This included a brush, plastic teats to allow sham-sucking, a PVC pipe filled with some dry molasses and a hanging rope to lick and chew (Figure 8.3). Locomotory play in these calves was increased by nearly 50% compared to controls. Control calves were in same-sized pens and received the same daily amount of milk and starter feed, but without these objects (Pempek et al. 2017).

Evidence for changes in play behaviour in animals exposed to ancestrally fitness-boosting or fitness-harming situations

In the wild, many observational studies report play levels decreasing in young animals as environmental conditions become fitness-threatening and increasing as conditions improve and become fitness-boosting (Test 3 in Chapter 2). Classic examples are food availability, ambient temperature and injury risk (reviewed in Held and Ŝpinka 2011). Controlled studies in domestic animals similarly show negative effects on play under conditions that would have been fitness-threatening in their ancestors (overview in Held 2017). For example, early weaning at 3–4 weeks of age depresses all types of play in piglets. This is weaning long before biologically evolved nutritional independence from the sow would be completed at 9–17 weeks (e.g. Donaldson et al. 2002; Hötzel et al. 2010; Colson et al. 2012; Franchi et al. 2022). Dairy calves are typically separated from their mothers within a few days after they are born and then fed replacer milk until they are about two months old. Thus, for these calves, unlike for farmed piglets, separation from the mother happens before they are fully weaned off milk, and all types of play are reduced in response to a loss of milk and (independently) a loss of mother (e.g. Krachun et al. 2010; Rushen and de Passillé 2012).

Poor health and nutritional status, hunger and reduced energy availability also reduce playfulness in animals. Größbacher et al. (2020), for example, reported that compared to healthy individuals, calves in poor health show less locomotor play in their home pens (other types of play not recorded). Health was assessed using diarrhoea, coughing/sneezing and increased respiratory rate as symptoms. Differences in birthweight alone can lead to differences in play levels in young animals as recently shown in piglets (Franchi et al. 2022); and calf studies have demonstrated that low vs high milk allowance reduces play in individuals of the same age, specifically locomotor play (Duve et al. 2012; Jensen et al. 2015). Furthermore, nutritional status provides the strongest evidence yet for a more incremental relationship with play levels: piglets with lower growth rates are less playful (Šilerová et al. 2010 Brown et al.; 2015, Franchi

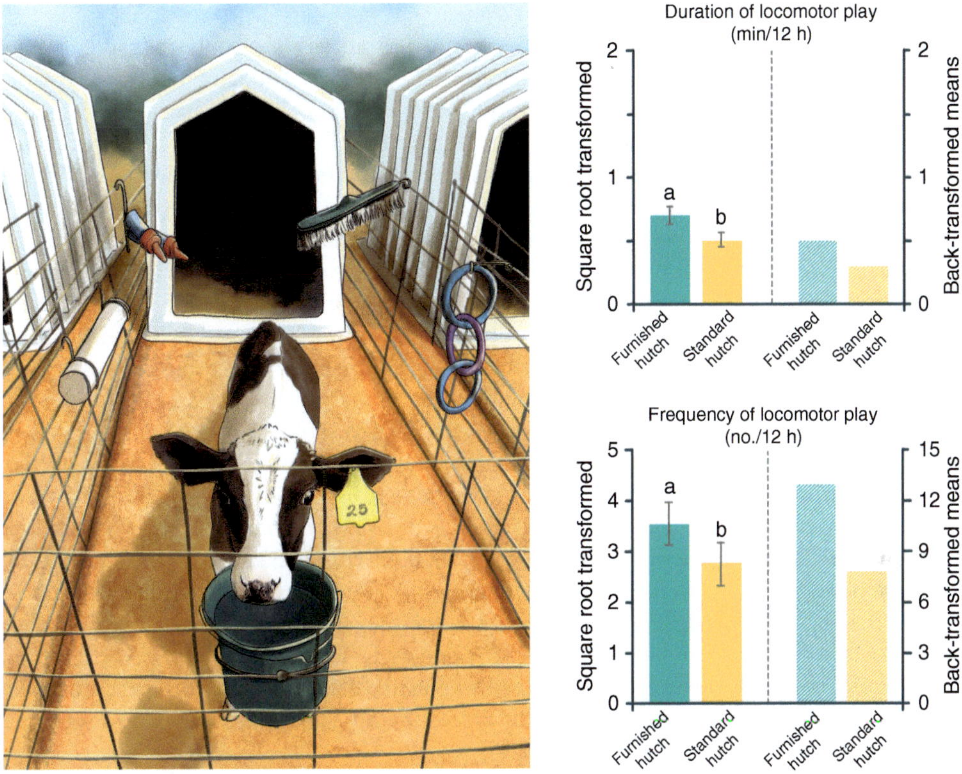

Figure 8.3 Duration (min/12h) and frequency (no./12h) of locomotory play in calves housed individually pre-weaning with (furnished hutch; left illustration) and without (standard hutch) objects in their home enclosures. In the histograms, bars to the right show back-transformed means. Source: Adapted from Pempek et al. (2017).

et al. 2023); total energy intake from milk and concentrate combined was positively correlated to locomotor play levels in yet another calf study (Rushen et al. 2016).

Evidence from affect-modulating drugs

Analgesic and anaesthetic drugs lessen or even reverse the effects of pain on playfulness in young animals (Test 4 in Chapter 2). Pain mitigation during disbudding provides some good examples. Calves given pain medication together with local anaesthesia before disbudding were more playful in a test arena three hours after disbudding than untreated controls (all play types combined; Mintline et al. 2013). In goat kids, pain medication given before disbudding similarly increases playfulness in the three hours after disbudding compared to controls, here recorded in the home pens (locomotory and social combined; Ajuda et al. 2020).

Direct effects of affect-modulating drugs on playfulness have been best studied for social play in rats, confirming that it is associated with positive affect. Opioid drugs and stimulation of reward centres in the brain have been consistently shown to increase social play in rats in neuropharmacological studies (see recent review by Achterberg and Vanderschuren, 2023).

Covariance with pre-validated measures

In rats, opioid stimulation of brain reward centres triggers not just social play but spontaneous tongue protrusion and paw licking (Peciña and Berridge 2000). These are untrained spontaneous, but consistent, behavioural responses shown by rats after tasting a sucrose solution. As such they are indicators of hedonic liking, a positive affective phenomenon (Peciña et al. 2006). Co-variation of social play and these behaviours after opioid stimulation is therefore evidence associated with Test 5 in Chapter 2.

The best evidence available for covariance with pre-validated measures is from vocalisations of rats during social play that are independently also emitted in other high-arousal positive affective states. When rats play with each other, they emit 50 kHz ultrasonic vocalisations as also mentioned in Chapter 4 (50 kHz USVs; Knutson et al. 1998; also e.g. Melotti et al. 2014; Hammond et al. 2019). Rats also produce these 50 kHz USVs during mating, in anticipation of rewards, and during stimulation of brain reward pathways (Sales, 1972; Burgdorf et al. 2008; Vanderschuren, 2010; Schwarting et al. 2018). Frequency modulated forms of 50 kHz vocalisations (especially trills) are most closely linked to positive affect, whereas 'flat' calls may also occur in contexts that appear aversive (e.g. aggression; Wright et al. 2010). In response to aversive stimuli, rats mostly produce different, namely 22 kHz, vocalisations (Knutson et al. 2002; see also recent review by Premoli et al. 2023). The trill-type 50 kHz vocalisations associated with rat social play thus specifically indicate positive affective states.

Note that many other species also vocalise when they play, again most notably during social play (see recent review by Winkler and Bryant 2021). However, these play vocalisations typically take their validation for being linked to positive affective states from their association with playfulness. They are therefore not independently validated as needed for our purposes here.

Responsiveness and selectivity

Factors affecting responsiveness of play as an indicator

Here, I consider what is currently known about situations where playfulness does *not* appear to reliably increase with positive affect or decline with negative affect.

Unknown quantitative relationships and potential changes in the nature of play

Little is known about the quantitative relationship between play levels and affective states, and whether it depends on play type. A welfare indicator should ideally be responsive to incremental changes in affective states such that any improvements or deterioration in welfare are reflected in changes in the indicator. If so, incremental changes in play levels would for example help us when ranking housing conditions in terms of their relative affective valence (Figure 8.4, lines B, C and D). They would also allow us to track relative improvement or deterioration in an individual's affective state, for example as it recovers from illness. However, whether the relationship is all-or-nothing (as in Figure 8.4, line A) or incremental (as in lines B, C and D) is not yet known. Line B can be seen as depicting a graded, incremental *increase* in playfulness with intensity of *positive* affective state, whereas line D reflects a gradual *decrease* in play as affective state becomes increasingly *negative*. There is some evidence for both of these, with line B aligning with findings that levels of adult human playfulness reflect levels of joy, and that levels of rat play are associated with 50 kHz calling rates; line D is supported by findings that poorer nutritional status correlates with decreasing play levels.

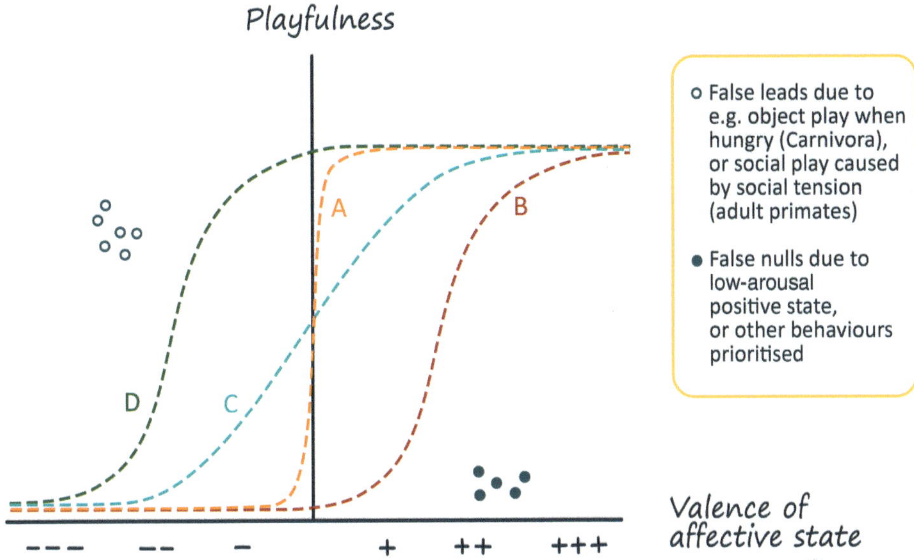

Figure 8.4 Heuristic graphs suggesting four potential quantitative relationships between affective state and play levels. Play levels are typically higher with positive affect and lower with negative affect but data are currently missing to strongly support one or other of the possible relationships depicted; for this reason, the lines are dashed. All four graphs taper off at the extremes, as floor and ceiling levels are reached. In addition, clusters of false nulls and leads are indicated. Lines A and B assume animals only play when they experience no negative affect. In line A, play levels do not reflect how much positive affect animals experience; whereas in line B, increasing play levels would indicate incremental, relative increases in positive affective state. Lines C and D both show graded quantitative relationships between playfulness and affect, but with play also occurring during negative affect. See text for more details. Source: Based on Ahloy-Dallaire et al. (2018).

Line C may thus be the most suitable model based on what is currently known since it combines lines B and D. Where line C crosses the vertical axis will differ between individuals, reflecting a baseline level of *a priori* playfulness.

Understanding quantitative relationships is important also for potential floor and ceiling effects because they could similarly lead to false nulls. They would show that no further improvements (ceiling) or deterioration (floor) in affective valence cause a further increase or decrease in playfulness (see Figure 8.4). In other words: the animal already plays so much that she cannot play more even if affect becomes more positive; or she has already stopped playing such that any worsening of her affective state cannot lead to less play.

The second point on poor responsiveness here is about potential changes in the nature of play in affect-relevant situations, that is changes in *how* the animal plays. If only play quantity is studied and found to be unchanged but the *types* of play do change, this could mean a false null. Changes in types of play may reflect affect in animals when overall levels do not, as mentioned earlier in the case of traumatised children.

When all is well, but play is not prioritised

When animals are well and their living conditions are good, they may stop playing in favour of behaviours that they are currently more strongly motivated to perform because of their greater immediate fitness consequences (Lawrence 1987; Fraser and Duncan 1998; review in Held and Ŝpinka 2011). Mating and feeding can serve as examples here: captive wolves reduce both solitary and social play during mating compared to non-mating periods, and during feeding versus post-feeding periods (Cordoni 2009); in harbour seals, solitary locomotor play was found to be reduced during the mating season (Renouf 1993). Following Test 3 arguments, as these are fitness-boosting situations, affective states are positive for successful individuals in these situations. However, other behaviours have taken priority over playing, leading to false nulls.

Low-arousal positive states may not induce play

Another main reason for false nulls is to do with arousal. Animal play typically happens with excitement and high arousal, and experimental evidence supports this: ACTH-administration directly increases social play in pair-housed rats (Arelis 2006; see also Chapter 5 on corticosteroids); the barks emitted by pigs during play also signal alarm, a high-arousal state (Chan 2011; see also Chapter 4); and as we saw earlier, the 50 kHz USVs that accompany social play in rats also characterise other high-arousal, positively-valenced situations such as mating (e.g. Bogacki-Rychlik et al. 2021) but not positive, low-arousal states such as satiation. Furthermore, the anticipation of positive rewards, such as food, in some species leads to 'anticipatory hyperactivity' with increased locomotory activity and behavioural transitions often including locomotor play components (Boissy et al. 2007; pigs: e.g. Dudink et al. 2006; rats: e.g. Hammond et al. 2019). Play behaviour is thus less sensitive to low-arousal positive affect such as contentment, satiation or other calm, relaxed states, following a two-dimensional model of affect as laid out in Chapter 1. We can also think of this as creating a threshold effect of arousal on playfulness.

Differences in play levels dependent on age, sex, breed/genetic line or litter of origin

Playfulness changes with age but in a predictable manner. More difficult are challenges to responsiveness caused by differences in playfulness between animals of the same age. Examples here are sex, breed/genetic line and litter of origin effects. In many species, young males, for example, reliably play-fight more than females. This could lead to underestimating positive affect in females if only social play, including play-fighting, were measured. Breed/genetic line effects on playfulness are also known for several species (e.g. mice: Walker and Byers 1991; rats: Siviy et al. 2003; cats: Delgado and Hecht 2019; chicken: Lundén et al. 2022). In pigs, lower birth weights and growth rates in litter of origin also contribute to lower social and locomotor play levels before weaning; higher birth weights and growth rates before weaning, in turn, reduce playfulness after weaning, probably because of a greater post-weaning growth check (Brown et al. 2015; Brown et al. 2018). Similar litter effects are also reported for cats and feral dogs (Martin and Bateson 1985; Pal 2010). Whether or not these effects are affect-neutral is currently unknown.

Residual individual differences

Even when age, sex and breed/line are accounted for, individuals can still differ in playfulness in a given situation. Biben and Champoux (1999), for example, found large individual differences in playfulness between juvenile squirrel monkeys in response to the same social challenge. Such differences could be caused by some individuals being more sensitive to social stress and/or generally more fearful. Support for greater anxiety/fearfulness potentially being directly associated with lower individual playfulness comes from play tests with dogs, rats, mink and calves (overview in Ahloy-Dallaire et al. 2018). One example is the study by Rushen and de Passillé (2014) that reported a direct correlation between higher individual fearfulness and lower locomotor playfulness in calves in an arena test. On the other hand, Lampe and colleagues reported that individual differences in tests of anxiety-like states were *uncorrelated* to social playfulness in rats (Lampe et al. 2017): in pair tests with familiar cage mates, individual rats differed consistently in their social playfulness when tested at different ages and levels of play motivation (i.e. either with or without short preceding social isolation). These differences were independent of individual responses in a test designed to assess anxiety predisposition. However, the anxiety test was not specifically assessing *social* anxiety (Lampe et al. 2017). Consistently lower individual social playfulness might therefore still have been related to higher underlying social anxiety. In dogs and humans, playfulness appears as a stable predisposition or personality trait, consistent across different situations (Svartberg and Forkman 2002; Proyer 2012) that can be related to other personality traits, such as sociability (Jones and Gosling 2005). If negative correlations between *a priori* playfulness and fearfulness/anxiety (including social) can be shown to be robust, they are good examples of construct validity (Test 5 in Chapter 2). However, if reasons for individual differences in play levels are unknown and/or it is unclear whether and how they relate to affective valence, they then cannot be accounted for and could create false nulls as well as false leads.

Factors affecting selectivity of play as an indicator

Play in adults, social play, play-fighting and notes on object play

Social play can *increase* during socially stressful situations, as seen in adult primates most likely towards reducing social tension and aggression before feeding, or when a new male arrives in an established group; and in adult horses, in groups with poor cohesion and stability (Palagi et al. 2006; Antonacci et al. 2010; Norscia and Palagi 2011; Blois-Heulin et al. 2015). Likewise, Yamanashi et al. (2018) reported that social play (including play-fighting) increased in adult chimpanzees before feeding, and in all-male groups, and that it was positively correlated with aggression levels. Asensio et al. (2022) similarly found increased play-fighting among adult howler monkeys as food competition increased. In adult wolves, individuals close in social rank are more likely to engage in play-fighting, which is flexibly used to assess the competitiveness of other pack members, manage group cohesiveness and reduce aggression (reviewed in Cordoni and Palagi 2019). That play-fighting can be part of asserting or assessing rank has been demonstrated also in dogs, pigs and rats (e.g. Bauer and Smuts 2007; Newberry et al. 1988; Smith et al. 1999). In all these situations, social playfulness may thus reflect negative affect caused by social tension, not just positive affect (Figure 8.4).

Object play is fiendishly variable in carnivores and primates, where it is most common and can be associated with hunger before feeding (see recent review in Pellis et al. 2019) or with maternal deprivation (see later). It is thus potentially associated with negative as well as positive valence.

In human adults, play activities generally, not just social play, can also increase spontaneously during times of boredom best evidenced for boredom at work (e.g. Game 2007, Wan et al. 2014). Boredom is characterised as an aversive low-stimulation, low-arousal state (Burn 2017). It is biologically plausible that animals may similarly experience boredom as aversive, and that play in animals may work to endogenously stimulate brains to cope with stress or avoid boredom (Pellis and Pellis 2009; Held and Špinka 2011; Burn 2017), potentially as part of an adaptive increase in information gathering in response to low environmental stimulation (Burn 2017).

Maternal deprivation

False leads can also arise in young mammals where playfulness can increase before weaning as a response to reduced or absent maternal care and/or inadequate milk supply. Disrupted maternal care and/or milk provision in young mammals would have been fitness-threatening in their ancestors, suggesting negative affective states following Test 3 in Chapter 2. Bateson and colleagues showed that separation from the mother, maternal undernourishment and disrupted milk production in the mother all increased object play in kittens (Bateson et al. 1981, 1990; Bateson and Young 1981). Similar findings have since been reported in other species, suggesting a more generalised effect with delayed adaptive benefits (Bateson et al. 1981, 1990; e.g. horses: Cameron et al. 2008; rhesus monkeys: Devinney et al. 2003).

Motivational rebound and positive contrast

Finally, there are motivational rebound and positive contrast situations. They are particularly important to take on board because of how tricky they are to deal with in terms of background affective state (Held and Ŝpinka 2011; Ahloy-Dallaire et al. 2018). Motivational rebound is the phenomenon that a behaviour can increase above baseline levels after a period of restriction (e.g. Nicol 1987; Jensen 1999). It is interpreted as indicating a prior restriction of ancestrally adaptive behaviours (Vestergaard 1982). Following the arguments of Tests 2 and 3 in Chapter 2, a motivational rebound would then indicate a negative affective state before the rebound.

In animals of all ages, a temporary increase in playfulness can reflect a rebound after environmental under-stimulation, social or spatial restriction. Many studies testify to this effect with good examples from chickens, and again from young pigs and calves. Broiler chicks in barren pens show more play than chicks in enriched pens, in 'worm-running' and free-space tests in their home pens (Liu et al. 2020). Piglets from barren pens play more than piglets from same-sized enriched pens when given access to extra space (Wood-Gush et al. 1990). Calves, similarly, show more locomotor play after a period of confinement than do unconfined controls (e.g. Jensen 1999). Young calves will also show more locomotor play when their home pens are temporarily made slightly bigger, or they are moved to a substantially bigger test arena, with the largest rebound seen in calves from the smallest home pens (Jensen and Kyhn 2000; see also Rushen and de Passillé 2014; Valníčková et al. 2015).

While the increase in play might arguably indicate 'relief' or a 'positive contrast' at the time at which the play is being expressed, it does represent a false lead in that subjects with the poorest *prior* welfare are playing the most (and see Chapter 7 on judgement biases for similar interpretive problems). If so, play in these contexts may accurately reflect short-term, in-the-moment emotional states but not longer-term mood.

Final considerations for use of play as an indicator of affective state

Considerations and recommendations for current use in practice

An ideal indicator would have a reliably monotonic relationship with affective valence (as in Figure 2.6 in Chapter 2), and playfulness taken as a whole does not (see Figure 8.4). But that is not insurmountable when it comes to using it in practice. Where it is spontaneous and observed, we 'simply' need to pay attention to and account for the context of any changes in play levels. Where possible in experimental studies, we can also aim to control the factors known to affect the responsiveness and selectivity of this indicator.

Young animals are predictably highly motivated to play in stress-free conditions. Thus, for them, reduced playful levels will reliably indicate more negative affective states and thus poor or worsened welfare (with the possible exception of object play; see below). In adults, in contrast, we need to be aware that increased social play can flag situations of social tension and instability, and since playing can increase with boredom in human adults, it might also do so in animals. This and other mounting evidence furthermore suggest that play types differ in their motivational bases: they may thus respond differently to changes in affective valence and should ideally be specified and considered separately. Social play can reflect positive or negative affect depending on context and age. Added to that, play-fighting is often included and recorded under 'social play', can escalate into aggressive fighting and be difficult to distinguish from it. This can raise definitional uncertainty: when does social play stop being play? Where are the boundaries? Reciprocity (taking turns) and play markers can help with the distinction. In young animals where it can often be cleanly distinguished from aggressive fighting, play-fighting is a good welfare indicator. In adolescent and adult animals, however, this distinction is more difficult, making social playfulness a less reliable indicator of positive affect. This further cautions against the indiscriminate use of social play in older animals as indicating relatively good welfare, at least at our current state of knowledge (i.e. avoid where possible for now).

Object play similarly is highly context-dependent in species where it is common and may be associated with hunger (i.e. avoid where possible for now). A conservative suggestion for use in current practice is to focus on locomotor play (i.e. embrace where possible) because at our current state of knowledge it has higher selectivity as a welfare indicator than social and object play.

At all ages and for all play types, playful levels increasing above baseline can indicate a motivational play rebound and/or a positive short-term emotion after being released from restricting or otherwise aversive conditions. In these contexts, it could hence be a good indicator of current affective state but a poor, maybe even inverse, indicator, of preceding longer-term 'mood' state. Play rebound can be controlled for by not using experimental designs where opportunities to play are changed between home and test environments.

We also need to bear in mind that individuals can differ in playfulness for reasons unrelated to treatment. Sex, age, litter of origin and breed are easy sources of individual differences to account for. Trickier are residual differences in play levels caused by dissimilarities in fearfulness or sociability, and any *a priori* individual differences in playfulness as an underlying personality trait. Where welfare assessment is aimed at the individual, it needs to be established whether being less (*a priori*) playful as an independent trait means being generally in a less good welfare state. For example, applying Test 1 from Chapter 2 could reveal whether people, who differ in the playfulness trait are more or less 'happy' as measured independently. To add to this complexity, play may help individuals to cope, meaning that spontaneously high players may either be better at coping, or have more need of coping. That said, for current use in practice, *in situ* welfare assessment is typically aimed at groups of animals, often to assess their treatment. If play behaviour is included in the assessment in such cases, it can be quantified for a representative sample to account for potential pre-existing individual differences: increased locomotor and social play in young ungulates and social play in young rats can indicate conditions that are relatively good for welfare at the group level. But within groups of identically treated animals, residual individual differences in play are currently far harder to use in welfare assessment and should be avoided unless corroborated by strong data.

Considerations and recommendations for future use in practice and in research: avoiding circularity, and improving responsiveness and selectivity

Playing being rewarding and refining the 'play' category

One issue to consider in future work is that play does not only reflect but also improve affective states. Play behaviour is a self-rewarding activity and experienced as pleasurable. The evidence for this effect is robust. An excellent, validating study in the context of this book is by Rygula et al. (2012) who found a direct link between rats having engaged in social play showing more positive judgement bias in pre-validated tests, indicative of more positive affect than rats that had not. Other strong evidence comes from neuropharmacological studies of social play in rats (most recently reviewed by Achterberg and Vanderschuren 2023). Engaging in play being experienced as pleasurable means that at the time of playing an animal will be in a more positive emotional state compared to the same situation but not playing. As a side-note, it is this feature that may help explain why playfulness can increase with social stress and boredom: a lift in affective valence resulting from playing may help individuals cope with aversive situations and thus be biologically adaptive. In any case, play being rewarding and improving affective valence is clearly important for animal welfare, in particular for efforts to improve it. It is confusing however, for those wishing to use play for assessing background affective states present before playing occurred (e.g. moods).

More relevant to this book, play being rewarding is also one of the defining features in Burghardt's list for recognising play (Burghardt 2005, 2010). As mentioned earlier, another feature is that play is expressed in a 'relaxed field'. This created a circular trap for the scientific study of animal welfare that was not intended but has proved difficult to avoid. An improved understanding of play types in different species – how they are motivated and respond differently to affect-modulating conditions – should eventually lead to refining how we conceptualise play. Gajdoš Kmecová and colleagues, for example, recently proposed a refinement for deciding whether cat social interactions are play or not (Gajdoš Kmecová et al. 2021). Based on Panksepp (1998), they set out a framework that integrates 'psychological' (affect) and motivational factors to identify playful interactions – though not set in the context of play in welfare assessment and with hypothesised factors still to be validated (Gajdoš Kmecová et al. 2021). For welfare assessment, further validation studies are therefore needed with more refined play ethograms, to improve responsiveness and selectivity.

The second issue is how to deal with residual individual differences once we have accounted for the effects of sex, age, breed and so on. A starting point here is to be aware that stable, *a priori* predispositions may affect both how and how much an individual plays, all else being equal. Such individual predispositions can be for greater or lesser fearfulness or sociability, more active or passive responding in the face of challenge, or indeed for greater or lesser playfulness itself. All these, individually and in interaction, can influence an animal's play behaviour in a given situation. Research is therefore needed to clarify whether such *a priori* differences by themselves are affect-neutral or already carry differences in background affective states (moods). If the latter, we then need to learn how to interpret these differences with regard to changed play levels. Less sociable individuals, for example, more fearful, or generally less playful ones, may be observed to play less than others when offered less space or straw or toys or opportunity to play with others. Is that because their affective state worsened more than that of others under those poorer conditions? Or was their affective state equally negatively influenced, but they were less likely to play than others because of their different predispositions? Mixed model designs can help identify and account for such interactions between pre-existing individual predispositions and observed playfulness.

Does expression of play occur only during absolutely positive affective states? Does it increase incrementally as affective states improve?

For young animals, four potential positive quantitative relationships between playfulness and affective valence can be envisaged as laid out earlier and visualised in Figure 8.4. Research clarifying the quantitative relationship for different species and for different play types will improve the responsiveness of play as an indicator. Could it be that some types of play increase as in line A, others as in lines B, C or D? Such quantitative questions are important to tackle also for practical reasons. A new emphasis on positive experience in welfare assessments (e.g. Rault et al. 2025) demands measures that reflect when an animal's background affective state becomes absolutely or overwhelmingly positive, or indeed 'optimal' (as in Ahloy-Dallaire et al. 2018). While challenging to sort out, this could potentially be achieved by i) showering animals with abundance and/or ii) using indicators already known to change qualitatively with affective state, such as rat vocalisations (see Chapter 2, and Chapter 9 on reward reactivity).

At all ages, there is an additional risk of false nulls when arousal levels are too low, e.g. when an animal is satiated after a meal (false nulls cluster, Figure 8.4). For adult animals, there is an additional risk of both false leads and false nulls, as social play can occur under stressful situations, and play may be absent when the animal is otherwise occupied, such as during mating (Figure 8.4). Again, a better understanding of the quantitative relationships between playfulness and affective state for all play types will raise confidence in how to interpret adult playfulness in terms of background affective valence.

Changes in the nature of play

We know little about how the *quality* of play changes in animals as their affective states get better or worse. Work with children suggests we may improve the responsiveness and selectivity of play behaviour as a welfare indicator if we consider such changes also in animals. Based on these studies (see section on construct validity), promising starting points for further research are shifts from social to more solitary play types and play bouts becoming more fragmented, as affective states worsen.

Measuring playfulness

The study of playfulness in domestic species has typically been carried out in controlled setups, where play behaviour has been elicited in test situations. The main reason is that spontaneous play is a rare activity in terms of time spent per day even in young animals. Tests are typically conducted away from the home in test pens or arenas, where play is stimulated by providing extra space, social partners and/or new objects. Playfulness in test situations, however, can be affected by motivational play rebound, fear of novelty or unfamiliar play mates and thus difficult to interpret in terms of background affective state – as discussed previously. Good examples of playfulness most cleanly relatable to background affective valence are where novelty and rebound effects are reduced or eliminated by measuring spontaneous playfulness in the home environment such as in Vinke *et al.* (2005) and Pempek et al. (2017), or when play is elicited in the home environment with familiar and preferred stimuli straw (e.g. Bolhuis *et al.* 2005; Jensen et al. 1998) and different play types are recorded separately.

Further development of remote recording of locomotor play using accelerometers will also help promote the use of spontaneous locomotor play in home environments (e.g. Rushen and de Passillé 2012; Luu et al. 2013; Winder *et al.* 2017; Gladden et al. 2020; Grössbacher et al. 2020; Carslake et al. 2023; Larsen et al. 2023).

Missing positively-valenced, low-arousal affective states

A final big challenge for improving responsiveness is the association of playfulness with high arousal. As discussed, an ideal welfare indicator would be responsive across a broad range of positive and negative welfare states. However, when an animal is calm, sleepy or otherwise un-excited, it may not show spontaneous play, but that does not mean no positive affect is experienced. The fact that playing is a rewarding activity, experienced as pleasurable, may help here. Humans with depression, chronic pain, and a range of other conditions associated with negative affective valence are less likely to seek pleasurable activities and/or derive pleasure from engaging in them (see Chapter 9 on reward reactivity). Reduced reward reactivity is thus commonly used in translational animal models of depression (Wang et al. 2021); and reduced playfulness is already a known indicator of depression in children (see construct validation section). However, in some other negatively-valenced situations, responsiveness to rewards in humans is high, suggesting an inverted C-shaped relationship between affective valence and reward reactivity (Figure 9.2 in Chapter 9). A potential case in point for play behaviour could be where playfulness increases with boredom in human adults as mentioned earlier as a false lead. With play being a known rewarding activity it may be fruitful to explore whether, and quantitatively how, playfulness relates to independently validated measures of reward responding in animals. We may then be able to predict and test motivation to engage in play against a background of low-arousal affective states.

In conclusion

With play a central component in current positive welfare frameworks, and its biological complexity no longer underestimated, care must be taken for now in its application. That said, with our current knowledge, we can confidently use spontaneous play in young animals, especially locomotor and social play, to indicate welfare at group level. Individual differences in play under identical conditions and adult play need more investigation, as do object and adult social play. We also need to better understand the quantitative and qualitative relationships between playfulness and affective valence; and what unites and distinguishes different play types as reflecting affective states, that is ultimately how better to typify 'play'.

References

Achterberg, E.M. and Vanderschuren, L.J. (2023) *The neurobiology of social play behaviour: Past, present and future. Neuroscience and Biobehavioral Reviews*, 105319. https://doi.org/10.1016/j.neubiorev.2023.105319

Ahloy Dallaire, J. and Mason, G.J. (2016) *Play in juvenile mink: Litter effects, stability over time, and motivational heterogeneity. Developmental Psychobiology*, 58(8), 945–957. https://doi.org/10.1002/dev.21425

Ahloy-Dallaire, J., Espinosa, J. and Mason, G. (2018) *Play and optimal welfare: Does play indicate the presence of positive affective states? Behavioural Processes*, 156, 3–15. https://doi.org/10.1016/j.beproc.2017.11.011

Ajuda, I., Battini, M., Mattiello, S., Arcuri, C. and Stilwell, G. (2020) *Evaluation of pain mitigation strategies in goat kids after cautery disbudding. Animals*, 10(2), 277. https://doi.org/10.3390/ani10020277

Antonacci, D., Norscia, I. and Palagi, E. (2010) *Stranger to familiar: Wild strepsirhines manage xenophobia by playing. PLoS One*, 5(10), e13218. https://doi.org/10.1371/journal.pone.0013218

Arelis, C.L. (2006) *Stress and the power of play*. PhD thesis,. University of Lethbridge, Faculty of Arts and Science, Alberta, Canada.

Asensio, N., Zandona, E., Dunn, J.C. and Cristóbal-Azkarate, J. (2022) *Socioecological correlates of social play in adult mantled howler monkeys. Animal Behaviour*, 186, 219–229. https://doi.org/10.1016/j.anbehav.2022.01.017

Bailly-Caumette, E., Bertelsen, M. and Jensen, M.B. (2023) *Social and locomotor play behavior of dairy calves kept with the dam either full time or half time in straw-bedded pens. JDS communications*, 4(4), 278–283. https://doi.org/10.3168/jdsc.2022-0337

Barnett, L.A. (1991) *The playful child: Measurement of a disposition to play. Play and Culture*, 4(1), 51–74.

Bateson, P. and Martin, P. (2013) *Play, playfulness, creativity and innovation*. Cambridge University Press, Cambridge, United Kingdom.

Bateson, P., Martin, P. and Young, M. (1981) *Effects of interrupting cat mothers' lactation with bromocriptine on the subsequent play of their kittens. Physiology & Behavior*, 27(5), 841–845. https://doi.org/10.1016/0031-9384(81)90051-2

Bateson, P., Mendl, M. and Feaver, J. (1990) *Play in the domestic cat is enhanced by rationing of the mother during lactation. Animal Behaviour*, 40(3), 514–525. https://doi.org/10.1016/s0003-3472(05)80532-9

Bateson, P. and Young, M. (1981) *Separation from the mother and the development of play in cats. Animal Behaviour*, 29(1), 173–180. https://doi.org/10.1016/s0003-3472(81)80163-7

Bauer, E.B. and Smuts, B.B. (2007) *Cooperation and competition during dyadic play in domestic dogs, Canis familiaris. Animal Behaviour*, 73, 489–499. https://doi.org/10.1016/j.anbehav.2006.09.006

Beattie, V.E., Walker, N. and Sneddon, I.A. (1998) *Preference testing of substrates by growing pigs. Animal Welfare*, 7(1), 27–34. https://doi.org/10.1017/s0962728600020236

Berger, P., Bitsch, F., Bröhl, H. and Falkenberg, I. (2018) *Play and playfulness in psychiatry: A selective review. International Journal of Play*, 7(2), 210–225. https://doi.org/10.1080/21594937.2017.1383341

Bertelsen, M. and Jensen, M.B. (2019) *Does dairy calves' motivation for social play behaviour build up over time? Applied Animal Behaviour Science*, 214, 18–24. https://doi.org/10.1016/j.applanim.2019.02.017

Biben, M. and Champoux, M. (1999) *Play and stress: Cortisol as a negative correlate of play in Saimiri. Play & Culture Studies*, 2, 191–208.

Blois-Heulin, C., Rochais, C., Camus, S., Fureix, C., Lemasson, A., Lunel, C., Bezard, E. and Hausberger, M. (2015) *Animal welfare: Could adult play be a false friend? Animal Behavior and Cognition*, 2(2), 156–185. https://doi.org/10.12966/abc.05.04.2015

Bogacki-Rychlik, W., Rolf, M. and Bialy, M. (2021) *Anticipatory 50-kHz precontact ultrasonic vocalizations and sexual motivation: Characteristic pattern of ultrasound subtypes in an individual analyzed profile. Frontiers in Behavioral Neuroscience*, 15, 722456. https://doi.org/10.3389/fnbeh.2021.722456

Boissy, A., Manteuffel, G., Jensen, M.B., Moe, R.O., Spruijt, B., Keeling, L.J., Winckler, C., Forkman, B., Dimitrov, I., Langbein, J., Bakken, M., Veissier, V. and Aubert, A. (2007) *Assessment of positive emotions in animals to improve their welfare. Physiology & Behavior*, 92(3), 375–397. https://doi.org/10.1016/j.physbeh.2007.02.003

Bolhuis, J.E., Schouten, W.G., Schrama, J.W. and Wiegant, V.M. (2005) *Behavioural development of pigs with different coping characteristics in barren and substrate-enriched housing conditions. Applied Animal Behaviour Science*, 93(3–4), 213–228. https://doi.org/10.1016/j.applanim.2005.01.006

Brown, S.M., Klaffenböck, M., Nevison, I.M. and Lawrence, A.B. (2015) *Evidence for litter differences in play behaviour in pre-weaned pigs. Applied Animal Behaviour Science*, 172, 17–25. https://doi.org/10.1016/j.applanim.2015.09.007

Brown, S.M., Peters, R., Nevison, I.M. and Lawrence, A.B. (2018) *Playful pigs: Evidence of consistency and change in play depending on litter and developmental stage. Applied Animal Behaviour Science*, 198, 36–43. https://doi.org/10.1016/j.applanim.2017.09.018

Bučková, K., Moravcsíková, Á., Šárová, R., Rajmon, R., and Špinka, M. (2022) *Indication of social buffering in disbudded calves. Scientific Reports*, 12(1), 13348. https://doi.org/10.1038/s41598-022-15919-8

Burgdorf, J., Kroes, R.A., Moskal, J.R., Pfaus, J.G., Brudzynski, S.M. and Panksepp, J. (2008) *Ultrasonic vocalizations of rats (Rattus norvegicus) during mating, play, and aggression: Behavioral concomitants, relationship to reward, and self-administration of playback. Journal of Comparative Psychology*, 122(4), 357. https://doi.org/10.1037/a0012889

Burghardt, G.M. (1998) The evolutionary origins of play revisited: Lessons from turtles. In Bekoff, M., Byers, J. A. (eds) *Animal play: Evolutionary, comparative, and ecological perspectives*. Cambridge University Press, Cambridge, pp. 1–26.

Burghardt, G.M. (2005) *The Genesis of Animal Play: Testing the Limits*. MIT Press, Cambridge, Massachusetts.

Burghardt, G.M. (2010) Defining and recognizing play. In Nathan, P., Pellegrini, A. D. (eds) *Oxford Handbook of the Development of Play*. Oxford University Press, Oxford, England, pp. 9–18. https://doi.org/10.1093/oxfordhb/9780195393002.013.0002

Burn, C.C. (2017) *Bestial boredom: A biological perspective on animal boredom and suggestions for its scientific investigation. Animal Behaviour*, 130, 141–151. https://doi.org/10.1016/j.anbehav.2017.06.006

Cameron, E.Z., Linklater, W.L., Stafford, K.J. and Minot, E.O. (2008) *Maternal investment results in better foal condition through increased play behaviour in horses. Animal Behaviour*, 76(5), 1511–1518. https://doi.org/10.1016/j.anbehav.2008.07.009

Carslake, C. (2022) *Sensor-Based Approaches to Monitoring the Behaviour, Health, and Welfare of Dairy Calves*. PhD Thesis, University of Nottingham, United Kingdom.

Chan, W.Y. (2011) The meaning of barks: Vocal communication of fearful and playful affective states in pigs. PhD thesis, Washington State University, USA.

Colson, V., Martin, E., Orgeur, P. and Prunier, A. (2012) *Influence of housing and social changes on growth, behaviour and cortisol in piglets at weaning. Physiology & Behavior*, 107(1), 59–64. https://doi.org/10.1016/j.physbeh.2012.06.001

Cordoni, G. (2009) *Social play in captive wolves (Canis lupus): Not only an immature affair. Behaviour*, 146(10), 1363–1385. https://psycnet.apa.org/doi/10.1163/156853909x427722

Cordoni, G. and Palagi, E. (2019) *Back to the future: A glance over wolf social behavior to understand dog–human relationship. Animals*, 9(11), 991. https://doi.org/10.3390/ani9110991

Delgado, M. and Hecht, J. (2019) *A review of the development and functions of cat play, with future research considerations. Applied Animal Behaviour Science*, 214, 1–17. https://doi.org/10.1016/j.applanim.2019.03.004

Devinney, B.J., Berman, C.M. and Rasmussen, K.L.R. (2003) *Individual differences in response to sibling birth among free-ranging yearling rhesus monkeys (Macaca mulatta) on Cayo Santiago. Behaviour*, 140, 899–924. https://psycnet.apa.org/doi/10.1163/156853903770238373

Donaldson, T.M., Newberry, R.C., Špinka, M. and Cloutier, S. (2002) *Effects of early play experience on play behaviour of piglets after weaning. Applied Animal Behaviour Science*, 79(3), 221–231. https://doi.org/10.1016/s0168-1591(02)00138-7

Dudink, S., Simonse, H., Marks, I., de Jonge, F.H. and Spruijt, B.M. (2006) *Announcing the arrival of enrichment increases play behaviour and reduces weaning-stress-induced behaviours of piglets directly after weaning. Applied Animal Behaviour Science*, 101(1–2), 86–101. https://doi.org/10.1016/s0168-1591(02)00138-7

Duve, L.R., Weary, D.M., Halekoh, U. and Jensen, M.B. (2012) *The effects of social contact and milk allowance on responses to handling, play, and social behavior in young dairy calves. Journal of Dairy Science*, 95(11), 6571–6581. https://doi.org/10.3168/jds.2011-5170

Fagen, R. (1981) *The Significance of Play: Animal Play Behavior*. Oxford University Press, New York, p. 684.

Fantuzzo, J., Sutton-Smith, B., Atkins, M., Meyers, R., Stevenson, H., Coolahan, K., Weiss, A. and Manz, P. (1996) *Community-based resilient peer treatment of withdrawn maltreated preschool children. Journal of Consulting and Clinical Psychology*, 64(6), 1377. https://psycnet.apa.org/doi/10.1037/0022-006X.64.6.1377

Fife-Cook, I. and Franks, B. (2019) *Positive welfare for fishes: Rationale and areas for future study. Fishes*, 4(2), 31. https://doi.org/10.3390/fishes4020031

Franchi, G.A., Larsen, M.L., Kristoffersen, I.H., Winters, J.F., Pedersen, L.J. and Jensen, M.B. (2023) *Play behaviour positively relates to weight gain, feeding behaviour and drinking behaviour in weaner pigs (Sus scrofa). Applied Animal Behaviour Science*, 259, 105836. https://doi.org/10.1016/j.applanim.2023.105836

Franchi, G.A., Larsen, M.L., Winters, J.F., Jensen, M.B. and Pedersen, L.J. (2022) *Investigating the effects of two weaning methods and two genetic hybrids on play behavior in weaner pigs (Sus scrofa). Frontiers in Animal Science*, 3, 909038. https://doi.org/10.3389/fanim.2022.909038

Fraser, D. and Duncan, I.J. (1998) *'Pleasures', 'pains' and animal welfare: Toward a natural history of affect. Animal Welfare*, 7(4), 383–396. https://doi.org/10.1017/s0962728600020935

Fredrickson, B.L. (2001) *The role of positive emotions in positive psychology: The broaden-and-build theory of positive emotions. American Psychologist*, 56(3), 218. https://doi.org/10.1037//0003-066x.56.3.218–226.

Gajdoš Kmecová, N., Pet'ková, B., Kottferová, J., Skurková, L. and Mills, D.S. (2021) *Are these cats playing? A closer look at social play in cats and proposal for a psychobiological approach and standard terminology. Frontiers in Veterinary Science*, 8, 712310. https://doi.org/10.3389/fvets.2021.712310

Game, A.M. (2007) *Workplace boredom coping: Health, safety, and HR implications. Personnel Review*, 36(5), 701–721. https://doi.org/10.1108/00483480710774007

Gariépy, N. and Howe, N. (2003) *The therapeutic power of play: Examining the play of young children with leukaemia. Child: Care, Health and Development*, 29(6), 523–537. https://doi.org/10.1046/j.1365-2214.2003.00372.x

Gladden, N., Cuthbert, E., Ellis, K. and McKeegan, D. (2020) *Use of a tri-axial accelerometer can reliably detect play behaviour in newborn calves. Animals*, 10(7), 1137. https://doi.org/10.3390/ani10071137

Graves, P.L. (1976) *Nutrition, infant behavior, and maternal characteristics: A pilot study in West Bengal, India. The American Journal of Clinical Nutrition*, 29(3), 305–319. https://doi.org/10.3390/ani10071137

Größbacher, V., Bučková, K., Lawrence, A.B., Špinka, M. and Winckler, C. (2020) *Discriminating spontaneous locomotor play of dairy calves using accelerometers. Journal of Dairy Science*, 103(2), 1866–1873. https://doi.org/10.3168/jds.2019-17005

Hammond, T., Bombail, V., Nielsen, B.L., Meddle, S.L., Lawrence, A.B. and Brown, S.M. (2019) *Relationships between play and responses to tickling in male juvenile rats. Applied Animal Behaviour Science*, 221, 104879. https://doi.org/10.1016/j.applanim.2019.104879

Held, S.D.E. (2017) Play behaviour. In Jensen, P. (ed.) *Ethology of Domestic Animals: An Introductory Text*, 3rd ed. CABI Publishing, Wallingford, pp. 90–103.

Held, S.D.E. and Špinka, M. (2011) *Animal play and animal welfare. Animal Behaviour*, 81(5), 891–899. https://doi.org/10.1016/j.anbehav.2011.01.007

Hötzel, M.J., Machado Filho, L.C.P., Irgang, R. and Alexandre Filho, L. (2010) *Short-term behavioural effects of weaning age in outdoor-reared piglets. Animal*, 4(1), 102–107. https://doi.org/10.1017/s1751731109990875

Jensen, M.B. (1999) *Effects of confinement on rebounds of locomotor behaviour of calves and heifers, and the spatial preferences of calves. Applied Animal Behaviour Science*, 62(1), 43–56. https://doi.org/10.1016/s0168-1591(98)00208-1

Jensen, M.B., Duve, L.R. and Weary, D.M. (2015) *Pair housing and enhanced milk allowance increase play behavior and improve performance in dairy calves. Journal of Dairy Science*, 98(4), 2568–2575. https://doi.org/10.3168/jds.2014-8272

Jensen, M.B. and Kyhn, R. (2000) *Play behaviour in group-housed dairy calves, the effect of space allowance. Applied Animal Behaviour Science*, 67(1–2), 35–46. https://doi.org/10.1016/s0168-1591(99)00113-6

Jensen, M.B., Vestergaard, K.S. and Krohn, C.C. (1998) *Play behaviour in dairy calves kept in pens: The effect of social contact and space allowance. Applied Animal Behaviour Science*, 56, 97–108. https://doi.org/10.1016/s0168-1591(97)00106-8

Jones, A.C. and Gosling, S.D. (2005) *Temperament and personality in dogs (Canis familiaris): A review and evaluation of past research. Animal Behaviour Science*, 95(1–2), 1–53. https://doi.org/10.1016/j.applanim.2005.04.008

Jones, J.B., Burgess, L.R., Webster, A.J.F. and Wathes, C.M. (1996) *Behavioural responses of pigs to atmospheric ammonia in a chronic choice test. Animal Science*, 63(3), 437–445. https://doi.org/10.1017/s1357729800015332

Kellman, J. and Radwan, K. (2022) *Towards an expanded neuroscientific understanding of social play. Neuroscience & Biobehavioral Reviews*, 132, 884–891. https://doi.org/10.1016/j.neubiorev.2021.11.005

Knutson, B., Burgdorf, J. and Panksepp, J. (1998) *Anticipation of play elicits high frequency ultrasonic vocalizations in young rats. Journal of Comparative Psychology*, 112, 65–73. https://doi.org/10.1037/0735-7036.112.1.65

Knutson, B., Burgdorf, J. and Panksepp, J. (2002) *Ultrasonic vocalizations as indices of affective states in rats. Psychological Bulletin*, 128(6), 961. https://doi.org/10.1037/0033-2909.128.6.961

Krachun, C., Rushen, J. and de Passillé, A.M. (2010) *Play behaviour in dairy calves is reduced by weaning and by a low energy intake. Applied Animal Behaviour Science*, 122(2–4), 71–76. https://doi.org/10.1016/j.applanim.2009.12.002

Lampe, J.F., Burman, O., Würbel, H. and Melotti, L. (2017) *Context-dependent individual differences in playfulness in male rats. Developmental Psychobiology*, 59(4), 460–472. https://doi.org/10.1002/dev.21509

Larsen, M.L., Wang, M., Willems, S., Liu, D. and Norton, T. (2023) *Automatic detection of locomotor play in young pigs: A proof of concept. Biosystems Engineering*, 229, 154–166. https://doi.org/10.1016/j.biosystemseng.2023.03.006

Lawrence, A.B. (1987) *Consumer demand theory and the assessment of animal welfare. Animal Behaviour*, 35, 293–295. https://doi.org/10.1016/s0003-3472(87)80236-1

Lawrence, A.B., Vigors, B. and Sandøe, P. (2019) *What is so positive about positive animal welfare? - A critical review of the literature. Animals*, 9(10), 783. 10.3390/ani9100783

Leung C.L.R. (2014) Adult playfulness and its relationship with humor, subjective happiness and depression: A comparative study of Hong Kong and mainland China. Bachelor Thesis in Social Sciences in Psychology, City University of Hong Kong, Institutional Repository. http://dspace.cityu.edu.hk/handle/2031/7525

Liu, Z., Torrey, S., Newberry, R.C. and Widowski, T. (2020) *Play behaviour reduced by environmental enrichment in fast-growing broiler chickens. Applied Animal Behaviour Science*, 232, 105098. https://doi.org/10.1016/j.applanim.2020.105098

Lundén, G., Oscarsson, R., Hedlund, L., Gjøen, J. and Jensen, P. (2022) *Play ontogeny in young chickens is affected by domestication and early stress. Scientific Reports*, 12(1), 13576. https://doi.org/10.1038/s41598-022-17617-x

Luu, J., Johnsen, J.F., de Passillé, A.M. and Rushen, J. (2013) *Which measures of acceleration best estimate the duration of locomotor play by dairy calves? Applied Animal Behaviour Science*, 148(1–2), 21–27. https://doi.org/10.1016/j.applanim.2013.07.004

Mandel, R., Whay, H.R., Klement, E. and Nicol, C.J. (2016) *Invited review: Environmental enrichment of dairy cows and calves in indoor housing. Journal of Dairy Science*, 99(3), 1695–1715. https://doi.org/10.3168/jds.2015-9875

Martin, P. and Bateson, P. (1985) *The ontogeny of locomotor play behaviour in the domestic cat. Animal Behaviour*, 33, 502–510. https://doi.org/10.1016/s0003-3472(85)80073-7

Martin, P. and Caro, T.M. (1985) *On the functions of play and its role in behavioral development. Advances in the Study of Behavior*, 15, 59–103. https://doi.org/10.1016/s0065-3454(08)60487-8

Mason, G.J., Cooper, J. and Clarebrough, C. (2001) *Frustrations of fur-farmed mink. Nature*, 410(6824), 35–36. https://doi.org/10.1038/35065157

McGrath, P.J., Beyer, J., Cleeland, C., Eland, J., McGrath, P.A. and Portenoy, R. (1990) *Report of the subcommittee on assessment and methodologic issues in management of pain in childhood cancer. Pediatrics*, 86, 814–817. https://doi.org/10.1542/peds.86.5.814

Mellor, D.J. and Beausoleil, N.J. (2015) *Extending the 'Five Domains' model for animal welfare assessment to incorporate positive welfare states. Animal Welfare*, 24(3), 241–253. https://doi.org/10.7120/09627286.24.3.241

Melotti, L., Bailoo, J.D., Murphy, E.M., Burman, O. and Würbel, H. (2014) *Play in rats: Association across contexts and types, and analysis of structure. Animal Behavior and Cognition*, 1(4), 489–501. https://doi.org/10.12966/abc.11.06.2014

Mintline, E.M., Stewart, M., Rogers, A.R., Cox, N.R., Verkerk, G.A., Stookey, J.M. and Tucker, C.B. (2013) *Play behavior as an indicator of animal welfare: Disbudding in dairy calves. Applied Animal Behaviour Science*, 144(1–2), 22–30. https://doi.org/10.1016/j.applanim.2012.12.008

Mintline, E.M., Wood, S.L., de Passillé, A.M., Rushen, J. and Tucker, C.B. (2012) *Assessing calf play behavior in an arena test. Applied Animal Behaviour Science*, 141(3–4), 101–107. https://doi.org/10.1016/j.applanim.2012.08.006

Mol Lous, A., De Wit, C.A., De Bruyn, E.E. and Riksen-Walraven, J.M. (2002) *Depression markers in young children's play: A comparison between depressed and nondepressed 3-to 6-year-olds in various play situations. Journal of Child Psychology and Psychiatry*, 43(8), 1029–1038. https://doi.org/10.1111/1469-7610.00230

Mol Lous, A., de Wit, C.A., de Bruyn, E.E., Riksen-Walraven, J.M. and Rost, H. (2000) *Depression and play in early childhood: Play behavior of depressed and nondepressed 3-to 6-year-olds in various play situations. Journal of Emotional and Behavioral Disorders*, 8(4), 249–260. https://doi.org/10.1177/106342660000800405

Monto, A.S., Bryan, E.R. and Ohmit, S. (1987) *Rhinovirus infections in Tecumseh, Michigan: Frequency of illness and number of serotypes. Journal of Infectious Diseases*, 156(1), 43–49. https://doi.org/10.1093/infdis/156.1.43

Newberry, R.C., Wood-Gush, D.G.M. and Hall, J.W. (1988) *Playful behaviour of piglets. Behavioural Processes*, 17(3), 205–216. https://doi.org/10.1016/0376-6357(88)90004-6

Nicol, C.J. (1987) *Behavioural responses of laying hens following a period of spatial restriction. Animal Behaviour*, 35(6), 1709–1719. https://doi.org/10.1016/s0003-3472(87)80063-5

Nijhof, S.L., Vinkers, C.H., van Geelen, S.M., Duijff, S.N., Achterberg, E.M., Van Der Net, J., Veltkamp, R.C., Grootenhuis, M.A., van de Putte, E.M., Hillegers, M.H. and van der Brug, A.W. (2018) *Healthy play, better coping: The importance of play for the development of children in health and disease. Neuroscience & Biobehavioral Reviews*, 95, 421–429. https://doi.org/10.1016/j.neubiorev.2018.09.024

Norscia, I. and Palagi, E. (2011) *When play is a family business: Adult play, hierarchy, and possible stress reduction in common marmosets. Primates*, 52, 101–104. https://doi.org/10.1007/s10329-010-0228-0

O'Connor, E.A., Parker, M.O., McLeman, M.A., Demmers, T.G., Lowe, J.C., Cui, L., Davey, E.L., Owen, R.C., Wathes, C.M. and Abeyesinghe, S.M. (2010) *The impact of chronic environmental stressors on growing pigs, Sus scrofa (Part 1): Stress physiology, production and play behaviour. Animal*, 4(11), 1899–1909. https://doi.org/10.1017/S1751731110001072

Pal, S.K. (2010) *Play behaviour during early ontogeny in free-ranging dogs (Canis familiaris). Applied Animal Behaviour Science*, 126, 140–153. https://doi.org/10.1016/j.applanim.2010.06.005

Palagi, E., Burghardt, G.M., Smuts, B., Cordoni, G., Dall'Olio, S., Fouts, H.N., Řeháková-Petrů, M., Siviy, S.M. and Pellis, S.M. (2016) *Rough-and-tumble play as a window on animal communication. Biological Reviews*, 91, 311–327. https://doi.org/10.1111/brv.12172

Palagi, E., Paoli, T. and Tarli, S.B. (2006) *Short-term benefits of play behavior and conflict prevention in Pan paniscus. International Journal of Primatology*, 27, 1257–1270.https://doi.org/10.1007/s10764-006-9071-y

Panksepp, J. (1998) *Affective Neuroscience: The Foundations of Human and Animal Emotions.* Oxford University Press, New York, NY.

Parker, M.O., O'Connor, E.A., McLeman, M.A., Demmers, T.G.M., Lowe, J.C., Owen, R.C., Davey, E.L., Wathes, C.M. and Abeyesinghe, S.M. (2010) *The impact of chronic environmental stressors on growing pigs, Sus scrofa (Part 2): social behaviour. Animal*, 4(11), 1910–1921. https://doi.org/10.1017/S1751731110001084

Patenaude, A.F. and Kupst, M.J. (2005) *Psychosocial functioning in pediatric cancer. Journal of Pediatric Psychology*, 30(1), 9–27. https://doi.org/10.1093/jpepsy/jsi012

Peciña, S. and Berridge, K.C. (2000) *Opioid site in nucleus accumbens shell mediates eating and hedonic 'liking' for food: Map based on microinjection Fos plumes. Brain Research*, 863(1–2), 71–86. https://doi.org/10.1016/s0006-8993(00)02102-8

Peciña, S., Smith, K.S. and Berridge, K.C. (2006) *Hedonic hot spots in the brain. The Neuroscientist*, 12(6), 500–511. https://doi.org/10.1177/1073858406293154

Pellis, S.M. and Pellis, V.C. (1996) *On knowing it's only play: The role of play signals in play fighting. Aggression and Violent Behavior*, 1(3), 249–268. https://doi.org/10.1016/1359-1789(95)00016-x

Pellis, S.M. and Pellis, V.C. (2009) *The Playful Brain.* Oneworld Publications, Oxford.

Pellis, S.M. and Pellis, V.C. (2010) Play. In Mills, D. (ed.) *Encyclopedia of Applied Animal Behaviour and Welfare.* CAB International, Wallingford, pp. 477–479.

Pellis, S.M., Pellis, V.C., Pelletier, A. and Leca, J.B. (2019) *Is play a behavior system, and, if so, what kind? Behavioural Processes*, 160, 1–9. https://doi.org/10.1016/j.beproc.2018.12.011

Pempek, J.A., Eastridge, M.L. and Proudfoot, K.L. (2017) *The effect of a furnished individual hutch pre-weaning on calf behavior, response to novelty, and growth. Journal of Dairy Science*, 100(6), 4807–4817. https://doi.org/10.3168/jds.2016-12180

Pinquart, M. and Shen, Y. (2011) *Depressive symptoms in children and adolescents with chronic physical illness: An updated meta-analysis. Journal of Pediatric Psychology*, 36(4), 375–384. https://doi.org/10.1093/jpepsy/jsq104

Premoli, M., Pietropaolo, S., Wöhr, M., Simola, N. and Bonini, S.A. (2023) *Mouse and rat ultrasonic vocalizations in neuroscience and neuropharmacology: State of the art and future applications. European Journal of Neuroscience*, 57(12), 2062–2096. https://doi.org/10.1111/ejn.15957

Proyer, R.T. (2012) *A psycho-linguistic study on adult playfulness: Its hierarchical structure and theoretical considerations. Journal of Adult Development*, 12, 141–149. https://doi.org/10.1007/s10804-012-9141-7

Rault, J.L., Bateson, M., Boissy, A., Forkman, B., Grinde, B., Gygax, L., Harfeld, J.L., Hintze, S., Keeling, L.J., Kostal, L., Lawrence, A.B., Mendl, M.T., Miele, M., Newberry, R.C., Sandøe, P., Špinka, M., Taylor, A.H., Webb, L.E., Whalin, L. and Jensen, M.B. (2025) *A consensus on the definition of positive animal welfare. Biology Letters*, 21, 20240382. https://doi.org/10.1098/rsbl.2024.0382

Reimert, I., Bolhuis, J.E., Kemp, B. and Rodenburg, T.B. (2013) *Indicators of positive and negative emotions and emotional contagion in pigs. Physiology & Behavior*, 109, 42–50. https://doi.org/10.1016/j.physbeh.2012.11.002

Renouf, D. (1993) *Play in a captive breeding colony of harbour seals (Phoca vitulina): Constrained by time or by energy? Journal of Zoology*, 231(3), 351–363. https://doi.org/10.1111/j.1469-7998.1993.tb01923.x

Rooney, N.J., Bradshaw, J.W. and Robinson, I.H. (2000) *A comparison of dog–dog and dog–human play behaviour. Applied Animal Behaviour Science*, 66(3), 235–248. https://doi.org/10.1016/s0168-1591(99)00078-7

Rushen, J. and de Passillé, A.M. (2012) *Automated measurement of acceleration can detect effects of age, dehorning and weaning on locomotor play of calves. Applied Animal Behaviour Science*, 139(3–4), 169–174. https://doi.org/10.1016/j.applanim.2012.04.011

Rushen, J. and de Passillé, A.M. (2014) *Locomotor play of veal calves in an arena: Are effects of feed level and spatial restriction mediated by responses to novelty? Applied Animal Behaviour Science*, 155, 34–41. https://doi.org/10.1016/j.applanim.2014.03.009

Rushen, J., Wright, R., Johnsen, J.F., Mejdell, C.M. and de Passillé, A.M. (2016) *Reduced locomotor play behaviour of dairy calves following separation from the mother reflects their response to reduced energy intake. Applied Animal Behaviour Science*, 177, 6–11. https://doi.org/10.1016/j.applanim.2016.01.023

Rygula, R., Pluta, H. and Popik, P. (2012) *Laughing rats are optimistic. PLoS One*, 7(12), e51959. https://doi.org/10.1371/journal.pone.0051959

Sales, G.D. (1972) *Ultrasound and mating behaviour in rodents with some observations on other behavioural situations. Journal of Zoology*, 168, 149–164. https://doi.org/10.1111/j.1469-7998.1972.tb01345.x

Schwarting, R.K., Kisko, T.M. and Wöhr, M. (2018) Playback of ultrasonic vocalizations to juvenile and adult rats: Behavioral and neuronal effects. In *Handbook of Behavioral Neuroscience*, vol 25, pp. 357–369. https://doi.org/10.1016/b978-0-12-809600.0.0.034-2

Sigman, M., Neumann, C., Baksh, M., Bwibo, N. and McDonald, M.A. (1989) *Relationship between nutrition and development in Kenyan toddlers. The Journal of Pediatrics*, 115(3), 357–364. https://doi.org/10.1016/s0022-3476(89)80832-7

Šilerová, J., Špinka, M., Šárová, R. and Algers, B. (2010) *Playing and fighting by piglets around weaning on farms, employing individual or group housing of lactating sows. Applied Animal Behaviour Science*, 124, 83–89. https://doi.org/10.1016/j.applanim.2010.02.003

Siviy, S.M., Love, N.J., DeCicco, B.M., Giordano, S.B. and Seifert, T.L. (2003) *The relative playfulness of juvenile Lewis and Fischer-344 rats. Physiology & Behavior*, 80, 385–394. https://doi.org/10.1016/j.physbeh.2003.09.002

Smith, J.H., Wathes, C.M. and Baldwin, B.A. (1996) *The preference of pigs for fresh air over ammoniated air. Applied Animal Behaviour Science*, 49, 417–424. https://doi.org/10.1016/0168-1591(96)01048-9

Smith, L.K., Fantella, S.L. and Pellis, S.M. (1999) *Playful defensive responses in adult male rats depend on the status of the unfamiliar opponent. Aggressive Behavior*, 25(2), 141–152. https://doi.org/10.1016/j.neubiorev.2023.105037

Spinka, M., Newberry, R.C. and Bekoff, M. (2001) *Mammalian play: Training for the unexpected. The Quarterly Review of Biology*, 76(2), 141–168. https://doi.org/10.1086/393866

Sutherland, M., Worth, G., Cameron, C. and Verbeek, E. (2019) *Effect of morphine administration on social and non-social play behaviour in calves. Animals*, 9(2), 56. https://doi.org/10.3390/ani9020056

Svartberg, K. and Forkman, B. (2002) *Personality traits in the domestic dog (Canis familiaris). Applied Animal Behaviour Science*, 79(2), 133–155. https://doi.org/10.1016/s0168-1591(02)00121-1

Thornton, P.D. and Waterman-Pearson, A.E. (2002) *Behavioural responses to castration in lambs. Animal Welfare*, 11(2), 203–212. https://doi.org/10.1017/s0962728600028153

Trevlas, E., Grammatikopoulos, V., Tsigilis, N. and Zachopoulou, E. (2003) *Evaluating playfulness: Construct validity of the Children's Playfulness Scale. Early Childhood Education Journal*, 31, 33–39. https://doi.org/10.1023/a:1025132701759

Valníčková, B., Stěhulová, I., Šárová, R. and Špinka, M. (2015) *The effect of age at separation from the dam and presence of social companions on play behavior and weight gain in dairy calves. Journal of Dairy Science*, 98(8), 5545–5556. https://doi.org/10.3168/jds.2014-9109

Vanderschuren, L.J. (2010) *How the brain makes play fun. American Journal of Play*, 2(3), 315–337.

Versluys, B. (2017) *Adults with an anxiety disorder or with an obsessive-compulsive disorder are less playful: A matched control comparison. The Arts in Psychotherapy*, 56, 117–128. https://doi.org/10.1016/j.aip.2017.06.003

Vestergaard, K. (1982) *Dust-bathing in the domestic fowl—diurnal rhythm and dust deprivation. Applied Animal Ethology*, 8(5), 487–495. https://doi.org/10.1016/0304-3762(82)90061-x

Vinke, C.M., Van Leeuwen, J. and Spruijt, B.M. (2005) *Juvenile farmed mink (Mustela vison) with additional access to swimming water play more frequently than animals housed with a cylinder and platform, but without swimming water. Animal Welfare*, 14(1), 53–60. https://doi.org/10.1017/s096272860002892x

Walker, C. and Byers, J.A. (1991) *Heritability of locomotor play in house mice, Mus domesticus. Animal Behaviour*, 42(6), 891–897. https://doi.org/10.1016/s0003-3472(05)80141-1

Wan, H.C., Downey, L.A. and Stough, C. (2014) *Understanding non-work presenteeism: Relationships between emotional intelligence, boredom, procrastination and job stress. Personality and Individual Differences*, 65, 86–90. https://doi.org/10.1016/j.paid.2014.01.018

Wang, S., Leri, F. and Rizvi, S.J. (2021) *Anhedonia as a central factor in depression: Neural mechanisms revealed from preclinical to clinical evidence. Progress in Neuro-Psychopharmacology and Biological Psychiatry*, 110, 110289. https://doi.org/10.1016/j.pnpbp.2021.110289

Webb, L.E., Veenhoven, R., Harfeld, J.L. and Jensen, M.B. (2019) *What is animal happiness? Annals of the New York Academy of Sciences*, 1438(1), 62–76.

Williams, L.A. (2021) *From human wellbeing to animal welfare. Neuroscience & Biobehavioral Reviews*, 131, 941–952. https://doi.org/10.1016/j.neubiorev.2021.09.014

Winder, C.B., LeBlanc, S.J., Haley, D.B., Lissemore, K.D., Godkin, M.A. and Duffield, T.F. (2017) *Clinical trial of local anaesthetic protocols for acute pain associated with caustic paste disbudding in dairy calves. Journal of Dairy Science*, 100(8), 6429–6441. https://doi.org/10.3168/jds.2017-12724

Winkler, S.L. and Bryant, G.A. (2021) *Play vocalisations and human laughter: A comparative review. Bioacoustics*, 30(5), 499–526. https://doi.org/10.1080/09524622.2021.1905065

Wood-Gush, D.G.M., Vestergaard, K. and Petersen, H.V. (1990) *The significance of motivation and environment in the development of exploration in pigs. Biology of Behaviour*, 15, 39–52.

Wright, J.M., Gourdon, J.C. and Clarke, P.B.S. (2010) *Identification of multiple call categories within the rich repertoire of adult rat 50-kHz ultrasonic vocalizations: Effects of amphetamine and social context. Psychopharmacology*, 211(1), 1–13. https://doi.org/10.1007/s00213-010-1859-y

Yamanashi, Y., Nogami, E., Teramoto, M., Morimura, N. and Hirata, S. (2018) *Adult-adult social play in captive chimpanzees: Is it indicative of positive animal welfare? Applied Animal Behaviour Science*, 199, 75–83. https://doi.org/10.1016/j.applanim.2017.10.006

Yeates, J.W. and Main, D.C. (2008) *Assessment of positive welfare: A review. The Veterinary Journal*, 175(3), 293–300. https://doi.org/10.1016/j.tvjl.2007.05.009

Reactivity to Rewards

9

Rebecca K. Meagher[1] and Carole Fureix[2]

[1]Department of Animal Science and Aquaculture, Dalhousie University, Truro, Nova Scotia, Canada
[2]Bristol Veterinary School, University of Bristol, Bristol, UK

Introduction

In this chapter, we examine how motivations for rewards link with affect, and the caveats to consider when using reactivity to reward as an indicator of affective states. As we will see, either increased or decreased reward reactivity can be observed in cases of negative affective states. Hence, from a dimensional view of affect (e.g. valence-arousal, see Chapter 1), reward reactivity is not a valid indicator of affective states overall. However, measuring reactivity to reward is fundamental when assessing some *specific discrete* affective states where responsiveness to reward is crucial for operationalising them. We therefore discuss the construct validity of reactivity to reward as an indicator of such discrete affective states. Here we use depression and boredom as examples, yet the rationale we present can be extended to other discrete affective states of relevance. But first, we start with discussing how to define a reward, and how to measure reward reactivity.

What is a reward?

'Rewards' are stimuli or situations which generate pleasure (i.e. are positively-valenced, not merely 'less negative'). The discussion of how to identify a reward is still open. In this chapter, we consider a stimulus or a situation to be rewarding when an individual is typically motivated to opportunistically approach or consume the stimulus, even in the absence of meeting an immediate biological need (e.g. Fraser and Duncan 1998; Mason 2010). An example would be eating a second serving of pudding/dessert, even if already feeling satiated. We also expect animals to prefer rewarding stimuli/situations over other less rewarding options (see Chapter 3 on preference tests). Furthermore, a stimulus is deemed to be rewarding when it activates neurobiological pathways known to be activated by pleasurable situations such as sexual behaviour (at least in species in which we know sexual behaviour can be pleasurable, e.g. in self-reporting humans) and taking some recreational

Box 9.1 Sucrose as a reward . . . or not, and beyond

Sucrose ticks many boxes for its consumption to be judged as rewarding in several species. However, although the preference for sweet compounds is common to several species, not all species display a preference for sweet stimuli, due to variations in their taste receptors' specificity and structures (e.g. Li et al. 2009; Shirazi-Beechey 2018). For instance, felids lack a functional Tas1r2 gene and appear insensitive to sweeteners (Li et al. 2009), rendering the use of a sweet compound consumption test irrelevant, and highlighting the importance of using species-relevant rewards. Moreover, depressed people, although prone to reduced reactivity to reward in general, often rate sweet tastes as equally pleasant as non-depressed people (Amsterdam et al. 1987; Berlin et al. 1998; Dichter et al. 2010; Swiecicki et al. 2015; Pluciute et al. 2021; although see Bedwell et al. 2019 reporting that anhedonic people rate sweet as less pleasant). Lastly, individuals from the same species may vary in their hedonic responses to sweet taste (*e.g.* sweet likers and sweet dislikers in humans, e.g. Iatridi et al. 2019). Therefore, reactions to other reward types (i.e. non-sucrose-related) should ideally also be included when intending to assess an animal's motivation for rewards.

drugs (Lowe and Butryn 2007; Olsen 2011; Pratt et al. 2012; see also Chapter 1). Such neurobiological pathways involve dopamine activity in the medial prefrontal cortex in mammals (or its equivalent nidopallium caudolateral in birds), the orbitofrontal cortex, the nucleus accumbens and other areas such as the ventral tegmental area (reviewed by e.g. Ironside et al. 2018; Smeets et al. 2000; Kubikova and Košťál 2010; Rolls et al. 2020), and/or opioid activity, some of which happens in overlapping brain regions (Berridge et al. 2009; Ironside et al. 2018). Sucrose, for instance, is a good example of a reward in mice and rats (e.g. Scheggi et al. 2018). Mice and rats under normal circumstances will consume sucrose even when satiated and prefer sucrose over non-sweetened options (e.g. Willner 2017a), and sucrose intake also engages reward pathways involved in certain recreational drugs (Olsen 2011; Pratt et al. 2012; but see also Box 9.1). With most other putative rewards, evidence of all three criteria is not yet available. While such evidence strengthens the conclusion that a stimulus or situation is a reward (see Box 9.2), we will use examples where not all lines of evidence are available, keeping this caveat in mind.

How is reactivity to reward measured?

Reward reactivity includes components such as 'wanting' (motivation to get access to the reward) and 'liking' (the hedonic response, i.e. the assumed feeling of pleasure, e.g. Der-Avakian et al. 2016; Berridge et al. 2009, see also Chapter 3). Several behavioural and physiological measures have been developed to assess animals' reactions to reward (Figure 9.1). Unless specified otherwise, the measures listed later can reflect both wanting and liking processes. First, reactivity to reward is typically measured by assessing the animal's simple *consumption of*, and/or *preference*[1] *for* rewards such as sucrose

[1] Here referring to an action, such as consuming more, when the animal is exposed to freely choosing to interact with at least two stimuli.

Box 9.2 Why is the *convergence* of criteria necessary to define a stimulus or situation as rewarding?

Preference is a necessary condition to define a stimulus as being rewarding, but not a sufficient one. Preference tests can indeed be used to identify stimuli that animals like and want, following the assumption that animals seek what induces positive affect. As discussed in Chapter 3, however, some preference might instead reflect avoidance of an alternative. For instance, rodents under normal circumstances prefer darker over bright environments (e.g. Lockard 1963; Sousa et al. 2006). This preference for darkness, however, may reflect the avoidance of brighter areas as anti-predation strategies, rather than demonstrating that being in darker environments is pleasure-generating in itself.

That the animal is willing to pay a cost to get access to a stimulus can also be expected if the stimulus is rewarding (Chapter 3). Whether we can use such motivational strength (i.e. how much the animal is willing to 'work') as a defining feature for reward is still open to discussion, however. Following Fraser and Duncan's framework (1998), pleasure-driven behaviours are expected to be expressed at times where the fitness costs of doing so are sufficiently low. Put differently, an animal would be expected to spend some energy to display the pleasure-driven behaviour, but not to pay a cost in terms of not being able to fulfil some other needs (e.g. when actions are required to cope with threat to survival). For example, in a maximum price paid test, it is reasonable to regard a resource ranked as high as the top benchmark (e.g. food) as being essential for the animal (i.e. fulfilling a need). However, it might be difficult in practice to set-up a cut-off of motivational strength that would reflect a threshold for identifying an opportunistic 'motivated but not too motivated' pleasure-driven behaviour.

Regarding neurobiological pathways, some of the relevant neurotransmitters and brain regions can also be activated during non-rewarding situations. For instance, dopamine release may also happen in response to strong aversive stimuli such as tail shocks, restraint or receiving an anxiogenic drug (Horvitz 2000). While activation of certain opioid receptors is more strongly linked with positive affect or reward (e.g. Barenz et al. 2024), many animal studies do not measure activity in specific pathways enough to definitively say that the animal is experiencing pleasure. Hence convergence of the criteria, rather than evidence of each in itself, is important to define a stimulus or a situation as rewarding.

(e.g. Scheggi et al. 2018). Second, *latency to approach / to consume* and/or *time spent investigating* the reward can also be used (e.g. Meagher and Mason 2012). Third, reward reactivity may also be assessed through formal tests of motivation, with animals working or paying a price to access the reward (Chapter 3). Fourth, the strength of *conditioned place preference*, in which animals develop a preference for places specifically associated with natural or artificial reinforcers (e.g. Papp et al. 1991, 1992; Chapter 2), can also be used for assessing reactivity to reward. Fifth, *anticipatory behaviours*, typically assessed in classical conditioning paradigms where the measure is behaviour during the period before an expected reward is delivered, have been tested in animals (Anderson et al. 2020). Sixth, evaluating the animal's response to an unexpected decrease in anticipated reward provides a measure of their *sensitivity to reward loss* (e.g. Burman et al. 2008; Luo et al. 2020). Seventh, *specific facial expressions*, apparently conserved across several mammal species, are also suggested to reflect 'liking,' i.e. a pleasurable response to a substance in contact

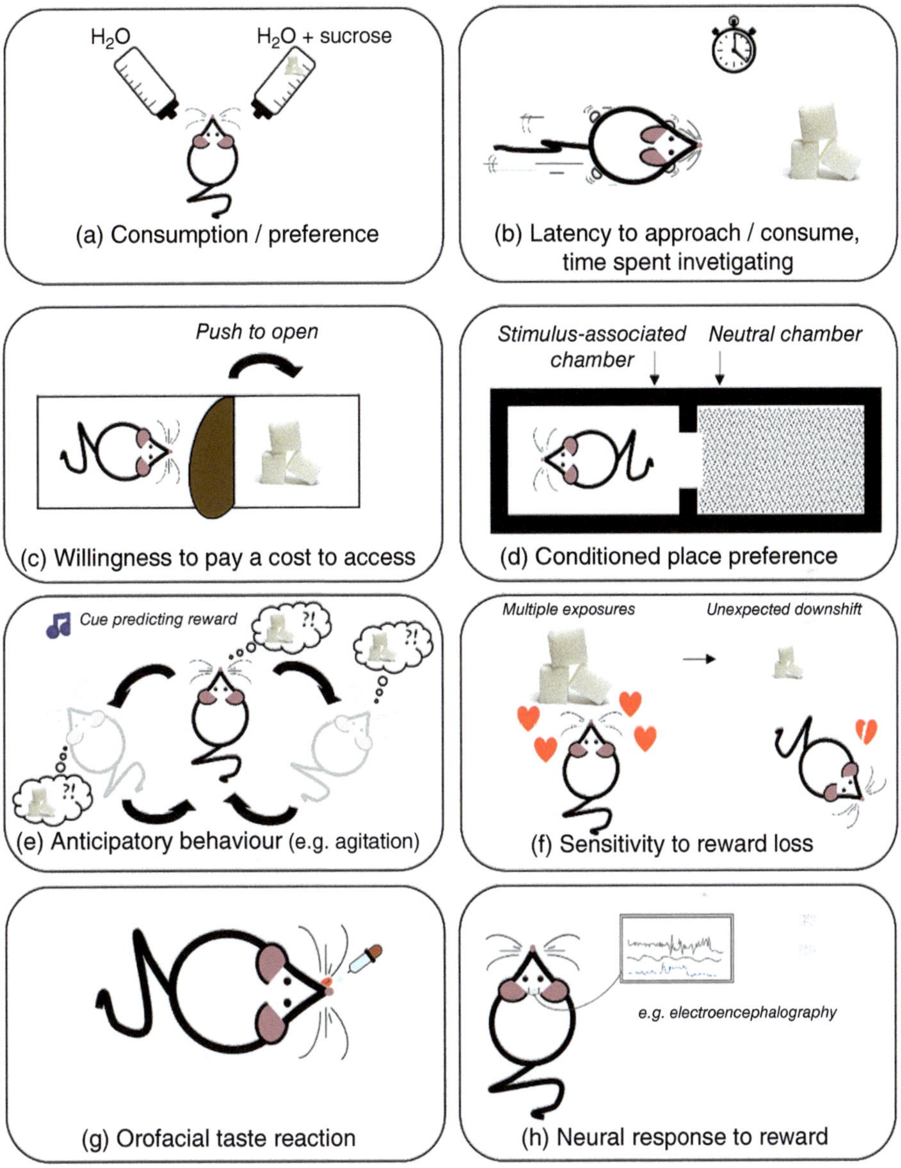

Figure 9.1 Schematic representation of some of the tests used to assess reactivity to reward. Tests are schematically represented here with rodents (most commonly studied), but their overall rationale can be applied to other species. Source: Humusak / Pixabay.

with the mouth (Berridge 2009; Berridge et al. 2009; Dwyer 2012; Clarkson et al. 2018; Yamamoto et al. 2021). This is the case for rhythmic tongue protrusion in response to sucrose or, specifically in sodium deprived rats, to salty solutions (Berridge et al. 2009). Additional measures (although more technically challenging, Planchez et al. 2019) can include assessing the magnitude of neural responses to a reward *via* EEG or neural imaging techniques (e.g. Milyavskaya et al. 2019).

How is reward reactivity related to pre-existing moods?

Whilst the valid assessment of affective valence is a key focus of most chapters in this book, there is substantial evidence, as mentioned earlier, that reward reactivity shows poor selectivity in this regard. Indeed, reactivity to reward can either increase or decrease in humans during pre-existing states of negative mood (Figure 9.2). For instance, reduced reward reactivity (e.g. reporting reduced experience of positive feelings, loss of preference) can be observed in people with depression (American Psychiatric Association APA 2013), chronic pain (Becker et al. 2012; Trøstheim et al. 2020), sickness (De La Garza 2005), post-traumatic stress disorders (Vinograd et al. 2021; APA 2013), burnout (Bianchi et al. 2015) and substance use disorders (Trøstheim et al. 2020). Conversely, reactivity to reward can increase with boredom (Milyavskaya et al. 2019), and with exposure to adverse life events that may trigger drug-taking relapse (Weiss 2005; Shalev et al. 2010) and 'comfort eating' of high-fat, high-sugar or high-calorie food (e.g. Klatzkin et al. 2019; Gemesi et al. 2022). Therefore, this inverted C-shaped relationship between reward reactivity and valence makes its use as a welfare indicator risky when taking a dimensional view of affect (see Chapter 2). However, assessing reactivity to reward is crucial for certain <u>specific discrete</u> affective states, like depression and boredom, where reward reactivity is part of the operationalisation of the states. We discuss later how to assess each in a way that best avoids false nulls arising from poor responsiveness of the indicator (e.g. Type II errors) and false leads caused by poor selectivity (e.g. Type I errors).

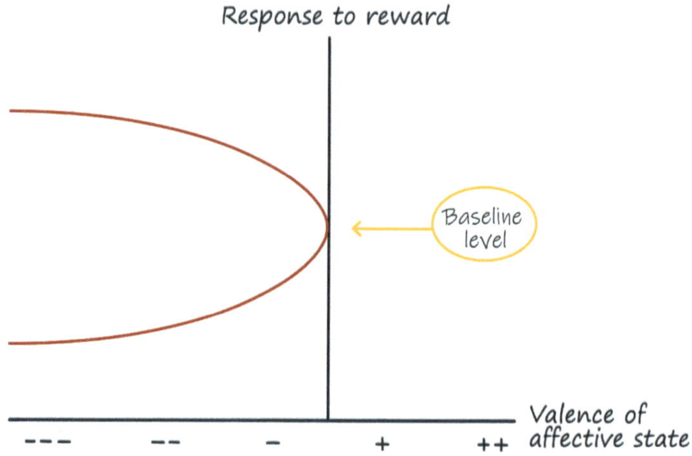

Figure 9.2 The inverted C-shaped relationship between reactivity to reward and negative mood when considering a dimensional (valence-arousal) view of affect. Reward reactivity may increase or decrease depending on the individual, or on the particular circumstances at stake, making its use as an indicator of affective state difficult. Note that the baseline position on the *y*-axis will differ between individuals.

Assessing reactivity to reward when focusing on specific discrete affective states: The example of depression

In humans, clinical depression[2] is a prevalent debilitating mood disorder, diagnosed by the co-occurrence of affective, cognitive, behavioural and homeostasis symptoms (World Health Organisation [WHO] 2022; APA 2013). Adverse life events commonly trigger depression, especially in people with predisposing genetic factors, and/or who were exposed to early-life aversive experiences (e.g. Anderson et al. 2022; Kendall et al. 2021; Nelson et al. 2017). Animals can be exposed to situations similar to those triggering depression in humans and can show depression-like symptoms (MacLellan et al. 2021). Considering the likely negative welfare implications of depression-like states for affected animals, it is thus important to be able to identify such states. Crucially, reduced reactivity to reward is one of the core criteria often used for operationalizing depression-like states (e.g. Fureix et al. 2015), as we now discuss.

Reduced reward reactivity in depression – construct validation

In the subsequent sections, we will apply three of the five tests listed in Chapter 2 used for construct validation of indicators of affective states. As a reminder, these are tests to assess to what extent an indicator accurately tracks self-reported affective states in humans (**Test 1**); reliably changes in animals (i) exposed to stimuli that they prefer/avoid (**Test 2**), (ii) exposed to ancestrally fitness-boosting or fitness-harming situations (**Test 3**) or (iii) given affect-modulating drugs (**Test 4**) and/or reliably covaries with pre-validated affect indicators (**Test 5**). We will focus here on Tests 1, 2, and 4.

Reduced reactivity to reward tracks self-reported affective states in depressed humans (Test 1).

Loss of pleasure (also termed anhedonia) is core to human clinical depression diagnosis (WHO 2022; APA 2013). Indeed, according to *Diagnostic Manual of Mental Disorders,* 5th *edition* (*DSM-5*), at least one of the diagnostic symptoms must be either depressed mood or loss of interest or pleasure (APA 2013). Anhedonia can predict greater severity of depression (but see Indicator Responsiveness section), while not all other symptoms do (Pelizza and Ferrari 2009; Lin et al. 2022; Zimmerman et al. 2018; Faravelli et al. 1996; Tolentino and Schmidt 2018; Fawcett et al. 1983). Anhedonia can also characterise specific subtypes of the illness. For instance, a near-complete absence of the capacity for pleasure (not simply a diminution) is one of the characteristics of the melancholic subtype of clinical depression (e.g. Planchez et al. 2019). Furthermore, inability to anticipate pleasure is also useful in distinguishing clinical depression from grieving response to a significant loss, which may resemble a depressive episode (APA 2013).

Reduced reactivity to reward is observed in animals subjected to non-preferred conditions (Test 2)

Laboratory mice and rats housed in non-preferred conditions, such as single-housing or barren cages, show reduced preference for sucrose compared to their counterparts housed in

[2] By which we mean major depressive disorder in the Diagnostic Manual of Mental Disorders – 5th edition DSM-5 (APA 2013), or depressive episodes in the International Statistical Classification of Diseases and Related Health Problems ICD-11 (WHO 2022).

preferred conditions (Mileva and Bielajew 2015; Fernández-Teruel et al. 2002; Solberg et al. 1999).[3] Furthermore, a classical biomedical procedure used to induce depression-like features in rodents, the 'chronic unpredictable stress' (CUS) paradigm, reduces reward reactivity in rats and mice. CUS involves exposing the animals to various aversive so-called 'micro'-events, usually once or twice daily for 3–4 hours over several weeks (Antoniuk et al. 2019; Willner 2017a; 2017b). Aversive events may include tilting the cages, unexpectedly modifying light cycles, wetting cages bedding, etc.[4] (Willner 2017b). CUS induces signs of poor welfare, such as negative judgment bias (Chapter 7, Harding et al. 2004) and chromodacryorrhea (ocular porphyrin 'red tears', Young et al. 2024) in rats. CUS robustly decreases rats' and mice's intake and/or preference for sucrose, even controlling for weight changes when observed, and although not necessarily affecting overall food intake (discussed in e.g. Planchez et al. 2019; Willner 1997, 2017a, 2017b). CUS-exposed rodents also show an impaired conditioned preference for places associated to food and amphetamine or morphine reinforcers (Papp et al. 1991,1992; Willner 2017a; Antoniuk et al. 2019; Wang et al. 2017). CUS-exposed male rats also show reduced mounting sexual behaviour compared to non-CUS exposed control rats (D'Aquila et al. 1994).

Outside of laboratory environments, aversive husbandry procedures can also induce reduced reactivity to sucrose. For instance, an atypical (flat neck) posture has been evidenced in certain riding school horses kept under restricted, non-preferred conditions (Fureix et al. 2015). That atypical posture co-occurs with decreased reactivity to environmental stimulation, and lower level of cortisol (compared to control individuals not displaying the posture), which can occur in depressed people (discussed in e.g. Fureix et al. 2012). When tested in standardised sucrose intake tests, horses displaying this atypical posture the most also consumed the least sucrose, as predicted if affected by depression-like states (Fureix et al. 2015).

Antidepressant drugs restore response to reward (Test 4)

In rodents, a CUS-induced reduction in reward reactivity overall shows specific and selective responsiveness to antidepressant treatments (reviewed in Willner 2017a). Indeed, sucrose consumption is restored after chronic (but not acute) administration of a wide range of established antidepressants (e.g. Venlafaxine, Escitalopram, Imipramine), but not by other drugs that are clinically ineffective in humans as antidepressants. Moreover, experimental antidepressant strategies that shows encouraging results in clinical investigations (e.g. repetitive transcranial magnetic stimulation, deep brain stimulation in depressed people) appear efficient at normalising rodents' CUS hedonic behaviour (e.g. Willner 2017a).

Problems with reduced reactivity to reward as a responsive measure of depression

Clinically, depression is diagnosed by the co-occurrence of several symptoms, of which at least one must be *either* depressed mood or loss of interest or pleasure (APA 2013). This means that some people can be diagnosed with depression without showing anhedonia. Moreover, when

[3] But this effect is not shown in all studies (e.g. Sparling et al. 2020, MacLellan et al. 2021).

[4] To our knowledge, no formal preference / avoidance studies have been conducted that aim to demonstrate that rodents would specifically avoid the micro-events used in practice in CUS. These micro-events are nonetheless likely to induce physical or social discomfort. It thus seems reasonable to predict that rodents would avoid exposures to these micro-events if they were provided with the option to do so.

defining the severity of clinical depression in the *DSM-5*, all symptoms are weighted equally (e.g. Zimmerman et al. 2018). Indeed, *DSM-5* depression severity is based on the number of symptoms, the level of distress caused by their intensity and the degree of social and occupational impairment of functioning (*DSM-5*). In other words, a patient can be clinically diagnosed with either mild, moderate or severe depression, with or without exhibiting anhedonia in each case. Furthermore, anhedonia is defined in the *DSM-5* as a '*markedly diminished interest or pleasure in* all, *or* almost all, *activities most of the day, nearly every day*' (APA 2013, p. 160; our emphasis underlined). This suggests that among depressed people who exhibit anhedonia, pleasure reactivity is already near zero ('diminished' for all previously pleasurable activities), and thus anhedonia may not be sensitive to differences in degree of depression. However, in depression research, continuous scales exist that measure severity of either depression (e.g. Hamilton Depression Rating Scale 'HDRS', Clinical Global Impression Severity scale) or anhedonia (e.g. Physical Anhedonia Scale, Social Anhedonia Scale). When using such continuous measurement tools, anhedonia generally positively correlates with, or predicts, greater severity of depression (Pelizza and Ferrari 2009; Lin et al. 2022; Zimmerman et al. 2018; Faravelli et al. 1996; Tolentino and Schmidt 2018; Fawcett et al. 1983; but see Schrader 1997 reporting an absence of correlation between anhedonia and severity of depression). Note that the observed relationship between depression severity and anhedonia is unlikely to be due to depression severity scales and anhedonia scales measuring the same things, since only a few items in the scales assessing the severity of depression specifically measure presence of anhedonia (e.g. only one out of the 21 HDRS items relate to loss of interest). This demonstrates some responsiveness of anhedonia to the severity of depression in humans, but only when assessing the phenomena with more sensitive methodologies.

Less research has addressed this question in animals. Nevertheless, in laboratory mice and rats, variations in CUS intensity sensitively predict reduction in sucrose intake. Indeed, more intense CUS regimes decrease sucrose intake more (Strekalova and Steinbusch 2010; Muscat and Willner 1992). In these studies, CUS intensity was defined by features such as the duration of exposure to aversive events, number of events and/or the nature of individual aversive events (e.g. exposing the animals to predator cues, to social challenges, applying events directly within the cage or more distant to the cage). Reduced reactivity to reward hence appears to be a responsive measure of putative depression-like states in animals, although more research is needed in animals before being able to rely firmly on reward reactivity as an indicator of the severity of depression-like states.

Problems with reduced reward reactivity as a selective measure of depression

Anhedonia is a core symptom of depression but does not occur selectively in depression alone. For instance, anhedonia can be present in other negative discrete affective states, such as chronic pain (e.g. Trøstheim et al. 2020). There are known interactions between pain and reward processing, in both humans and animals (Liu et al. 2019), unsurprisingly given that they affect some of the same brain structures and neurotransmitter systems (i.e. dopamine and opioids; Becker et al. 2012). Therefore, a reduction in reward reactivity alone cannot distinguish between chronic pain and depression. Further, some depressed people also show *increased* reactivity to reward on some measures, such as an increased motivation to approach preferred snack foods reward, although they have a decreased motivation to approach monetary and social rewards (Fussner et al. 2017). Depression with a seasonal pattern (previously called seasonal affective disorder, also known as winter depression), may also induce a craving

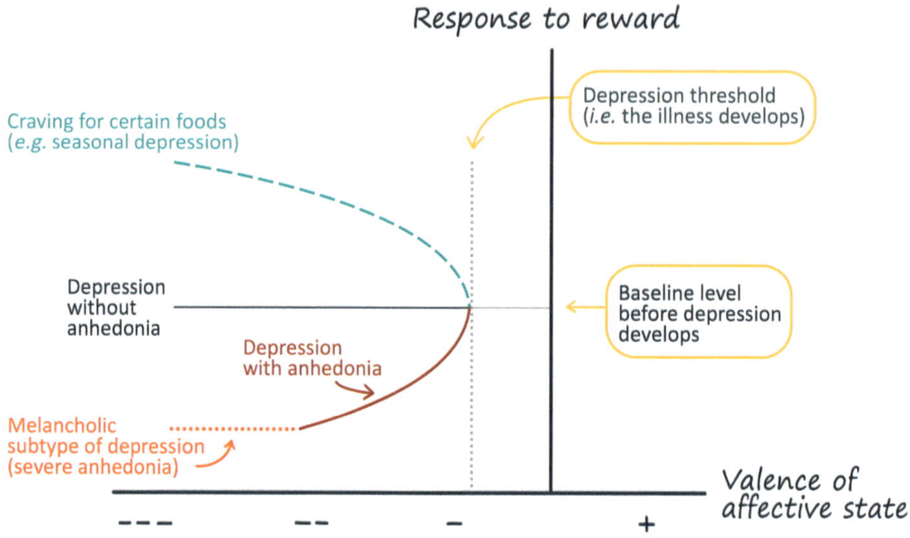

Figure 9.3 The relationship between reactivity to reward and depression. Note that the baseline position on the *y*-axis will differ between individuals. In clinically depressed humans with anhedonia (red, solid, thick line), reward reactivity may decrease as the severity of depression increases. In the melancholic subtype of depression (dotted orange line), general capacity for pleasure is near zero (floor effect). However, clinical depression, irrespective of its severity, can also be diagnosed in the absence of anhedonia. In such a case (dark grey thin line), reactivity to reward is expected to remain at its baseline level. Lastly, reactivity to reward can also increase for some measures as the illness develops (dashed turquoise line), although the exact slope of the curve representing this relationship (shown as near linear on the graph) remains to our knowledge to be investigated. For instance, higher intake of specific foods such as carbohydrates and sweets can be observed in seasonal depression.

for, and higher intake of, specific foods such as carbohydrates and sweets (Swiecicki et al. 2015; APA 2013). Figure 9.3 shows the relationship between reactivity to reward and depression. We now discuss how to deal with this lack of selectivity when measuring response to reward in putatively depressed animals.

Use in practice: reduced reactivity to reward as a measure of depression-like states in animals

Reduced reward reactivity is, as we have seen, not specific to depression. Therefore, when measuring reactivity to reward in putatively depressed animals, first it is crucial to concomitantly assess anhedonia alongside other depression-like symptoms (Wang et al. 2017; see also MacLellan et al. 2021 who deeply discuss how to operationalise the range of depression symptoms for application to animals, which is beyond the scope of this chapter). Second, one may also test, when possible, whether antidepressant drugs restore response to reward. If so, this provides further evidence that the observed reduced response to reward reflects a depression-like state.[5] Using a Test 4 approach with affect-modulating drugs can also help

[5] Although response to antidepressant is not sufficient to demonstrate that the animal is in a depression-like state, as discussed in e.g. Fureix et al. 2022.

with ruling out other potential causes of anhedonia. For instance, if animals in certain circumstances show reduced reactivity to reward *and* are at risk of being in pain (e.g. in disbudded calves, Lecorps et al. 2020), one could also measure whether administering painkillers reduces anhedonia. If anhedonia *is* reduced by painkiller, this suggests that pain triggers the symptom, or that the two states are comorbid.[6] If anhedonia is *not* reduced by painkillers, this suggests that anhedonia reflects depression-like states.

Moreover, as already outlined, depressed mood may also enhance (instead of reducing) reactivity to sweets. Thus, reactivity to several rewards should ideally also be included to assess anhedonia (see also Box 9.1). Candidate rewarding stimuli could be preferred social partners (Ferdowsian et al. 2011) and cues from conspecifics signalling opportunities for mating (Malkesman et al. 2010). Social withdrawal and reduced libido may indeed occur in depressed people (WHO 2022; APA 2013). Moreover, captive primates kept in suboptimal conditions (e.g. confined, isolated) sometimes display an unusual, hunched posture (MacLellan et al. 2021). This hunched posture recalls psychomotor retardation in some depressed humans (Buyukdura et al. 2011; Burton et al. 2012), can be triggered by early-life aversive experiences (as human depression can be), may co-occur with biomarkers of depression (e.g. smaller hippocampal volumes; see Chapter 13) and can be reduced by antidepressant drugs (reviewed in MacLellan et al. 2021). Interestingly, such hunched postures have also been linked with reduced sexual and parental behaviour (Li et al. 2013). Social stimuli thus appear to be sensible candidates to test anhedonia in research on depression-like states, although more data are needed first to demonstrate their rewarding properties.

Lastly, anhedonia may relate to reduced liking ('consummatory' anhedonia) and/or wanting ('motivational' anhedonia). Anhedonia has mostly been studied as consummatory in both humans (Treadway and Zald 2011) and animals (MacLellan et al. 2021), but clinically depressed people can also be affected by motivational anhedonia (e.g. Rizvi et al. 2016, Slaney et al. 2022). Therefore, future work can explore incorporating paradigms or measures that allow measurement of both of these components where possible.

Reward reactivity can also be affected by other factors, including non-affective ones. For instance, individuals vary in their general appetite levels. That variation may confound the consumption of sucrose, irrespectively of sucrose hedonic properties. In other words, individuals with smaller general appetites may consume less sucrose during food-based anhedonia testing than those with greater appetites, while all may be perceiving sucrose as equally pleasurable. For instance, in the aforementioned study in which riding school horses displaying an atypical posture were tested for signs of anhedonia, lower hay consumption levels (used as a proxy for general appetite levels) predict lower sucrose intakes overall during anhedonia testing (Fureix et al. 2015). In calves, disbudding reduces sucrose intake, but pre-disbudding *baseline* sucrose intakes also strongly predict sucrose intakes following disbudding (Lecorps et al. 2020). Such non-mood-related individual differences in sucrose intake may thus, if not controlled for, confound mood-related anhedonia results. When using sucrose to measure reactivity to reward, we therefore strongly recommend to control for general appetite levels, e.g. by measuring *preference for* rather than merely *consumption of* sucrose (i.e. by exposing the animal to one bottle filled with sweetened solution and one bottle filled with water, and express sucrose intake as a ratio), or by adding a relevant baseline measure of general appetite/water intake as co-variate in the statistical model.

[6] Comorbid states will be suspected if the animal also presents with other depression-like symptoms, and/or antidepressants also reduce anhedonia.

Another aspect that can influence reactivity to reward is its novelty. Indeed, a neophobic reaction to a novel reward may induce a reduced reward consumption when consumption is assessed over a restricted period of time, due to delayed latency to start consuming. For instance, in horses tested for anhedonia with sugar blocks that were new to them, higher latency to eat their normal pellets scented with novel aromatics (i.e. measuring neophobia towards novel food) predicts lower sucrose intake over the 30 hours of exposure to the sugar blocks (Fureix et al. 2015). Similarly, in mice (which usually prefer, and are keen to pay a price to get access to, enriched cages), the most neophobic animals in novel object tests and when first exposed to sugar also used enriched cages least (Walker and Mason 2011). A reduced reward consumption in neophobic animals may therefore be wrongly interpreted as a sign of anhedonia. Controlling for neophobic reactions is thus crucial, and may include habituating the animals to the rewards and to the testing environment prior to testing and/or adding relevant measures in the statistical model. However, while the novelty of the reward can be a confounding factor here in depression research, it is a key element when operationalising some other affective states, such as boredom, which we discuss in the next section.

Assessing reactivity to reward when focusing on specific discrete affective states: The example of boredom

Boredom has been defined as 'the aversive experience of wanting, but being unable, to engage in satisfying activity' (Eastwood et al. 2012). In humans, it can have a range of negative consequences for physical and mental health, and it can reasonably be expected that some of these might also be experienced by other animals (Burn 2017; Meagher 2019).

Increased reactivity to reward in boredom – construct validation

As described previously in the section on depression, we will apply three of the five tests listed in Chapter 2 used for construct validation of indicators of affective states. In this case, we assess increased reactivity to reward as an indicator of boredom and focus on Tests 1–3.

Increased reactivity to reward tracks self-reported affective states in bored humans (Test 1)

Human research suggests that boredom results in increased motivation for novel or varying stimuli; these stimuli can be rewarding but also range from this to mildly aversive or frightening (reviewed by Eastwood et al. 2012; Bench and Lench 2019). This motivation for arousing stimuli, which can include any stimuli that require more cognitive processing, is otherwise known as sensation-seeking motivation (Zuckerman 1994). Although most studies have thus investigated general sensation-seeking and not reward-seeking specifically, it has been suggested that sensation-seekers are especially attracted to rewarding stimuli (Barbee et al. 1987; Norbury and Husain 2015). Indeed, novelty itself could be considered to be rewarding in the context of boredom, given this motivation to seek it even though there is no clear biological need to be addressed. Furthermore, there is evidence that novelty can activate dopamine systems in the midbrain that also respond to reward (e.g. the ventral tegmental area), at least in relatively exploratory species (e.g. humans: Wittmann et al. 2007; rodents: reviewed in Tapper and Molas 2020).

Boredom in these human studies is typically identified based on some form of self-report (Eastwood et al. 2012). A few studies have directly linked reactivity to rewards with self-report of current boredom. Havermans et al. (2015), for example, induced boredom through a monotonous film and confirmed this through people's ratings of how they felt before and after watching. Participants in the treatment subjected to this boring film ate more chocolate candies on average than those who watched an affectively neutral film clip. Similar results were obtained in a second study where a boredom-inducing treatment increased eating of sweets which were chosen over a healthier food option (Moynihan et al. 2015). These authors also found that self-reported experiences of boredom over the course of a week predicted food consumption including fat and carbohydrates, further suggesting that boredom can increase motivation for foods typically considered enjoyable and not necessarily because of nutritional needs. These studies mirror work looking retrospectively at people's overall reactivity to reward: self-reported reward reactivity (measured by 'Sensitivity to Punishment and Sensitivity to Reward' – questionnaires) correlates at an individual level with self-reported 'emotional eating' and expectation that eating will alleviate boredom (Hennegan et al. 2013).

Increased reactivity to rewards and novel stimuli is observed in animals subjected to conditions predicted to induce boredom (Tests 2 and 3)

Humans typically report feeling bored in monotonous, constrained situations, or otherwise when lacking in things to do that engage the person's attention (see e.g. Toohey 2011; Eastwood et al. 2012). Animal work on boredom has made use of these types of condition: being confined to non-enriched or minimally enriched housing, compared to an enriched condition. This began with studies in mink, where the enriched housing condition that incorporated some change of enrichment over time had previously been demonstrated to be preferred over the 'boring', monotonous housing (Meagher and Mason 2012), thus providing the preferred and avoided conditions needed to address Test 2 (Chapter 2). The literature on exploratory behaviour suggests that this behaviour of seeking and approaching novel stimuli or environments increases fitness in the long-term through gathering of useful information that can help the animal to exploit future opportunities (Geana et al. 2016). Environments that allow for this type of behaviour could thus potentially be considered ancestrally fitness-boosting conditions, making this also relevant to Test 3 (Chapter 2).

Two studies of these conditions suggest that boredom-like states will increase reward reactivity. First, Meagher and Mason (2012) showed that compared to animals in enriched housing, animals in 'boring', non-enriched housing ate more food treats, and for one of two 'rewards' that did not involve consumption of food (specifically, stimuli associated with prey or hunting behaviour, judged to be rewarding since they elicit approach responses), exhibited higher exploratory responses (assessed via time oriented to and in contact with the stimulus). Evidence from parrots also supports the hypothesis that long-term housing in barren environments increases interest in novelty, as indicated by longer durations spent exploring new objects (Meehan and Mench 2002). Again, if novelty in general is considered rewarding, at least for neophilic species such as parrots, this is consistent with the birds having increased reward reactivity linked to boredom in these housing conditions.

Problems with increased reactivity to reward as a sensitive measure of boredom

While the limited animal research conducted so far has consistently found increases in reactions to novel stimuli in conditions hypothesised to induce boredom, this effect has not always been apparent if looking only at stimuli that would typically be expected to be rewarding for the species. Although the first mink study discussed earlier showed such an effect, this

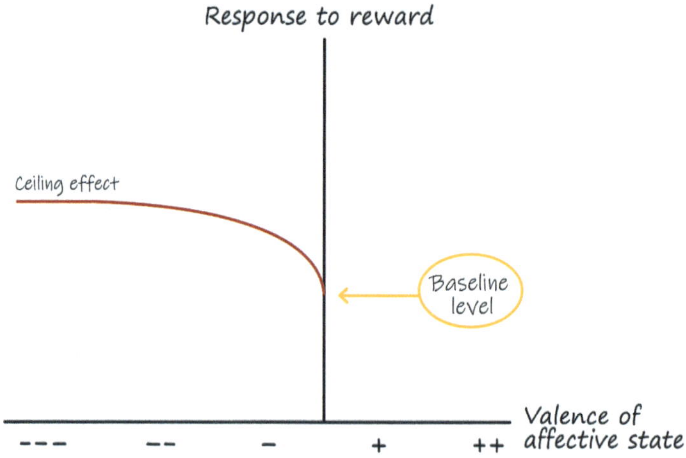

Figure 9.4 The relationship between reactivity to reward and boredom. Note that again, the baseline position on the *y*-axis will differ between individuals. As boredom develops (thus a negative emotional valence), reactivity is predicted to increase. However, evidence from non-human animals to date suggests this effect may be relatively small and prone to ceiling effects.

effect did not hold up when data were pooled across three studies (Polanco et al. 2021). Similarly with ferrets, Burn et al. (2020) reported differences in reactions to three aversive or ambiguous stimuli but not those expected to be rewarding (mouse bedding, because of its ecological significance as a prey-related cue, and a ball as a preferred toy that elicits play). This less consistent difference in reactivity to stimuli classified as rewards for the species (see Introduction criteria to define a reward), compared to novel ambiguous or aversive stimuli, is likely due to ceiling effects (i.e. all values close to the maximum; Meagher et al. 2017; see Figure 9.4).

In addition to this limitation to sensitivity that seems to affect all measures used if the stimuli are very rewarding, not all measures of response have been equally successful in detecting an effect of 'boring' housing. For instance in mink, latency to approach stimuli often did not show significant housing effects, even with novel stimuli not traditionally considered 'rewards' (or shown to meet all of the criteria for identifying rewards earlier), whereas duration of orientation towards and exploratory contact with stimuli did increase in 'boring' housing (Polanco et al. 2021). A further example of how the measure chosen may influence whether an effect of 'boring' housing can be detected comes from using anticipatory behaviour. In rats, increased behavioural transitions during the period before a predicted reward is delivered are used as an indicator of anticipation and have been interpreted as evidence of increased responsiveness to rewards (e.g. van der Harst and Spruijt 2007, but see Anderson et al. 2020 for a critical discussion of this interpretation). This increase in behavioural transitions in anticipation of a sweet food reward appears to be higher in rats housed in standard laboratory conditions, which would be hypothesised to induce boredom, than in more naturalistic housing (van der Harst et al. 2003; Makowska and Weary 2016). However, these studies were not explicitly investigating boredom and thus did not measure other indicators to validate the anticipatory behaviour as an indicator of this state. Furthermore, similar work in mink (also not explicitly investigating boredom) has in contrast *not* found such differences in anticipatory behaviour between housing conditions (Vinke et al. 2004, 2006) – again potentially reflecting poor sensitivity (responsiveness) of the measure to boredom, at least in this species or when the reward used is too exciting to the species.

Finally, conflicting effects of other affective states experienced at or near the same time might reduce the extent to which reactivity measures detect boredom, although there is less direct evidence of this issue to date. For example, animals reared in barren environments often show increased fear of novelty (e.g. Jones 1982), and it has long been recognised that fear typically reduces or at least slows exploration (see e.g. Russell 1973). With the parrot example mentioned earlier (Meehan and Mench 2002), this could be seen through increased latencies to explore the novel objects in the barren-housed birds, despite this group eventually showing elevated overall time exploring the stimuli. However, novelty is important in testing for boredom. With short tests, fear-induced hesitation could prevent such housing effects on overall motivation from being detected. As discussed earlier, depression can also reduce reactivity to rewards, thus having the opposite predicted effect on exploration of rewards or novel stimuli. However, boredom and depression may sometimes co-occur (e.g. people's proneness to boredom can correlate with depressive symptoms; Farmer and Sundberg 1986). Co-occurrence of two qualitatively different negative affective states, or short-term interruption of one by another (fear), may interfere with assessing specific states *via* reward responses.

Problems with increased reactivity to reward as a specific measure of boredom

Although increased motivation for reward (and other stimuli) is seen in boredom, it can also result from other negative states. As mentioned earlier, increased reactivity to certain rewards can be observed in depression with seasonal pattern. Chronic exposure to aversive events or conditions predominantly blunts reward reactivity, producing anhedonia, but *acute* stress can increase motivation for rewards through activity in specific dopaminergic pathways (reviewed by Baik 2020), e.g. triggering non-boredom-related drug-taking relapse or 'comfort eating' (see Introduction). Moreover, early-life challenges can influence development of these pathways and increase motivation for putative rewards such as ethanol in rodents (reviewed by Hanson et al. 2021). The aversive events used in such studies, such as maternal separation in rodents' early lives, are not associated with boredom. Thus, some paradigms for assessing reward reactivity might detect increased reactivity to rewards in the absence of boredom (although potentially still reflecting negative affect).

Use in practice: increased reactivity to reward as a measure of boredom

Clearly, increased reward reactivity is an indicator that can provide supportive evidence of boredom but must be used in combination with other indicators, including increased preference for or exploration of novel, typically aversive, stimuli. For instance, if an animal displays anticipatory behaviour prior to the regular meal, one will need to investigate whether this anticipatory behaviour also predicts greater motivation to interact with a broad range of stimuli (including novel and slightly aversive objects) before interpreting this anticipatory behaviour itself as a sign of boredom. Currently, this is an emerging area of research and thus such work has been conducted in a limited range of species. To be applied in a new species, any measures must be validated through use of well-understood treatments consistent with known causes of boredom from the human literature and known to affect the welfare of the species. Contexts that typically trigger boredom in humans can help distinguish boredom from other states that might increase reward reactivity. As research in this field develops, it is hoped that other boredom indicators will be established so that correlations among them can help with validation.

There are a number of areas for future validation work to explore. One is the testing of additional measures of reactivity to reward for use in this context. For example, using an EEG measure called feedback negativity[7], Milyavskaya et al. (2019) showed that people assigned to a boring task showed a higher magnitude of response to both a preferred food (candy) and non-food (monetary) reward than did control participants who conducted no prior task. Such EEG measures have not yet been used in animals to assess changes in reward reactivity as related to emotional states. However, feedback-related negativity has also been measured in non-human primates and rodents (Warren et al. 2015), indicating feasibility of use in other species. In general, human work specifically linking boredom with neurobiological measures of reward reactivity is limited, so currently there can be less reliance on homologies in neurobiological responses to infer boredom in animals (but see links between reactivity to psychostimulant rewards and novelty or sensation seeking at the trait level, e.g. Arenas et al. 2016). However, with the increased interest in boredom in recent years leading to an increasing evidence base, the use of neurobiological markers of boredom may become possible. Other physiological measures might also become better established for humans and thus useful in other species. For example, a proposed 'psychophysiological signature' of boredom in humans of increasing heart rate, low skin conductance and high cortisol has been identified (Merrifield and Danckert 2014). Although these physiological measures are likely not specific enough to be used as evidence of boredom on their own since linked to general arousal, future work could test whether these co-vary with reward reactivity to provide corroborating evidence.

Pharmacological validation in animals is a promising avenue for future validation of reward reactivity measures of boredom. Methylphenidate, widely used to treat attention-deficit/hyperactivity disorder (ADHD), is thought to act, at least in part, through blocking reuptake of dopamine and thus increasing its levels in areas of the brain relevant to reward reactivity (Furukawa et al. 2020; Jaeschke et al. 2021). Although its effects on boredom have not been a major focus, some studies report that both ADHD patients (Golubchik et al. 2021) and healthy volunteers doing a learning task (Klinge et al. 2018) experience less boredom when treated with methylphenidate. This further points to a direct link between reward reactivity and boredom. In animals, there has been no attempt as of yet to validate reward reactivity as a measure of boredom using methylphenidate or other drugs.

Some refinements can be recommended to the existing measures of reward reactivity in boredom (orientation to and direct exploration of relevant stimuli). One is to continue using a range of potential rewards of different intensity but look more explicitly at within-individual variation in reactivity to these different stimuli. Franks et al. (2023) proposed that 'curiosity', which has been linked to positive well-being in humans (e.g. Kashdan and Steger 2007), will lead to targeted but sustained interest in specific stimuli the animal chooses to explore, whereas showing uniform exploration of all stimuli would be more consistent with negative states such as boredom. This focus on sustained interest towards specific stimuli in curiosity but not boredom also suggests that relatively long test durations might be useful. It would make detection of differential responses possible across stimuli, allowing researchers to distinguish the different states and identify more specific measures. Such increased test durations (as long as they do not physically fatigue the animals) might also help to reduce ceiling effects that can lead to failure to detect an effect of 'boring' conditions, reducing the responsiveness problems associated with measures of reward reactivity. An additional way to help identify

[7] Feedback negativity is a component of the brain's immediate response to a stimulus, which is known to co-vary with approach motivation and self-reported interest in rewards.

true differences in reactivity to rewards might be to include rewards that are only moderately preferred, such that individuals who are not experiencing boredom are more likely to stop responding before they reach physical limits and therefore ceiling effects do not occur. Lastly, including rewards that are not truly novel might help to prevent transient fear counteracting housing effects on motivation for the stimuli, as also recommended for assessing depression earlier. While novel stimuli are typically the focus of boredom investigations, novelty does not seem to be necessary for stimuli that are rewarding in other ways.

Conclusions

Reactivity to reward is influenced by a range of moods, making them of interest for assessing animal affect. Since negatively-valenced mood states are associated with both increases and decreases in reactivity to reward, these measures cannot be used to reliably indicate affective valence. However, we have presented evidence that they are useful, and indeed sometimes key, indicators of certain discrete affective states, including depression and boredom, for which validations of reward reactivity measures have been published. We encourage similar work in relation to other discrete emotions that are hypothesised to influence motivation for reward. In conjunction with other measures, indicators of reward reactivity may thus be useful markers of specific discrete affective states.

References

American Psychiatric Association (2013) *Diagnostic and Statistical Manual of Mental Disorders*, fifth ed. American Psychiatric Association, Arlington, VA.

Amsterdam, J.D., Settle, R.G., Doty, R.L., Abelman, E. and Winokur, A. (1987) *Taste and smell perception in depression. Biological Psychiatry*, 22, 1481–1485. https://doi.org/10.1016/0006-3223(87)90108-9

Anderson, C., von Keyserlingk, M.A.G., Lidfors, L.M. and Weary, D.M. (2020) *Anticipatory behaviour in animals: A critical review. Animal Welfare*, 29, 231–238. https://doi.org/10.7120/09627286.29.3.231

Anderson, L.R., Monden, C.W.S. and Bukodi, E. (2022) *Stressful Life Events, Differential Vulnerability, and Depressive Symptoms: Critique and New Evidence. Journal of Health and Social Behavior*, 63(2), 283–300. https://doi.org/10.1177/00221465211055993

Antoniuk, S., Bijata, M., Ponimaskin, E. and Wlodarczyk, J. (2019) *Chronic unpredictable mild stress for modeling depression in rodents: Meta-analysis of model reliability. Neuroscience & Biobehavioral Reviews*, 99, 101–116. https://doi.org/10.1016/j.neubiorev.2018.12.002

Arenas, M.C., Aguilar, M.A., Montagud-Romero, S., Mateos-García, A., Navarro-Francés, C.I., Miñarro, J. and Rodríguez-Arias, M. (2016) *Influence of the novelty-seeking endophenotype on the rewarding effects of psychostimulant drugs in animal models. Current Neuropharmacology*, 14, 87–100. https://doi.org/10.2174/1570159x13666150921112841

Baik, J.-H. (2020) *Stress and the dopaminergic reward system. Experimental & Molecular Medicine*, 52(12), 1879–1890. https://doi.org/10.1038/s12276-020-00532-4

Barbee, A., Hammock, G. and Richardson, D. (1987) *Are sensation seekers attracted to intense or positive stimuli? The importance of "sensation" to research on sensation seeking. Personality and Individual Differences*, 8, 441–443. https://doi.org/10.1016/0191-8869(87)90048-1

Barenz, J., O'Donnell, M. and Smith, J. (2024) The endogenous opioid system as a pathway of positive emotions. In Kerr, P. L., Sirbu, C., Gregg, J. M. (eds) *Endogenous Opioids. Advances in Neurobiology*, vol 35. Springer, Cham. https://doi.org/10.1007/978-3-031-45493-6_12

Becker, S., Gandhi, W. and Schweinhardt, P. (2012) *Cerebral interactions of pain and reward and their relevance for chronic pain. Neuroscience Letters*, 520, 182–187. https://doi.org/10.1016/j.neulet.2012.03.013

Bedwell, J.S., Spencer, C.C., Chirino, C.A. and O'Donnell, J.P. (2019) *The sweet taste test: Relationships with anhedonia subtypes, personality traits, and menstrual cycle phases. Journal of Psychopathology and Behavioral Assessment*, 41(2), 235–248. https://doi.org/10.1007/s10862-019-09717-2

Bench, S.W. and Lench, H.C. (2019) *Boredom as a seeking state: Boredom prompts the pursuit of novel (even negative) experiences. Emotion*, 19, 242–254. https://psycnet.apa.org/doi/10.1037/emo0000433

Berlin, I., Givry-Steiner, L., Lecrubier, Y. and Puech, A.J. (1998) *Measures of anhedonia and hedonic responses to sucrose in depressive and schizophrenic patients in comparison with healthy subjects. European Psychiatry*, 13(6), 303–309. https://doi.org/10.1016/S0924-9338(98)80048-5

Berridge, K.C. (2009) *'Liking' and 'wanting' food rewards: Brain substrates and roles in eating disorders. Physiology & Behavior*, 97(5), 537–550. https://doi.org/10.1016/j.physbeh.2009.02.044

Berridge, K.C., Robinson, T.E. and Aldridge, J.W. (2009) *Dissecting components of reward: 'liking', 'wanting', and learning. Current Opinion in Pharmacology*, 9, 65–73. https://doi.org/10.1016/j.coph.2008.12.014

Bianchi, R., Schonfeld, I.S. and Laurent, E. (2015) *Burnout–depression overlap: A review. Clinical Psychology Review*, 36, 28–41. https://doi.org/10.1016/j.cpr.2015.01.004

Burman, O.H.P., Parker, R.M.A., Paul, E.S. and Mendl, M. (2008) *Sensitivity to reward loss as an indicator of animal emotion and welfare. Biology Letters*, 4(4), 330. https://doi.org/10.1098/rsbl.2008.0113–333.

Burn, C.C. (2017) *Bestial boredom: A biological perspective on animal boredom and suggestions for its scientific investigation. Animal Behaviour*, 130, 141–151. https://doi.org/10.1016/j.anbehav.2017.06.006

Burn, C.C., Raffle, J. and Bizley, J.K. (2020) *Does "playtime" reduce stimulus-seeking and other boredom-like behaviour in laboratory ferrets? Animal Welfare*, 29, 19–26. https://doi.org/10.71.0/09627286.29.1.019

Burton, C., McKinstry, B., Szentagotai Tătar, A., Serrano-Blanco, A., Pagliari, C. and Wolters, M. (2012) *Activity monitoring in patients with depression: A systematic review. Journal of Affective Disorders*, 145(1), 21–28. https://doi.org/10.1016/j.jad.2012.07.001

Buyukdura, J.S., McClintock, S.M. and Croarkin, P.E. (2011) *Psychomotor retardation in depression: Biological underpinnings, measurement, and treatment. Progress in Neuro-Psychopharmacology & Biological Psychiatry*, 35(2), 395–409. https://doi.org/10.1016/j.pnpbp.2010.10.019

Clarkson, J., Dwyer, D., Flecknell, P., Leach, M. and Rowe, C. (2018) *Handling method alters the hedonic value of reward in laboratory mice. Scientific Reports*, 8, 2448. https://doi.org/10.1038/s41598-018-20716-3

D'Aquila, P.S., Brain, P. and Willner, P. (1994) *Effects of chronic mild stress on performance in behavioural tests relevant to anxiety and depression. Physiology & Behavior*, 56(5), 861–867. https://doi.org/10.1016/0031-9384(94)90316-6

De La Garza, R. (2005) *Endotoxin- or pro-inflammatory cytokine-induced sickness behavior as an animal model of depression: Focus on anhedonia. Neuroscience & Biobehavioral Reviews*, 29(4), 761–770. https://doi.org/10.1016/j.neubiorev.2005.03.016

Der-Avakian, A., Barnes, S.A., Markou, A. and Pizzagalli, D.A. (2016) *Translational Assessment of Reward and Motivational Deficits in Psychiatric Disorders. Current Topics in Behavioral Neurosciences*, 28, 231–262. https://doi.org/10.1007/7854_2015_5004

Dichter, G.S., Smoski, M.J., Kampov-Polevoy, A.B., Gallop, R. and Garbutt, J.C. (2010) *Unipolar depression does not moderate responses to the Sweet Taste Test. Depression and Anxiety*, 27(9), 859–863. https://doi.org/10.1002/da.20690

Dwyer, D. (2012) *Licking and liking: The assessment of hedonic responses in rodents. Quarterly Journal of Experimental Psychology*, 65(3), 371–394. https://doi.org/10.1080/17470218.2011.652969

Eastwood, J.D., Frischen, A., Fenske, M.J. and Smilek, D. (2012) *The Unengaged Mind. Perspectives on Psychological Science*, 7, 482–495. https://doi.org/10.1177/1745691612456044

Faravelli, C., Servi, P., Arends, J.A. and Strik, W.K. (1996) *Number of symptoms, quantification, and qualification of depression. Comprehensive Psychiatry*, 37(5), 307–315. https://doi.org/10.1016/S0010-440x(96)90011-5

Farmer, R. and Sundberg, N.D. (1986) *Boredom Proneness—The Development and Correlates of a New Scale. Journal of Personality Assessment*, 50(1), 4–17. https://doi.org/10.1207/s15327752jpa5001_2

Fawcett, J.C., Clark, D.C., Scheftner, W.A. and Gibbons, R.D. (1983) *Assessing anhedonia in psychiatric patients. The pleasure scale. Archives of General Psychiatry*, 40(1), 79–84. https://doi.org/10.1001/archpsyc.1983.01790010081010

Ferdowsian, H.R., Durham, D.L., Kimwele, C., Kranendonk, G., Otali, E., Akugizibwe, T., Mulcahy, J.B., Ajarova, L. and Johnson, C.M. (2011) *Signs of mood and anxiety disorders in chimpanzees. PLoS One*, 6(6), e19855. https://doi.org/10.1371/journal.pone.0019855

Fernández-Teruel, A., Driscoll, P., Gil, L., Aguilar, R., Tobeña, A. and Escorihuela, R.M. (2002) *Enduring effects of environmental enrichment on novelty seeking, saccharin and ethanol intake in two rat lines (RHA/Verh and RLA/Verh) differing in incentive-seeking behavior. Pharmacology Biochemistry and Behavior*, 73(1), 225–231. https://doi.org/10.1016/s0091-3057(02)00784-0

Franks, B., Gaffney, L.P., Graham, C. and Weary, D.M. (2023) *Curiosity in zebrafish (Danio rerio)? Behavioral responses to 30 novel objects. Frontiers in Veterinary Science*, 9, 1062420–1062420. https://doi.org/10.3389/fvets.2022.1062420

Fraser, D. and Duncan, I.J.H. (1998) *'Pleasures', 'pains' and animal welfare: Toward a natural history of affect. Animal Welfare*, 7(4), 383–396. https://doi.org/10.1017/s0962728600020935

Fureix, C., Beaulieu, C., Argaud, S., Rochais, C., Quinton, M., Henry, S., Hausberger, M. and Mason, G. (2015) *Investigating anhedonia in a non-conventional species: do some riding horses Equus caballus display symptoms of depression? Applied Animal Behaviour Science*, 162, 26–36. https://doi.org/10.1016/j.applanim.2014.11.007

Fureix, C., Jego, P., Henry, S., Lansade, L. and Hausberger, M. (2012) *Towards an Ethological Animal Model of Depression? A Study on Horses. PLoS One*, 7(6). https://doi.org/10.1371/journal.pone.0039280

Fureix, C., Trevarthen, A.C., Finnegan, E.M., Bučková, K., Paul, E.S. and Mendl, M.T. (2022) *Do greater levels of in-cage waking inactivity in laboratory mice reflect a spontaneous depression-like symptom? A pharmacological investigation. Pharmacology, Biochemistry, and Behavior*, 212, 173311–173311. https://doi.org/10.1016/j.pbb.2021.173311

Furukawa, E., da Costa, R.Q.M., Bado, P., Hoefle, S., Vigne, P., Monteiro, M., Wickens, J.R., Moll, J., Tripp, G. and Mattos, P. (2020) *Methylphenidate modifies reward cue responses in adults with ADHD: An fMRI study. Neuropharmacology*, 162, 107833. https://doi.org/10.1016/j.neuropharm.2019.107833

Fussner, L.M., Mancini, K.J. and Luebbe, A.M. (2017) *Depression and approach motivation: Differential relations to monetary, social, and food reward. Journal of Psychopathology and Behavioral Assessment*, 40(1), 117–129. https://doi.org/10.1007/s10862-017-9620-z

Geana, A., Wilson, R.C., Daw, N. and Cohen, J.D. (2016) Boredom, Information-Seeking and Exploration. *38th Annual Meeting of the Cognitive Science Society*, Philadelphia, Pennsylvania, August 10th-13th, 2016

Gemesi, K., Holzmann, S.L., Kaiser, B., Wintergerst, M., Lurz, M., Groh, G., Böhm, M., Krcmar, H., Gedrich, K., Hauner, H. and Holzapfel, C. (2022) *Stress eating: an online survey of eating behaviours, comfort foods, and healthy food substitutes in German adults. BMC Public Health*, 22(1), 391–391. https://doi.org/10.1186/s12889-022-12787-9

Golubchik, P., Schoen, G. and Weizman, A. (2021) *Mood and proneness to boredom are associated with poorer continuous performance test results, which may improve with methylphenidate treatment, in Attention-Deficit/Hyperactivity Disorder. Journal of Child and Adolescent Psychopharmacology*, 31, 310–314. https://doi.org/10.1089/cap.2021.0001

Hanson, J.L., Williams, A.V., Bangasser, D.A. and Peña, C.J. (2021) *Impact of early life stress on reward circuit function and regulation. Frontiers in Psychiatry*, 12, 744690. https://doi.org/10.3389/fpsyt.2021.744690

Harding, E., Paul, E. and Mendl, M. (2004) *Cognitive bias and affective state. Nature*, 427, 312. https://doi.org/10.1038/427312a

Havermans, R.C., Vancleef, L., Kalamatianos, A. and Nederkoorn, C. (2015) *Eating and inflicting pain out of boredom. Appetite*, 85, 52–57. https://doi.org/10.1016/j.appet.2014.11.007

Hennegan, J.M., Loxton, N.J. and Mattar, A. (2013) *Great expectations. Eating expectancies as mediators of reinforcement sensitivity and eating. Appetite*, 71, 81–88. https://doi.org/10.1016/j.appet.2013.07.013

Horvitz, J.C. (2000) *Mesolimbocortical and nigrostriatal dopamine responses to salient non-reward events. Neuroscience*, 96(4), 651–656. https://doi.org/10.1016/s0306-4522(00)00019-1

Iatridi, V., Hayes, J.E. and Yeomans, M.R. (2019) *Reconsidering the classification of sweet taste liker phenotypes: A methodological review. Food Quality and Preference*, 72, 56–76. https://doi.org/10.1016/j.foodqual.2018.09.001

Ironside, M., Kumar, P., Kang, M.S. and Pizzagalli, D.A. (2018) *Brain mechanisms mediating effects of stress on reward sensitivity. Current Opinion in Behavioral Sciences, Apathy and Motivation*, 22, 106–113. https://doi.org/10.1016/j.cobeha.2018.01.016

Jaeschke, R.R., Sujkowska, E. and Sowa-Kućma, M. (2021) *Methylphenidate for attention-deficit/hyperactivity disorder in adults: A narrative review. Psychopharmacology*, 238, 2667–2691. https://doi.org/10.1007/s00213-021-05946-0

Jones, R.B. (1982) *Effects of early environmental enrichment upon open-field behavior and timidity in the domestic chick. Developmental Psychobiology*, 15(2), 105–111. https://doi.org/10.1002/dev.420150203

Kashdan, T.B. and Steger, M.F. (2007) *Curiosity and pathways to well-being and meaning in life: Traits, states, and everyday behaviors. Motivation and Emotion*, 31(3), 159–173. https://doi.org/10.1007/s110.1-007-9068-7

Kendall, K.M., Van Assche, E., Andlauer, T.F.M., Choi, K.W., Luykx, J.J., Schulte, E.C. and Lu, Y. (2021) *The genetic basis of major depression. Psychological Medicine*, 51(13), 2217–2230. https://doi.org/10.1017/s0033291721000441

Klatzkin, R.R., Dasani, R., Warren, M., Cattaneo, C., Nadel, T., Nikodem, C. and Kissileff, H.R. (2019) *Negative affect is associated with increased stress-eating for women with high perceived life stress. Physiology & Behavior*, 210, 112639–112639. https://doi.org/10.1016/j.physbeh.2019.112639

Klinge, C., Shuttleworth, C., Muglia, P., Nobre, A.C., Harmer, C.J. and Murphy, S.E. (2018) *Methylphenidate enhances implicit learning in healthy adults. Journal of Psychopharmacology*, 32, 70–80. https://doi.org/10.1177/0269881117731472

Kubikova, L. and Košt'ál, L. (2010) *Dopaminergic system in birdsong learning and maintenance. Journal of Chemical Neuroanatomy*, 39, 112. https://doi.org/10.1016/j.jchemneu.2009.10.004–123.

Lecorps, B., Nogues, E., von Keyserlingk, M.A.G. and Weary, D.M. (2020) *Pessimistic dairy calves are more vulnerable to pain-induced anhedonia. PLoS One*, 15(11), e0242100. https://doi.org/10.1371/journal.pone.0242100

Li, X., Glaser, D., Li, W., Johnson, W.E., O'Brien, S.J., Beauchamp, G.K. and Brand, J.G. (2009) *Analyses of sweet receptor gene (Tas1r2) and preference for sweet stimuli in species of Carnivora. Journal of Heredity*, 100(suppl-1), S90–S100. https://doi.org/10.1093/jhered/esp015

Li, X., Xu, F., Xie, L., Ji, Y., Cheng, K., Zhou, Q., Wang, T., Shively, C., Wu, Q., Gong, W., Fang, L., Zhan, Q., Melgiri, N.D. and Xie, P. (2013) *Depression-like behavioral phenotypes by social and social plus visual isolation in the adult female Macaca fascicularis. PLoS One*, 8(9), e73293–e73293. https://doi.org/10.1371/journal.pone.0073293

Lin, J., Su, Y., Rizvi, S.J., Jagoda, J., Li, J., Wu, Y., Dai, Y., Zhang, Y., Kennedy, S.H. and Si, T. (2022) *Define and characterize the anhedonia in major depressive disorder: An explorative study. Journal of Affective Disorders*, 313, 235–242. https://doi.org/10.1016/j.jad.2022.06.082

Liu, X., Wang, N., Gu, L., Guo, J., Wang, J. and Luo, F. (2019) *Reward processing under chronic pain from the perspective of "liking" and "wanting": a narrative review. Pain Research and Management*, 2019. https://doi.org/10.1155/2019/6760121

Lockard, R.B. (1963) *Some effects of light upon the behavior of rodents. Psychological Bulletin*, 60(6), 509–529. https://doi.org/10.1037/h0046113

Lowe, M.R. and Butryn, M.L. (2007) *Hedonic hunger: A new dimension of appetite? Physiology & Behavior*, 91(4), 432–439. https://doi.org/10.1016/j.physbeh.2007.04.006

Luo, L., Reimert, I., Graat, E.A.M., Smeets, S., Kemp, B. and Bolhuis, J.E. (2020) *Effects of early life and current housing on sensitivity to reward loss in a successive negative contrast test in pigs. Animal Cognition*, 23(1), 121–130. https://doi.org/10.1007/s10071-019-01322-w

MacLellan, A., Fureix, C., Polanco, A. and Mason, G. (2021) *Can animals develop depression? An overview and assessment of 'depression-like' states. Behaviour*, 158(14–15), 1303–1353. https://doi.org/10.1163/1568539x-bja10132

Makowska, I.J. and Weary, D.M. (2016) *Differences in anticipatory behaviour between rats (Rattus norvegicus) housed in standard versus semi-naturalistic laboratory environments. PLoS One*, 11, e0147595. https://doi.org/10.1371/journal.pone.0147595

Malkesman, O., Scattoni, M.L., Paredes, D., Tragon, T., Pearson, B., Shaltiel, G., Chen, G., Crawley, J.N. and Manji, H.K. (2010) *The female urine sniffing test: A novel approach for assessing reward-seeking behavior in rodents. Biological Psychiatry*, 67(9), 864–871. https://doi.org/10.1016/j.biopsych.2009.10.018

Mason, G.J. (2010) *Species differences in responses to captivity: Stress, welfare and the comparative method. Trends in Ecology & Evolution*, 25(12), 713–721. https://doi.org/10.1016/j.tree.2010.08.011

Meagher, R.K. (2019) *Is boredom an animal welfare concern? Animal Welfare*, 28(1), 21–32. https://doi.org/10.71.0/09627286.28.1.021

Meagher, R.K., Campbell, D.L.M. and Mason, G.J. (2017) *Boredom-like states in mink and their behavioural correlates: A replicate study. Applied Animal Behaviour Science*, 197, 112–119. https://doi.org/10.1016/j.applanim.2017.08.001

Meagher, R.K. and Mason, G.J. (2012) *Environmental enrichment reduces signs of boredom in caged mink. PLoS One*, 7, e49180. https://doi.org/10.1371/journal.pone.0049180

Meehan, C.L. and Mench, J.A. (2002) *Environmental enrichment affects the fear and exploratory responses to novelty of young Amazon parrots. Applied Animal Behaviour Science*, 79, 75–88. https://doi.org/10.1016/s0168-1591(02)00118-1

Merrifield, C. and Danckert, J. (2014) *Characterizing the psychophysiological signature of boredom. Experimental Brain Research*, 232, 481–491. https://doi.org/10.1007/s00221-013-3755-2

Mileva, G.R. and Bielajew, C. (2015) *Environmental manipulation affects depressive-like behaviours in female Wistar-Kyoto rats. Behavioural Brain Research*, 293, 208–216. https://doi.org/10.1016/j.bbr.2015.07.035

Milyavskaya, M., Inzlicht, M., Johnson, T. and Larson, M.J. (2019) *Reward sensitivity following boredom and cognitive effort: A high-powered neurophysiological investigation. Neuropsychologia, Cognitive Effort*, 123, 159–168. https://doi.org/10.1016/j.neuropsychologia.2018.03.033

Moynihan, A.B., Tilburg, W.A.P., van Igou, E.R., Wisman, A., Donnelly, A.E. and Mulcaire, J.B. (2015) *Eaten up by boredom: Consuming food to escape awareness of the bored self. Frontiers in Psychology*, 6. https://doi.org/10.3389/fpsyg.2015.00369

Muscat, R. and Willner, P. (1992) *Suppression of sucrose drinking by chronic mild unpredictable stress: A methodological analysis. Neuroscience & Biobehavioral Reviews*, 16, 507–517. https://doi.org/10.1016/S0149-7634(05)80192-7

Nelson, J., Klumparendt, A., Doebler, P. and Ehring, T. (2017) *Childhood maltreatment and characteristics of adult depression: Meta-analysis. British Journal of Psychiatry*, 210(2), 96–104. https://doi.org/10.1192/bjp.bp.115.180752

Norbury, A. and Husain, M. (2015) *Sensation-seeking: Dopaminergic modulation and risk for psychopathology. Behavioural Brain Research*, 288, 79–93. https://doi.org/10.1016/j.bbr.2015.04.015

Olsen, C.M. (2011) *Natural rewards, neuroplasticity, and non-drug addictions. Neuropharmacology*, 61(7), 1109–1122. https://doi.org/10.1016/j.neuropharm.2011.03.010

Papp, M., Lappas, S., Muscat, R. and Willner, P. (1992) *Attenuation of place preference conditioning but not place aversion conditioning by chronic mild stress. Journal of Psychopharmacology*, 6(3), 352–356. https://doi.org/10.1177/026988119200600302

Papp, M., Willner, P. and Muscat, R. (1991) *An animal model of anhedonia – Attenuation of sucrose consumption and place preference conditioning by chronic unpredictable mild stress. Psychopharmacology*, 104(2), 255–259. https://doi.org/10.1007/bf02244188

Pelizza, L. and Ferrari, A. (2009) *Anhedonia in schizophrenia and major depression: State or trait? Annals of General Psychiatry*, 8(1), 22–22. https://doi.org/10.1186/1744-859x-8-22

Planchez, B., Surget, A. and Belzung, C. (2019) *Animal models of major depression: Drawbacks and challenges. Journal of Neural Transmission*, 126, 1383–1408. https://doi.org/10.1007/s00702-019-02084-y

Pluciute, A., Adomaitienė, V., Steibliene, V., Jarutiene, L., Bartkiene, E., Lele, V., Cernauskas, D., Klupsaite, D., Žadeikė, D. and Juodeikiene, G. (2021) *Differences in the preferred food tastes characteristics in patients with depressive disorder and healthy subjects. European Psychiatry*, 64(S1), S325–S326. https://doi.org/10.1192/j.eurpsy.2021.874

Polanco, A., Meagher, R.K. and Mason, G.J. (2021) *Boredom-like states in farmed mink are rapidly reduced by providing environmental enrichment. Applied Animal Behaviour Science*, 238, 105323. https://doi.org/10.1016/j.applanim.2021.105323

Pratt, W.E., Choi, E. and Guy, E.G. (2012) *An examination of the effects of subthalamic nucleus inhibition or mu-opioid receptor stimulation on food-directed motivation in the non-deprived rat. Behavioural Brain Research*, 230(2), 365–373. https://doi.org/10.1016/j.bbr.2012.02.031

Rizvi, S.J., Pizzagalli, D.A., Sproule, B.A. and Kennedy, S.H. (2016) *Assessing anhedonia in depression: Potentials and pitfalls. Neuroscience and Biobehavioral Reviews*, 65, 21–35. https://doi.org/10.1016/j.neubiorev.2016.03.004

Rolls, E.T., Cheng, W. and Feng, J. (2020) *The orbitofrontal cortex: Reward, emotion and depression. Brain Communications*, 2(2), fcaa196. https://doi.org/10.1093/braincomms/fcaa196

Russell, P.A. (1973) *Relationships between exploratory behaviour and fear: A review. British Journal of Psychology*, 64(3), 417–433. https://doi.org/10.1111/j.2044-8295.1973.tb01369.x

Scheggi, S., De Montis, M.G. and Gambarana, C. (2018) *Making Sense of Rodent Models of Anhedonia. International Journal of Neuropsychopharmacology*, 21(11), 1049–1065. https://doi.org/10.1093/ijnp/pyy083

Schrader, G.D. (1997) *Does anhedonia correlate with depression severity in chronic depression? Comprehensive Psychiatry*, 38(5), 260–263. https://doi.org/10.1016/s0010-440X(97)90057-2

Shalev, U., Erb, S. and Shaham, Y. (2010) *Role of CRF and other neuropeptides in stress-induced reinstatement of drug seeking. Brain Research*, 1314, 15–28. https://doi.org/10.1016/j.brainres.2009.07.028

Shirazi-Beechey, S. (2018) *Taste perception throughout life and evolution. Journal of Animal Science*, 96(Supplement S3), 353–354. https://doi.org/10.1093/jas/sky404.778

Slaney, C., Perkins, A.M., Davis, R., Penton-Voak, I., Munafò, M.R., Houghton, C.J. and Robinson, E.S.J. (2022) *Objective measures of reward sensitivity and motivation in people with high v. low anhedonia. Psychological Medicine*, 53(10), 1–9. https://doi.org/10.1017/S0033291722001052

Smeets, W.J.A.J., Marin, O. and Gonzalez, A. (2000) *Evolution of the basal ganglia: New perspectives through a comparative approach. Journal of Anatomy*, 196, 501–517. https://doi.org/10.1046/j.1469-7580.2000.19640501.x

Solberg, L.C., Horton, T.H. and Turek, F.W. (1999) *Circadian rhythms and depression: Effects of exercise in an animal model. American Journal of Physiology. Regulatory, Integrative and Comparative Physiology*, 276(1), R152–R161. https://doi.org/10.1152/ajpregu.1999.276.1.r152

Sousa, N., Almeida, O.F.X. and Wotjak, C.T. (2006) *A hitchhiker's guide to behavioral analysis in laboratory rodents. Genes, Brain and Behavior*, 5(s2), 5–24. https://doi.org/10.1111/j.1601-183x.2006.00228.x

Sparling, J.E., Barbeau, K., Boileau, K. and Konkle, A.T.M. (2020) *Environmental enrichment and its influence on rodent offspring and maternal behaviours, a scoping style review of indices of depression and anxiety. Pharmacology, Biochemistry, and Behavior*, 197, https://doi.org/10.1016/j.pbb.2020.172997

Strekalova, T. and Steinbusch, H.W.M. (2010) *Measuring behavior in mice with chronic stress depression paradigm. Progress in Neuro-Psychopharmacology & Biological Psychiatry*, 34(2), 348–361. https://doi.org/10.1016/j.pnpbp.2009.12.014

Swiecicki, L., Scinska, A., Bzinkowska, D., Torbinski, J., Sienkiewicz-Jarosz, H., Samochowiec, J. and Bienkowski, P. (2015) *Intensity and pleasantness of sucrose taste in patients with winter depression. Nutritional Neuroscience*, 18(4), 186–191. https://doi.org/10.1179/1476830514y.0000000115

Tapper, A.R. and Molas, S. (2020) *Midbrain circuits of novelty processing. Neurobiology of Learning and Memory*, 176, 107323. https://doi.org/10.1016/j.nlm.2020.107323

Tolentino, J.C. and Schmidt, S.L. (2018) *DSM-5 criteria and depression severity: Implications for clinical practice. Frontiers in Psychiatry*, 9, 450–450. https://doi.org/10.3389/fpsyt.2018.00450

Toohey, P. (2011) *Boredom: A Lively History*. Yale University Press, London.

Treadway, M.T. and Zald, D.H. (2011) *Reconsidering anhedonia in depression: Lessons from translational neuroscience. Neuroscience & Biobehavioral Reviews*, 35(3), 537–555. https://doi.org/10.1016/j.neubiorev.2010.06.006

Trøstheim, M., Eikemo, M., Meir, R., Hansen, I., Paul, E., Kroll, S.L., Garland, E.L. and Leknes, S. (2020) *Assessment of anhedonia in adults with and without mental illness: A systematic review and meta-analysis. JAMA Network Open*, 3(8), e2013233–e2013233. https://doi.org/10.1016/j.neubiorev.2010.06.006

van der Harst, J. and Spruijt, B.M. (2007) *Tools to measure and improve animal welfare: Reward-related behaviour. Animal Welfare*, 16(S1), 67–73. https://doi.org/10.1017/s0962728600031742

van der Harst, J.E., Baars, A.M. and Spruijt, B.M. (2003) *Standard housed rats are more sensitive to rewards than enriched housed rats as reflected by their anticipatory behaviour. Behavioural Brain Research*, 142(1–2), 151–156. https://doi.org/10.1016/s0166-4328(02)00403-5

Vinke, C.M., Houx, B.B., Van Den Bos, R. and Spruijt, B.M. (2006) *Anticipatory behaviour and stereotypical behaviour in farmed mink (Mustela vison) in the presence, absence and after the removal of swimming water. Applied Animal Behaviour Science*, 96, 129–142. https://doi.org/10.1016/j.applanim.2005.04.022

Vinke, C.M., Van Den, R.B. and Spruijt, B.M. (2004) *Anticipatory activity and stereotypical behaviour in American mink (Mustela vison) in three housing systems differing in the amount of enrichments. Applied Animal Behaviour Science*, 89, 145–161. https://doi.org/10.1016/j.applanim.2004.06.002

Vinograd, M., Stout, D.M., Risbrough, V.B. and Pizzagalli, D.A. (2021) Anhedonia in posttraumatic stress disorder: Prevalence, phenotypes, and neural circuitry. In Pizzagalli, D. A. (ed.) *Anhedonia: Preclinical, Translational, and Clinical Integration*. Springer Nature, Cham, Switzerland, pp. 185–200.

Walker, M.D. and Mason, G.J. (2011) *Female C57BL/6 mice show consistent individual differences in spontaneous interaction with environmental enrichment that are predicted by neophobia. Behavioural Brain Research*, 224(1), 207–212. https://doi.org/10.1016/j.bbr.2011.06.003

Wang, Q., Timberlake, M.A., Prall, K. and Dwivedi, Y. (2017) *The recent progress in animal models of depression. Progress in Neuro-Psychopharmacology and Biological Psychiatry*, 77, 99–109. https://doi.org/10.1016/j.pnpbp.2017.04.008

Warren, C.M., Hyman, J.M., Seamans, J.K. and Holroyd, C.B. (2015) *Feedback-related negativity observed in rodent anterior cingulate cortex. Journal of Physiology-Paris*, 109, 87–94. https://doi.org/10.1016/j.jphysparis.2014.08.008

Weiss, F. (2005) *Neurobiology of craving, conditioned reward and relapse. Current Opinion in Pharmacology*, 5(1), 9–19. https://doi.org/10.1016/j.coph.2004.11.001

Willner, P. (1997) *Validity, reliability and utility of the chronic mild stress model of depression: A 10-year review and evaluation. Psychopharmacology*, 134, 319–329. https://doi.org/10.1007/s002130050456

Willner, P. (2017a) *The chronic mild stress (CMS) model of depression: History, evaluation and usage. Neurobiology of Stress*, 6(C), 78–93. https://doi.org/10.1016/j.ynstr.2016.08.002

Willner, P. (2017b) *Reliability of the chronic mild stress model of depression: A user survey. Neurobiology of Stress*, 6(C), 68–77. https://doi.org/10.1016/j.ynstr.2016.08.001

Wittmann, B.C., Bunzeck, N., Dolan, R.J. and Düzel, E. (2007) *Anticipation of novelty recruits reward system and hippocampus while promoting recollection. NeuroImage*, 38, 194–202. https://doi.org/10.1016/j.neuroimage.2007.06.038

World Health Organisation (2022) *ICD-11 for Mortality and Morbidity Statistics - Depressive Disorders*. ICD-11 for Mortality and Morbidity Statistics (who.int): World Health Organisation. Accessible at: https://icd.who.int/browse/2024-01/mms/en#1563440232 (Accessed: 17th December 2024).

Yamamoto, T., Mizuta, H. and Ueji, K. (2021) *Analysis of facial expressions in response to basic taste stimuli using artificial intelligence to predict perceived hedonic ratings. PLoS One*, 16(5), e0250928. https://doi.org/10.1371/journal.pone.0250928

Young, L.E., McCallum, R.T., Perreault, M.L. and Mason, G.J. (2024) *Opportunistically using a Chronic Unpredictable Stress study to investigate 'inactive-but-awake' behaviour as a potential welfare indicator in laboratory rats. Applied Animal Behaviour Science*, 274, 106249. https://doi.org/10.1016/j.applanim.2024.106249

Zimmerman, M., Balling, C., Chelminski, I. and Dalrymple, K. (2018) *Understanding the severity of depression: Which symptoms of depression are the best indicators of depression severity? Comprehensive Psychiatry*, 87, 84–88. https://doi.org/10.1016/j.comppsych.2018.09.006

Zuckerman, M. (1994) *Sensation seeking: Beyond the optimal level of arousal*, 2nd ed. Lawrence Erlbaum Associates, Hillsdale, NJ.

10 Dehydroepiandrosterone (DHEA) and Glucocorticoids-to-DHEA(S) Ratio

Jessica C. Whitham and Lance J. Miller
Brookfield Zoo Chicago, Brookfield, Illinois, USA

Introduction

In recent years, welfare scientists have made concerted efforts to non-invasively measure biomarkers that provide insight into affective states (Pressman and Cohen 2005; Staley et al. 2018; Whitham et al. 2020). There is mounting evidence that dehydroepiandrosterone (DHEA) and its sulphated metabolite (DHEA-S) – collectively referred to as DHEA(S) – can help us better understand the long-term affective states (moods) of humans and likely non-human animals. For humans, it has been suggested that DHEA-S represents a more stable indicator of long-term affective states than DHEA (Kamin and Kertes 2017). Furthermore, DHEA(S), which has antioxidative, anti-inflammatory and neuroprotective properties, also exerts widespread positive effects on immune function, as well as behaviours associated with reproduction and survival (Muehlenbein et al. 2003; Guilliams and Edwards 2010; Kamin and Kertes 2017). In this chapter, we consider whether DHEA(S), as well as the ratio of glucocorticoids (GCs) to DHEA(S), can be used to non-invasively assess long-term affective states.

Both DHEA(S) and GCs can be measured for diverse species using a variety of non-invasive sample types, including saliva (Fels et al. 2019), urine (Behringer et al. 2012), faeces (Takeshita et al. 2014, 2018), hair (Peric et al. 2017; Bergamin et al. 2019; Montillo et al. 2020) and even nails (Warnock et al. 2010). We focus our discussion on mammals, as physiology varies across classes of animals. DHEA(S) has been investigated across diverse mammalian species from ungulates to primates to cetaceans (e.g. Takeshita et al. 2014; O'Brien et al. 2017; Fels et al. 2019; Placci et al. 2020; Arena et al. 2021). In most, DHEA(S) is released by adrenal glands, triggered by activation of the hypothalamic–pituitary–adrenal (HPA) axis (Raison and Miller 2003; Guilliams and Edwards 2010; see Chapter 5 and Figure 10.1). DHEA,

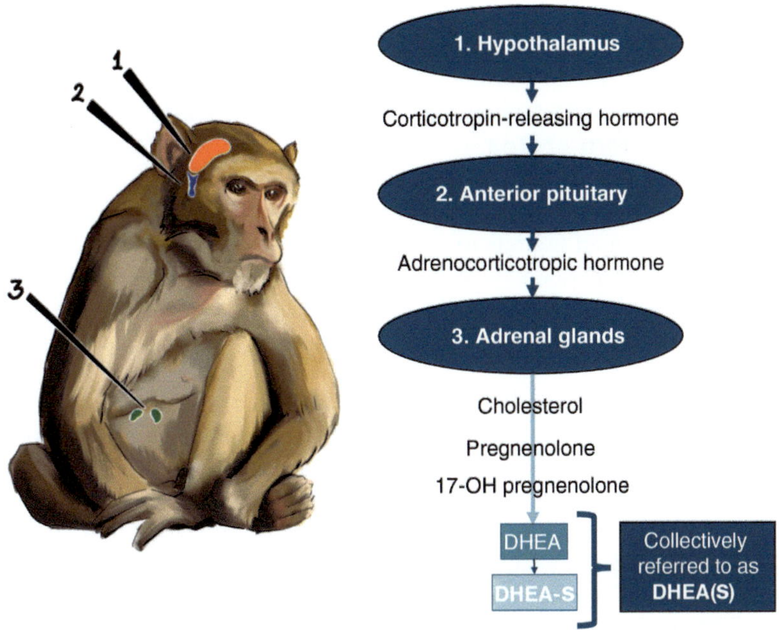

Figure 10.1 Synthesis of DHEA and DHEA-S in the adrenal glands in most mammals.

which is secreted by the *zona reticularis* of the adrenal cortex, is both an anabolic steroid and a precursor to sex steroids (Leowattana 2004; Maninger et al. 2009). There is no evidence that DHEA exerts feedback on the HPA axis, and dysregulation of the axis may be associated with a reduction in DHEA(S) (Leowattana, 2004; Guilliams and Edwards 2010; Kamin and Kertes 2017). DHEA(S) can also originate from sources other than the adrenal gland (e.g. placenta, gonads, liver, brain; Klinge et al. 2018; Gabai et al. 2020).

Here, we investigate the validity of DHEA(S) and the GC:DHEA(S) ratio as a means to assess long-term affective states. While acute negative events may be associated with a temporary increase in DHEA(S), individuals experiencing long-term negative affect generally exhibit reduced levels (Morgan et al. 2004; Izawa et al. 2008; Lennartsson et al. 2015; Fustini et al. 2017; Kamin and Kertes 2017; Dutheil et al. 2021). In this chapter, we discuss the construct validity, limitations and potential applications of both DHEA(S) and GC:DHEA(S) as indicators of mood in mammals. While only a handful of studies have examined how DHEA(S) and GC:DHEA(S) may be linked to affective states in non-human mammals, we argue that – based on preliminary studies as well as the human literature – these biomarkers have the potential to provide insight into an individual's welfare status when assessed over time.

Construct validity

In this section, we use the framework set out in Chapter 2 to critically investigate the evidence that DHEA(S), as well as the ratio of GCs to DHEA(S), can serve as valid, non-invasive indicators of long-term affective states for both humans and non-human mammals (e.g. Cawood and Bancroft 1996; Van Niekerk et al. 2001; Schmidt et al. 2005; Alhaj et al. 2006; Placci et al. 2020; Arena et al. 2021).

Humans as a model for using DHEA(S) and GC:DHEA(S) as indicators of long-term affective states

We start by examining the human literature. In Chapter 2, validation Test 1 involves determining whether the variable of interest accurately tracks self-reported affective states in humans: the ability of human subjects to self-report allows us to consider how DHEA(S) and GC:DHEA(S) may be associated with well-being and subjective states such as pain, misery and malaise. In this section, we investigate how these biomarkers are related to long-term affective states and psychological well-being in humans.

Mental health and psychological well-being

Numerous studies have examined the relationship between DHEA(S) and depression, anxiety, mood and psychological well-being. In a mood induction study that involved collecting saliva samples before and after a writing task, the level of positive affect exhibited in the writing task was predictive of increases in both DHEA and cortisol (Williams et al. 2008). There is also correlational evidence that high DHEA(S) is associated with long-term positive affect. For instance, in a study of women aged 40–60, plasma DHEA was directly related to positive affect and well-being (Cawood and Bancroft 1996). In addition, there is ample evidence that low DHEA(S) is linked to negative affect and depressive symptoms (e.g. Susman et al. 1991; Cawood and Bancroft 1996; Goodyer et al. 1996; Van Niekerk et al. 2001; Wong et al. 2011), and that elevated cortisol:DHEA(S) is associated with symptoms of dissociation, depression, anxiety, negative mood and angry temperament (e.g. Markopoulou et al. 2009; Morgan et al. 2009; Rasmusson et al. 2004; Leff-Gelman et al. 2020). For example, Goodyer et al. (1996) reported that among adolescents, high evening salivary cortisol and low morning salivary DHEA were independently associated with major depressive disorder. A separate study of adolescents discovered an inverse relationship between serum DHEA levels and subsequent negative affect (Susman et al. 1991). Among older men, higher evening salivary DHEA was correlated with lower negative mood in the morning and lower anxiety (Van Niekerk et al. 2001). In postmenopausal women, plasma DHEA-S was inversely associated with depressed mood (Barrett-Connor et al. 1999). Finally, within patients with schizophrenia – who generally have elevated DHEA levels and lower cortisol:DHEA ratios than healthy controls – the cortisol:DHEA ratio was found to be inversely correlated with hippocampal volume (Ji et al. 2021; see also Chapter 13 on hippocampal plasticity). Furthermore, the cortisol:DHEA-S ratio was positively associated with severity of depression, angry temperament, anxiety and hostility in patients with schizophrenia (Ritsner et al. 2004).

How aversive experiences are associated with DHEA(S)

There is evidence that aversive experiences and stressful life events can have a lasting impact on DHEA(S). In older adults (65+ years old), those who experienced higher stress severity for life events in the preceding year demonstrated significantly higher cortisol:DHEA ratios (despite no change in cortisol) than subjects who had experienced little stress (Heaney et al. 2014). This is consistent with the finding that older adults who experienced bereavement and its associated emotional distress within the prior two months had significantly higher cortisol:DHEA-S ratios than age-matched, non-bereaved controls (Khanfer et al. 2011). Similarly, healthy caregivers of patients with Alzheimers disease had lower salivary DHEA-S levels and increased cortisol:DHEA-S ratios (yet similar cortisol levels) compared to healthy controls (Jeckel et al. 2010). Finally, subjects who reported higher perceived stress at work exhibited an attenuated DHEA-S response and higher cortisol:DHEA-S ratio during an acute psychosocial stress test than individuals reporting low perceived stress at work (Lennartsson et al. 2013). However, in contrast, post-traumatic stress disorder (PTSD) is associated with higher DHEA, as we review below.

The relationship between mood and DHEA(S)

DHEA(S) also often influences mood, having both anxiolytic and antidepressant properties (Wolkowitz et al. 1995; Alhaj et al. 2006; Maninger et al. 2009; Kamin and Kertes 2017). For example, DHEA may be protective in subjects with PTSD, mitigating the severity of PTSD symptoms and negative mood (Rasmusson et al. 2004). Indeed, although patients with PTSD generally have elevated DHEA compared to controls (like people with schizophrenia, as outlined earlier), *within* this group, those individuals with relatively low cortisol:DHEA ratios were found to have the fewest negative mood symptoms (Rasmusson et al. 2004).

Several studies also report that administering DHEA can often improve mood and psychological well-being. For those diagnosed with depression, DHEA administration can reduce depressive symptoms and improve mood (e.g. Wolkowitz et al. 1995, 1999; Strous et al. 2003; Schmidt et al. 2005). DHEA administration can also benefit subjects who have not been diagnosed with a mental disorder. For instance, DHEA treatment enhanced mood and/or psychological well-being in studies of healthy young men (Alhaj et al. 2006), women with adrenal insufficiency (Arlt et al. 1999, 2000), adults with Addison's disease (Hunt et al. 2000), postmenopausal women (Labrie et al. 1997) and adults of advanced age (Morales et al. 1994).

In sum, low DHEA(S) and a high cortisol:DHEA(S) ratio are often indicative of negative affect and mood, sometimes even in cases in which cortisol remains unchanged (Guilliams and Edwards 2010; Jeckel et al. 2010; Heaney et al. 2014), and even in groups with conditions paradoxically associated with elevated DHEA. But why? It is possible that acute aversive events influence DHEA(S) and that negative affect plays a role in the synthesis/release of these hormones. However, there is also evidence that DHEA(S) can downregulate negative emotions, thereby influencing welfare (as well as being influenced by welfare).

Physical health

A considerable amount of research has investigated DHEA(S)'s immune-enhancing properties and relation to physical health. DHEA(S) increases the production of cytokines responsible for promoting white blood cell activity and inhibits the production of inflammatory cytokines (e.g. Hazeldine et al. 2010; Alves et al. 2016; Cao et al. 2020). Thus, there is evidence that DHEA supplementation can enhance health, including decreasing disease activity, severity or symptoms (e.g. systemic lupus erythematosus: Van Vollenhoven and McGuire 1996; AIDS: Centurelli and Abate 1997; Crohn's disease: Andus et al. 2003; pulmonary hypertension: Dumas de La Roque et al. 2012). Some studies also investigated the role of DHEA(S) in the aging process. For example, Bauer (2008) described how replacement therapy can enhance immune function and overall well-being in aging humans, while Morales et al. (1994) found that DHEA-treated senior subjects reported improved physical well-being (e.g. deeper sleep, increased energy).

In humans, high cortisol:DHEA(S) is also often associated with physical illness and infection. While these studies do not include measures of affect, the conditions involved all have the potential to negatively impact subjective state (Beishuizen et al. 2002; Butcher et al. 2005; Phillips et al. 2010). For instance, Beishuizen et al. (2002) discovered that among patients experiencing septic shock and trauma, the most critically ill subjects had the highest cortisol levels and lowest DHEA-S levels. A positive relationship was even found between the cortisol:DHEA-S ratio and various types of mortality over the next 15 years (Phillips et al. 2010). Overall, the causality may be bidirectional, with low DHEA(S) both being an outcome of poor health status and also negatively impacting physical well-being.

Construct validity of DHEA(S) and GC:DHEA(S) as indicators of long-term affective states in studies of non-human mammals

The idea of using DHEA(S) to investigate long-term affect in non-human mammals is still in its infancy. To evaluate the construct validity of DHEA(S) and GC:DHEA(S) for welfare research, we primarily focus on non-clinical studies that consider how DHEA(S) may be related to long-term affective states.

We sifted through the literature to apply Test 2 from Chapter 2 – i.e. assessing whether endocrine variables are reliably associated with stimuli that individuals prefer/avoid. Nejad et al. (2021) determined that dairy cows allowed to graze at pasture in alpine areas for 24 hours per day had higher hair DHEA, lower hair cortisol, lower plasma cortisol:DHEA ratios and lower hair cortisol:DHEA ratios than control cows (i.e. cows living indoors) or those only allowed to graze for 12 hours per day. In horses, those who spent little or no time out in paddocks (a preferred environment: Lee et al. 2011) and displayed oral and locomotor stereotypies had similar cortisol levels to control horses who spent more time outside and did not display stereotypic behaviour, but they had significantly higher plasma cortisol:DHEA (Arena et al. 2021). In a similar study, horses exposed to more natural management practices (e.g. more outdoor access, socializing allowed in paddocks, pasture feeding allowed) also had significantly higher hair DHEA concentrations, and lower cortisol:DHEA concentrations (as well as lower cortisol concentrations), compared to horses who experienced a traditional stable management style (Placci et al. 2020). Likewise, piglets housed under conditions designed to enhance welfare via fewer aversive procedures and more preferred resources (e.g. no tail-docking, outdoor access, access to straw) exhibited higher salivary DHEA and lower cortisol:DHEA ratios (as well as lower salivary cortisol) than piglets who faced more negative experiences (Fels et al. 2019). Furthermore, there is evidence that transportation (generally regarded as aversive: EFSA Panel on Animal Health and Welfare 2022) is associated with a decrease in plasma DHEA and increased cortisol:DHEA in young bulls (Buckham Sporer et al. 2008), as well as elevated hair cortisol:DHEA in sows (Trevisan et al. 2017).

If diseases are assumed to threaten fitness and cause malaise as they do in humans, then further evidence linking low DHEA to negative affect in animals (validation Test 3 in Chapter 2) comes from some examples given earlier, plus studies finding low DHEA levels in diseased seals (Gundlach et al. 2018), cats infected by feline leukaemia virus and feline immunodeficiency virus (Tejerizo et al. 2012) and lame cows with abnormal posture, visible lesions and reduced appetites (Almeida et al. 2008).

In addition, turning to Test 5 in Chapter 2 (i.e. relationship with other welfare indicators), as well as the human mortality findings presented earlier, a study on female macaques reported a link between passive, depression-like behaviour (combined with avoiding the experimenters) and both low DHEA-S concentrations and higher cortisol:DHEA-S ratios, despite no significant difference in cortisol alone (Goncharova et al. 2010).

Overall, this sums up preliminary evidence that in mammals, higher DHEA(S) and a lower GC:DHEA(S) ratio is associated with better well-being when considered in the long term (non-acute). However, despite this, these biomarkers have only recently been integrated into welfare studies, and more research is certainly needed. Furthermore, these biomarkers are prone to both false leads (i.e. when the indicator changes, but affective valence has not changed) and false nulls (i.e. when affective state has changed but the indicator has not picked this up), as we review later.

Responsiveness and selectivity

For those hoping to incorporate DHEA(S) or the GC:DHEA(S) ratio into welfare research, it is crucial to consider how responsive and selective these biomarkers would be as indicators of long-term affective states. Here, we review a number of relevant factors (summarised in Table 10.1) and attempt to illustrate potential false leads (i.e. Type I errors) and false nulls (i.e. Type II errors) that could arise if we were to always assume a simple relationship between a relatively high GC:DHEA(S) ratio and reduced welfare (Figure 10.2).

Overall, DHEA(S) and the GC:DHEA(S) ratio do not appear to be highly prone to false nulls in which they fail to respond to negative moods. However, in one study of elderly humans, depressed subjects and controls did not differ in DHEA or cortisol:DHEA (Moraes et al. 2016). Similarly, in captive Japanese macaques (*Macaca fuscata*), GC:DHEA-S did not differ between females housed singly indoors and those housed socially outdoors (Takeshita et al. 2014), despite evidence that the latter conditions are better for welfare (e.g. Polanco 2021). Indeed, it should be noted that those housed singly indoors had higher faecal DHEA-S concentrations, which is contrary to what would be expected if DHEA-S was reflecting long-term affective state (see discussion of false leads later).

Furthermore, there is scope for false nulls (as well as false leads, as we discuss later) if potential sources of 'noise' are ignored. For example, some researchers reported that exogenous DHEA did not result in improved mood and/or psychological well-being, especially for humans post middle age (e.g. Wolf et al. 1997; Flynn et al. 1999; Barnhart et al. 1999; Arlt et al. 2001; Nair et al. 2006). It is also crucial to consider diurnal rhythms when discussing both the responsiveness and selectivity of DHEA(S). DHEA(S) peaks in the evening for some nocturnal species, like golden hamsters (Pieper and Lobocki 2000). In humans, circulating DHEA is highest in the morning (Hucklebridge et al. 2005), but DHEA-S does not vary considerably from one day to the next or have a strong diurnal rhythm, due to its long half-life and slow rate of metabolic clearance (Leowattana 2004). Welfare researchers must also take seasonal variation into account, as seasonal fluctuations of DHEA(S) may obscure or mimic welfare-related changes (e.g. Boonstra et al. 2008; Soma et al. 2008). In stallions, for example, the hair accumulation values for DHEA-S and cortisol varied seasonally (Olvera-Maneu et al. 2021). Reproductive status is important, as well. For example, DHEA(S) plays a role in maintaining pregnancy and parturition, with DHEA(S) increasing at the end of pregnancy as the foetus' adrenal gland develops (Rainey et al. 2004; Takeshita et al. 2016, 2019; Gabai et al. 2020; da Silva et al. 2021). Finally, there may be sex differences in DHEA(S) and the GC:DHEA(S) ratio (e.g. Klinge et al. 2018; Gabai et al. 2020; Pieper and Lobocki 2000), as well as age differences (e.g. Muehlenbein et al. 2003; Goncharova et al. 2010; O'Brien et al. 2017). Such factors could cause both false nulls and false leads, if not acknowledged and controlled for properly.

Furthermore, in terms of selectivity of the indicator, welfare scientists must consider factors/situations that can increase the likelihood of false leads, in which presumed *negative* states or conditions are associated with *high* DHEA(S) (e.g. as in some cases of human PTSD, depression, and intimate partner violence: Pico-Alfonso et al. 2004, Rasmusson et al. 2004; Morrison et al. 2011). One reason, in some cases, is that there is overwhelming evidence that

Table 10.1 Examples of factors that may influence the responsiveness and selectivity of DHEA(S) as an indicator of affective state, with potential errors if low DHEA(S) or high GC:DHEA(S) is always assumed to indicate more negative affect.

Examples of potential false leads where neutral states or negative affect is associated with increased rather than decreased DHEA(S)	
Acute psychological stressors	Humans: First-time parachute jumps (Oberbeck et al. 1998) Psychosocial stress tests (Izawa et al. 2008; Lennartsson et al. 2012)
Acute aversive events	Non-humans: Overstocking (Fustini et al. 2017) Castration (Ayala and Martos 2013)
Housing/environmental conditions	Non-humans: Female Japanese macaques housed singly indoors had higher DHEA-S than those housed socially outdoors (Takeshita et al. 2014).
Pregnancy	Some species exhibit an increase in DHEA(S) towards the end of pregnancy (Rainey et al. 2004; Takeshita et al. 2016, 2019; Gabai et al. 2020; da Silva et al. 2021).
Inflammation and some infections/forms of illness	Non-humans: • Sepsis (Dembek et al. 2017) • Metritis (Gundlach et al. 2017) • Sole ulcers (O'Driscoll et al. 2015) • Inflammatory mammary carcinoma (Illera et al. 2006) • Ehrlichiosis (Rondelli et al. 2015) • Endometrial oedema (Falomo et al. 2020) Humans: • Post-traumatic stress disorder, PTSD (Rasmusson et al. 2004) • Schizophrenia (Ji et al. 2021) • Depressive symptoms (Morrison et al. 2011)
Others	Sex, season, time of day: see text
Examples of potential false nulls where negative affect is not associated with decreased DHEA(S)	
Housing/environmental conditions	Non-humans: GC:DHEA-S did not differ between female Japanese macaques housed singly indoors and those housed socially outdoors (Takeshita et al. 2014).
Some forms of illness	In elderly humans, depressed subjects and controls did not differ in DHEA or cortisol:DHEA (Moraes et al. 2016).
Others	Sex, season and time of day: see text.

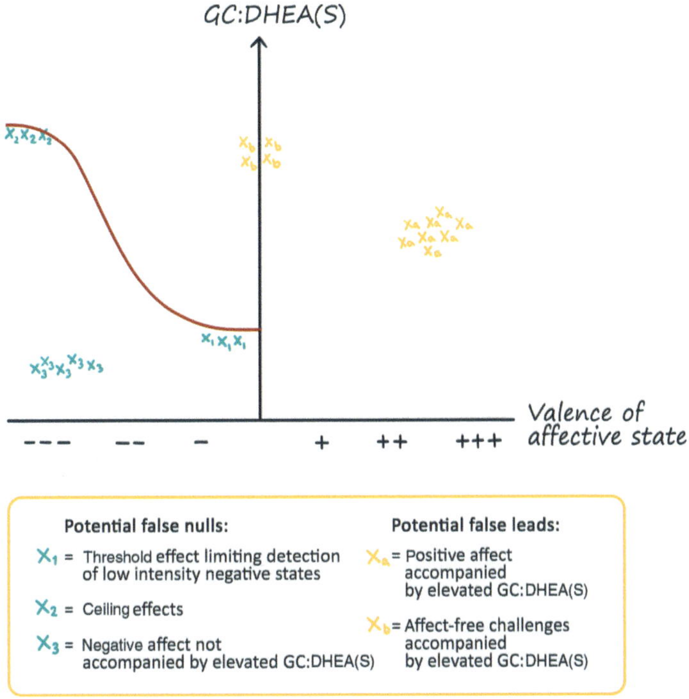

Figure 10.2 GC:DHEA(S) responses while experiencing positive and negative affective states including potential false nulls and false leads. No data are currently available for the area to the right of the vertical axis.

acute psychological stressors are associated with *increases* in DHEA(S). For example, an increase in plasma DHEA was found for pregnant dairy cows experiencing overstocking, a short-term negative event (Fustini et al. 2017). Likewise, Dutheil et al.'s (2021) meta-analysis found that acute mental stress in humans (e.g. first-time parachute jumps, undergoing psychosocial stress tests) results in a short-term increase in DHEA(S) (Oberbeck et al. 1998; Lennartsson et al. 2012). Several researchers have postulated that such secretion of DHEA(S) following acute mental stress plays a protective role, including antagonizing the effects of cortisol (Morgan et al. 2004; Hechter et al. 1997). Indeed, Izawa et al. (2008) reported that lower DHEA and an elevated cortisol:DHEA ratio during a psychosocial stress test were associated with increased negative affect during and shortly after this test.

In addition, the relationship between DHEA(S) and affective state can be influenced by factors such as cytokine release, with IL-8 potentially increasing DHEA(S) levels (Gabai et al. 2020). The source of DHEA can play a role, too. For example, Illera et al. (2006) discovered that plasma DHEA was *elevated* in female dogs with inflammatory mammary carcinomas, and that serum DHEA was highly correlated with tumour tissue DHEA, suggesting that the tumours themselves were producing DHEA.

Perhaps these issues explain why some studies on non-human mammals report that negative events, diseases and chronic conditions are associated with *elevated* rather than reduced levels of DHEA(S). For example, high DHEA(S) concentrations have been reported in animals experiencing medical conditions: newborn foals with sepsis (e.g. Dembek et al. 2017), domestic dogs with canine ehrlichiosis (Rondelli et al. 2015), domestic dogs with inflammatory mammary carcinomas (Illera et al. 2006), mares with a high degree of endometrial oedema (Falomo et al. 2020), cows diagnosed with metritis and leukopenia (Gundlach et al. 2017) and cows with sole ulcers and elevated inflammatory biomarkers (O'Driscoll et al. 2015). Elevated DHEA(S) was also found in horses following castration (Ayala and Martos 2013), though it is possible that this was influenced by the loss of testosterone.

Finally, it is important to recognise that when comparing studies that integrate DHEA(S) and/or GC:DHEA(S), the methodologies (e.g. sample type, sample collection) may vary greatly. Furthermore, eliciting stimuli/conditions may differ in their nature, persistency (acute vs. chronic) and/or aetiology, and may therefore affect DHEA(S) release (from various sources) differently (O'Driscoll et al. 2017; Falomo et al. 2020; Gabai et al. 2020).

Use in practice

Welfare scientists working in agricultural settings, companion animal shelters, laboratories, zoological facilities and conservation centres could possibly benefit by employing DHEA(S) and the GC:DHEA(S) ratio to evaluate an individual's welfare status. These biomarkers have the potential to inform management decisions and interventions designed to improve individual animal welfare. In the future, perhaps DHEA supplementation could even become a novel tool for improving welfare. However, as the previous section has shown, there are still a considerable number of caveats. At this time, more research is needed to gain a better understanding of how DHEA(S) and GC:DHEA(S) can be applied across diverse taxa.

Moving forward, we encourage welfare scientists to validate DHEA(S) and the GC:DHEA(S) ratio across a variety of taxonomic groups. This requires researchers to conduct both biochemical validations and to determine whether these biomarkers have construct validity as indicators of affect by (i) assessing individuals as they are exposed to stimuli that they both prefer and avoid (Test 2 in Chapter 2; Whitham et al. 2020) and (ii) comparing DHEA(S) and GC:DHEA(S) to other validated indicators of affect (Test 5 in Chapter 2), such as heart rate variability, data from judgement bias tests (see Chapter 7) and abnormal behaviour (see Chapter 11).

Before integrating DHEA(S) into a particular welfare study, the researcher must first consider whether it is an appropriate biomarker for the species of interest. If both DHEA and DHEA-S are present, the researcher will have to determine which is more appropriate for investigating chronic versus acute affective states, as they may have different half-lives or rates of metabolic clearance (Kamin and Kertes 2017). For example, while DHEA-S appears to be better for examining long-term affective states or mood in humans, it is reported to be undetectable in some animals (e.g. in studies that examined the adrenals of cycling sows and the plasma of nonpregnant gilts), to circulate at nanomolar levels in some species (e.g. dogs, pigs) and to circulate at micromolar levels in various primate species (Kaminska et al. 2000; Rege et al. 2019; Gabai et al. 2020). Moreover, it is important to keep in mind that DHEA(S) can originate from sources other than the adrenals (Klinge et al. 2018; Gabai et al. 2020).

Ultimately, welfare researchers must now determine (i) which of DHEA, DHEA-S or the GC:DHEA(S) ratio is most sensitive and (ii) most specific to changes in affect, and (iii) whether adrenal DHEA(S) has greater validity as an indicator of affect, as compared to DHEA(S) that originates from other organs. It is also vital for researchers to examine how DHEA(S) varies by sex, reproductive status, health status and age, as well as how these hormones fluctuate across seasons/time of day. This will allow researchers to assess the degree to which such variables may alter levels independently of affective state, potentially impacting both the responsiveness and selectivity of DHEA(S) as an indicator of mood. It may even be possible to determine whether DHEA(S) plays a role in helping animals cope with adversity by investigating whether the individuals who fare better when exposed to the same acute challenge exhibit higher DHEA(S), as is the case with humans facing short-term psychological stressors/aversive events, and also with humans suffering from PTSD and schizophrenia. We do, however, caution researchers against using DHEA(S) as an indicator when an animal is known to have a disease or condition that may impact this biomarker. Furthermore, one must consider whether the individual has recently experienced any negative acute events, as we know that these can be associated with temporary increases in DHEA(S).

Finally, there are numerous methodological considerations that must be taken into account when designing studies that involve DHEA(S). For example, welfare scientists must determine which sample type is most appropriate. Overall, a case can be made for using faeces and urine, as non-invasive sample types that represent a cumulative, pooled concentration may be preferable when investigating long-term affective states. Whitham et al. (2020) provide a more detailed discussion of methodological considerations, including overall study design, sample type selection and sample processing (see also da Silva et al. (2021) for factors to consider when analysing faecal samples).

Conclusions

As discussed throughout this book, a multifaceted approach that integrates valid physiological biomarkers with other measures is crucial when assessing animal affective states. Combining DHEA(S) and the GC:DHEA(S) ratio with other welfare indicators allows for a more thorough assessment of mental, physical and emotional health (Whitham et al. 2020). Overall, there is mounting evidence that when considering mood: (i) high DHEA(S) is associated with positive well-being, and (ii) low DHEA(S) and a high cortisol:DHEA(S) ratio are linked to negative affect in the long-term. While DHEA(S) is prone to false leads under a variety of circumstances, it seems that DHEA(S) is not as susceptible to false nulls as baseline GCs. Furthermore, the GC:DHEA(S) ratio in particular may be a more responsive indicator than GCs alone when assessing long-term affective states. Ultimately, welfare scientists must proceed with caution before integrating DHEA(S) and the GC:DHEA(S) ratio into their welfare toolkits.

References

Alhaj, H.A., Massey, A.E. and McAllister-Williams, R.H. (2006) *Effects of DHEA administration on episodic memory, cortisol and mood in healthy young men: A double-blind, placebo-controlled study*. *Psychopharmacology*, 188, 541–551. https://doi.org/10.1007/s00213-005-0136-y

Almeida, P.E., Weber, P.S.D., Burton, J.L. and Zanella, A.J. (2008) *Depressed DHEA and increased sickness response behaviors in lame dairy cows with inflammatory foot lesions*. *Domestic Animal Endocrinology*, 34, 89–99. https://doi.org/10.1016/j.domaniend.2006.11.006

Alves, V.B.F., Basso, P.J., Nardini, V., Silva, A., Chica, J.E.L. and de Barros Cardoso, C.R. (2016) *Dehydroepiandrosterone (DHEA) restrains intestinal inflammation by rendering leukocytes hyporesponsive and balancing colitogenic inflammatory responses. Immunobiology*, 221, 934–943. https://doi.org/10.1016/j.imbio.2016.05.013

Andus, T., Klebl, F., Rogler, G., Bregenzer, N., Scholmerich, J. and Straub, R.H. (2003) *Patients with refractory Crohn's disease or ulcerative colitis respond to dehydroepiandrosterone: A pilot study. Alimentary Pharmacology & Therapeutics*, 17, 409–414. https://doi.org/10.1046/j.1365-2036.2003.01433.x

Arena, I., Marliani, G., Sabioni, S., Gabai, G., Bucci, D. and Accorsi, P.A. (2021) *Assessment of horses' welfare: Behavioral, hormonal, and husbandry aspects. Journal of Veterinary Behavior*, 41, 82–90. https://doi.org/10.1016/j.jveb.2021.01.006

Arlt, W., Callies, F. and Allolio, B. (2000) *DHEA replacement in women with adrenal insufficiency-Pharmacokinetics, bioconversion and clinical effects on well-being, sexuality and cognition. Endocrine Research*, 26, 505–511. https://doi.org/10.3109/07435800009048561

Arlt, W., Callies, F., Koehler, I., van Vlijmen, J.C., Fassnacht, M., Strasburger, C.J., Seibel, M.J., Huebler, D., Ernst, M., Oettel, M., Reincke, M., Schulte, H.M. and Allolio, B. (2001) *Dehydroepiandrosterone supplementation in healthy men with an age-related decline of dehydroepiandrosterone secretion. The Journal of Clinical Endocrinology & Metabolism*, 86, 4686–4692. https://doi.org/10.1210/jcem.86.10.7974

Arlt, W., Callies, F., van Vlijmen, J.C., Koehler, I., Reincke, M., Bidlingmaier, M., Huebler, D., Oettel, M., Ernst, M., Schulte, H.M. and Allolio, B. (1999) *Dehydroepiandrosterone replacement in women with adrenal insufficiency. The New England Journal of Medicine*, 341, 1013–1020. https://doi.org/10.1056/nejm199909303411401

Ayala, I. and Martos, N. (2013) *Changes in serum dehydroepiandrosterone, androstenedione, testosterone, and 17β-oestradiol levels associated with disease and surgery in the horse. Acta Veterinaria Brno*, 82, 91–96. https://doi.org/10.2754/avb201382010091

Barnhart, K.T., Freeman, E., Grisso, J.A., Rader, D.J., Sammel, M., Kapoor, S. and Nestler, J.E. (1999) *The effect of dehydroepiandrosterone supplementation to symptomatic perimenopausal women on serum endocrine profiles, lipid parameters, and health-related quality of life. The Journal of Clinical Endocrinology & Metabolism*, 84, 3896–3902. https://doi.org/10.1210/jcem.84.11.6153

Barrett-Connor, E., von Muhlen, D., Laughlin, G.A. and Kripke, A. (1999) *Endogenous levels of dehydroepiandrosterone sulfate, but not other sex hormones, are associated with depressed mood in older women: The Rancho Bernardo Study. Journal of the American Geriatrics Society*, 47, 685–691. https://doi.org/10.1111/j.1532-5415.1999.tb01590.x

Bauer, M.E. (2008) *Chronic stress and immunosenescence: A review. Neuroimmunomodulation*, 15, 241–250. https://doi.org/10.1159/000156467

Behringer, V., Hohmann, G., Stevens, J.M., Weltring, A. and Deschner, T. (2012) *Adrenarche in bonobos (Pan paniscus): Evidence from ontogenetic changes in urinary dehydroepiandrosterone-sulfate levels. Journal of Endocrinology*, 214, 55–65. https://doi.org/10.1530/joe-12-0103

Beishuizen, A., Thijs, L.G. and Vermes, I. (2002) *Decreased levels of dehydroepiandrosterone sulphate in severe critical illness: A sign of exhausted adrenal reserve? Critical Care*, 6, 434–438. https://doi.org/10.1186/cc1530

Bergamin, C., Comin, A., Corazzin, M., Faustini, M., Peric, T., Scollo, A., Gottardo, F., Montillo, M. and Prandi, A. (2019) *Cortisol, DHEA, and sexual steroid concentrations in fattening pigs' hair. Animals*, 9, 345. https://doi.org/10.3390/ani9060345

Boonstra, R., Lane, J.E., Boutin, S., Bradley, A., Desantis, L., Newman, A.E. and Soma, K.K. (2008) *Plasma DHEA levels in wild, territorial red squirrels: Seasonal variation and effect of ACTH. General and Comparative Endocrinology*, 158, 61–67. https://doi.org/10.1016/j.ygcen.2008.05.004

Buckham Sporer, K.R., Weber, P.S.D., Burton, J.L., Earley, B. and Crowe, M.A. (2008) *Transportation of young beef bulls alters circulating physiological parameters that may be effective biomarkers of stress. Journal of Animal Science*, 86, 1325–1334. https://doi.org/10.2527/jas.2007-0762

Butcher, S.K., Killampalli, V., Lascelles, D., Wang, K., Alpar, E.K. and Lord, J.M. (2005) *Raised cortisol: DHEAS ratios in the elderly after injury: Potential impact upon neutrophil function and immunity. Aging Cell*, 4, 319–324. https://doi.org/10.1111/j.1474-9726.2005.00178.x

Cao, J., Zhang, H., Yang, Z., Zhao, J. and Ma, H. (2020) *Effect of dehydroepiandrosterone on the immune response and gut microbiota in dextran sulfate sodium-induced colitis mice. Molecular Immunology*, 118, 60–72. https://doi.org/10.1016/j.molimm.2019.12.008

Cawood, E.H. and Bancroft, J. (1996) *Steroid hormones, the menopause, sexuality and well-being of women. Psychological Medicine*, 26, 925–936. https://doi.org/10.1017/S0033291700035261

Centurelli, M.A. and Abate, M.A. (1997) *The role of dehydroepiandrosterone in AIDS. Annals of Pharmacotherapy*, 31, 639–642.

da Silva, G.P., de Melo, J.T., Monteiro, F.O.B., Ferreira, A.K.P., Carneiro, L.A. and Takeshita, R.S.C. (2021) *Validation of a dehydroepiandrosterone-sulfate assay in three platyrrhine primates (Alouatta caraya, Aotus azarae infulatus, and Sapajus apella). International Journal of Primatology*, 42, 722–736. https://doi.org/10.1007/s10764-021-00239-x

Dembek, K.A., Timko, K.J., Johnson, L.M., Hart, K.A., Barr, B.S., David, B., Burns, T.A. and Toribio, R.E. (2017) *Steroids, steroid precursors, and neuroactive steroids in critically ill equine neonates. The Veterinary Journal*, 225, 42–49. https://doi.org/10.1016/j.tvjl.2017.05.009

Dumas de La Roque, E., Savineau, J.P., Metivier, A.C., Billes, M.A., Kraemer, J.P., Doutreleau, S., Jougon, J., Marthan, R., Moore, N., Fayon, M., Baulieu, E.-E. and Dromer, C. (2012) *Dehydroepiandrosterone (DHEA) improves pulmonary hypertension in chronic obstructive pulmonary disease (COPD): A pilot study. Annales d'Endocrinologie*, 73, 20–25. https://doi.org/10.1016/j.ando.2011.12.005

Dutheil, F., de Saint Vincent, S., Pereira, B., Schmidt, J., Moustafa, F., Charkhabi, M., Bouillon-Minois, J.-B. and Clinchamps, M. (2021) *DHEA as a biomarker of stress: A systematic review and meta-analysis. Frontiers in Psychiatry*, 12, 1–14. https://doi.org/10.3389/fpsyt.2021.688367

EFSA Panel on Animal Health and Welfare (AHAW), Nielsen, S.S., Alvarez, J., Bicout, D.J., Calistri, P., Canali, E., Drewe, J.A., Garin-Bastuji, B., Gonzales Rojas, J.L., Gortázar Schmidt, C., Michel, V., Miranda Chueca, M.Á., Padalino, B., Pasquali, P., Roberts, H.C., Spoolder, H., Stahl, K., Velarde, A., Viltrop, A., Winckler, C., Earley, B., Edwards, S., Faucitano, L., Marti, S., de La Lama, G.C.M., Costa, L.N., Thomsen, P.T., Ashe, S., Mur, L., Van der Stede, Y. and Herskin, M. (2022) *Welfare of cattle during transport. EFSA Journal*, 20(9), e07442. https://doi.org/10.2903/j.efsa.2022.7442

Falomo, M.E., Del Re, B., Rossi, M., Giaretta, E., Da Dalt, L. and Gabai, G. (2020) *Relationship between postpartum uterine involution and biomarkers of inflammation and oxidative stress in clinically healthy mares (Equus caballus). Heliyon*, 6, e03691. https://doi.org/10.1016/j.heliyon.2020.e03691

Fels, M., Rauterberg, S., Schwennen, C., Ligges, U., Herbrandt, S., Kemper, N. and Schmicke, M. (2019) *Cortisol/dehydroepiandrosterone ratio in saliva: Endocrine biomarker for chronic stress in pigs? Livestock Science*, 222, 21–24. https://doi.org/10.1016/j.livsci.2019.01.022

Flynn, M.A., Weaver-Osterholtz, D., Sharpe-Timms, K.L., Allen, S. and Krause, G. (1999) *Dehydroepiandrosterone replacement in aging humans. The Journal of Clinical Endocrinology & Metabolism*, 84, 1527–1533. https://doi.org/10.1210/jcem.84.5.5672

Fustini, M., Galeati, G., Gabai, G., Mammi, L.E., Bucci, D., Baratta, M., Accorsi, P.A. and Formigoni, A. (2017) *Overstocking dairy cows during the dry period affects dehydroepiandrosterone and cortisol secretion. Journal of Dairy Science*, 100, 620–628. https://doi.org/10.3168/jds.2016-11293

Gabai, G., Mongillo, P., Giaretta, E. and Marinelli, L. (2020) *Do dehydroepiandrosterone (DHEA) and its sulfate (DHEAS) play a role in the stress response in domestic animals? Frontiers in Veterinary Science*, 7, 588835. https://doi.org/10.3389/fvets.2020.588835

Goncharova, N.D., Marenin, V.Y. and Oganyan, T.E. (2010) *Aging of the hypothalamic-pituitary-adrenal axis in nonhuman primates with depression-like and aggressive behavior. Aging*, 2, 854–866. https://doi.org/10.18632/aging.100227

Goodyer, I.M., Herbert, J., Altham, P.M., Pearson, J., Secher, S.M. and Shiers, H.M. (1996) *Adrenal secretion during major depression in 8- to 16-year-olds, I. Altered diurnal rhythms in salivary cortisol and dehydroepiandrosterone (DHEA) at presentation. Psychological Medicine*, 26, 245–256. https://doi.org/10.1017/S0033291700034644

Guilliams, T.G. and Edwards, L. (2010) *Chronic stress and the HPA axis: Clinical assessment and therapeutic considerations. Standard*, 9, 1–12.

Gundlach, N.H., Feldmann, M., Gundelach, Y., Gil, M.A., Siebert, U., Hoedemaker, M. and Schmicke, M. (2017) *Dehydroepiandrosterone and cortisol/dehydroepiandrosterone ratios in dairy cattle with postpartum metritis. Research in Veterinary Science*, 115, 530–533. https://doi.org/10.1016/j.rvsc.2017.09.024

Gundlach, N.H., Schmicke, M., Ludes-Wehrmeister, E., Ulrich, S.A., Araujo, M.G. and Siebert, U. (2018) *New approach to stress research in phocids—Potential of dehydroepiandrosterone and cortisol/dehydroepiandrosterone ratio as markers for stress in harbor seals (Phoca vitulina) and gray seals (Halichoerus grypus). Journal of Zoo and Wildlife Medicine*, 49, 556–563. https://doi.org/10.1638/2017-0191.1

Hazeldine, J., Arlt, W. and Lord, J. (2010) *Dehydroepiandrosterone as a regulator of immune cell function. Journal of Steroid Biochemistry and Molecular Biology*, 120, 127–136. https://doi.org/10.1016/j.jsbmb.2009.12.016

Heaney, J.L.J., Carroll, D. and Phillips, A.C. (2014) *Physical activity, life events stress, cortisol, and DHEA: Preliminary findings that physical activity may buffer against the negative effects of stress. Journal of Aging and Physical Activity*, 22, 465–473. https://doi.org/10.1123/japa.2012-0082

Hechter, O., Grossman, A. and Chatterton, R.T. Jr. (1997) *Relationship of dehydroepiandrosterone and cortisol in disease. Medical Hypotheses*, 49, 85–91. https://doi.org/10.1016/s0306-9877(97)90258-9

Hucklebridge, F., Hussain, T., Evans, P. and Clow, A. (2005) *The diurnal patterns of the adrenal steroids cortisol and dehydroepiandrosterone (DHEA) in relation to awakening. Psychoneuroendocrinology*, 30, 51–57. https://doi.org/10.1016/j.psyneuen.2004.04.007

Hunt, P.J., Gurnell, E.M., Huppert, F.A., Richards, C., Prevost, A.T., Wass, J.A., Herbert, J. and Chatterjee, V.K. (2000) *Improvement in mood and fatigue after dehydroepiandrosterone replacement in Addison's disease in a randomized, double blind trial. The Journal of Clinical Endocrinology and Metabolism*, 85, 4650–4656. https://doi.org/10.1210/jcem.85.12.7022

Illera, J.C., Pérez-Alenza, M.D., Nieto, A., Jiménez, M.A., Silvan, G., Dunner, S. and Peña, L. (2006) *Steroids and receptors in canine mammary cancer. Steroids*, 71, 541–548. https://doi.org/10.1016/j.steroids.2005.11.007

Izawa, S., Sugaya, N., Shirotsuki, K., Yamada, K.C., Ogawa, N., Ouchi, Y., Nagano, Y., Suzuki, K. and Nomura, S. (2008) *Salivary dehydroepiandrosterone secretion in response to acute psychosocial stress and its correlations with biological and psychological changes. Biological Psychology*, 79, 294–298. https://doi.org/10.1016/j.biopsycho.2008.07.003

Jeckel, C.M.M., Lopes, R.P., Berleze, M.C., Luz, C., Feix, L., de Lima Argimon, I.I., Stein, L.M. and Bauer, M.E. (2010) *Neuroendocrine and immunological correlates of chronic stress in "strictly healthy" populations. Neuroimmunomodulation*, 17, 9–18. https://doi.org/10.1159/000243080

Ji, E., Weickert, C.S., Purves-Tyson, T., White, C., Handelsman, D.J., Desai, R., O'Donnell, M., Liu, D., Galletly, C., Lenroot, R. and Weickert, T.W. (2021) *Cortisol-dehydroepiandrosterone ratios are inversely associated with hippocampal and prefrontal brain volume in schizophrenia. Psychoneuroendocrinology*, 123, 104916. https://doi.org/10.1016/j.psyneuen.2020.104916

Kamin, H.S. and Kertes, D.A. (2017) *Cortisol and DHEA in development and psychopathology. Hormones and Behavior*, 89, 69–85. https://doi.org/10.1016/j.yhbeh.2016.11.018

Kaminska, B., Opalka, M., Ciereszko, R.E. and Dusza, L. (2000) *The involvement of prolactin in the regulation of adrenal cortex function in pigs. Domestic Animal Endocrinology*, 19, 147–157. https://doi.org/10.1016/s0739-7240(00)00076-x

Khanfer, R., Lord, J.M. and Phillips, A.C. (2011) *Neutrophil function and cortisol: DHEAS ratio in bereaved older adults. Brain, Behavior, and Immunity*, 25(6), 1182–1186. https://doi.org/10.1016/j.bbi.2011.03.008

Klinge, C.M., Clark, B.J., Prough, R.A. (2018). *Dehydroepiandrosterone research: Past, current, and future. Vitamins and Hormones* 108: 1–28. https://doi.org/10.1016/bs.vh.2018.02.002, New York, Elsevier

Labrie, F., Diamond, P., Cusan, L., Gomez, J.L., Bélanger, A. and Candas, B. (1997) *Effect of 12-month dehydroepiandrosterone replacement therapy on bone, vagina, and endometrium in postmenopausal women. The Journal of Clinical Endocrinology and Metabolism*, 82(10), 3498–3505. https://doi.org/10.1210/jcem.82.10.4306

Lee, J., Floyd, T., Erb, H. and Houpt, K. (2011) *Preference and demand for exercise in stabled horses. Applied Animal Behaviour Science*, 130(3–4), 91–100. https://doi.org/10.1016/j.applanim.2011.01.001

Leff-Gelman, P., Flores-Ramos, M., Carrasco, A.E.Á., Martínez, M.L., Takashima, M.F.S., Coronel, F.M.C., Labonne, B.F., Dosal, J.A.Z., Chávez Peón, P.B., Morales, S.G. and Camacho-Arroyo, I. (2020) *Cortisol and DHEA-S levels in pregnant women with severe anxiety. BMC Psychiatry*, 20, 1–14. https://doi.org/10.1186/s12888-020-02788-6

Lennartsson, A.K., Kushnir, M.M., Bergquist, J. and Jonsdottir, I.H. (2012) *DHEA and DHEA-S response to acute psychosocial stress in healthy men and women. Biological Psychology*, 90, 143–149. https://doi.org/10.1016/j.biopsycho.2012.03.003

Lennartsson, A.K., Sjörs, A. and Jonsdottir, I.H. (2015) *Indication of attenuated DHEA-s response during acute psychosocial stress in patients with clinical burnout. Journal of Psychosomatic Research*, 79, 107–111. https://doi.org/10.1016/j.jpsychores.2015.05.011

Lennartsson, A.K., Theorell, T., Rockwood, A.L., Kushnir, M.M. and Jonsdottir, I.H. (2013) *Perceived stress at work is associated with lower levels of DHEA-S. PLoS One*, 8(8), e72460. https://doi.org/10.1371/journal.pone.0072460

Leowattana, W. (2004) *DHEAS as a new diagnostic tool. Clinica Chimica Acta*, 341, 1–15. https://doi.org/10.1016/j.cccn.2003.10.031

Maninger, N., Wolkowitz, O.M., Reus, V.I., Epel, E.S. and Mellon, S.H. (2009) *Neurobiological and neuropsychiatric effects of dehydroepiandrosterone (DHEA) and DHEA sulfate (DHEAS). Frontiers in Neuroendocrinology*, 30, 65–91. https://doi.org/10.1016/j.yfrne.2008.11.002

Markopoulou, K., Papadopoulos, A., Juruena, M.F., Poon, L., Pariante, C.M. and Cleare, A.J. (2009) *The ratio of cortisol/DHEA in treatment resistant depression. Psychoneuroendocrinology*, 34, 19–26. https://doi.org/10.1016/j.psyneuen.2008.08.004

Montillo, M., Nodari, S.R., Peric, T., Polloni, A., Corazzin, M., Bergamin, C., Balestrieri, A., Prandi, A. and Comin, A. (2020) *Steroids in pig hair and welfare evaluation systems: combined approaches to improve management in pig breeding? Veterinaria Italiana*, 56, 177–184. https://doi.org/10.12834/vetit.1974.11885.1

Moraes, H., Deslandes, A., Maciel-Pinheiro, P.D.T., Corrêa, H. and Laks, J. (2016) *Cortisol, DHEA, and depression in the elderly: The influence of physical capacity. Arquivos de Neuro-Psiquiatria*, 74, 456–461. https://doi.org/10.1590/0004-282x20160059

Morales, A.J., Nolan, J.J., Nelson, J.C. and Yen, S.S.C. (1994) *Effects of a replacement dose of dehydroepiandrosterone in men and women of advancing age. The Journal of Clinical Endocrinology and Metabolism*, 78, 1360–1367. https://doi.org/10.1210/jcem.78.6.7515387

Morgan, C., Rasmusson, A., Pietrzak, R., Coric, V. and Southwick, S.M. (2009) *Relationships among plasma dehydroepiandrosterone and dehydroepiandrosterone sulfate, cortisol, symptoms of dissociation, and objective performance in humans exposed to underwater navigation stress. Biological Psychiatry*, 66, 334–340. https://doi.org/10.1016/j.biopsych.2009.04.004

Morgan, C., Southwick, S., Hazlett, G., Rasmusson, A., Hoyt, G., Zimolo, Z. and Charney, N. (2004) *Relationships among plasma dehydroepiandrosterone sulfate and cortisol levels, symptoms of dissociation, and objective performance in humans exposed to acute stress. Archives of General Psychiatry*, 61, 819–825. https://doi.org/10.1001/archpsyc.61.8.819

Morrison, M.F., Freeman, E.W., Lin, H. and Sammel, M.D. (2011) *Higher DHEA-S (dehydroepiandrosterone sulfate) levels are associated with depressive symptoms during the menopausal transition: Results from the PENN Ovarian Aging Study. Archives of Women's Mental Health*, 14, 375–382. https://doi.org/10.1007/s00737-011-0231-5

Muehlenbein, M.P., Campbell, B.C., Richards, R.J., Svec, F., Phillippi-Falkenstein, K.M., Murchison, M.A. and Myers, L. (2003) *Dehydroepiandrosterone-sulfate as a biomarker of senescence in male non-human primates. Experimental Gerontology*, 38, 1077–1085. https://doi.org/10.1016/j.exger.2003.07.001

Nair, K.S., Rizza, R.A., O'Brien, P., Dhatariya, K., Short, K.R., Nehra, A., Vittone, J.L., Klee, G.G., Basu, A., Basu, R., Cobelli, C., et al. (2006) *DHEA in elderly women and DHEA or testosterone in elderly men. New England Journal of Medicine*, 355, 1647–1659. https://doi.org/10.1056/nejmoa054629

Nejad, J.G., Lee, B.H., Kim, J.Y., Chemere, B., Sung, K.I. and Lee, H.G. (2021) *Effect of alpine grazing on plasma and hair cortisol, serotonin, and DHEA in dairy cows and its welfare impact. Domestic Animal Endocrinology*, 75, 106581. https://doi.org/10.1016/j.domaniend.2020.106581

Oberbeck, R., Benschop, R.J., Jacobs, R., Hosch, W., Jetschmann, J.U., Schürmeyer, T.H., Schmidt, R.E. and Schedlowski, M. (1998) *Endocrine mechanisms of stress-induced DHEA-secretion. Journal of Endocrinological Investigation*, 21, 148–153. https://doi.org/10.1007/bf03347293

O'Brien, J.K., Steinman, K.J., Fetter, G.A. and Robeck, T.R. (2017) *Androgen and glucocorticoid production in the male killer whale (Orcinus orca): Influence of age, maturity, and environmental factors. Andrology*, 5, 180–190. https://doi.org/10.1111/andr.12254

O'Driscoll, K., McCabe, M. and Earley, B. (2015) *Differences in leukocyte profile, gene expression, and metabolite status of dairy cows with or without sole ulcers. Journal of Dairy Science*, 98, 1685–1695. https://doi.org/10.3168/jds.2014-8199

O'Driscoll, K., McCabe, M. and Earley, B. (2017) *Leukocyte profile, gene expression, acute phase response, and metabolite status of cows with sole hemorrhages. Journal of Dairy Science*, 100, 9382–9391. https://doi.org/10.3168/jds.2017-13106

Olvera-Maneu, S., Carbajal, A., Gardela, J. and Lopez-Bejar, M. (2021) *Hair cortisol, testosterone, dehydroepiandrosterone sulfate and their ratios in stallions as a retrospective measure of hypothalamic–pituitary–adrenal and hypothalamic–pituitary–gonadal axes activity: exploring the influence of seasonality. Animals*, 11, 2202. https://doi.org/10.3390/ani11082202

Peric, T., Corazzin, M., Romanzin, A., Bovolenta, S., Prandi, A., Montillo, M. and Comin, A. (2017) *Cortisol and DHEA concentrations in the hair of dairy cows managed indoor or on pasture. Livestock Science*, 202, 39–43. https://doi.org/10.1016/j.livsci.2017.05.020

Phillips, A., Carroll, D., Gale, C. and Lord, J. (2010) *Cortisol, DHEA sulphate, their ratio, and all-cause and cause-specific mortality in the Vietnam Experience Study. European Journal of Endocrinology*, 163, 285–292. https://doi.org/10.1530/eje-10-0299

Pico-Alfonso, M.A., Garcia-Linares, M.I., Celda-Navarro, N., Herbert, J. and Martinez, M. (2004) *Changes in cortisol and dehydroepiandrosterone in women victims of physical and psychological intimate partner violence. Biological Psychiatry*, 56(4), 233–240. https://doi.org/10.1016/j.biopsych.2004.06.001

Pieper, D.R. and Lobocki, C.A. (2000) *Characterization of serum dehydroepiandrosterone secretion in golden hamsters. Proceedings of the Society for Experimental Biology and Medicine*, 224, 278–284. https://doi.org/10.1046/j.1525-1373.2000.22432.x

Placci, M., Marliani, G., Sabioni, S., Gabai, G., Mondo, E., Borghetti, P., De Angelis, E. and Accorsi, P.A. (2020) *Natural horse boarding vs traditional stable: A comparison of hormonal, hematological and immunological parameters. Journal of Applied Animal Welfare Science*, 23, 366–377. https://doi.org/10.1080/10888705.2019.1663737

Polanco, A. (2021). Validating markers of cumulative experience in laboratory rhesus monkeys (Macaca mulatta). Doctoral dissertation, University of Guelph, Canada.

Pressman, S.D. and Cohen, S. (2005) *Does positive affect influence health? Psychological Bulletin*, 131, 925–971. https://doi.org/10.1037/0033-2909.131.6.925

Rainey, W.E., Rehman, K.S. and Carr, B.R. (2004) *The human fetal adrenal: Making adrenal androgens for placental estrogens. Seminars in Reproductive Medicine*, 22, 327–336. https://doi.org/10.1055/s-2004-861549

Raison, C. and Miller, A. (2003) *When not enough is too much: The role of insufficient glucocorticoid signaling in the pathophysiology of stress-related disorders. American Journal of Psychiatry*, 160, 1554–1565. https://doi.org/10.1176/appi.ajp.160.9.1554

Rasmusson, A., Vasek, J., Lipschitz, D.S., Vojvoda, D., Mustone, M.E., Shi, Q., Gudmundsen, G., Morgan, C.A., Wolfe, J. and Charney, D.S. (2004) *An increased capacity for adrenal DHEA release is associated with decreased avoidance and negative mood symptoms in women with PTSD. Neuropsychopharmacology*, 29, 1546–1557. https://doi.org/10.1038/sj.npp.1300432

Rege, J., Garber, S., Conley, A.J., Elsey, R.M., Turcu, A.F., Auchus, R.J. and Rainey, W.E. (2019) *Circulating 11-oxygenated androgens across species. Journal of Steroid Biochemistry and Molecular Biology*, 190, 242–249. https://doi.org/10.1016/j.jsbmb.2019.04.005

Ritsner, M., Maayan, R., Gibel, A., Strous, R.D., Modai, I. and Weizman, A. (2004) *Elevation of the cortisol/dehydroepiandrosterone ratio in schizophrenia patients. European Neuropsychopharmacology*, 14, 267–273. https://doi.org/10.1016/j.euroneuro.2003.09.003

Rondelli, M.C.H., Munhoz, T.D., Catandi, P.B., Freschi, C.R., Junior, R.P., Machado, R.Z. and Tinucci-Costa, M. (2015) *Serum DHEA-S increases in dogs naturally infected with Ehrlichia canis. Research in Veterinary Science*, 100, 18–20. https://doi.org/10.1016/j.rvsc.2015.04.010

Schmidt, P.J., Daly, R.C., Bloch, M., Smith, M.J., Danaceau, M.A.S., Clair, L.S., Murphy, J.H., Haq, N. and Rubinow, D.R. (2005) *Dehydroepiandrosterone monotherapy in midlife-onset major and minor depression. Archives of General Psychiatry*, 62, 154–162. https://doi.org/10.1001/archpsyc.62.2.154

Soma, K.K., Scotti, M.A.L., Newman, A.E., Charlier, T.D. and Demas, G.E. (2008) *Novel mechanisms for neuroendocrine regulation of aggression. Frontiers in Neuroendocrinology*, 29, 476–489. https://doi.org/10.1016/j.yfrne.2007.12.003

Staley, M., Conners, M.G., Hall, K. and Miller, L.J. (2018) *Linking stress and immunity: ImmunoglobulinA as a non-invasive physiological biomarker in animal welfare studies. Hormones and Behavior*, 102, 55–68. https://doi.org/10.1016/j.yhbeh.2018.04.011

Strous, R.D., Maayan, R., Lapidus, R., Stryjer, R., Lustig, M., Kotler, M. and Weizman, A. (2003) *Dehydroepiandrosterone augmentation in the management of negative, depressive, and anxiety symptoms in schizophrenia. Archives of General Psychiatry*, 60, 133–141. https://doi.org/10.1001/archpsyc.60.2.133

Susman, E.J., Dorn, L.D. and Chrousos, G.P. (1991) *Negative affect and hormone levels in young adolescents: Concurrent and predictive perspectives. Journal of Youth and Adolescence*, 20, 167–190. https://doi.org/10.1007/bf01537607

Takeshita, R.S., Bercovitch, F.B., Huffman, M.A., Mouri, K., Garcia, C., Rigaill, L. and Shimizu, K. (2014) *Environmental, biological, and social factors influencing fecal adrenal steroid concentrations in female Japanese macaques (Macaca fuscata). American Journal of Primatology*, 76, 1084–1093. https://doi.org/10.1002/ajp.22295

Takeshita, R.S., Huffman, M.A., Mouri, K., Shimizu, K. and Bercovitch, F.B. (2016) *Dead or alive? Predicting fetal loss in Japanese macaques (Macaca fuscata) by fecal metabolites. Animal Reproduction Science*, 175, 33–38. https://doi.org/10.1016/j.anireprosci.2016.10.006

Takeshita, R.S.C., Bercovitch, F.B., Huffman, M.A. and Kinoshita, K. (2018) *Development and validation of an enzyme immunoassay for fecal dehydroepiandrosterone sulfate in Japanese macaques (Macaca fuscata). International Journal of Primatology*, 39, 208–221. https://doi.org/10.1007/s10764-018-0026-x

Takeshita, R.S.C., Mendonça, R.S., Bercovitch, F.B. and Huffman, M.A. (2019) *Developmental changes in the endocrine stress response in orangutans (Pongo pygmaeus). Journal of Comparative Physiology*, 189, 659–672. https://doi.org/10.1007/s00360-019-01235-7

Tejerizo, G., Doménech, A., Illera, J.C., Silván, G. and Gómez-Lucía, E. (2012) *Altered plasma concentrations of sex hormones in cats infected by feline immunodeficiency virus or feline leukemia virus. Domestic Animal Endocrinology*, 42, 113–120. https://doi.org/10.1016/j.domaniend.2011.11.001

Trevisan, C., Montillo, M., Prandi, A., Mkupasi, E.M., Ngowi, H.A. and Johansen, M.V. (2017) *Hair cortisol and dehydroepiandrosterone concentrations in naturally Taenia solium infected pigs in Tanzania. General and Comparative Endocrinology*, 246, 23–28. https://doi.org/10.1016/j.ygcen.2017.03.007

Van Niekerk, J.K., Huppert, F.A. and Herbert, J. (2001) *Salivary cortisol and DHEA: Association with measures of cognition and well-being in normal older men, and effects of three months of DHEA supplementation. Psychoneuroendocrinology*, 26, 591–612. https://doi.org/10.1016/s0306-4530(01)00014-2

Van Vollenhoven, R.F. and McGuire, J.L. (1996) *Studies of dehydroepiandrosterone (DHEA) as a therapeutic agent in systemic lupus erythematosus. Annales de Médecine Interne (Paris)*, 147, 290–296.

Warnock, F., McElwee, K., Seo, R.J., McIsaac, S., Seim, D., Ramirez-Aponte, T., Macritchie, K.A.N. and Young, A.H. (2010) *Measuring cortisol and DHEA in fingernails: A pilot study. Neuropsychiatric Disease and Treatment*, 6, 1–7.

Whitham, J.C., Bryant, J.L. and Miller, L.J. (2020) *Beyond glucocorticoids: Integrating dehydroepiandrosterone (DHEA) into animal welfare research. Animals*, 10, 1381. https://doi.org/10.3390/ani10081381

Williams, R., Laffoon, A., Warnick, J. E. (2008). Positive emotion displayed in a writing task predicts salivary biochemical changes. Psychobiology in the Sun Belt Conference 5: 42–43.

Wolf, O.T., Neumann, O., Hellhammer, D.H., Geiben, A.C., Strasburger, C.J., Dressendörfer, R.A., Pirke, K.M. and Kirschbaum, C. (1997) *Effects of a two-week physiological dehydroepiandrosterone substitution on cognitive performance and well-being in healthy elderly women and men. The Journal of Clinical Endocrinology & Metabolism*, 82, 2363–2367. https://doi.org/10.1210/jcem.82.7.4056

Wolkowitz, O.M., Reus, V.I., Keebler, A., Nelson, N., Friedland, M., Brizendine, L. and Roberts, E. (1999) *Double-blind treatment of major depression with dehydroepiandrosterone. American Journal of Psychiatry*, 156, 646–649. https://doi.org/10.1176/ajp.156.4.646

Wolkowitz, O.M., Reus, V.I., Roberts, E., Manfredi, F., Chan, T., Ormiston, S., Johnson, R., Canick, J., Brizendine, L. and Weingartner, H. (1995) *Antidepressant and cognition-enhancing effects of DHEA in major depression. Annals of the New York Academy of Sciences*, 774, 337–339. https://doi.org/10.1111/j.1749-6632.1995.tb17403.x-i1

Wong, S.Y., Leung, J.C., Kwok, T., Ohlsson, C., Vandenput, L., Leung, P.C. and Woo, J. (2011) *Low DHEAS levels are associated with depressive symptoms in elderly Chinese men: Results from a large study. Asian Journal of Andrology*, 13, 898–902. https://doi.org/10.1038/aja.2011.116

Part 3

Indicators of Cumulative Affective Experience

Abnormal Repetitive Behaviours

11

Georgia J. Mason and Lindsey Kitchenham

Department of Integrative Biology & Campbell Centre for the Study of Animal Welfare, University of Guelph, Guelph, Ontario, Canada

Introduction

Tigers and polar bears repeatedly pacing worn down tracks in a zoo enclosure, pigs in intensive farms making chewing movements until their saliva froths, mice in biomedical research facilities running in circles in their cages – abnormal repetitive behaviours (*ARBs*) like these are common. One conservative estimate is that over 70 million zoo, farm and lab animals a year perform these behaviours (Mason and Latham 2004). ARBs are sometimes divided into *stereotypic behaviours* – defined by either their highly predictable appearance or the role of frustration or brain dysfunction in their aetiology (Mason 2006) – and more variable-looking behaviours such as fur- or feather-plucking, or (in humans) obsessive-compulsive rituals. And along with being diverse in appearance, ARBs are diverse mechanistically too. Some apparently derive from normal active responses to frustration or motivational conflict (as described in any ethological textbook): escape attempts, or displacement, redirected or sham activities (e.g. Keiper 1969a; Rushen and de Passillé 1995; Wiedenmayer 1997), which in monotonous environments become repetitive and predictable (Figure 11.1, pathways 1a and 3). But neurological research and insights from human clinical populations also highlight roles for captivity-induced, potentially dysfunctional, neurological changes (Figure 11.1, pathways 1c, 4 and 5). These may reflect lasting developmental harms from raising young animals in impoverished environments and/or more reversible, state-related effects such as neuroinflammation, induced by low moods, chronic stress and even (as we will see) infection. Together these disinhibit normal behavioural control to make subjects more compulsive, impulsive and/or persistent and behaviourally inflexible, depending on which brain regions are affected (e.g. Kitchenham and Mason 2021; Muehlmann and Lewis 2012; Wilkes and Lewis 2018).

ARBs have long been linked with poor welfare, deemed '*a sure sign of wrong treatment*' (Hediger 1950, p. 75) and '*amongst the most important indicators of long-term-welfare problems*' (Broom and Johnson 1993, p. 79). Yet as research on ARBs expanded, their relationships with welfare started looking more nuanced, leading to statements like '*not all stereotypies*

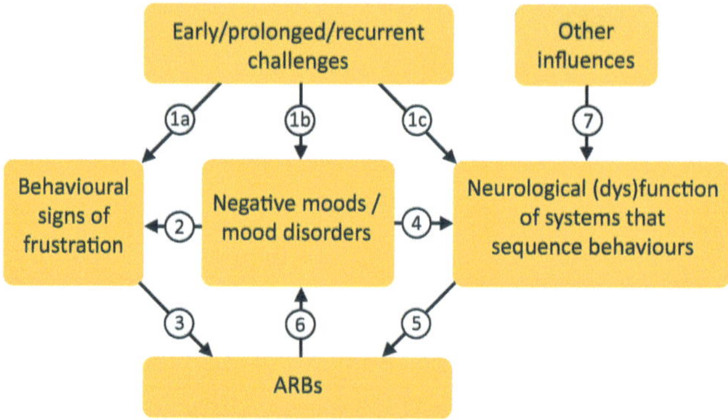

Figure 11.1 The aetiology of ARBs (abnormal repetitive behaviours). Source: Adapted from Mason and Kitchenham (2026).

1) *Challenges cause (a) motivational frustration, leading to escape attempts, and/or vacuum/redirected/displacement activities; (b) negative moods or even mood disorders and (c) altered development of neurological systems that sequence behaviour (e.g. prefrontal cortex and basal ganglia in mammals).*
2) *Negative moods may enhance behavioural signs of frustration (e.g. by increasing escape motivations), while mood disorders may promote agitation and restlessness.*
3) *Behavioural signs of frustration are repeated, perhaps due to positive reinforcement from beneficial consequences of performance (coping effects), and become predictable due to both environmental monotony and normal processes of routine formation.*
4) *Negative moods/mood disorders cause or correlate with state-dependent neurological changes (perhaps involving neuroinflammation) in brain regions involved in behavioural control.*
5) *Effects 1c) and 4) cause behavioural disinhibition and inflexibility (impulsive, compulsive or persistent behavioural selection).*
6) *ARB performance may itself further lower mood by compromising adaptive behaviour (e.g. normal social interactions).*

Other influences on the function of brain regions involved in behavioural control may be unrelated to affective states (e.g. reflecting genetic effects, individual differences, stimulant drug effects, etc.: see* ***Scope* *section for details).*

are sensitive indicators of current stress or frustration' (Mason and Latham 2004, p. S64); '*there is no one to one correspondence between stereotypy and welfare*' (Ashley 2007, p. 216), and even '*stereotypy is not a valid welfare correlate on its own. Some stereotypies indicate poor welfare. Others show the opposite*' (Dawkins 2021, p. 55). Here, we therefore tackle the often-paradoxical connections between ARBs and welfare (to end up – spoiler alert – finding that those early commentators were right after all). To do this, like other chapters, we first present evidence (sometimes from experiments that, we warn, are very distressing) for construct validity via **Tests 1–5** described in Chapter 2. This will include examples where ARBs show good responsiveness to differing degrees of poor welfare. We then discuss how changes in overall activity levels can generate both false null and false lead errors, before exploring other causes of false nulls (this welfare indicator's main weakness), as well as some ongoing puzzles that need more research. We end with some simple rules of thumb for using ARBs accurately in welfare assessment. But first, we cover three topics that determine this chapter's scope and eliminate some causes of false nulls and false leads from the get-go: what to include as *ARBs*; why ARBs should not be used to infer moment-to-moment emotion; and why ARBs should only be used to compare the welfare impacts of different environments (and not, for instance, different individuals).

Clarifying this chapter's scope

What we are terming an ARB

Just like forms of play (Chapter 8), ARBs are heterogeneous, and the precise borders between them and other activities are not always clear. The term *abnormal repetitive behaviour* has long been applied to diverse behaviours that concern and/or mystify (e.g. Mills and Luescher 2006) and are perhaps rare or absent in free-living populations, or appear inappropriate or excessive (e.g. Garner 2005). This chapter deliberately casts a wide net, consciously excluding only two candidates. One was coprophagia by chimpanzees: distasteful to humans, but a natural behaviour occurring in the wild (e.g. Hopper et al. 2016). The other was wheel-running by rodents: very plausibly an ARB (Mason and Würbel 2016), yet highly motivated, and reliant on 'enrichments' whose use we don't wish to accidentally discourage. These aside, we collated a vast, diverse array of ARBs, from highly predictable pacing to much more variable responses, all united in being studied in ways permitting the application of **Tests 1–5**. (And we revisit this knotty issue of definitions and categorisation at the end of the chapter.)

Why ARBs should not be used to infer moment-to-moment changes in emotion

This chapter straddles the 'mood' and 'cumulative negative experience' sections because, while ARB proneness indicates long-term poor welfare (as we will see), the onset and offset of ARB bouts do *not* reliably reflect moment-to-moment changes in current negative emotions (Mason and Kitchenham 2026). For one, aversive stimuli often fail to trigger ARB bouts, making them prone to *false nulls* as indicators of emotion. For example, being newly exposed to hunger (e.g. Cabib and Bonaventura 1997; Duncan and Wood-Gush 1972), newly confined (e.g. Deaton et al. 1977) or newly isolated from companions (e.g. Meehan et al. 2004), typically does not cause ARBs to develop despite being negative. And in animals with already-developed ARBs, frightening stimuli likewise often fail to trigger ARB performance (e.g. Poirier et al. 2019).

Furthermore, ARB bouts may be performed during moments of calm or even positive valence. Human children may perform bouts while watching television or falling asleep (e.g. Oakley et al. 2015; Tröster 1994), and adults with developmental conditions like autistic spectrum disorder may perform bouts during moments of excitement or intense focus that are emotionally neutral or even positive (e.g. Lanzarini et al. 2021; Willemsen-Swinkels et al. 1998). Similar patterns occur in animals. In some dogs, food (along with reported happiness and excitement) is a trigger for tail-chasing (Tiira et al. 2012); while pacing and weaving movements often occur just before expected meals (e.g. in broiler breeder chickens [Kostal et al. 1992], horses [McAfee et al. 2002; Wickens 2009], and diverse Carnivora species [Mason and Mendl 1997]). Ingesting food can also elicit ARB bouts, as seen in sows (e.g. Lawrence and Terlouw 1993), calves (e.g. de Passillé et al. 1992, 1993; Rushen and de Passillé 1995) and broiler breeder (e.g. Kostal et al. 1992), especially if this experience seems rather positive: larger meals elicit more post-prandial object-pecking by chickens (Savory and Mann 1999), while regurgitation-and-re-ingestion by chimpanzees is often triggered by eating highly palatable treats (e.g. Baker and Easley 1996). ARB bouts are thus also prone to *false leads* as indicators of negative emotion. Contributing yet further to this problem, performing ARBs can sometimes be immediately calming or stress-reducing. For people, hair plucking and skin picking are reportedly gratifying/stress-relieving (e.g. Chamberlain 2007; Kłosowska et al. 2021); bouts of ARB performance can reduce heart-rates (e.g. Heathers et al. 2019); and gum-chewing can reduce anxiety and cortisol during acute stress (Smith 2016). Similar physiological data show that in calves, amongst others, repetitive tongue-playing also predicts reduced heart rates (Seo et al. 1998).

This chapter therefore avoids using ARB performance to infer short-term emotions: focusing on the triggers and momentary affective correlates of bouts risks making welfare assessments inaccurate. Instead, we consider why ARBs slowly emerge in the first place, and why they come to reach certain levels in terms of population prevalences or average time budgets.

Why ARBs should only be used to compare the welfare impacts of events or environments

Even taking this long-term view, *some* factors that promote or inhibit the emergence of ARB are *still* unrelated to welfare (see pathway 7 in Figure 11.1). This is because (i) the neurological systems responsible for behavioural sequencing are affected by many factors, not all welfare-related; and (ii) ARBs rely on subjects being behaviourally active and persistent, yet some are not. Such complexities risk generating myriad false nulls and false leads in welfare assessment. A solution is to only use ARBs to make certain comparisons: *of events or environments* external to the animal. How and why should become clear after illustrating the problem.

One illustration comes from drugs that mimick dopamine or stimulating its release. These specific drugs often induce ARBs: repetitive licking, gnawing and sniffing are common in animals, while humans may repetitively manipulate objects (e.g. taking items apart and putting them together again). But they do not lower mood (far from it): animals seek them out (e.g. Sanchis-Segura and Spanagel 2006), and humans also take them voluntarily (either therapeutically for Parkinson's disease or recreationally for their euphoric effects). These drug-induced ARBs thus do not indicate poor welfare. Instead they simply reflect normal behavioural control being disinhibited pharmacologically by drugs that interact with neurological systems responsible for behavioural sequencing (Everitt and Robbins 2005; Shukla and Pandey 2020; Mason and Kitchenham 2026).

These neurological systems are affected by genetic background too. Consequently, some inherited neurological predispositions can promote or inhibit ARB development for reasons unrelated to welfare. Data from different mouse strains illustrate this. Faced with similar challenges (e.g. small barren cages or prolonged food deprivation), mice from DBA/2 and C58/J strains show far more ARB than C57Bl/6 animals. These differences reflect strain differences in brain systems responsible for behavioural sequencing: in stressed C57Bl/6s, their prefrontal cortices strongly inhibit dopamine activity in the basal ganglia (Cabib 2006; Cabib et al. 1998, 2020, 2021), basal ganglial regions that inhibit behaviour are more active (e.g. Lewis et al. 2018), and consequently C57Bl/6 mice are both more behaviourally flexible than the other strains (e.g. Whitehouse et al. 2017) and more inactive too (Fureix et al. 2016). In contrast, these differences in ARB do *not* reflect strain differences in welfare: C57Bl/6 mice seem no less anxious than the other strains (Cabib et al. 2021; Moy et al. 2007) and indeed are worryingly prone to passive helplessness in tests for depression-like states (Fureix et al. 2016).

If it would be inaccurate to infer strain differences in welfare from strain differences in ARB, then by the same logic, it would be risky to infer welfare differences from ARB differences between species or sexes.[1] And this issue should seem familiar: a common theme in this book has been that genetic differences can affect the expression of welfare indicators quite independently of animals' subjective well-being. We would hesitate to use, say, Chapter 4's vocalisations, Chapter 5's corticosteroid levels, or (coming up next) Chapter 13's hippocampal volumes to compare welfare across species or sexes; and the same principle applies to ARBs.

Furthermore, for ARBs this same principle applies to individual differences too. When identically treated age-matched, sex-matched, strain-/species-matched individuals differ in how much ARB they perform, this again seems to not reliably reflect differences in well-being. Data on other welfare indicators often reveal no significant differences between spontaneously high- and low-ARB subjects faced with the same challenges, and sometimes they even indicate that higher ARB subjects are better off (Mason and Latham 2004, Mason and Kitchenham 2026). One likely reason is the type of personality pre-disposing subjects to ARBs (perhaps shaped, at least partly, by variability in the neurological systems discussed earlier). For both humans and animals, it has been suggested that active, assertive individuals are more likely to develop ARBs in sub-optimal conditions, while others instead become still and passive (Deaton et al. 1977; Ijichi et al. 2013). Furthermore, the potentially beneficial consequences of ARB-performance outlined earlier may even help those individuals who develop ARBs to cope better with sub-optimal conditions. Whatever the reason, one thing is evident: when faced with the same challenges, individuals who display high ARB cannot be assumed to have poorer welfare than otherwise similar individuals who display little or none.

Together, this flags some comparisons to avoid. Using ARBs to compare welfare across subjects differentially exposed to drugs that influence behavioural sequencing or across subjects differing in ARB for endogenous reasons (e.g. strain or individual predispositions) can lead to inaccurate conclusions (both false leads and false nulls). This chapter therefore focuses only on how ARB-proneness is affected by exposing groups of otherwise similar animals to different events or environments.

Evidence for construct validity, including responsiveness to degrees of welfare impact

For every test, more details and examples are given in Mason and Kitchenham (2026).

Test 1: links between ARB proneness and negative experiences in humans

Non-clinical human ARBs

ARBs in children, including potentially harmful forms like air-swallowing and skin-picking, emerge or increase during challenges like periods of stress at home or school (e.g. Câmara-Souza et al. 2023; Devanarayana and Rajindrajith 2012; Hwang et al. 2007), after being moved into residential care (Tröster 1994), or if their parents are abusive or indifferent

[1] This is not because species differences in ARB *never* indicate species differences in welfare: they can. Thus in zoos, wide-ranging, semi-nomadic Carnivora species are both more prone to stereotypic route-tracing than naturally more 'stay-at-home' species, and show reduced fecundity and higher infant mortality (Bandeli et al. 2023; Miller et al. 2019); while naturally browsing Ungulate species are both more prone to oral ARB than naturally grazing species, and have shorter lifespans (Mellor and Mason 2023). But importantly, some species (e.g. crab-eating macaques, Norway rats, donkeys) apparently rarely show ARB even when sub-optimally treated, probably becoming inactive instead; it is crucial not to assume such animals are fine.

(Egeland et al. 1983). Body-rocking – common in infants but normally grown out of – also persists and even increases in some children when they start attending school (Rafaeli-Mor et al. 1999). In adults, one WWII study suggested that ARBs increase in students facing the threat of war (Young 1947). Imprisonment – frightening, boring, and a cause of anxiety and depression (e.g. Haney 2003; Morin 2016) – also dramatically increases ARBs: pacing becomes prevalent (e.g. Vivar 2014), rituals like repeatedly folding clothes may develop (Anonymous 2014) and harmful ARBs like self-cutting may emerge (e.g. Haines et al. 1995; Anonymous 2014). Being isolated when captive has further effects (e.g. Metcalf 2020): after four weeks of solitary confinement, most prisoners of war report pacing, repetitively exercising or performing ritualistic activities (Deaton et al. 1977).

Illustrating good responsiveness, such ARBs may increase with the degree of challenge experienced. In juvenile residential homes, certain forms (e.g. self-scratching) are most prevalent in children believed to have experienced or witnessed abuse, and more kinds of ARBs emerge the longer children are in care (Tröster 1994). Nail-biting severity is also greater in adolescents who experienced more traumas as children (e.g. death, divorce, abuse, illness: Yasir et al. 2021). Perhaps reflecting such differential exposure, ARBs also positively covaried with anxiety in one study of school children (Keating et al. 2023), as did tooth-grinding severity in college students (Câmara-Souza et al. 2023). Turning to adults, in prisoners, levels of pacing likewise increase the longer they are in or exposed to solitary confinement (Deaton et al. 1977, Vivar 2014), while *lifetime stress scores*, a tally of all past negative events, predict tooth-grinding severity in Type A personality (i.e. neurotic) subjects (Pingitore et al. 1991).

Children who would have been neurotypical, but for exposure to challenge

In Romania in the 1980s, many thousands of otherwise neurotypical children were raised in so-called orphanages where they faced extreme neglect: barren conditions often involving confinement in cribs; little social contact or affection from caregivers (sometimes even abuse) and malnourishment. This caused physical health problems (Beckett et al. 2003), anxiety (Ellis et al. 2004; Nelson et al. 2014), shortened telomeres (Drury et al. 2012; see Chapter 12), and prevalent ARBs (e.g. Carlson and Earls 1997): 50–80% of children rocked back and forth or shifted rhythmically from foot to foot while holding their cribs' railings (Bos and Van Balen 2010; Fisher et al. 1997). Some ARBs then persisted once children were adopted into normal family homes (Beckett et al. 2002; Fisher et al. 1997), as predicted by two welfare-relevant variables: post-adoption ARBs were more prevalent the longer the child had been institutionalised (Beckett et al. 2002; Bos and Van Balen 2010), and the less time spent in normal family homes (Fisher et al. 1997; Hoksbergen et al. 2005).

More recently, it has emerged that some children also develop ARBs after *Streptococcus* infections (e.g. those causing sore throats), thanks to an autoimmune syndrome sometimes termed *Pediatric Autoimmune Neuropsychiatric Disorders Associated with Streptococcal Infections* (PANDAS). This causes the sudden onset of obsessions, compulsions and tics (Leonardi et al. 2024; Swedo et al. 2024), as well as signs of poor welfare: anxiety, irritability, mood swings, suicidality, and sleep disturbance (National Institute of Mental Health 2025; PANDAS Network 2025; Wang et al. 2024). Risk factors include genetic predispositions as well as exposure to chronic stress (Gagliano et al. 2025) and recurrent infections (Murphy et al. 2007). Stress exposure after onset (e.g. lock-down during the COVID19 pandemic) can also exacerbate symptoms (e.g. Guido et al. 2021).

Neurological changes – including in brain systems responsible for behavioural sequencing – are implicated in tragic cases like these. Effects in institutionalised children included reduced cortical volume and activity, and altered cortico-basal ganglia connectivity (e.g. Behen et al. 2009; Chugani et al. 2001; Sheridan et al. 2012); and children showing persistent ARBs post-adoption had the greatest signs of neurological dysfunction (e.g. poorer cognitive abilities; Beckett et al. 2002; Bos and Van Balen 2010). Turning to PANDAS, this syndrome seems to reflect neurological inflammation (Leonardi et al. 2024), with cross-reactive autoantibodies attacking neuronal targets in the basal ganglia (ARBs therefore co-occurring with other behavioural changes like altered motor skills and hyperactivity).

Humans with clinical conditions characterised by ARBs

Several clinical conditions are *defined* by ARBs. These range from the predictable (e.g. tics in Tourette's syndrome; stereotypies in Rett's syndrome, autism spectrum disorder and stereotyped movement disorders) to the more flexible (e.g. rituals in obsessive-compulsive disorder [OCD]; body-focused hair plucking and skin-picking in trichotillomania and excoriation disorders). These ARBs are thus diverse, looking different to observers, as well as feeling different to performers (and being heterogeneous even within each condition). Nevertheless, they share commonalities including altered cortico-basal ganglia functioning (e.g. Langen et al. 2011; Shukla and Pandey 2020) and often being exacerbated by stress (e.g. Shulman et al. 1996).

Thus challenges like being bullied, exposed to trauma or relationship problems are often associated with, and likely cause, increased ARB. For example, adults with excoriation disorder (Özten et al. 2015; Spitzer et al. 2022) and trichotillomania (Lochner et al. 2002; Özten et al. 2015) are more likely to have experienced childhood traumas (e.g. abuse or neglect) than age- and sex-matched controls. They are also more likely to have experienced traumatic events in adulthood (excoriation disorder: Kłosowska et al. 2021; Spitzer et al. 2022; trichotillomania: Özten et al. 2015), as are adults with OCD (Horesh et al. 2018; Sarkhel et al. 2011; Vidal-Ribas et al. 2015).

Furthermore, *within* affected populations, the degree of challenge experienced can increase ARB severity in a dose-response-like. In children with autism, those who are bullied more show more anxiety and more ARB (Cappadocia et al. 2012); while children and adolescents with Tourette's syndrome show more severe tics, the more major stressors they have experienced (Horesh et al. 2018, Steinberg et al. 2013). Similarly, across adults with OCD or Tourette's syndrome, more severe compulsions (Boger et al. 2020) or tics (Yang et al. 2022) are predicted by having faced more severe or numerous adverse experiences in childhood. In adults with OCD or trichotillomania, more severe ARB is predicted by previously experiencing any major stressor in their lifetime (OCD: Cromer et al. 2007; Imthon et al. 2020; trichotillomania: Houghton et al. 2016); the presence of a traumatic life event the year before onset can predict more chronic OCD (Goldberg et al. 2015), while Lin et al. (2010) find that prior psychosocial stress and past *Streptococcus* infections have additive effects on both OCD and Tourettic tic severity. Furthermore a cumulative experience –more severe, numerous or diverse adverse experiences over the whole lifespan predict more severe OCD (Cromer et al. 2007; Sarkhel et al. 2011; Vidal-Ribas et al. 2015) and more prolonged hair-plucking in trichotillomania (Gershuny et al. 2006).

Correspondingly, increased ARB in such populations is often associated with more negative moods or even mood disorders (e.g. Grant et al. 2017; Oakley et al. 2015; Özten et al. 2015; Spitzer et al. 2022). Indeed, across subjects, ARB severity covaries in graded, dose-response type ways with scores for anxiety, distress or depression in populations affected by OCD, Tourette's syndrome, autism spectrum disorder, Rett's syndrome, and excoriation disorder and trichotillomania (reviewed in Mason and Kitchenham 2026).

Test 2: effects of aversive experiences on animals' ARBs

Here we cover three types of well-studied aversive experience: barren housing; threatened or actual physical harm; and food deprivation.

Effects of barren housing

ARBs are often promoted by long-term housing conditions lacking 'enrichments': specific resources that, for at least 10 species (see Table 11.1), studies of motivation show are preferred (see Chapter 3).

Table 11.1 Resources that animals are motivated to interact with and their impact on ARB (abnormal repetitive behaviour) if absent.

Species	Preferred resource	Effect on ARB in animals housed without this
Chickens	Nest boxes Foraging/dust-bathing substrates Perches for sleeping on at night	Increased pacing before egg-laying More prevalent, more time-consuming feather pecking More prevalent feather pecking
Parrots (*assuming similarities across species*)	Opportunities to forage (chew and manipulate) – studied in orange-winged Amazon parrots	Reduced feather-plucking/ chewing – studied in African grey parrots
Chinchillas	Dust-bathing substrates	More prevalent fur-chewing (trend)
Mice	Nesting material Diverse resources including more space, diverse forms of shelter and nesting material, running wheels, hammocks, and chewable cardboard	More cage-top twirling More prevalent, time-consuming ARB including backflipping, route-tracing, bar-mouthing, cage-top twirling and fur-plucking
Bank voles	Loose hay and twigs as cover	More prevalent repetitive jumping
Gerbils	Opaque artificial burrows (chambers plus tunnels)	Increased stereotypic digging in cage corners
American mink	Water in which animals can swim/head-dip Running wheels Diverse resources including space, flowing water and novel or manipulable objects	Locomotor ARBs (e.g. pacing) are more prevalent and time-consuming (in 2/3 studies) Increased locomotor ARBs (e.g. pacing) and stationary whole-body ARBs Pacing, head-twirling and scrabbling are more prevalent and time-consuming
'Blue' (Arctic) foxes	Dirt floors for digging	More time spent in oral ARBs
Pigs	Straw to manipulate forage in	Increased prevalence of tail-biting
Cattle	Brushes to self-scratch and groom with	More time spent bar-licking

Adapted from Mason and Kitchenham (2026).

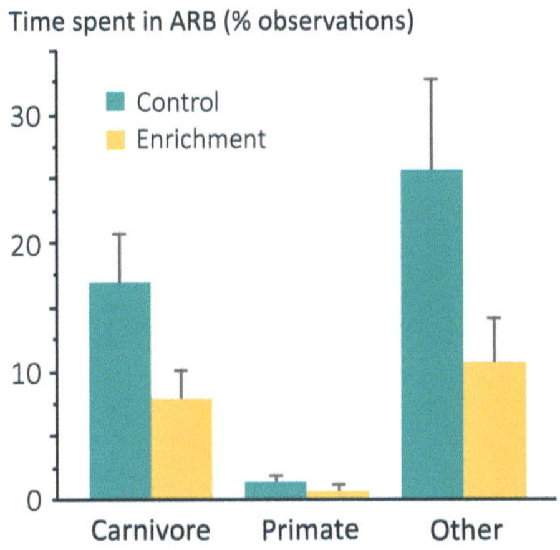

Figure 11.2 A summary of 22 zoo studies on the effects on ARB (abnormal repetitive behaviour) of 'enrichments' (alterations to the enclosure, or additions of objects or stimuli, made with the aim of increasing welfare). This intervention approximately halved the time devoted to ARB. Target species were mainly mammalian. Source: Reproduced from Swaisgood and Shepherdson (2006) / with permission of CABI.

Assuming that more barren environments are similarly aversive to other species whose housing- or resource-preferences remain unknown, this impact is widespread, increasing ARBs in Japanese quail (Laurence et al. 2014), and multiple species kept in in zoos (Shyne 2006; Swaisgood and Shepherdson 2005, 2006; see Figure 11.2).

Furthermore, such effects can be graded, with ARBs declining the better-resourced environments are. In American mink, locomotor ARBs like pacing are not reduced by merely adding a ball and chew-chain to farm cages (Meagher et al. 2014); but are approximately halved by providing suspended tunnels as well as chewing objects (Hansen et al. 2007), or highly valued running wheels (Malmkvist et al. 2024), and reduced by three-quarters by climbing structures and highly motivating large cages full of preferred resources (Díez-León et al. 2013, 2016). Similarly, the more motivated animals are to access particular resources, the more effectively these can reduce ARB. Thus, laboratory mice moved from well-resourced 'enriched' to barren cages showed more route tracing *and* greater motivations to escape (pushing heavier weights to do so: Latham and Mason 2010); and middle-aged barren-raised mice who were most motivated to reach enriched cages also showed the largest ARB reductions when rehomed there (Tilly et al. 2010). Durations of barren housing can have similar graded effects, suggesting cumulative impacts over time. In chickens, feather-pecking increases steadily over three months without litter, while not doing so if litter is present (Blokhuis and Arkes 1984); and in young orange-winged Amazon parrots (Meehan et al. 2004), ARBs similarly increase over eight months in barren housing, while barely doing so if given foraging, perching, climbing and exploratory opportunities.

Effects of actual or threatened physical harm

Repeated threats or physical harms can exacerbate ARBs. For example, maternally deprived rhesus monkeys exposed to 4–8 weeks of chronic unpredictable stress (CUS; the paradigm we first met in Chapter 2 meant to induce depression-like states, and here includ restraint and threat from humans brandishing nets), developed increased levels of pacing, rocking, bouncing and hair-pulling (Zhang et al. 2016). Similarly, chickens exposed to CUS (including restraint and social disruption) display more feather-pecking (Mindus et al. 2021). And across

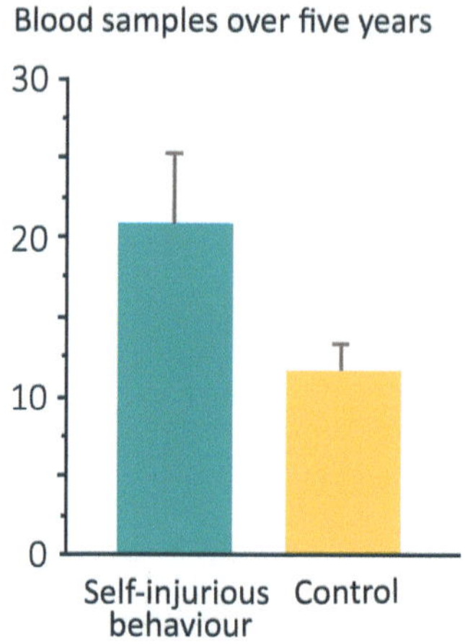

Figure 11.3 Total number of blood sample collections per individual over the preceding 5-year period for 27 single-caged rhesus monkeys who displayed self-injurious behaviour (SIB; largely self-biting) versus 161 who did not. Differences are statistically significant, with SIB animals having experienced nearly twice as many blood draws. Source: Reproduced from Novak (2003) / with permission of John Wiley & Sons.

dozens of single-housed rhesus monkeys in a research facility, experiencing more blood draws in recent years was found to be a risk factor for both self-injurious behaviour, largely self-biting (Novak 2003; Lutz et al. 2003; Figure 11.3) and repetitive 'saluting' (Lutz et al. 2003). Studies of such threats also reveal graded, dose-response-like effects. In chimpanzees, time budgets of self-directed ARBs and nodding covaried with the proportion of life spent in research projects (Warniment and Brent 1997); and in rhesus macaques, the number of research projects a monkey was used in over their lifetime helped predict levels of *motor stereotypy* (a broad category including pacing, twisting and bouncing: Gottlieb et al. 2013).

Moving to harms from conspecifics, levels of route-tracing, jumping and similar ARBs in mice are higher in barren cages that also elevate aggression between cage-mates, compared to well-resourced cages (e.g. Nip et al. 2019); and when well-resourced, if 'enrichments' are clustered and reduced in number, levels of ARB and aggression both increase in correlated, graded ways (Akre et al. 2011).

Effects of food restriction

Being fed less food than they would prefer promotes object-pecking, pacing and feather-pecking in chickens (Kostal et al. 1992; Savory et al. 1993; Duncan and Wood-Gush 1972; Savory and Mann 1999; Morrissey et al. 2014), chain and bar manipulation in pigs (Spoolder et al. 1995) and non-nutritive sucking in calves (e.g. Rushen and de Passillé 1995). Assuming that food restriction is also aversive to other animals (not least because impactful on fitness see **Test 3**), then it is also relevant to note that it exacerbates oral ARBs in cows (Redbo et al. 1996); rocking and swaying in chimpanzees, along with regurgitation and reingestion (Berkson and Mason 1964; Baker and Easley 1996); pacing in captive Carnivora (e.g. mink: Bildsøe et al. 1991; Mason and Mendl 1997) and repetitive jumping to the cage lid in DBA/2 mice (Cabib and Bonaventura 1997).

Furthermore, varying degrees of deprivation can have dose-response-like effects, including oral ARBs in sows (Figure 11.4) and object-pecking in chickens (Savory et al. 1993). Pacing increases in food-thwarted hens relative to the degree and duration of feed restriction (Duncan and Wood-Gush 1972).

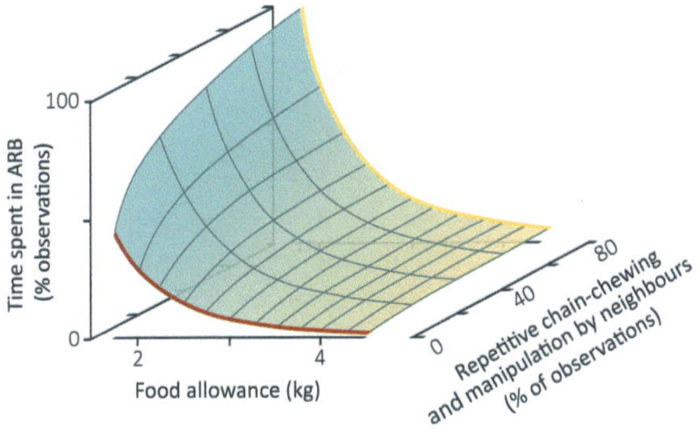

Figure 11.4 Relationship between total ARB (abnormal repetitive behaviour; primarily oral movements like bar-chewing) and daily food allowance across 76 sows (also showing the influence of neighbours' behaviour). Source: Reproduced from Appleby et al. (1989) / with permission of Elsevier.

Test 3: the effects of (ancestral) threats to fitness on animals' ARBs

Other threats to ancestral fitness promote ARB performance too. Again, we survey three well-documented examples: illness and infection; early parental loss; and broken or absent social bonds.

Effects of illness and infection

Infection can cause, or at least predict, elevated levels of ARB in diverse species. For example, rodents show ARBs like head-bobbing and excessive self-grooming after repeated *Streptococcus* exposure: a PANDAS-like effect (Mora et al. 2020; Wilson et al. 2023). Reflecting similar syndromes, and/or pain and discomfort, studies of horses and pigs find associations between oral ARB performance and gastritis, gastric ulcers and equine colic (Arias-Esquivel et al. 2024; Hemmings et al. 2007; Scantlebury et al. 2015). And in dogs, clinical conditions can also potentiate ARBs. Most dogs prone to repetitive surface-licking had gastrointestinal disorders, and treatment with hypoallergenic diets and antacid/anti-nausea medication reduced this ARB in over half (Bécuwe-Bonnet et al. 2012). In another study, dogs showing repetitive 'fly snapping' all had gastric or duodenal problems, and medical treatment reduced ARB in most of these dogs (Frank et al. 2012). Finally, although causal links are uncertain, comparisons across farms reveal that in chickens, feather-pecking is more likely in flocks affected by egg peritonitis and infectious bronchitis (Green et al. 2000), while in pigs, tail- and ear-biting is more likely in herds with high levels of respiratory disease (Boyle et al. 2022).

Effects of early parental loss

Reduced parental care in infancy, in ways that would increase mortality in the wild, often increases ARB too. Feather-pecking by hens is more prevalent in birds raised without mothers (Perré et al. 2002), and adult zoo parrots reared by humans perform more ARB than parent-raised birds (Williams et al. 2017). Surveys of pet parrots similarly report more feather-plucking in human-reared birds (Costa et al. 2016), with those removed earliest from the nest developing the most ARB (Schmid et al. 2006). Other owner surveys indicate that kittens who suck body parts are more likely to be orphans than those who do not (Delgado et al. 2020); cats who overgroom as adults, as well as tail-chasing dogs, were removed earlier from their mothers than non-affected conspecifics (Ahola et al. 2017; Tiira et al. 2012); and equine facilities that separate mothers from foals, rather than allowing natural weaning, have more prevalent ARBs (Parker et al. 2008).

Experiments on mammals reveal similar patterns, primates being particularly well researched. One study removed rhesus monkeys from their mothers immediately after birth (isolating them in barren cages). They soon developed digit-sucking, self-biting, and rhythmic rocking, the latter two ARBs persisting into adulthood (when new forms like pacing also developed; Cross and Harlow 1965; Harlow and Harlow 1962). In another, juvenile squirrel monkeys separated from their mothers for five days responded by pacing (Jones and Clark 1973). In today's research primates, premature maternal separation via nursery rearing is also a risk factor for ARBs like hair-pulling, SIB and motor stereotypies, displayed years later in adulthood (e.g. Bellanca and Crockett 2002; Gottlieb et al. 2013; Lutz et al. 2003, 2007). Furthermore, degree of parental deprivation may predict the magnitudes of such effects: across chimpanzees, there was an inverse linear relationship between adults' ARB levels and the ages at which, years earlier, they had been removed from their mothers (Warniment and Brent 1997).

In other mammals, ARBs are more time-consuming in subjects removed from their mothers at younger rather than older ages, as seen in CD-1 mice (Würbel and Stauffacher 1997) and African striped mice (Jones et al. 2010b). Striped mice raised by mothers alone, instead of biparentally as would happen naturally, also display more ARB in adulthood (Jones et al. 2010a). Tail-biting is more prevalent, and pacing is more time-consuming, in adult mink separated from their mothers at younger rather than older ages (Mason 1994; Jeppesen et al. 2000). Oral responses (object- and allo-sucking) soon emerge in calves and lambs prematurely weaned from their mothers (e.g. Napolitano et al. 2008); and in piglets, premature weaning also triggers repetitive belly-nosing ARBs that, in a graded way, are more prevalent and time-consuming the younger the animals were when losing their mothers (e.g. Faccin et al. 2020; Weary et al. 1999; Widowski et al. 2008).

Effects of social isolation

In species for whom group living is evolutionarily adaptive, being isolated or living in smaller groups than in the wild promotes ARB development. Avian examples include canaries (Keiper 1970; Sargent and Keiper 1967) and several parrot species (Meehan et al. 2003; Polverino et al. 2015; Williams et al. 2017). Similar effects occur in mammals (albeit often confounded with space allowance), including rats (Hurst et al. 1997, 1998), mice (Lewis and Hurst 2004), various primates (e.g. rhesus monkeys: Gottlieb et al. 2013; chimpanzees: Brent et al. 1989), horses (e.g. Visser et al. 2008; Flamand et al. 2025), dholes (a wild canid: Maisch et al. 2017), dogs (e.g. Hubrecht et al. 1992; Corsetti et al. 2023), calves (e.g. Leruste et al. 2014) and perhaps pigs (though here social contact was confounded with straw provision; e.g. Broom et al. 1995; Karlen et al. 2007).

Furthermore, graded effects may occur. For elephants (African and Asian), how much time they spend alone predicts time spent swaying, head-bobbing and other ARBs (Greco et al. 2016); while for zoo-housed primates, across 24 species the difference between average group sizes in the wild and those in captivity (typically smaller) predicts levels of self-plucking (Pomerantz et al. 2013). The duration of challenge can also have dose-response-like effects suggest cumulative impacts over time. In Broom et al.'s (1995) study, single-housed sows' levels of sham-chewing and bar-biting increased four- to fivefold between their first and fourth parities (while group-housed animals' stayed low); in orange-winged Amazon parrots (Meehan et al. 2003), young birds steadily developed more ARB over 12 months of single housing (while no such effects occurred in pair-housed peers); and in deer mice from well-resourced cages, 12 hours of isolation in barren test cages every few days steadily increased the ARB they displayed when in the barren cage (Figure 11.5).

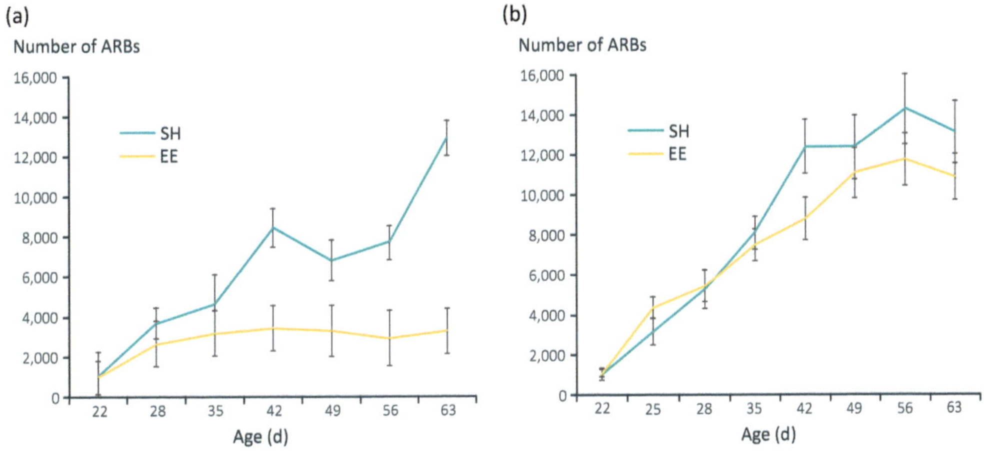

Figure 11.5 The development of ARB (abnormal repetitive behaviour) in differentially housed deer mice, and the impact of regular isolation in a barren test cage. (a) **Control group**: Each mouse is placed alone in a barren test cage (for automated ARB assessment) just once at one of the eight ages shown. Data are from mice reared in environmentally 'enriched' (EE; *n* = 93) housing or barren standard housing (SH; *n* = 123). (b) **Recurrent isolation group**: Each mouse is placed alone in a barren assessment cage eight times at the ages shown, i.e. approximately every five days. Data again are from mice reared in environmentally 'enriched' (EE; *n* = 40) housing or barren standard housing (SH; *n* =19). Error bars are ± SEM. Source: Figure redrawn from Bechard et al. (2017).

Past isolation can have lasting effects too. In Hawaiian crows, for instance, being reared without conspecifics increases rates of multiple ARBs in adulthood (Harvey et al. 2002). And again, effects of degrees of past isolation or social disruption can have graded effects. Across rescued apes in a chimpanzee sanctuary, for example, levels of abnormal behaviour (including ARB) increased with age in animals who had lived alone prior to rescue (while not increasing in previously socially housed subjects, whose levels were always low: Chen et al. 2025). Likewise, extents of past single housing help to steadily increase risks of current ARB in laboratory rhesus (Gottlieb et al. 2013; Lutz et al. 2003; Rommeck et al. 2009; Vandeleest et al. 2011) and pig-tailed macaques (Figure 11.6). In rhesus monkeys, being paired with multiple different cage-mates over their lifetime (thence more social disruption) also steadily increases risks of current ARB (Gottlieb et al. 2013). And in elephants, current ARB levels were predicted by the number of previous transfers between zoos: something typically involving broken social bonds since animals are generally moved on their own (Figure 11.7).

Test 4: the effects of drugs that reduce negative affect

In humans, mood-enhancing drugs can reduce some ARBs (effectiveness varying with aetiology). Tricyclic antidepressants can reduce ARBs in neurotypical people (e.g. Hwang et al. 2007) and be effective for stereotyped movement disorder and trichotillomania (reviewed by Shukla and Pandey 2020). Anxiolytic benzodiazepines may help reduce tics in Tourette's, especially for anxious subjects (Singer 2010). And selective serotonin reuptake inhibitors (SSRIs) like fluoxetine (a.k.a. Prozac) are a common treatment for OCD (e.g. Stein 2002; Hugo et al. 2003; Singer 2010; Lutz 2014).

In the earliest such study of animals, benzodiazapines reduced spot-picking and route-tracing in caged canaries (Keiper 1969b). These also reduced repetitive pacing in

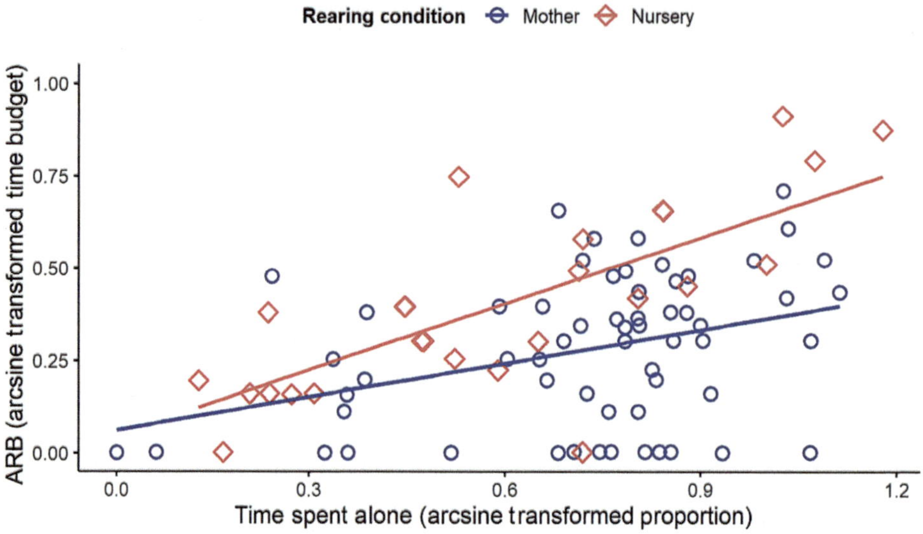

Figure 11.6 In adult male pigtailed macaques, the current proportion of time spent in total abnormal behaviour (including ARB) is predicted by the proportion of time spent singly housed before 48 months of age and by whether mother-reared or nursery-reared without the mother. Re-analysis also reveals a synergistic interaction between the two ($p<0.05$): having been nursery-reared promotes ARB more in monkeys who spent more time singly caged than in peers who spent more time socially housed; while spending time alone also promoted ARB more in nursery-reared than in mother-reared animals. Source: Bellanca and Crockett (2002) / with permission of John Wiley & Sons.

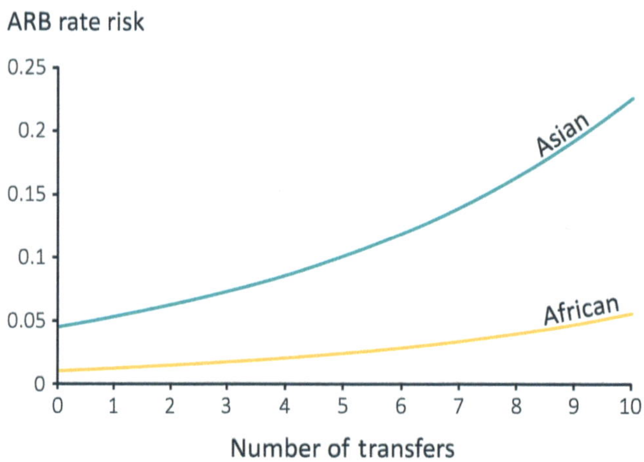

Figure 11.7 Risks for rates of ARB (abnormal repetitive behaviour) like swaying and bobbing, as predicted by inter-zoo transfers for 96 African and 84 Asian elephants. ARB rates were calculated for each elephant by dividing the counts of each behaviour by the total number of active visible samples recorded. Baseline risks for performing an ARB at or above observed rates are thus about 5%, increasing to about 10% after 5 transfers. Statistically, this corresponds to a 17.5% increase in risk for every additional transfer experienced. Source: Greco et al. (2016) / PUBLIC LIBRARY OF SCIENCE (PLOS) / CC BY 4.0.

food-thwarted hungry hens (even *preventing* it, if given before pacing emerged; Duncan and Wood-Gush 1972); and reduced time spent backflipping and route-tracing by conventionally caged mice exposed to CUS (Olsson and Sherwin 2006). Anti-depressants like SSRIs, and less commonly tricyclics, can also be effective (e.g. Mills and Luescher 2006; Poroshinska et al. 2024). SSRIs have reduced self-injurious behaviours in rhesus monkeys (Fontenot et al. 2005), somersaulting and weaving in vervet monkeys (Hugo et al. 2003), route-tracing and repetitive jumping in single-caged deer mice (Korff et al. 2008, 2009; Wolmarans et al. 2013), ARBs in highly stereotypic bank voles (Meers and Ödberg 2005) and fur-chewing in farmed chinchillas (Galeano et al. 2013). Fluoxetine has also reduced, sometimes even eliminated, pacing in zoo-housed bears of various species (e.g. Jeong et al. 2019; Poulsen and Cambell Teskey 2006), and a tiger (Baker 2002), as well as self-licking, tail-chasing and 'compulsive sucking' in pet dogs and cats (e.g. Mosallanejad et al. 2015; Rapoport et al. 1992; Tynes and Sinn 2014). Similarly, tricyclics have reduced feather-picking in pet cockatoos (van Zeeland 2018).

Test 5: ARBs in environments that induce other signs of negative affect

Data from experiments

Environments or events that make ARBs prevalent and time-consuming often induce other signs of poor welfare. Indeed, a tally across publications on 'stereotypies' found that across 196 comparisons of environments that affected this subset of ARBs, those that increased it induced other signs of poor welfare in 68% of cases: significantly above chance (Figure 11.8).

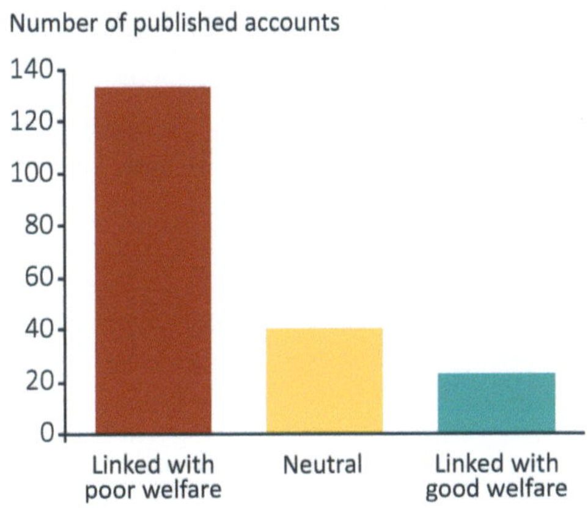

Figure 11.8 A tally of 196 studies comparing animals exposed to different environments (e.g. housing styles, stocking densities, weaning ages and so on), showing how those that elevate ARB (abnormal repetitive behaviour) affect other welfare indicators. Of these, 133 reported other signs of decreased welfare when ARBs increased, 23 reported improved welfare and 40 found no significant differences. Note, however, that some studies used indicators we would now question as signs of poor welfare, such as elevated baseline corticosteroids or non-validated judgement bias tests. Source: Adapted from Mason and Latham (2004).

Many subsequent experiments have mirrored this pattern. For example, aversive ARB-promoting environments that are barren or lack specific preferred resources (relevant to **Test 2**): decrease spleen and thymus weight, male testosterone levels and female fecundity in mink, while increasing their fear of humans and skeletal (developmental) asymmetry (Díez-León et al. 2013, 2016; Meagher et al. 2014); reportedly lead to more reproductive failure in dholes (Maisch et al. 2017); increase behavioural and neurological signs of depression and anxiety, decrease resilience to disease and shorten lifespans in mice (Cait et al. 2022); reduce growth rates in calves (Zhang et al. 2021); and reduce body condition in hamsters (Hauzenberger et al. 2006). Likewise, coyotes housed in small kennels rather than large enclosures not only pace more but also show greater heart rate and hiding responses to human proximity (Brummer et al. 2010); while rhesus macaques moved to a new building (and given new neighbours) not only show more SIB but also have elevated cortisol and disrupted sleep (Davenport et al. 2008).

Turning to studies of fitness threats (relevant to **Test 3**), for social animals the following ARB-promoting environments also have the following effects: single-caging for parrots and primates can shorten telomeres (Aydinonat et al. 2014), reduce hippocampal neurogenesis (e.g. Cinini et al. 2014) and impair health and survival (e.g. Lewis et al. 2000; Capitanio et al. 2019); and disrupting social bonds can increase rhesus monkeys' vulnerability to viral disease (e.g. Capitanio et al. 2019) and elephants' mortality rates (Clubb et al. 2008). Turning to rearing animals without mothers, for chicks this (potentially) increases both feather-pecking and also fearfulness (Rodenburg et al. 2009); for rhesus monkeys this both increases ARBs and shortens telomeres (Schneper et al. 2016), increases rates of physical illness (Conti et al. 2012), curtails lifespans (Lewis et al. 2000), and promotes significant weight loss if faced with ARB-inducing CUS (Zhang et al. 2016); and for chimpanzees who also have poor bonds to their human carers, this increases proneness to both stereotypic rocking *and* respiratory infections (Clay et al. 2015).

Varying the degrees of welfare impact may even reveal graded relationships between group-level ARBs and other welfare indicators. For brown capuchins exposed to a series of 'enrichments', the more effectively these reduced abnormal behaviour (including ARBs), the more effectively they reduced alarm reactions (Boinski et al. 1999). Likewise, in piglets, being weaned at progressively younger ages predicted progressively lower growth rates a progressively greater belly-nosing (Faccin et al. 2020).

Data from epidemiological studies

Turning to 'real world', non-experimental comparisons where varying environments are confounded with endogenous effects (e.g. individual differences), across flocks of chickens, feather-pecking severity (inferred from plumage scores) and fear of humans covaried (De Haas et al. 2014). Across stabled horses varying in background and age, more ARB-prone subjects slept less (Hausberger et al. 2007), showed more depression-like withdrawn behaviour (Hausberger et al. 2007), had more prevalent gastric ulceration (Patiño et al. 2020) and were less fertile (Benhajali et al. 2014). Across 6 zoos, female white rhinoceroses without oestrous cycles were rated as higher in ARB (Carlstead and Brown 2005); and across 52 Felidae individuals spanning 13 species and 5 zoos, keeper ratings for being 'stereotypical' ('fixed and oversimplified in behavior') were mirrored by high ratings for fearfulness and anxiety (O'Connor and Vonk 2022). Similarly, owner surveys on pet cats from diverse homes and experiences indicate that fearfulness and over-grooming are related (Mikkola et al. 2022), as are 'shyness' and tail-chasing in pet dogs (Tiira et al. 2012).

Summary of construct validity and responsiveness

Overall, a substantial body of research indicates that prolonged or repeated challenges (potentially including disease), negative events early in life (e.g. during infancy) and even cumulative exposure to challenges over a whole lifespan can cause the emergence of ARBs and potentiate their performance. Indeed, several well-studied forms of animal ARB (pacing by Carnivora, repetitive jumping and circling in caged rodents, feather-plucking by Psittacines and self-injurious behaviour in primates) passed *all* of **Tests 2–5**. Correspondingly, in humans, the development of ARBs and increases in their severity often reflect negative moods or even mood disorders (anxiety and depression). Consistent with a central role for mood, antidepressants and anxiolytics often reduce ARBs in both humans and animals. Furthermore, suggesting good responsiveness as indicators, relationships between ARBs and long-lasting negative moods, mood disorders, or cumulative negative experience are often graded, showing positive relationships between levels of ARB and the severity of negative experiences or self-reported mood, the duration of negative experiences and sometimes the number of challenges endured.

Now, there is a longstanding hypothesis that ARBs help performers to cope with stress: an idea sometimes used to downgrade ARBs as welfare indicators (e.g. Dawkins 2021; Mason et al. 2007). But our collated evidence shows that *despite* performing ARBs, people affected by chronic stress, imprisonment or OCD, autism spectrum disorder and related clinical conditions are still vulnerable to low moods and affective disorders. And in animals, environments that promote ARB performance are still aversive and still cause other signs of chronic stress. Thus, even if ARBs sometimes have immediate positive consequences (as outlined in the Introduction), the welfare of those who perform them clearly remains very poor.

Overall, the emergence of ARBs therefore has strong construct validity as a welfare indicator; and their levels (population prevalence and/or average time budgets) also have the potential to be *responsive* as indicators too. Reflecting early, repeated or long-term challenges, they allow the assessment of how different environments or events influence moodeven mood disorders or cause cumulative negative affect over a lifespan.

However, as with every indicator in this book, ARBs are not perfect. We found many instances where they either failed to respond to challenges or increased during neutral or even improved conditions. We review these, and the reasons for them, next.

Problems with the responsiveness and selectivity of ARBs as welfare indicators

ARBs often seemed to fail validation tests, especially through subjects not becoming more ARB-prone when challenged. For example, children with neglectful parents do not perform more ARBs than well-parented controls (unlike those with depressed or abusive parents), despite still seeming unhappy (Egeland et al. 1983). SSRIs do not tackle the ARBs of autism spectrum disorder (e.g. Maneeton et al. 2024; Williams et al. 2011). Negative experiences also do not cause the onset of ARBs in autism spectrum disorder, and several studies find similar for OCD, trichotillomania and excoriation (reviewed in Mason and Kitchenham 2026). Furthermore, roughly 70% of people worldwide experience at least one traumatic event in their lifetime (Benjet et al. 2016), yet most do not respond by developing ARB. Similarly, for animals, aversive or fitness-threatening challenges do not always impact ARBs. Some studies find no apparent effects of motherless rearing on feather-pecking in hens

(Angevaare et al. 2012; Rodenburg et al. 2009) or feather-plucking in parrots (e.g. Ebisawa et al. 2021; Schmid et al. 2006); while others fail to find that early maternal separation increases ARB levels in mice (Bailoo et al. 2020; Bechard et al. 2012). CUS seems not to increase ARB in mice either (Olsson and Sherwin 2006) despite being well-validated as reducing rodent welfare; and in cynomolgus monkeys (aka crab-eating macaques), sustained social threat or social isolation causes depression-like forms of inactivity rather than ARB (Li et al. 2013; Willard and Shively 2012, Shively et al. 2005). Finally, apparently mood-enhancing drugs may have no effects on animals' ARBs, or even be counter-productive (e.g. Burke et al. 2022; Olsson and Sherwin 2006).

Some of these anomalies may just be stochastic statistical errors, with Type I errors causing false leads and Type II errors false nulls. However, as we review next, certain consistent patterns emerge that help such results make sense (Again, more details and examples are given in Mason and Kitchenham 2026).

When challenges reduce animals' activity levels, ARBs can generate both false null and false lead results

If relatively positive environments increase overall activity levels, relatively negative ones decrease them, effects on ARB may be masked causing false nulls. To illustrate, coyotes in small pens hid more, and their heart rates recovered more slowly after human disturbance, compared to coyotes in larger pens, yet their ARB levels seemed similar in the two conditions (8–9% observations in the former, 6–7% in the latter: Brummer et al. 2010). However, overall activity was also lower in smaller pens. When we accounted for this, a housing effect emerged in the expected direction: ARB as a proportion of active time was nearly doubled in the smaller enclosures (now 17–19% compared to 8–10%).

Such effects may also cause false leads, whereby relatively positively treated subjects show *more* ARB. To illustrate, lab-born African striped mice display much more ARB than wild-caught conspecifics, despite being significantly less stressed and less anxious (Jones et al. 2011); and single-caged rats show more ARB when undisturbed than if exposed daily to CUS (Young et al. 2024). Likewise, farmed mink show more ARB if undisturbed than if restrained daily for an hour in tiny catching cages (something they avoid if they can: Bildsøe et al. 1991); more ARB if healthy than if sick (G. Mason, personal observation); and yet more ARB if supplied with simple toy-like enrichments to play with (e.g. balls: Meagher et al. 2014). In every instance, the elevated ARB in the better treatments was accompanied by elevated overall activity. Furthermore, in the two cases where we could statistically correct for this, doing so caused these false lead results to disappear (Meagher et al. 2014; Mason unpubl. reanalysis of Young et al. 2024). This still left results arguably yielding false nulls. But as we will see below, false nulls are common when using ARB to assess welfare.

Further problems with responsiveness

In the myriad instances of ARBs failing to change in sub-optimal conditions, many of these reflected a confounding effect of overall activity, either because activity levels were changed as above or because some genotypes generally display inactive responses to challenge instead of active ones (as we saw for C57Bl/6 mice, rats and cynomolgus monkeys). But there were other recurring themes too, especially that ARB emergence could be inhibited by apparent threshold effects; that increases in ARB performance could level out at apparent ceilings; and that, reflecting their origins in frustrated normal behaviour, some ARBs showed motivational selectivity, being increased only by specific challenges only.

When ARB emergence seems inhibited by threshold effects

As outlined in the *Scope* section earlier, food depriving or isolating animals for a day or two, or placing them in barren conditions for a day or two, seems insufficient to cause ARBs to emerge: onset may require days or even weeks of such treatment. Such delays could reflect the time needed for negative emotions to integrate into negative moods (see Chapter 1); the time needed for relevant neurological changes to occur (see Figure 11.1); and/or that negative affect must cross a threshold of severity before ARBs emerge.

Supporting this threshold idea, several studies indicate that a single challenge, even if prolonged, may not be enough to promote ARB: not unless combined with others. For example, housing mink in small cages that also lack swimming water potentiates ARB performance, yet each challenge *on its own* (small cage alone; no swimming water alone) had negligible effects (Hansen and Jeppesen 2000). In Japanese quail, a week of barren caging was found to have little effect on ARBs, unless birds were also subjected to CUS; and likewise a week of CUS had little effect on ARB, unless birds were also barren-caged (Laurence et al. 2014). In another CUS study, this time on young rhesus macaques, CUS had little effect on ARB unless animals were also maternally deprived; and even then, this was not manifest after one week of CUS, but only if CUS was more prolonged (Zhang et al. 2016). Thus, some challenges merely *prime* subjects for ARB: they have ARB-potentiating effects that stay hidden or latent unless animals are stressed yet further.

When increases in ARB performance seem prevented by ceiling effects

Once ARB has developed, several studies suggest ceiling effects on performance, even with increasing exposure to challenge. For instance, in parrots (Meehan et al. 2004) and mice (Nip 2018), ARBs increase over the first weeks or months of barren housing, yet average levels then plateau. Likewise, repeatedly isolating well-housed deer mice in test chambers caused their ARB to increase, but only up to the levels shown by barren-housed conspecifics; and furthermore, doing the same to barren-housed deer mice had no apparent effects at all, as if their levels were already at a maximum (see Figure 11.5). The previous section's two CUS studies also suggest ceiling effects: the high ARB of quails exposed to either barren housing or CUS for *two* weeks did not seem to elevate if both challenges were combined; while the elevated ARB levels of maternally deprived monkeys was no higher after eight weeks of CUS than after four weeks. In all these cases, ARBs thus level out at apparent maxima, not responding to increasingly prolonged or severe negative affect. Speculatively, such effects could reflect constraints on how much time or energy animals can devote to ARB; or perhaps even competition from increasing levels of inactivity (if fearfulness or depression-like apathy start to emerge as responses).

Motivational selectivity: when ARBs reflect specific negative experiences only

Finally, some cases of poor responsiveness highlight ARBs' motivational heterogeneity, some reflecting only very specific forms of frustration while being insensitive to others. For example, classic work on canaries by Keiper (1969a) showed that reducing cage sizes did not increase spot-pecking, despite increasing route-tracing; while providing less naturalistic diets did not increase route-tracing, despite increasing spot-pecking. Similarly, early loss of the mother has no obvious influence on tail- and ear-biting in growing pigs (reviewed by Boyle et al. 2022), despite causing belly-nosing (a direct response to maternal loss seemingly derived from normal udder massage). Likewise, for male mink, scrabbling at cage partitions (a conspecific-directed ARB reduced by removing male neighbours) is only minimally affected by housing animals in small barren cages, even though such cages are aversive and promote whole body ARBs (e.g. pacing) and repetitive head movements (Polanco et al. 2018); and nor is scrabbling increased by denying mink running wheels, despite these being valued resources that again help reduce pacing and head-twirling/nodding (Malmkvist et al. 2024).

Thus, frustrating animals may sometimes have little effect on particular ARBs: some challenges impact motivationally related ARBs only, such that not every ARB responds to all possible causes of negative mood.

Some outstanding puzzles

Before wrapping up, we flag three puzzles that cannot yet be interpreted: fascinating and potentially concerning topics that need more research using well-validated indicators.

First, the effects on ARBs of drugs aimed at enhancing mood are rather inconsistent. However, their effects on moods themselves are also inconsistent, especially for SSRIs: human research shows that these antidepressants can fail, take weeks or months to work or even be counterproductive (e.g. Fergusson et al. 2005; Kverno and Mangano 2021; Nierenberg et al. 2000). For example, for people with agitated depression, SSRIs typically make restlessness and other symptoms *worse* (Sohel et al. 2025). This reminds us how **Test 4** crucially relies on knowing a drug's true affective impact: an understanding we still lack for many animal species.

Second, apparent improvements in welfare can fail to reduce ARBs in some individuals, and it is unclear whether these are false nulls or true ones. Thus just as ARBs may persist in long-institutionalised children adopted from poor-quality orphanages into normal family homes (**Test 1**), so too can ARBs become hard to alleviate with well-resourced housing in animals who have been barren-housed for long periods of time (e.g. Tilly et al. 2010). It has also been suggested that mood-enhancing drugs become less effective on ARBs that have been performed for long periods of time (e.g. Duncan and Wood-Gush 1974; Jeong et al. 2019; Mills and Luescher 2006). Such persistent ARBs could perhaps reflect lasting impairments to the neurological systems that sequence behaviour: ones that thus still disinhibit behaviour despite improvements to well-being. If correct, these persistent ARBs are failing to track welfare: yet another case of poor responsiveness. However, just as in the drug examples above, another explanation is that treatments we *hope* are improving welfare are *not* actually doing so. If correct (and only findings from other welfare indicators will tell), then these treatment-resistant ARBs are worrying *true* nulls reflecting treatment-resistant negative affect. Alternatively, perhaps ARBs more validly reflect cumulative affective experience *over a lifetime* than they do just current mood. If this conception is correct, then persistent ARBs in aged or long-deprived subjects are again true nulls (not false ones), reflecting that small improvements late in life do little for cumulative lifelong well-being.

Third, as illustrated in Figure 11.4 (for pigs), many species' ARBs are exacerbated by ARB performance in neighbours. But despite this being a common effect, again we do not understand its welfare implications. Are such elevations merely false leads – increases due to increased activity, or perhaps even social learning, that do not indicate declines in welfare? Or, more worryingly, could they be true signs of prolonged negative affect, perhaps reflecting the aversiveness of being surrounded by abnormally behaving peers?

Use in practice: rules of thumb for the valid use of ARBs in welfare assessment

We found strong evidence that ARBs indicate long-term negative states: negative moods, sometimes even mood disorders. If environmental conditions cause the emergence of ARBs, or their increase, this is compelling evidence that such conditions compromise welfare.

Furthermore, ARBs may reflect cumulative negative affect ('cumulative stress'), suggesting that ARBs could represent new ways to assess animals' integrative lifetime welfare, and supporting recent suggestions that some ARBs in farm animals are so-called *iceberg indicators* integrating multiple welfare challenges (Cronin and Glatz 2021; EFSA AHAW Panel 2023; Šemrov and Patt 2024). We therefore applaud how ARBs are often used in welfare assessment and even policymaking. Thus for pigs, it is common to use them as welfare indicators, both in research studies and policy (e.g. Gonyou et al. 2012; Hoorweg et al. 2017; Wallgren et al. 2016; Zhang et al. 2017). The same holds for poultry (EFSA AHAW Panel 2023). It is also standard practice to use ARBs to assess welfare in laboratory animals (e.g. CCAC 2021; NC3Rs 2025) and zoo animals (e.g. Fazio et al. 2020; Hoorweg et al. 2017; Shyne and Block 2010; Swaisgood and Shepherdson 2005). We hope our demonstration of construct validity increases the impacts of these efforts, motivates those pushing for them and encourages those working with other species to do the same.

To help refine the accuracy with which this important indicator is used, this penultimate section therefore distils our review into some practical rules of thumb, helped – of course – by a heuristic graph (see Figure 11.9).

As laid out at the start of this chapter, our first rule of thumb is to not rely on ARBs to assess variation in welfare between similarly treated individuals, strains or species, but instead to recognise they are best suited for welfare comparisons between events or environments, especially if these occur early in life, are prolonged or are repeated. (This is why the *x*-axis of our heuristic graph is labelled 'affective valence of environment'.) Our second rule of thumb is not to be distracted by the triggers and momentary affective correlates of ARB bouts, but to focus on long-term effects only: why ARBs emerge in the first place and why their performance reaches certain levels. The most useful welfare-relevant question is thus not '*why is that animal performing an ARB right now?*' but '*why does that animal ever perform ARBs at all?*'. Our Figure 11.9 follows both these rules of thumb, to avoid depicting the myriad false leads and false nulls that would otherwise result.

A third rule of thumb is that ARBs often reflect not *just* the affective impact an animal's *current* environment but *also* challenges experienced in the past (potentially including disease). These might have simple additive effects, but alternatively, they might have more complex synergistic ones (like those of Figure 11.6), or even priming effects where the impact of one challenge on ARB-proneness is not detectable (cf. the graph's threshold effects) unless animals are challenged yet further.

A fourth rule of thumb is that changes in general activity can inflate or attenuate ARB performance in potentially misleading ways. Recognising certain forms of inactivity as competing responses to sub-optimal treatment, and also general activity as a potential confound that can be corrected for statistically, are both strategies likely to improve the accuracy of ARB in welfare assessment, so helping avoid both false leads (the solid dots in Figure 11.9) and false nulls (such as the X_3s in Figure 11.9).

A final rule of thumb concerns motivational specificity: recognise that not every challenge will elevate every ARB (another cause of Figure 11.9's Xs, if only specific forms are being assessed). Inferring the motivational origins of particular forms from their morphologies, and extrapolating from similar ARBs performed by related, better studied species, are two ways to narrow down the challenges likely to be particularly relevant to a given specific form of ARB.

Following these five rules of thumb helps ARBs achieve rather impressive selectivity to poor welfare: whenever events or environments cause ARB-prone phenotypes, negative affect can then very reliably be inferred; and events or environments that cause high levels of ARB

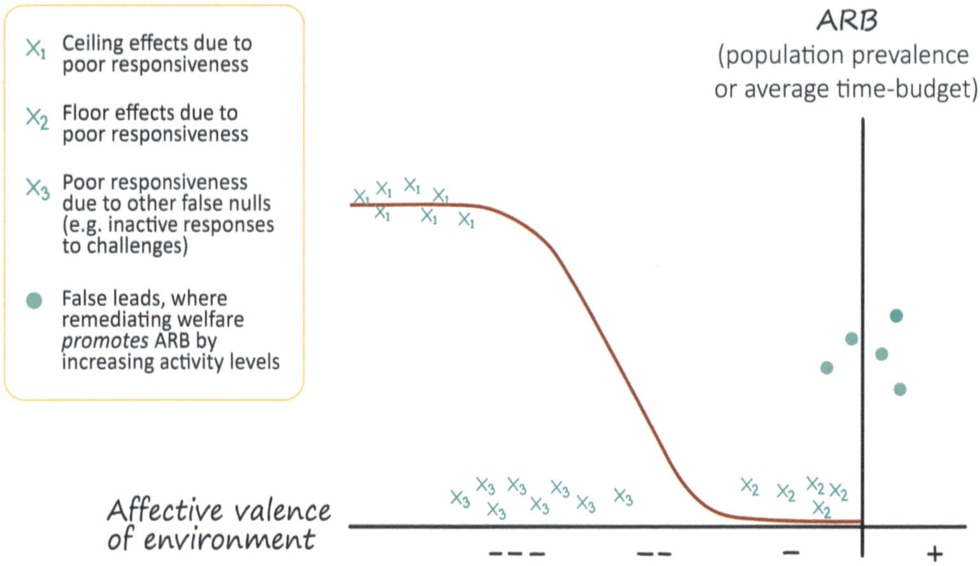

Figure 11.9 The strengths and limitations of ARBs as welfare indicators. The *x*-axis in this heuristic graph differs from those in other chapters as it specifically depicts the impact of *events or environments* on affective state. This is to follow our first rule of thumb (see chapter text). The *y*-axis represents ARB-proneness: population prevalence or average time budget (to follow our second rule-of-thumb: see text). The curve's flat portions are regions of poor responsiveness, with other markers indicating further false nulls as well as false leads (the chapter text providing further rules of thumb that augment this graph).

can reliably be inferred as being worse for welfare than those causing lower levels. However, the absence of ARBs is clearly far less revealing: this need *not* demonstrate good welfare. Furthermore, once developed, ARBs can also be somewhat blunt instruments in welfare assessment, not always covarying in nice, graded ways with degrees of exposure to challenge primarily due to ceiling effects. In such cases, ARBs will need supplementing or supplanting with other, more responsive validated indicators of negative mood or cumulative negative affect (e.g. judgement bias or telomere length).

To end

To this we add just two final notes. The first is a recognition that, as for all authors in this book, our review was narrative rather than systematic: our search process is non-replicable as well as non-exhaustive (despite being exhausting!), and our collation or interpretation of papers could therefore have been skewed by unconscious biases. Furthermore, because ARBs are heterogeneous in their properties and underlying mechanisms, what seems true for *most* ARBs need not be true for *every* ARB. This means that scientifically, our rules of thumb are actually *hypotheses*; and for greater rigour, each really needs testing (via experiment and/or meta-analysis), especially for forms of ARB that have as yet been little studied.

The second concerns definitions. Our loose definition of ARB allowed us to survey both those predictable forms traditionally termed 'stereotypies' and a variety of others from feather-plucking to tail-biting. Supporting previous hunches (Mason 2006), our review then revealed some fundamental commonalities across all of these diverse forms: a central role for poor welfare. This raises the question, should ARBs be *defined* as indicating negative affect? We opened this chapter by comparing ARBs with play. And we do this again here: just as being reduced by poor welfare is seen as a key, even defining characteristic of play, perhaps being promoted by poor welfare should be seen as a key, even defining characteristic of ARBs. But whatever terminology and definitions are adopted moving forward, one thing is clear: ARBs should never be ignored, because they flag serious welfare concerns.

References

Ahola, M.K., Vapalahti, K. and Lohi, H. (2017) *Early weaning increases aggression and stereotypic behaviour in cats. Scientific Reports*, 1, 10412. https://doi.org/10.1038/s41598-017-11173-5

Akre, A.K., Bakken, M., Hovland, A.L., Palme, R. and Mason, G.J. (2011) *Clustered environmental enrichments induce more aggression and stereotypic behaviour than do dispersed enrichments in female mice. Applied Animal Behaviour Science*, 131, 145–152. https://doi.org/10.1016/j.applanim.2011.01.010

Angevaare, M.J., Prins, S., van der Staay, F.J. and Nordquist, R.E. (2012) *The effect of maternal care and infrared beak trimming on development, performance and behavior of Silver Nick hens. Applied Animal Behaviour Science*, 140, 70–84. https://doi.org/10.1016/j.applanim.2012.05.004

Anonymous (2014) OCD inside. Prison UK – an insider's view. https://prisonuk.blogspot.com/2014/07/ocd-inside.html (accessed December 2025)

Appleby, M.C., Lawrence, A.B. and Illius, A.W. (1989) *Influence of neighbours on stereotypic behaviour of tethered sows. Applied Animal Behaviour Science*, 24, 137–146. https://doi.org/10.1016/0168-1591(89)90041-5

Arias-Esquivel, A.M., Vasco, A.C.C. de M., Lance, J., Warren, L.K., Rodriguez-Campos, L.A., Lee, M.C., Rodriguez, C.N. and Wickens, C.L. (2024) *Investigating the gastrointestinal physiology of mature horses with and without a history of cribbing behavior in response to feeding a digestive support supplement. Journal of Equine Veterinary Science*, 132, 104964. https://doi.org/10.1016/j.jevs.2023.104964

Ashley, P.J. (2007) *Fish welfare: Current issues in aquaculture. Applied Animal Behaviour Science*, 104, 199–235. https://doi.org/10.1016/j.applanim.2006.09.001

Aydinonat, D., Penn, D.J., Smith, S., Moodley, Y., Hoelzl, F., Knauer, F. and Schwarzenberger, F. (2014) *Social isolation shortens telomeres in African grey parrots (Psittacus erithacus erithacus). PLoS One*, 9, 2–6. https://doi.org/10.1371/journal.pone.0093839

Bailoo, J.D., Voelkl, B., Varholick, J., Novak, J., Murphy, E., Rosso, M., Palme, R. and Würbel, H. (2020) *Effects of weaning age and housing conditions on phenotypic differences in mice. Scientific Reports*, 10, 1–13. https://doi.org/10.1038/s41598-020-68549-3

Baker, D.G. (2002) *Combination therapy for footpad lesions in a captive Bengal tiger (Panthera tigris tigris). Journal of Zoo and Wildlife Medicine*, 33, 389–391. https://doi.org/10.1638/1042-72.0(2.02.033[0389:ctffli]2.0.co;2

Baker, K.C. and Easley, S.P. (1996) *An analysis of regurgitation and reingestion in captive chimpanzees. Applied Animal Behaviour Science*, 49, 403–415. https://doi.org/10.1016/0168-1591(96)010.1-1

Bandeli, M., Mellor, E.L., Kroshko, J., Maherali, H. and Mason, G.J. (2023) *The welfare problems of wide-ranging Carnivora reflect naturally itinerant lifestyles. Royal Society Open Science*, 10. https://doi.org/10.1098/rsos.230437

Bechard, A.R., Nicholson, A. and Mason, G.J. (2012) *Litter size predicts adult stereotypic behavior in female laboratory mice. Journal of the American Association for Laboratory Animal Science*, 51, 407–411.

Beckett, C., Bredenkamp, D., Castle, J., Groothues, C., O'Connor, T.G. and Rutter, M. (2002) *Behavior patterns associated with institutional deprivation: A study of children adopted from Romania. Journal of Developmental & Behavioral Pediatrics*, 23, 297–303. https://doi.org/10.1097/00004703-200210000-00001

Beckett, C., Castle, J., Groothues, C., O'Connor, T. and Rutter, M. (2003) *Health problems in children adopted from Romania: Association with duration of deprivation and behavioural problems. Adoption and Fostering*, 27, 19–29. https://doi.org/10.1177/030857590302700404

Bécuwe-Bonnet, V., Bélanger, M.C., Frank, D., Parent, J. and Hélie, P. (2012) *Gastrointestinal disorders in dogs with excessive licking of surfaces. Journal of Veterinary Behavior: Clinical Applications and Research* 7, 194–204. https://doi.org/10.1016/j.jveb.2011.07.003

Behen, M.E., Muzik, O., Saporta, A.S.D., Wilson, B.J., Pai, D., Hua, J. and Chugani, H.T. (2009) *Abnormal fronto-striatal connectivity in children with histories of early deprivation: A diffusion tensor imaging study. Brain Imaging and Behavior*, 3, 292–297. https://doi.org/10.1007/s11682-009-9071-6

Bellanca, R.U. and Crockett, C.M. (2002) *Factors predicting increased incidence of abnormal behavior in male pigtailed macaques. American Journal of Primatology*, 58, 57–69. https://doi.org/10.1002/ajp.10052

Benhajali, H., Ezzaouia, M., Lunel, C., Charfi, F. and Hausberger, M. (2014) *Stereotypic behaviours and mating success in domestic mares. Applied Animal Behaviour Science*, 153, 36–42. https://doi.org/10.1016/j.applanim.2014.01.002

Benjet, C., Bromet, E., Karam, E.G., Kessler, R.C., Mclaughlin, K.A., Ruscio, A.M., Shahly, V., Stein, D.J., Petukhova, M., Hill, E., Alonso, J., Atwoli, L., Bunting, B., De Girolamo, G., Florescu, S., Gureje, O., Lepine, J.P., Kawakami, N., Kovess-Masfety, V., Piazza, M., Scott, K.M., Shalev, A., Slade, T., Torres, Y., Viana, M.C., Zarkov, Z. and Koenen, K.C. (2016) *The epidemiology of traumatic event exposure worldwide: Results from the World Mental Health Survey Consortium. Psychological Medicine*, 46, 327–343. https://doi.org/10.1017/s0033291715001981

Berkson, G. and Mason, W.A. (1964) *Stereotyped behaviors of chimpanzees: Relation to general arousal and alternative activities. Perceptual and Motor Skills*, 19, 635–662. https://doi.org/10.2.66/pms.1964.19.2.635

Bildsøe, M., Heller, K.E. and Jeppesen, L.L. (1991) *Effects of immobility stress and food restriction on stereotypies in low and high stereotyping female ranch mink. Behavioural Processes*, 25, 179–189. https://doi.org/10.1016/0376-6357(91)90020-z

Blokhuis, H.J. and Arkes, J.G. (1984) *Some observations on the development of feather-pecking in poultry. Applied Animal Behaviour Science*, 12, 145–157. https://doi.org/10.1016/0168-1591(84)90104-7

Boger, S., Ehring, T., Schwarzkopf, W. and Werner, G.G. (2020) *Potential mediators of the association between childhood maltreatment and obsessive-compulsive disorder in adulthood. Journal of Obsessive-Compulsive and Related Disorders*, 27, 100587. https://doi.org/10.1016/j.jocrd.2020.100587

Boinski, S., Gross, T.S. and Davis, J.K. (1999) *Terrestrial predator alarm vocalizations are a valid monitor of stress in captive brown capuchins (Cebus apella). Zoo Biology*, 18, 295–312. https://doi.org/10.1002/(sici)1098-2361(1999)18:4<295::aid-zoo4>3.0.co;2-5

Bos, H. and Van Balen, F. (2010) *Children of the new reproductive technologies: Social and genetic parenthood. Patient Education and Counseling*, 81, 429–435. https://doi.org/10.1016/j.pec.2010.09.012

Boyle, L.A., Edwards, S.A., Bolhuis, J.E., Pol, F., Šemrov, M.Z., Schütze, S., Nordgreen, J., Bozakova, N., Sossidou, E.N. and Valros, A. (2022) *The evidence for a causal link between disease and damaging behavior in pigs. Frontiers in Veterinary Science*, 8. https://doi.org/10.3389/fvets.2021.771682

Brent, L., Lee, D.R. and Eichberg, J.W. (1989) *The effects of single caging on chimpanzee behavior. Laboratory Animal Science*, 39, 345–346.

Broom, D., Mendl, M. and Zanella, A. (1995) *A comparison of the welfare of sows in different housing conditions. Animal Science*, 61, 369–385. https://doi.org/10.1017/s1357729800013928

Broom, D.M. and Johnson, K.G. (1993) *Stress and Animal Welfare*. Chapman & Hall, London.

Brummer, S.P., Gese, E.M. and Shivik, J.A. (2010) *The effect of enclosure type on the behavior and heart rate of captive coyotes. Applied Animal Behaviour Science*, 125, 171–180. https://doi.org/10.1016/j.applanim.2010.04.012

Burke, J.T., Mograbi, D.C. and Wolmarans, D.W. (2022) *Behavioral restriction, lorazepam, and escitalopram uniquely influence the expression of naturalistic stereotypy in deer mice: Perspectives on anxiety- and compulsive-like behavior. Frontiers in Behavioral Neuroscience*, 16. https://doi.org/10.3389/fnbeh.2022.1071157

Cabib, S. (2006) The neurobiology of stereotypy. II: The role of stress. In Mason, G. J., Rushen, J. (eds) *Stereotypic Animal Behaviour Funadamentals and Applications to Welfare*. CABI, Wallingford, pp. 227–255.

Cabib, S. and Bonaventura, N. (1997) *Parallel strain-dependent susceptibility to environmentally-induced stereotypies and stress-induced behavioral sensitization in mice. Physiology & Behavior*, 61, 499–506. https://doi.org/10.1016/s0031-9384(96)00463-5

Cabib, S., Campus, P., Conversi, D., Orsini, C. and Puglisi-Allegra, S. (2020) *Functional and dysfunctional neuroplasticity in learning to cope with stress. Brain Sciences*, 10. https://doi.org/10.3390/brainsci10020127

Cabib, S., Campus, P., Latagliata, E.C., Orsini, C. and Tarmati, V. (2021) *Repetitive and inflexible active coping and addiction-like neuroplasticity in stressed mice of a helplessness – resistant inbred strain. Behavioral Science*, 11, 174. https://doi.org/10.3390/bs11120174

Cabib, S., Giardino, L., Calza, L., Zanni, M., Mele, A. and Puglisi-Allegra, A. (1998) *Stress promotes major changes in dopamine receptor densities within the mesoaccumbens and nigrostraital systems. Neuroscience*, 84, 193–200. https://doi.org/10.1016/s0306-4522(97)00468-5

Cait, J., Cait, A., Scott, R.W., Winder, C.B. and Mason, G.J. (2022) *Conventional laboratory housing increases morbidity and mortality in research rodents: Results of a meta-analysis. BMC Biology*, 20, 1–22. https://doi.org/10.1186/s12915-021-01184-0

Câmara-Souza, M.B., Carvalho, A.G., Figueredo, O.M.C., Bracci, A., Manfredini, D. and Rodrigues Garcia, R.C.M. (2023) *Awake bruxism frequency and psychosocial factors in college preparatory students. Cranio*, 41, 178–184. https://doi.org/10.1080/08869634.2020.1829289

Canadian Council on Animal Care (2021) CCAC guidelines: Animal welfare assessment. https://ccac.ca/Documents/Standards/Guidelines/CCAC_guidelines-Animal_welfare_assessment.pdf (accessed November 2025)

Capitanio, J.P., Cacioppo, S. and Cole, S.W. (2019) *Loneliness in monkeys: Neuroimmune mechanisms. Current Opinion in Behavioral Sciences*, 28, 51–57. https://doi.org/10.1016/j.cobeha.2019.01.013

Cappadocia, M.C., Weiss, J.A. and Pepler, D. (2012) *Bullying experiences among children and youth with autism spectrum disorders. Journal of Autism and Developmental Disorders*, 42, 266–277. https://doi.org/10.1007/s10803-011-1241-x

Carlson, M. and Earls, F. (1997) *Psychological and neuroendocrinological sequelae of early social deprivation in institutionalized children in Romania. Annals of the New York Academy of Sciences*, 807, 419–428. https://doi.org/10.1111/j.1749-6632.1997.tb51936.x

Carlstead, K. and Brown, J.L. (2005) *Relationships between patterns of fecal corticoid excretion and behavior, reproduction, and environmental factors in captive black (Diceros bicornis) and white (Ceratotherium simum) rhinoceros. Zoo Biology*, 24, 215–232. https://doi.org/10.1002/zoo.20050

Chamberlain, S. (2007) *Lifting the veil on trichotillomania. American Journal of Psychiatry*, 164, 568. https://doi.org/10.1176/appi.ajp.164.4.568

Chen, E., Pipolo, G., Crailsheim, D. and Morimoto, J. (2025) *The lasting impact of social isolation: Behavioral insights from former pet and entertainer chimpanzees in a sanctuary in Spain. American Journal of Primatology*, 87. https://doi.org/10.1002/ajp.23715

Chugani, H.T., Behen, M.E., Muzik, O., Juhá, C., Nagy, F. and Chugani, D.C. (2001) *Local brain functional activity following early deprivation: A study of postinstitutionalized Romanian orphans. Neuroimage*, 14, 1290–1301. https://doi.org/10.1006/nimg.2001.0917

Cinini, S.M., Barnabe, G.F., Galvão-Coelho, N., de Medeiros, M.A., Perez-Mendes, P., Sousa, M.B.C., Covolan, L. and Mello, L.E. (2014) *Social isolation disrupts hippocampal neurogenesis in young non-human primates. Frontiers in Neuroscience*, 8, 1–9. https://doi.org/10.3389/fnins.2014.00045

Clay, A.W., Bloomsmith, M.A., Bard, K.A., Maple, T.L. and Marr, M.J. (2015) *Long-term effects of infant attachment organization on adult behavior and health in nursery-reared, captive chimpanzees (Pan troglodytes). Journal of Comparative Psychology*, 129, 145–159. https://doi.org/10.1037/a0038901

Clubb, R., Rowcliffe, M., Lee, P., Mar, K.U., Moss, C. and Mason, G.J. (2008) *Compromised survivorship in zoo elephants. Science*, 322(5908), 1649. https://doi.org/10.1126/science.1164298

Conti, G., Hansman, C., Heckman, J.J., Novak, M.F.X., Ruggiero, A. and Suomi, S.J. (2012) *Primate evidence on the late health effects of early-life adversity. Proceedings of the National Academy of Sciences of the United States of America*, 109, 8866–8871. https://doi.org/10.1073/pnas.1205340109

Corsetti, S., Natoli, E., Palme, R. and Viggiano, E. (2023) *Intraspecific interactions decrease stress affecting welfare in shelter dogs: A comparison of four different housing conditions. Animals*, 13, 1–12. https://doi.org/10.3390/ani13111828

Costa, P., Macchi, E., Tomassone, L., Ricceri, F., Bollo, E., Scaglione, F.E., Tarantola, M., De Marco, M., Prola, L., Bergero, D. and Schiavone, A. (2016) *Feather picking in pet parrots: Sensitive species, risk factor and ethological evidence. Italian Journal of Animal Science*, 15, 473–480. https://doi.org/10.1080/1828051x.2016.1195711

Cromer, K.R., Schmidt, N.B. and Murphy, D.L. (2007) *An investigation of traumatic life events and obsessive-compulsive disorder. Behaviour Research and Therapy*, 45, 1683–1691. https://doi.org/10.1016/j.brat.2006.08.018

Cronin, G.M. and Glatz, P.C. (2021) *Causes of feather pecking and subsequent welfare issues for the laying hen: A review. Animal Production Science*, 61, 990–1005. https://doi.org/10.1071/an19628

Cross, H.A. and Harlow, H.F. (1965) *Prolonged and progressive effects of partial isolation on the behaviour of macaque monkeys. Journal of Experimental Research in Personality*, 1, 39–49.

Davenport, M.D., Lutz, C.K., Tiefenbacher, S., Novak, M.A. and Meyer, J.S. (2008) *A rhesus monkey model of self-injury: Effects of relocation stress on behavior and neuroendocrine function. Biological Psychiatry*, 63, 990–996. https://doi.org/10.1016/j.biopsych.2007.10.025

Dawkins, M.S. (2021) *The Science of Animal Welfare: Understanding What Animals Want.* Oxford, UK: Oxford University Press. https://doi.org/10.1093/oso/9780198848981.001.0001

De Haas, E.N., Bolhuis, J.E., De Jong, I.C., Kemp, B., Janczak, A.M. and Rodenburg, T.B. (2014) *Predicting feather damage in laying hens during the laying period: Is it the past or is it the present? Applied Animal Behaviour Science*, 160, 75–85. https://doi.org/10.1016/j.applanim.2014.08.009

de Passillé, A.M., Christopherson, R. and Rushen, J. (1993) *Non-nutritive sucking by the calf and postprandial secretion of insulin, CCK and gastrin. Physiology & Behavior*, 54, 1069–1073. https://doi.org/10.1016/0031-9384(93)90326-b

de Passillé, A.M., Metz, J.H.M., Mekking, P. and Wiepkema, P.R. (1992) *Does drinking milk stimulate sucking in young calves? Applied Animal Behaviour Science*, 34, 23–36. https://doi.org/10.1016/s0168-1591(05)80054-1

Deaton, J.E., Berg, S.W., Richlin, M. and Litrownik, A.J. (1977) *Coping activities in solitary confinement of U.S. Navy POWs in Vietnam. Journal of Applied Social Psychology*, 7, 239–257. https://doi.org/10.1111/j.1559-1816.1977.tb00749.x

Delgado, M.M., Walcher, I. and Buffington, C.A.T. (2020) *A survey-based assessment of risk factors for cross-sucking behaviors in neonatal kittens, Felis catus. Applied Animal Behaviour Science*, 230, 105069. https://doi.org/10.1016/j.applanim.2020.105069

Devanarayana, N.M. and Rajindrajith, S. (2012) *Aerophagia among Sri Lankan schoolchildren: Epidemiological patterns and symptom characteristics. Journal of Pediatric Gastroenterology and Nutrition*, 54, 516–520. https://doi.org/10.1097/mpg.0b013e318236051d

Díez-León, M., Bowman, J., Bursian, S., Filion, H., Galicia, D., Kanefsky, J., Napolitano, A., Palme, R., Schulte-Hostedde, A., Scribner, K. and Mason, G.J. (2013) *Environmentally enriched male mink gain more copulations than stereotypic, barren-reared competitors. PLoS One*, 8, 1–12. https://doi.org/10.1371/journal.pone.0080494

Díez-León, M., Bursian, S., Galicia, D., Napolitano, A., Palme, R. and Mason, G.J. (2016) *Environmentally enriching American mink (Neovison vison) increases lymphoid organ weight and skeletal symmetry, and reveals differences between two sub-types of stereotypic behaviour. Applied Animal Behaviour Science*, 177, 59–69. https://doi.org/10.1016/j.applanim.2015.12.002

Drury, S.S., Theall, K., Gleason, M.M., Smyke, A.T., De Vivo, I., Wong, J.Y.Y., Fox, N.A., Zeanah, C.H. and Nelson, C.A. (2012) *Telomere length and early severe social deprivation: Linking early adversity and cellular aging. Molecular Psychiatry*, 17, 719–727. https://doi.org/10.1038/mp.2011.53

Duncan, I.J.H. and Wood-Gush, D.G.M. (1972) *Thwarting of feeding behaviour in the domestic fowl. Animal Behaviour*, 20, 444–451. https://doi.org/10.1016/s0003-3472(72)80007-1

Duncan, I.J.H. and Wood-Gush, D.G.M. (1974) *The effect of a rauwolfia tranquilizer on stereotyped movements in frustrated domestic fowl. Applied Animal Ethology*, 1, 67–76. https://doi.org/10.1016/0304-3762(74)90009-1

Ebisawa, K., Nakayama, S., Pai, C., Kinoshita, R. and Koie, H. (2021) *Prevalence and risk factors for feather-damaging behavior in psittacine birds: Analysis of a Japanese nationwide survey. PLoS One*, 16, 1–15. https://doi.org/10.1371/journal.pone.0254610

Egeland, B., Sroufe, L.A. and Erickson, M. (1983) *The developmental consequence of different patterns of maltreatment. Child Abuse & Neglect*, 7, 459–469. https://doi.org/10.1016/0145-2134(83)90053-4

Ellis, H.H., Fisher, P.A. and Zaharie, S. (2004) *Predictors of disruptive behavior, developmental delays, anxiety, and affective symptomatology among institutionally reared Romanian children. Journal of the American Academy of Child & Adolescent Psychiatry*, 43, 1283–1292. https://doi.org/10.1097/01.chi.0000136562.24085.160

Everitt, B.J. and Robbins, T.W. (2005) *Neural systems of reinforcement for drug addiction: From actions to habits to compulsion. Nature Neuroscience*, 8, 1481–1489. https://doi.org/10.1038/nn1579

Faccin, J.E.G., Laskoski, F., Hernig, L.F., Kummer, R., Lima, G.F.R., Orlando, U.A.D., Goncalves, M.A.D., Mellagi, A.P.G., Ulguim, R.R. and Bortolozzo, F.P. (2020) *Impact of increasing weaning age on pig performance and belly nosing prevalence in a commercial multisite production system. Journal of Animal Science*, 98, 1–8. https://doi.org/10.1093/jas/skaa031

Fazio, J.M., Barthel, T., Freeman, E.W., Garlick-Ott, K., Scholle, A. and Brown, J.L. (2020) *Utilizing camera traps, closed circuit cameras and behavior observation software to monitor activity budgets, habitat use, and social interactions of zoo-housed Asian elephants (Elephus maximus). Animals*, 10, 1–21. https://doi.org/10.3390/ani10112026

Fergusson, D., Doucette, S., Glass, K.C., Shapiro, S., Healy, D., Hebert, P. and Hutton, B. (2005) *Association between suicide attempts and selective serotonin reuptake inhibitors: Systematic review of randomised controlled trials. The BMJ*, 330, 1–7. https://doi.org/10.1136/bmj.330.7488.396

Fisher, L., Ackland, T., Carter, M., Ellwood, A.-L., Gilman, L., Mainemer, H., Morison, S., Thompson, S., Aitken, H., Blamey, J., Boecker, C., Brown, K., Carley, S., Drew, R., Dykman, L., Krygsveld, S., Lawes, P., Mann, S., Morum, B., Mulloy, R., Ogmundson, A., Olsen, T., Plante, K., Rodrigues, L., Rutherford, R., Sayani, S., Vigue, T., Wang, C., Wiebe, V., Wong, S., Ames, E.W., Chisholm, K. and Savoie, L. (1997) *Problems reported by parents of Romanian orphans adopted to British Columbia. International Journal of Behavioral Development*, 20, 67–82. https://doi.org/10.1080/016502597385441

Flamand, A., Robinet, L., Raskin, A., Braconnier, M., Bouhamidi, A., Derolez, G., Lochin, C., Helleu, C. and Petit, O. (2025) *The social dimension of equine welfare: Social contact positively affects the emotional state of stalled horses. Animal Behaviour*, 221, 123055. https://doi.org/10.1016/j.anbehav.2024.123055

Fontenot, M.B., Padgett, E.E., Dupuy, A.M., Lynch, C.R., De Petrillo, P.B. and Higley, J.D. (2005) *The effects of flouxetine and buspirone on self-injurious and stereotypic behavior in adult male rhesus macaques. Comparative Medicine*, 55, 67–74.

Frank, D., Bélanger, M.C., Bécuwe-Bonnet, V. and Parent, J. (2012) *Prospective medical evaluation of 7 dogs presented with fly biting. Canadian Veterinary Journal*, 53, 1279–1284.

Fureix, C., Walker, M., Harper, L., Reynolds, K., Saldivia-Woo, A. and Mason, G.J. (2016) *Stereotypic behaviour in standard non-enriched cages is an alternative to depression-like responses in C57BL/6 mice. Behavioural Brain Research*, 305, 186–190. https://doi.org/10.1016/j.bbr.2016.02.005

Gagliano, A., Cucinotta, F., Giunta, I., Di Modica, I., De Domenico, C., Costanza, C., Germanò, E. and Frankovich, J. (2025) *Immune/inflammatory underpinnings of neurodevelopmental disorders and pediatric acute-onset neuropsychiatric syndrome: A scoping review. International Journal of Molecular Sciences*, 26, 1–36. https://doi.org/10.3390/ijms26167767

Galeano, M.G., Ruiz, R.D., de Cuneo, M.F. and Ponzio, M.F. (2013) *Effectiveness of fluoxetine to control fur-chewing behaviour in the chinchilla (Chinchilla lanigera). Applied Animal Behaviour Science*, 146, 112–117. https://doi.org/10.1016/j.applanim.2013.04.003

Garner, J.P. (2005) *Stereotypies and other abnormal repetitive behaviors: Potential impact on validity, reliability, and replicability of scientific outcomes. Institute for Laboratory Animal Research (ILAR) Journal*, 46, 106–117. https://doi.org/10.1093/ilar.46.2.106

Gershuny, B.S., Keuthen, N.J., Gentes, E.L., Russo, A.R., Emmott, E.C., Jameson, M., Dougherty, D.D., Loh, R. and Jenike, M.A. (2006) *Current posttraumatic stress disorder and history of trauma in trichotillomania. Journal of Clinical Psychology*, 62, 1521–1529. https://doi.org/10.1002/jclp.20303

Goldberg, X., Soriano-Mas, C., Alonso, P., Segalàs, C., Real, E., López-Solà, C., Subirà, M., Via, E., Jiménez-Murcia, S., Menchón, J.M. and Cardoner, N. (2015) *Predictive value of familiality, stressful life events and gender on the course of obsessive-compulsive disorder. Journal of Affective Disorders*, 185, 129–134. https://doi.org/10.1016/j.jad.2015.06.047

Gonyou, H.W., Develliers, N., Faucitano, L., Friendship, R., Pasma, T., Widowski, T.M., Ringgenberg, N. and Possberg, F.L. (2012) *Code of practice for the care & handling of pigs: review of scientific research on priority issues*. NFACC: Pig Code of Practice Scientists' Committee.

Gottlieb, D.H., Capitanio, J.P. and Mccowan, B. (2013) *Risk factors for stereotypic behavior and self-biting in rhesus macaques (Macaca mulatta): Animal's history, current environment, and personality. American Journal of Primatology*, 75, 995–1008. https://doi.org/10.1002/ajp.22161

Grant, J.E., Redden, S.A., Medeiros, G.C., Odlaug, B.L., Curley, E.E., Tavares, H. and Keuthen, N.J. (2017) *Trichotillomania and its clinical relationship to depression and anxiety. International Journal of Psychiatry in Clinical Practice*, 21, 302–306. https://doi.org/10.1080/13651501.2017.1314509

Greco, B.J., Meehan, C.L., Hogan, J.N., Leighty, K.A., Mellen, J., Mason, G.J. and Mench, J.A. (2016) *The days and nights of zoo elephants: Using epidemiology to better understand stereotypic behavior of African elephants (Loxodonta africana) and Asian elephants (Elephas maximus) in North American zoos. PLoS One*, 11, e0144276. https://doi.org/10.1371/journal.pone.0144276

Green, L.E., Lewis, K., Kimpton, A., Nicol, C.J. and House, L. (2000) *Cross-sectional study of the prevalence of feather pecking in laying hens in alternative systems and its associations with management and disease. Veterinary Record*, 147, 233–238. https://doi.org/10.1136/vr.147.9.233

Guido, C.A., Loffredo, L., Zicari, A.M., Pavone, P., Savasta, S., Gagliano, A., Brindisi, G., Galardini, G., Bertolini, A. and Spalice, A. (2021) *The impact of the COVID-19 epidemic during the lockdown on children with the pediatric acute-onset neuropsychiatric syndrome (PANDAS/PANS): The importance of environmental factors on clinical conditions. Frontiers in Neurology*, 12, 1–14. https://doi.org/10.3389/fneur.2021.702356

Haines, J., Williams, C.L. and Brain, K.L. (1995) *The psychopathology of incarcerated self-mutilators. Canadian Journal of Psychiatry*, 40, 514–522. https://doi.org/10.1177/070674379504000903

Haney, C. (2003) *Mental health issues in long-term solitary and "supermax" confinement. Crime & Delinquency*, 49, 124–156. https://doi.org/10.1177/0011128702239239

Hansen, C.P.B. and Jeppesen, L.L. (2000) *Short term behavioural consequences of denied access to environmental facilities in mink. Agricultural and Food Science*, 9, 149–155. https://doi.org/10.23986/afsci.5656

Hansen, S.W., Malmkvist, J., Palme, R. and Damgaard, B.M. (2007) *Do double cages and access to occupational materials improve the welfare of farmed mink? Animal Welfare*, 16, 63–76. https://doi.org/10.1017/s0962728600030943

Harlow, H.F. and Harlow, M.K. (1962) *Social deprivation in monkeys. Scientific American*, 207, 136–150.

Harvey, N.C., Farabaugh, S.M. and Druker, B.B. (2002) *Effects of early rearing experience on adult behavior and nesting in captive Hawaiian crows (Corvus hawaiiensis). Zoo Biology*, 21, 59–75. https://doi.org/10.1002/zoo.10024

Hausberger, M., Gautier, E., Müller, C. and Jego, P. (2007) *Lower learning abilities in stereotypic horses. Applied Animal Behaviour Science*, 107, 299–306. https://doi.org/10.1016/j.applanim.2006.10.003

Hauzenberger, A.R., Gebhardt-Henrich, S.G. and Steiger, A. (2006) *The influence of bedding depth on behaviour in golden hamsters (Mesocricetus auratus). Applied Animal Behaviour Science*, 100, 280–294. https://doi.org/10.1016/j.applanim.2005.11.012

Heathers, J.A.J., Gilchrist, K.H., Hegarty-Craver, M., Grego, S. and Goodwin, M.S. (2019) *An analysis of stereotypical motor movements and cardiovascular coupling in individuals on the autism spectrum. Biological Psychology*, 142, 90–99. https://doi.org/10.1016/j.biopsycho.2019.01.004

Hediger, H. (1950) *Wild Animals in Captivity.* Butterworth Scientific Publications, London.

Hemmings, A., McBride, S.D. and Hale, C.E. (2007) *Perseverative responding and the aetiology of equine oral stereotypy. Applied Animal Behaviour Science*, 104, 143–150. https://doi.org/10.1016/j.applanim.2006.04.031

Hoksbergen, R., Ter Laak, J., Rijk, K., Van Dijkum, C. and Stoutjesdijk, F. (2005) *Post-institutional autistic syndrome in Romanian adoptees. Journal of Autism and Developmental Disorders*, 35, 615–623. https://doi.org/10.1007/s10803-005-0005-x

Hoorweg, F.A., Vermeer, H.M., Pedersen, L.J. and Spoolder, H.A. (2017) Review on hunger induced behaviours: Aggression and stereotypies. EU Reference Centre for Animal Welfare - Pigs. https://edepot.wur.nl/574232 (accessed December 2025)

Hopper, L.M., Freeman, H.D. and Ross, S.R. (2016) *Reconsidering coprophagy as an indicator of negative welfare for captive chimpanzees. Applied Animal Behaviour Science*, 176, 112–119. https://doi.org/10.1016/j.applanim.2016.01.002

Horesh, N., Shmuel-Baruch, S., Farbstein, D., Ruhrman, D., Milshtein, N.B.A., Fennig, S., Apter, A. and Steinberg, T. (2018) *Major and minor life events, personality and psychopathology in children with Tourette syndrome. Psychiatry Research*, 260, 1–9. https://doi.org/10.1016/j.psychres.2017.11.016

Houghton, D.C., Mathew, A.S., Twohig, M.P., Saunders, S.M., Franklin, M.E., Compton, S.N., Neal-Barnett, A.M. and Woods, D.W. (2016) *Trauma and trichotillomania: A tenuous relationship. Journal of Obsessive-Compulsive and Related Disorders*, 11, 91–95. https://doi.org/10.1016/j.jocrd.2016.09.003

Hubrecht, R.C., Serpell, J.A. and Poole, T.B. (1992) *Correlates of pen size and housing conditions on the behaviour of kennelled dogs. Applied Animal Behaviour Science*, 34, 365–383. https://doi.org/10.1016/S0168-1591(05)80096-6

Hugo, C., Seier, J., Mdhluli, C., Daniels, W., Harvey, B.H., Toit, D., Wolfe-Coote, S., Nel, D. and Stein, D.J. (2003) *Fluoxetine decreases stereotypic behavior in primates. Progress in Neuro-Psychopharmacology & Biological Psychiatry*, 27, 639–643. https://doi.org/10.1016/s0278-5846(03)00073-3

Hurst, J.L., Barnard, C.J., Nevison, C.M. and West, C.D. (1997) *Housing and welfare in laboratory rats: Welfare implications of isolation and social contact among caged males. Animal Welfare*, 6, 329–347. https://doi.org/10.1017/s0962728600020042

Hurst, J.L., Barnard, C.J., Nevison, C.M. and West, C.D. (1998) *Housing and welfare in laboratory rats: The welfare implications of social isolation and social contact among females. Animal Welfare*, 7, 121–136. https://doi.org/10.1017/s0962728600020455

Hwang, J.B., Kim, J.S., Ahn, B.H., Jung, C.H., Lee, Y.H. and Kam, S. (2007) *Clonazepam treatment of pathologic childhood aerophagia with psychological stresses. Journal of Korean Medical Science*, 22, 205–208. https://doi.org/10.3346/jkms.2007.22.2.205

Ijichi, C.L., Collins, L.M. and Elwood, R.W. (2013) *Evidence for the role of personality in stereotypy predisposition. Animal Behaviour*, 85, 1145–1151. https://doi.org/10.1016/j.anbehav.2013.03.033

Imthon, A.K., Caldart, C.A., Do Rosário, M.C., Fontenelle, L.F., Miguel, E.C. and Ferrão, Y.A. (2020) *Stressful life events and the clinical expression of obsessive–compulsive disorder (OCD): An exploratory study. Journal of Clinical Medicine*, 9, 1–19. https://doi.org/10.3390/jcm910.371

Jeong, D.H., Yang, J.J. and Yeon, S.C. (2019) *Fluoxetine therapy to decrease stereotypic behavior in the Asiatic black bear (Ursus thibetanus). Journal of Zoo and Wildlife Medicine*, 50, 718–722. https://doi.org/10.1638/2017-0160

Jeppesen, L.L., Heller, K.E. and Dalsgaard, T. (2000) *Effects of early weaning and housing conditions on the development of stereotypies in farmed mink. Applied Animal Behaviour Science*, 68, 85–92. https://doi.org/10.1016/s0168-1591(00)00099-x

Jones, B.C. and Clark, D.L. (1973) *Mother-infant separation in squirrel monkeys living in a group. Developmental Psychobiology*, 6, 259–269. https://doi.org/10.1002/dev.420060311

Jones, M.A., Mason, G. and Pillay, N. (2010a) *Early social experience influences the development of stereotypic behaviour in captive-born striped mice Rhabdomys. Applied Animal Behaviour Science*, 123, 70–75. https://doi.org/10.1016/j.applanim.2009.12.009

Jones, M.A., Mason, G.J. and Pillay, N. (2011) *Correlates of birth origin effects on the development of stereotypic behaviour in striped mice, Rhabdomys. Animal Behaviour*, 82, 149–159. https://doi.org/10.1016/j.anbehav.2011.04.010

Jones, M.A., Van Lierop, M., Mason, G.J. and Pillay, N. (2010b) *Increased reproductive output in stereotypic captive Rhabdomys females: Potential implications for captive breeding. Applied Animal Behaviour Science*, 123, 63–69. https://doi.org/10.1016/j.applanim.2009.12.013

Karlen, G.A.M., Hemsworth, P.H., Gonyou, H.W., Fabrega, E., David Strom, A. and Smits, R.J. (2007) *The welfare of gestating sows in conventional stalls and large groups on deep litter. Applied Animal Behaviour Science*, 105, 87–101. https://doi.org/10.1016/j.applanim.2006.05.014

Keating, J., Van Goozen, S., Uljarevic, M., Hay, D. and Leekam, S.R. (2023) *Restricted and repetitive behaviors and their developmental and demographic correlates in 4–8-year-old children: A transdiagnostic approach. Frontiers in Behavioral Neuroscience*, 17, 1–15. https://doi.org/10.3389/fnbeh.2023.1085404

Keiper, R.R. (1969a) *Causal factors of stereotypies in caged birds. Animal Behaviour*, 17, 1–4.

Keiper, R.R. (1969b) *Drug effects on canary stereotypies. Psychopharmacologia*, 16, 16–24. https://doi.org/10.1007/bf00405252

Keiper, R.R. (1970) *Studies of stereotypy in the canary (Serinus canarius). Animal Behaviour*, 18, 353–357. https://doi.org/10.1016/s0003-3472(70)80047-1

Kitchenham, L. and Mason, G.J. (2021) *The neurobiology of environmentally induced stereotypic behaviours in captive animals: Assessing the basal ganglia pathways and cortico-striatal-thalamo-cortical circuitry hypotheses. Behaviour*, 158, 1251–1302. https://doi.org/10.1163/1568539x-bja10084

Kłosowska, J., Antosz-Rekucka, R., Kałużna-Wielobób, A. and Prochwicz, K. (2021) *Dissociative experiences mediate the relationship between traumatic life events and types of skin picking. Findings from non-clinical sample. Frontiers in Psychiatry*, 12, 1–10. https://doi.org/10.3389/fpsyt.2021.698543

Korff, S., Stein, D.J. and Harvey, B.H. (2008) *Stereotypic behaviour in the deer mouse: Pharmacological validation and relevance for obsessive-compulsive disorder. Progress in Neuro-Psychopharmacology*, 32, 348–355. https://doi.org/10.1016/j.pnpbp.2007.08.032

Korff, S., Stein, D.J. and Harvey, B.H. (2009) *Cortico-striatal cyclic AMP-phosphodiesterase-4 signalling and stereotypy in the deer mouse: Attenuation after chronic fluoxetine treatment. Pharmacology, Biochemistry, and Behavior*, 92, 514–520. https://doi.org/10.1016/j.pbb.2009.01.025

Kostal, L., Savory, C.J. and Hughes, B.O. (1992) *Diurnal and individual variation in behaviour of restricted-fed broiler breeders. Applied Animal Behaviour Science*, 32, 361–374. https://doi.org/10.1016/s0168-1591(05)80028-0

Kverno, K.S. and Mangano, E. (2021) *Treatment-resistant depression: Approaches to treatment. Journal of Psychosocial Nursing and Mental Health Services*, 59, 7–11. https://doi.org/10.3928/02793695-20210816-01

Langen, M., Durston, S., Kas, M.J.H., van Engeland, H. and Staal, W.G. (2011) *The neurobiology of repetitive behavior: ...and men. Neuroscience and Biobehavioral Reviews*, 35, 356–365. https://doi.org/10.1016/j.neubiorev.2010.02.005

Lanzarini, E., Pruccoli, J., Grimandi, I., Spadoni, C., Angotti, M., Pignataro, V., Sacrato, L., Franzoni, E. and Parmeggiani, A. (2021) *Phonic and motor stereotypies in autism spectrum disorder: Video analysis and neurological characterization. Brain Sciences*, 11. https://doi.org/10.3390/brainsci11040431

Latham, N. and Mason, G.J. (2010) *Frustration and perseveration in stereotypic captive animals: Is a taste of enrichment worse than none at all? Behavioral Brain Research*, 211, 96–104. https://doi.org/10.1016/j.bbr.2010.03.018

Laurence, A., Lumineau, S., Calandreau, L., Arnould, C., Leterrier, C., Boissy, A. and Houdelier, C. (2014) *Short- and long-term effects of unpredictable repeated negative stimuli on Japanese quail's fear of humans. PLoS One*, 9, 1–8. https://doi.org/10.1371/journal.pone.0093259

Lawrence, A.B. and Terlouw, E.M.C. (1993) *A review of behavioral factors involved in the development and continued performance of stereotypic behaviors in pigs. Journal of Animal Science*, 71, 2815–2825. https://doi.org/10.2527/1993.71102815x

Leonardi, L., Perna, C., Bernabei, I., Fiore, M., Ma, M., Frankovich, J., Tarani, L. and Spalice, A. (2024) *Pediatric acute-onset neuropsychiatric syndrome (PANS) and pediatric autoimmune neuropsychiatric disorders associated with streptococcal infections (PANDAS): Immunological features underpinning controversial entities. Children*, 11. https://doi.org/10.3390/children110910.3

Leruste, H., Brscic, M., Cozzi, G., Kemp, B., Wolthuis-Fillerup, M., Lensink, B.J., Bokkers, E.A.M. and van Reenen, C.G. (2014) *Prevalence and potential influencing factors of non-nutritive oral behaviors of veal calves on commercial farms. Journal of Dairy Science*, 97, 7021–7030. https://doi.org/10.3168/jds.2014-7917

Lewis, M.H., Gluck, J.P., Petitto, J.M., Hensley, L.L. and Ozer, H. (2000) *Early social deprivation in nonhuman primates: Long-term effects on survival and cell-mediated immunity. Biological Psychiatry*, 47, 119–126. https://doi.org/10.1016/s0006-3223(99)00238-3

Lewis, M.H., Lindenmaier, Z., Boswell, K., Edington, G., King, M.A. and Muehlmann, A.M. (2018) *Subthalamic nucleus pathology contributes to repetitive behavior expression and is reversed by environmental enrichment. Genes, Brain and Behavior*, 17, e12468. https://doi.org/10.1111/gbb.12468

Lewis, R. and Hurst, J. (2004) *The assessment of bar chewing as an escape behaviour in laboratory mice. Animal Welfare*, 13, 19–25. https://doi.org/10.1017/s0962728600026610

Li, X., Xu, F., Xie, L., Ji, Y., Cheng, K., Zhou, Q., Wang, T., Shively, C., Wu, Q., Gong, W., Fang, L., Zhan, Q., Melgiri, N.D. and Xie, P. (2013) *Depression-like behavioral phenotypes by social and social plus visual isolation in the adult female Macaca fascicularis. PLoS One*, 8. https://doi.org/10.1371/journal.pone.0073293

Lin, H., Williams, K.A., Katsovich, L., Findley, D.B., Grantz, H., Lombroso, P.J., King, R.A., Bessen, D.E., Johnson, D., Kaplan, E.L., Landeros-Weisenberger, A., Zhang, H. and Leckman, J.F. (2010) *Streptococcal upper respiratory tract infections and psychosocial stress predict future tic and obsessive-compulsive symptom severity in children and adolescents with Tourette syndrome and obsessive-compulsive disorder. Biological Psychiatry*, 67, 684–691. https://doi.org/10.1016/j.biopsych.2009.08.020

Lochner, C., Du Toit, P.L., Zungu-Dirwayi, N., Marais, A., Van Kradenburg, J., Seedat, S., Niehaus, D.J.H. and Stein, D.J. (2002) *Childhood trauma in obsessive-compulsive disorder, trichotillomania, and controls. Depression and Anxiety*, 15, 66–68. https://doi.org/10.1002/da.10028

Lutz, C., Well, A. and Novak, M. (2003) *Stereotypic and self-injurious behavior in rhesus macaques: A survey and retrospective analysis of environment and early experience. American Journal of Primatology*, 60, 1–15. https://doi.org/10.1002/AJP.10075

Lutz, C.K. (2014) *Stereotypic behavior in nonhuman primates as a model for the human condition. Institute for Laboratory Animal Research (ILAR) Journal*, 55, 284–296. https://doi.org/10.1093/ilar/ilu016

Lutz, C.K., Davis, E.B., Ruggiero, A.M. and Suomi, S.J. (2007) *Early predictors of self-biting in socially-housed rhesus macaques (Macaca mulatta). American Journal of Primatology*, 69, 584–590. https://doi.org/10.1002/AJP.20370

Maisch, H., Vellinga, N.R. and Boersma, H. (2017) Best practice guideline dhole (*C. alpinus*) 1st edition, EAZA Best Practice Guidelines. https://strapi.eaza.net/uploads/Dhole_EEP_BPG_first_ed_1_e9d3dc973e.pdf (accessed November 2025)

Malmkvist, J., Díez-León, M. and Christensen, J.W. (2024) *Animals with various forms of abnormal behaviour differ in learning performance and use of enrichment: Evidence from farm mink. Applied Animal Behaviour Science*, 271, 106167. https://doi.org/10.1016/j.applanim.2024.10.167

Maneeton, P., Maneeton, B., Kienngam, N., Maneeton, N., Winichaikul, Y. and Kawilapat, S. (2024) *Updated systematic review and meta-analysis of randomized controlled trials: Antidepressants for restricted and repetitive behaviors in autism spectrum disorder. Neuropsychiatric Disease and Treatment*, 20, 1711–1723. https://doi.org/10.2147/ndt.s465611

Mason, G.J. (1994) *Tail-biting in mink (Mustela vison) is influenced by age at removal from the mother. Animal Welfare*, 3, 305–311. https://doi.org/10.1017/s0962728600017061

Mason, G.J. (2006) Stereotypic behaviour in captive animals: Fundamentals and implications for welfare and beyond. In Mason, G. J., Rushen, J. (eds) *Stereotypic Animal Behaviour Fundamentals and Applications to Welfare*. CABI, Wallingford, pp. 325–356.

Mason, G.J., Clubb, R., Latham, N. and Vickery, S. (2007) *Why and how should we use environmental enrichment to tackle stereotypic behaviour? Applied Animal Behaviour Science*, 102, 163–188. https://doi.org/10.1016/j.applanim.2006.05.041

Mason, G.J. and Kitchenham, K. (2026) Assessing the validity of abnormal repetitive behaviours as indicators of poor animal welfare: A narrative review. Preprints.org, *Preprints ID:* 191216. https://doi.org/10.20944/preprints202512.2180.v1

Mason, G.J. and Latham, N. (2004) *Can't stop, won't stop: Is stereotypy a reliable animal welfare indicator? Animal Welfare*, 13, 57–69. https://doi.org/10.1017/s096272860001438x

Mason, G.J. and Mendl, M. (1997) *Do the stereotypies of pigs, chickens and mink reflect adaptive species differences in the control of foraging? Applied Animal Behaviour Science*, 53, 45–58. https://doi.org/10.1016/s0168-1591(96)01150-1

Mason, G.J. and Würbel, H. (2016) *What can be learnt from wheel-running by wild mice, and how can we identify when wheel-running is pathological? Proceedings of the Royal Society B: Biological Sciences*, 283(1824), 20150738.

McAfee, L.M., Mills, D.S. and Cooper, J.J. (2002) *The use of mirrors for the control of stereotypic weaving behaviour in the stabled horse. Applied Animal Behaviour Science*, 78, 159–173. https://doi.org/10.1016/s0168-1591(02)00086-2

Meagher, R.K., Dallaire, A.A., Campbell, J.L.M., Ross, D. and Møller, M.H. (2014) *Benefits of a ball and chain: Simple environmental enrichments improve welfare and reproductive success in farmed American mink (Neovison vison). PLoS One*, 9, 110589. https://doi.org/10.1371/journal.pone.0110589

Meehan, C.L., Garner, J.P. and Mench, J.A. (2003) *Isosexual pair housing improves the welfare of young Amazon parrots. Applied Animal Behaviour Science*, 81, 73–88. https://doi.org/10.1016/s0168-1591(02)00238-1

Meehan, C.L., Garner, J.P. and Mench, J.A. (2004) *Environmental enrichment and development of cage stereotypy in orange-winged Amazon parrots (Amazona amazonica). Developmental Psychobiology*, 44, 209–218. https://doi.org/10.1002/dev.20007

Meers, L. and Ödberg, F.O. (2005) *Paradoxical rate-dependent effect of fluoxetine on captivity-induced stereotypies in bank voles. Progress in Neuro-Psychopharmacology & Biological Psychiatry*, 29, 964–971. https://doi.org/10.1016/j.pnpbp.2005.06.002

Mellor, E.L. and Mason, G.J. (2023) *Feeding, mating and animal wellbeing: New insights from phylogenetic comparative methods. Proceedings of the Royal Society of London, Series B: Biological Sciences*, 290, 20222571. https://doi.org/10.1098/rspb.2022.2571

Metcalf, J. (2020) A prisoner's three tips for beating boredom. Prison Writers. https://prisonwriters.com/boredom-in-prison/ (accessed December 2025)

Mikkola, S., Salonen, M., Hakanen, E. and Lohi, H. (2022) *Fearfulness associates with problematic behaviors and poor socialization in cats. iScience*, 25, 105265. https://doi.org/10.1016/j.isci.2022.105265

Miller, L.J., Ivy, J.A., Vicino, G.A. and Schork, I.G. (2019) *Impacts of natural history and exhibit factors on carnivore welfare. Journal of Applied Animal Welfare Science*, 22, 188–196. https://doi.org/10.1080/10888705.2018.1455582

Mills, D.S. and Luescher, A. (2006) Veterinary and pharmacological approaches to abnormal repetitive behaviour. In Mason, G. J., Rushen, J. (eds) *Stereotypic Animal Behaviour Fundamentals and Applications to Welfare*. CABI, Wallingford, pp. 286–324.

Mindus, C., van Staaveren, N., Bharwani, A., Fuchs, D., Gostner, J.M., Kjaer, J.B., Kunze, W., Mian, M.F., Shoveller, A.K., Forsythe, P. and Harlander-Matauschek, A. (2021) *Ingestion of Lactobacillus rhamnosus modulates chronic stress-induced feather pecking in chickens. Scientific Reports*, 11, 1–14. https://doi.org/10.1038/s41598-021-96615-x

Mora, S., Martín-González, E., Flores, P. and Moreno, M. (2020) *Neuropsychiatric consequences of childhood group A streptococcal infection: A systematic review of preclinical models. Brain, Behavior, and Immunity*, 86, 53–62. https://doi.org/10.1016/j.bbi.2019.02.027

Morin, K.M. (2016) *Carceral space: Prisoners and animals. Antipode*, 48, 1317–1336. https://doi.org/10.1111/anti.12239

Morrissey, K.L.H., Widowski, T., Leeson, S., Sandilands, V., Arnone, A. and Torrey, S. (2014) *The effect of dietary alterations during rearing on growth, productivity, and behavior in broiler breeder females. Poultry Science*, 93, 285–295. https://doi.org/10.3382/ps.2013-03265

Mosallanejad, B., Varzi, H.N., Avizeh, R., Pourmahdi, M. and Khalili, F. (2015) *Comparative evaluation between hypericin (hypiran) and fluoxetine in treatment of companion dogs with tail chasing. Veterinary Research Forum*, 6, 167–172.

Moy, S.S., Nadler, J.J., Young, N.B., Perez, A., Holloway, L.P., Barbaro, R.P., Barbaro, J.R., West, L.M., Threadgill, D.W., Lauder, J.M., Magnuson, T.R. and Crawley, J.N. (2007) *Mouse behavioral tasks relevant to autism: Phenotypes of 10 inbred strains. Behavioural Brain Research*, 176, 4–20. https://doi.org/10.1016/j.bbr.2006.07.030

Muehlmann, A.M. and Lewis, M.H. (2012) *Abnormal repetitive behaviours: Shared phenomenology and pathophysiology. Journal of Intellectual Disability Research*, 56, 427–440. https://doi.org/10.1111/j.1365-2788.2011.01519.x

Murphy, T.K., Snider, L.A., Mutch, P.J., Harden, E., Zaytoun, A., Edge, P.J., Storch, E.A., Yang, M.C.K., Mann, G., Goodman, W.K. and Swedo, S.E. (2007) *Relationship of movements and behaviors to Group A Streptococcus infections in elementary school children. Biological Psychiatry*, 61, 279–284. https://doi.org/10.1016/j.biopsych.2006.08.031

Napolitano, F., De Rosa, G. and Sevi, A. (2008) *Welfare implications of artificial rearing and early weaning in sheep. Applied Animal Behaviour Science*, 110, 58–72. https://doi.org/10.1016/j.applanim.2007.03.020

National Institute of Mental Health (2025) PANS and PANDAS: Questions and answers. United States Department of Health and Human Services National Institutes of Health. https://www.nimh.nih.gov/health/publications/pandas (accessed December 2025)

NC3Rs (2025) Behavioural indicators. https://macaques.nc3rs.org.uk/welfare-assessment/behavioural-indicators (accessed December 2025)

Nielsen, S.S., Alvarez, J., Bicout, D.J., Calistri, P., Canali, E., Drewe, J.A., Garin-Bastuji, B., Gonzales Rojas, J.L., Schmidt, C.G., Herskin, M., Miranda Chueca, M.A., Padalino, B., Pasquali, P., Roberts, H.C., Spoolder, H., Stahl, K., Velarde, A., Viltrop, A., Winckler, C., Tiemann, I., de Jong, I., Gebhardt-Henrich, S.G., Keeling, L., Riber, A.B., Ashe, S., Candiani,

D., García Matas, R., Hempen, M., Mosbach-Schulz, O., Rojo Gimeno, C., Van der Stede, Y., Vitali, M., Bailly-Caumette, E. and Michel, V. (2023) *Scientific opinion on the welfare of broilers on farm. EFSA Journal*, 21(2), 7788, 236 pp. https://doi.org/10.2903/j.efsa.2023.7788

Nelson, C.A., Fox, N.A. and Zeanah, C.H. (2014) *Romania's Abandoned Children: Deprivation, Brain Development, and the Struggle for Recovery*. Harvard University Press, Cambridge.

Nierenberg, A.A., Farabaugh, A.H., Alpert, J.E., Gordon, J., Worthington, J.J., Rosenbaum, J.F. and Fava, M. (2000) *Timing of onset of antidepressant response with fluoxetine treatment. American Journal of Psychiatry*, 157, 1423–1428. https://doi.org/10.1176/appi.ajp.157.9.1423

Nip, E. (2018) The long-term effects of environmental enrichment on agonism in female C57BL/6, BALB/c, and DBA/2 mice. University of Guelph. MSc thesis. https://atrium.lib.uoguelph.ca/items/db4ae808-766c-46c7-8714-77de9fd6ddff (accessed November 2025)

Nip, E., Adcock, A., Nazal, B., Maclellan, A., Niel, L., Choleris, E., Levison, L. and Mason, G.J. (2019) *Why are enriched mice nice? Investigating how environmental enrichment reduces agonism in female C57BL/6, DBA/2, and BALB/c mice. Applied Animal Behaviour Science*, 217, 73–82. https://doi.org/10.1016/j.applanim.2019.05.002

Novak, M.A. (2003) *Self-injurious behavior in rhesus monkeys: New insights into its etiology, physiology, and treatment. American Journal of Primatology*, 59, 3–19. https://doi.org/10.1002/ajp.10063

Oakley, C., Mahone, E.M., Morris-Berry, C., Kline, T. and Singer, H.S. (2015) *Primary complex motor stereotypies in older children and adolescents: Clinical features and longitudinal follow-up. Pediatric Neurology*, 52, 398–403.e1. https://doi.org/10.1016/j.pediatrneurol.2014.11.002

O'Connor, V.L. and Vonk, J. (2022) *Scaredy-cats don't succeed: Behavioral traits predict problem-solving success in captive felidae. PeerJ*, 10. https://doi.org/10.7717/peerj.14453

Olsson, I.A.S. and Sherwin, C.M. (2006) *Behaviour of laboratory mice in different housing conditions when allowed to self-administer an anxiolytic. Laboratory Animals*, 40, 392–399. https://doi.org/10.1258/002367706778476389

Özten, E., Sayar, G.H., Eryılmaz, G., Kağan, G., Işık, S. and Karamustafalıoğlu, O. (2015) *The relationship of psychological trauma with trichotillomania and skin picking. Neuropsychiatric Disease and Treatment*, 11, 1203–1210. https://doi.org/10.2147/ndt.s79554

PANDAS Network (2025) Understanding PANDAS and PANS: A comprehensive resource guide. https://pandasnetwork.org/understanding-pandas/ (accessed December 2025)

Parker, M., Goodwin, D. and Redhead, E.S. (2008) *Survey of breeders' management of horses in Europe, North America and Australia: Comparison of factors associated with the development of abnormal behaviour. Applied Animal Behaviour Science*, 114, 206–215. https://doi.org/10.1016/j.applanim.2008.02.003

Patiño, J.J., Vélez, S.A. and Martínez, J.R. (2020) *Ethological, endocrinological, and gastroscopic evaluation of crib-biting Colombian creole horses. Journal of Veterinary Behavior*, 40, 92–97. https://doi.org/10.1016/j.jveb.2020.10.005

Perré, Y., Wauters, A.-M. and Richard-Yris, M.-A. (2002) *Influence of mothering on emotional and social reactivity of domestic pullets. Applied Animal Behaviour Science*, 75, 133–146. https://doi.org/10.1016/s0168-1591(01)00189-7

Pingitore, G., Chrobak, V. and Petrie, J. (1991) *The social and psychological factors of bruxism. The Journal of Prosthetic Dentistry*, 65, 443–446. https://doi.org/10.1016/0022-3913(91)90240-w

Poirier, C., Oliver, C.J., Castellano Bueno, J., Flecknell, P. and Bateson, M. (2019) *Pacing behaviour in laboratory macaques is an unreliable indicator of acute stress. Scientific Reports*, 9, 7476. https://doi.org/10.1038/s41598-019-43695-5

Polanco, A., Díez-León, M. and Mason, G.J. (2018) *Stereotypic behaviours are heterogeneous in their triggers and treatments in the American mink, Neovison vison, a model carnivore. Animal Behaviour*, 141, 105–114. https://doi.org/10.1016/j.anbehav.2018.05.006

Polverino, G., Manciocco, A., Vitale, A. and Alleva, E. (2015) *Stereotypic behaviours in Melopsittacus undulatus: Behavioural consequences of social and spatial limitations. Applied Animal Behaviour Science*, 165, 143–155. https://doi.org/10.1016/j.applanim.2015.02.009

Pomerantz, O., Meiri, S. and Terkel, J. (2013) *Socio-ecological factors correlate with levels of stereotypic behavior in zoo-housed primates. Behavioural Processes*, 98, 85–91. https://doi.org/10.1016/j.beproc.2013.05.005

Poroshinska, O., Polishchuk, A., Shmayun, S., Kozii, N., Shahanenko, R., Chornozub, M., Stovbetska, L., Shahanenko, V. and Koziy, V. (2024) *The use of anxiolytic drugs for the correction of behavioral disorders in mammals. Regulatory Mechanisms in Biosystems*, 15, 42–48. https://doi.org/10.15421/022406

Poulsen, E.M.B. and Cambell Teskey, G. (2006) Pacing, prozac and a polar bear. In Mason, G. J., Rushen, J. (eds) *Stereotypic Animal Behaviour Fundamentals and Applications to Welfare*. CABI, Wallingford, p. 309.

Rafaeli-Mor, N., Foster, L. and Berkson, G. (1999) *Self-reported body rocking and other habits in college students. American Journal of Mental Retardation*, 104, 1–10. https://doi.org/10.1352.0895-8017(1999)104<0001:sbaohi>2.0.co;2

Rapoport, J.L., Ryland, D.H. and Kriete, M. (1992) *Drug treatment of canine acral lick: An animal model of obsessive-compulsive disorder. Archives of General Psychiatry*, 49, 517–521. https://doi.org/10.1001/archpsyc.1992.01820070011002

Redbo, I., Emanuelson, M., Oredsson, N. and Lundberg, K. (1996) *Feeding level and oral stereotypies in dairy cows. Animal Science*, 62, 199–206. https://doi.org/10.1017/s1357729800014491

Rodenburg, T.B., Uitdehaag, K.A., Ellen, E.D. and Komen, J. (2009) *The effects of selection on low mortality and brooding by a mother hen on open-field response, feather pecking and cannibalism in laying hens. Animal Welfare*, 18, 427–432. https://doi.org/10.1017/s096272860000083x

Rommeck, I., Anderson, K., Heagerty, A., Cameron, A. and McCowan, B. (2009) *Risk factors and remediation of self-injurious and self-abuse behavior in rhesus macaques. Journal of Applied Animal Welfare Science*, 12, 61–72. https://doi.org/10.1080/10888700802536798

Rushen, J. and de Passillé, A.M. (1995) *The motivation of non-nutritive sucking in calves, Bos taurus. Animal Behaviour*, 49, 1503–1510. https://doi.org/10.1016/0003-3472(95)90071-3

Sanchis-Segura, C. and Spanagel, R. (2006) *Behavioural assessment of drug reinforcement and addictive features in rodents: An overview. Addiction Biology*, 11, 2–38. https://doi.org/10.1111/j.1369-1600.2006.00012.x

Sargent, T.D. and Keiper, R.R. (1967) *Stereotypies in caged canaries. Animal Behaviour*, 15, 62–66. https://doi.org/10.1016/s0003-3472(67)80012-5

Sarkhel, S., Praharaj, S.K. and Sinha, V.K. (2011) *Role of life events in obsessive compulsive disorder. Israel Journal of Psychiatry and Related Sciences*, 48, 182–185. https://doi.org/10.3126/ajms.v15i2.57961

Savory, C.J. and Mann, J.S. (1999) *Stereotyped pecking after feeding by restricted-fed fowls is influenced by meal size. Applied Animal Behaviour Science*, 62, 209–217. https://doi.org/10.1016/s0168-1591(98)00228-7

Savory, C.J., Maros, K. and Rutter, S.M. (1993) *Assessment of hunger in growing broiler breeders in relation to a commercial restricted feeding programme. Animal Welfare*, 2, 131–152. https://doi.org/10.1017/s0962728600015669

Scantlebury, C.E., Archer, D.C., Proudman, C.J. and Pinchbeck, G.L. (2015) *Management and horse-level risk factors for recurrent colic in the UK general equine practice population. Equine Veterinary Journal*, 47, 202–206. https://doi.org/10.1111/evj.12276

Schmid, R., Doherr, M.G. and Steiger, A. (2006) *The influence of the breeding method on the behaviour of adult African grey parrots (Psittacus erithacus). Applied Animal Behaviour Science*, 98, 293–307. https://doi.org/10.1016/j.applanim.2005.09.002

Schneper, L.M., Brooks-Gunn, J., Notterman, D.A. and Suomi, S.J. (2016) *Early-life experiences and telomere length in adult rhesus monkeys: An exploratory study. Psychosomatic Medicine*, 78, 1066–1071. https://doi.org/10.1097/psy.0000000000000402

Šemrov, M.Z. and Patt, A. (2024) The tail as an iceberg indicator: Interrelationships with welfare problems. In Driscoll, K., Valros, A. (eds) *Tail Biting in Pigs: A Comprehensive Guide to Its Aetiology, Impact and Wider Significance in Pig Management*. Brill, Wageningen, pp. 351–326.

Seo, T., Sato, S., Kosaka, K., Sakamoto, N. and Tokumoto, K. (1998) *Tongue-playing and heart rate in calves. Applied Animal Behaviour Science*, 58, 179–182. https://doi.org/10.1016/s0168-1591(97)00062-2

Sheridan, M.A., Fox, N.A., Zeanah, C.H., McLaughlin, K.A. and Nelson, C.A. (2012) *Variation in neural development as a result of exposure to institutionalization early in childhood. Proceedings of the National Academy of Sciences of the United States of America (PNAS)*, 109, 12927–12932. https://doi.org/10.1073/pnas.1200041109

Shively, C.A., Register, T.C., Friedman, D.P., Morgan, T.M., Thompson, J. and Lanier, T. (2005) *Social stress-associated depression in adult female cynomolgus monkeys (Macaca fascicularis). Biological Psychology*, 69, 67–84. https://doi.org/10.1016/j.biopsycho.2004.11.006

Shukla, T. and Pandey, S. (2020) *Stereotypies in adults: A systematic review. Neurologia i Neurochirurgia Polska*, 54, 294–304. https://doi.org/10.5603/pjnns.a2020.0058

Shulman, L.M., Sanchez-Ramos, J.R. and Weiner, W.J. (1996) Defining features, clinical conditions, and theoretical constructs of stereotyped movements. In Sprague, R. L., Newell, K. M. (eds) *Stereotyped Movements – Brain & Behavior Relationships*. American Psychiatric Association, Washington, DC, pp. 17–24.

Shyne, A. (2006) *Meta-analytic review of the effects of enrichment on stereotypic behavior in zoo mammals. Zoo Biology*, 25, 317–337. https://doi.org/10.1002/zoo.20091

Shyne, A. and Block, M. (2010) *The effects of husbandry training on stereotypic pacing in captive African wild dogs (Lycaon pictus). Journal of Applied Animal Welfare Science*, 13, 56–65. https://doi.org/10.1080/10888700903372069

Singer, H.S. (2010) *Treatment of tics and Tourette syndrome. Current Treatment Options in Neurology*, 12, 539–561. https://doi.org/10.1007/s11940-010-0095-4

Smith, A.P. (2016) *Chewing gum and stress reduction. Journal of Clinical and Translational Research*, 2, 52–54. https://doi.org/10.18053/jctres.02.201602.002

Sohel, A.J., Shutter, M.C., Patel, P. and Molla, M. (2025) Fluoxetine. StatPearls. https://www.ncbi.nlm.nih.gov/books/nbk459223/ (accessed December 2025)

Spitzer, C., Lübke, L., Müller, S., Lindstädt, T. and Gallinat, C. (2022) *Childhood maltreatment, traumatic experiences, and posttraumatic stress disorder in pathological skin picking: An online case-control study. General Hospital Psychiatry*, 78, 9–13. https://doi.org/10.1016/j.genhosppsych.2022.06.002

Spoolder, H.A.M., Burbidge, J.A., Edwards, S.A., Howard Simmins, P. and Lawrence, A.B. (1995) *Provision of straw as a foraging substrate reduces the development of excessive chain and bar manipulation in food restricted sows. Applied Animal Behaviour Science*, 43, 249–262. https://doi.org/10.1016/0168-1591(95)00566-b

Stein, D.J. (2002) *Obsessive-compulsive disorder. Lancet*, 360, 397–405. https://doi.org/10.1016/s0140-6736(02)09620-4

Steinberg, T., Shmuel-Baruch, S., Horesh, N. and Apter, A. (2013) *Life events and Tourette syndrome. Comprehensive Psychiatry*, 54, 467–473. https://doi.org/10.1016/j.comppsych.2012.10.015

Swaisgood, R. and Shepherdson, D. (2006) Environmental enrichment as a strategy for mitigating stereotypies in zoo animals: A literature review and meta-analysis. In Mason, G. J., Rushen, J. (eds) *Stereotypic Animal Behaviour Fundamentals and Applications to Welfare*. CAB International, Oxford, pp. 256–285.

Swaisgood, R.R. and Shepherdson, D.J. (2005) *Scientific approaches to enrichment and stereotypies in zoo animals: What's been done and where should we go next? Zoo Biology*, 24, 499–518. https://doi.org/10.1002/zoo.20066

Swedo, S., Menendez, C.M. and Cunningham, M.W. (2024) Pediatric autoimmune neuropsychiatric disorder associated with streptococcal infection (PANDAS): A case series. In Ferretti, J. J., Stevens, D. L., Fischetti, V. A. (eds) *Streptococcus Pyogenes: Basic Biology to Clinical Manifestations*. University of Oklahoma Health Sciences Center, Oklahoma City, pp. 75–79. https://doi.org/10.4103/mjdrdypu.mjdrdypu_569_22

Tiira, K., Hakosalo, O., Kareinen, L., Thomas, A., Hielm-Björkman, A., Escriou, C., Arnold, P. and Lohi, H. (2012) *Environmental effects on compulsive tail chasing in dogs. PLoS One*, 7, e41684–e41684. https://doi.org/10.1371/journal.pone.0041684

Tilly, S.L.C., Dallaire, J. and Mason, G.J. (2010) *Middle-aged mice with enrichment-resistant stereotypic behaviour show reduced motivation for enrichment. Animal Behaviour*, 80, 363–373. https://doi.org/10.1016/j.anbehav.2010.06.008

Tröster, H. (1994) *Prevalence and functions of stereotyped behaviors in nonhandicapped children in residential care. Journal of Abnormal Child Psychology*, 22, 79–97. https://doi.org/10.1007/bf02169257

Tynes, V.V. and Sinn, L. (2014) *Abnormal repetitive behaviors in dogs and cats: A guide for practitioners. The Veterinary Clinics of North America. Small Animal Practice*, 44, 543–564. https://doi.org/10.1016/j.cvsm.2014.01.011

van Zeeland, Y. (2018) *Medication for behavior modification in birds. Veterinary Clinics of North America: Exotic Animal Practice*, 21, 115–149. https://doi.org/10.1016/j.cvex.2017.08.005

Vandeleest, J.J., Mccowan, B. and Capitanio, J.P. (2011) *Early rearing interacts with temperament and housing to influence the risk for motor stereotypy in rhesus monkeys (Macaca mulatta). Applied Animal Behaviour Science*, 132, 81–89. https://doi.org/10.1016/j.applanim.2011.02.010

Vidal-Ribas, P., Stringaris, A., Rück, C., Serlachius, E., Lichtenstein, P. and Mataix-Cols, D. (2015) *Are stressful life events causally related to the severity of obsessive-compulsive symptoms? A monozygotic twin difference study. European Psychiatry*, 30, 309–316. https://doi.org/10.1016/j.eurpsy.2014.11.008

Visser, E.K., Ellis, A.D. and Van Reenen, C.G. (2008) *The effect of two different housing conditions on the welfare of young horses stabled for the first time. Applied Animal Behaviour Science*, 114, 521–533. https://doi.org/10.1016/j.applanim.2008.03.003

Vivar, J. (2014) *Dispatches from the Canadian carceral state: The truth about provincial prisons. Journal of Prisoners on Prisons*, 23, 6–9.

Wallgren, T., Westin, R. and Gunnarsson, S. (2016) *A survey of straw use and tail biting in Swedish pig farms rearing undocked pigs. Acta Veterinaria Scandinavica*, 58, 1–11. https://doi.org/10.1186/s13028-016-0266-8

Wang, M., Ricardi, R. and Ritfeld, G.J. (2024) *PANDAS, a series of difficult decisions: A case report. BMC Psychiatry*, 24. https://doi.org/10.1186/s12888-024-06180-6

Warniment, A. and Brent, L. (1997) *Abnormal behavior in a captive chimpanzee colony. The Newsletter*, 8(3), 1–3.

Weary, D.M., Appleby, M.C. and Fraser, D. (1999) *Responses of piglets to early separation from the sow. Applied Animal Behaviour Science*, 63, 289–300. https://doi.org/10.1016/s0168-1591(99)00021-0

Whitehouse, C.M., Curry-Pochy, L.S., Shafer, R., Rudy, J. and Lewis, M.H. (2017) *Reversal learning in C58 mice: Modeling higher order repetitive behavior. Behavioural Brain Research*, 332, 372–378. https://doi.org/10.1016/j.bbr.2017.06.014

Wickens, C.L. (2009) Investigation of specific stereotypic behaviors in horses. PhD thesis, Michigan State University. htt1?ps://www.1?proquest.com/o1?penview/3a44b2b306474057fc1778af44dc16a2/1?1?pq-origsite=gscholar&cbl=18750 (accessed November 2025)

Widowski, T.M., Torrey, S., Bench, C.J. and Gonyou, H.W. (2008) *Development of ingestive behaviour and the relationship to belly nosing in early-weaned piglets. Applied Animal Behaviour Science*, 110, 109–127. https://doi.org/10.1016/j.applanim.2007.04.010

Wiedenmayer, C. (1997) *Causation of the ontogenetic development of stereotypic digging in gerbils. Animal Behaviour*, 53, 461–470. https://doi.org/10.1006/anbe.1996.0296

Wilkes, B.J. and Lewis, M.H. (2018) *The neural circuitry of restricted repetitive behavior: Magnetic resonance imaging in neurodevelopmental disorders and animal models. Neuroscience and Biobehavioral Reviews*, 92, 152–171. https://doi.org/10.1016/j.neubiorev.2018.05.022

Willard, S.L. and Shively, C.A. (2012) *Modeling depression in adult female cynomolgus monkeys (Macaca fascicularis). American Journal of Primatology*, 74, 528–542. https://doi.org/10.1002/ajp.21013

Willemsen-Swinkels, S.H.N., Buitelaar, J.K., Dekker, M. and Van Engeland, H. (1998) *Subtyping stereotypic behavior in children: The association between stereotypic behavior, mood, and heart rate. Journal of Autism and Developmental Disorders*, 28, 547–557. https://doi.org/10.1023/a:1026008313284

Williams, I., Hoppitt, W. and Grant, R. (2017) *The effect of auditory enrichment, rearing method and social environment on the behavior of zoo-housed psittacines (Aves: Psittaciformes); implications for welfare. Applied Animal Behaviour Science*, 186, 85–92. https://doi.org/10.1016/j.applanim.2016.10.013

Williams, K., Wheeler, D.M., Silove, N. and Hazell, P. (2011) *Cochrane review: Selective serotonin reuptake inhibitors (SSRIs) for autism spectrum disorders (ASD). Evidence-Based Child Health*, 6, 1044–1078. https://doi.org/10.1002/ebch.804

Wilson, C., Gattuso, J.J., Hannan, A.J. and Renoir, T. (2023) *Mechanisms of pathogenesis and environmental moderators in preclinical models of compulsive-like behaviours. Neurobiology of Disease*, 185, 106223. https://doi.org/10.1016/j.nbd.2023.106223

Wolmarans, D.W., Brand, L., Stein, D.J. and Harvey, B.H. (2013) *Reappraisal of spontaneous stereotypy in the deer mouse as an animal model of obsessive-compulsive disorder (OCD): Response to escitalopram treatment and basal serotonin transporter (SERT) density. Behavioural Brain Research*, 256, 545–553. https://doi.org/10.1016/j.bbr.2013.08.049

Würbel, H. and Stauffacher, M. (1997) *Age and weight at weaning affect corticosterone level and development of stereotypies in ICR-mice. Animal Behaviour*, 53, 891–900. https://doi.org/10.1006/anbe.1996.0424

Yang, K., Essa, A., Noriega, D., Yu, D., Osiecki, L., Gauvin, C.A., Illmann, C., Bortolato, M., Dunn, E.C., Mathews, C.A. and Scharf, J.M. (2022) *Relationship between adverse childhood experiences and symptom severity in adult men with Tourette syndrome. Journal of Psychiatric Research*, 155, 252–259. https://doi.org/10.1016/j.jpsychires.2022.08.024

Yasir, W., Sherin, M. and Kazmi, S.F. (2021) *The role of childhood trauma in developing nail biting and trichotillomania among adolescents. International Journal of Innovation, Creativity and Change*, 15, 1314–1332.

Young, F.M. (1947) *The incidence of nervous habits observed in college students. Journal of Personality*, 15, 309–320. https://doi.org/10.1111/j.1467-6494.1947.tb010.1.x

Young, L.E., McCallum, R.T., Perreault, M.L. and Mason, G.J. (2024) *Opportunistically using a chronic unpredictable stress study to investigate 'inactive-but-awake' behaviour as a potential welfare indicator in laboratory rats. Applied Animal Behaviour Science*, 274. https://doi.org/10.1016/j.applanim.2024.106249

Zhang, C., Juniper, D.T. and Meagher, R.K. (2021) *Effects of physical enrichment items and social housing on calves' growth, behaviour and response to novelty. Applied Animal Behaviour Science*, 237, 105295. https://doi.org/10.1016/j.applanim.2021.105295

Zhang, M.-Y., Li, X., Zhang, X.-H., Liu, H.-G., Li, J.-H. and Bao, J. (2017) *Effects of confinement duration and parity on stereotypic behavioral and physiological responses of pregnant sows. Physiology & Behavior*, 179, 369–376. https://doi.org/10.1016/j.physbeh.2017.07.015

Zhang, Z.-Y., Mao, Y., Feng, X.-L., Zheng, N., Lü, L.-B., Ma, Y.-Y., Qin, D.-D. and Hu, X.-T. (2016) *Early adversity contributes to chronic stress induced depression-like behavior in adolescent male rhesus monkeys. Behavioural Brain Research*, 306, 154–159. https://doi.org/10.1016/j.bbr.2016.03.040

Telomere Length

12

Melissa Bateson[1] and Luise A. Seeker[2]

[1]Faculty of Medical Sciences, Newcastle University, Newcastle upon Tyne, UK
[2]Department of Bioengineering, Stanford University, Stanford, California, USA

Introduction

"*Every stress leaves an indelible scar, and the organism pays for its survival after a stressful situation by becoming a little older*" (Selye 1956). This quote establishes the hypothesis that there are causal links between stressful experiences – defined here as states resulting in subjectively negative affective valence – and rates of biological ageing. In this chapter, we review accumulating evidence that telomere length (TL), a cellular biomarker that shortens with age in many species, provides a valid indicator of an individual's cumulative negative affective experience. We show that negative affective states are typically associated with more telomere shortening, while a healthy lifestyle and most positive affective experiences are reflected in less telomere shortening, and possibly even telomere lengthening.

We start by presenting the general argument for why biological age is a plausible indicator of cumulative affective experience. Next, we introduce TL as a biomarker of biological age and describe how it is measured. In the subsequent section, we review converging lines of evidence that TL is a valid indicator of cumulative affective experience, while in the following section, we highlight some issues relating to the responsiveness and selectivity of TL when used for this purpose. In the final section, we discuss the practicalities of using TL as a welfare indicator, illustrating its successful use. Since much of the literature on TL relates explicitly to health rather than affective state, we conclude with a note on the tensions between health and affective views of welfare. While there is strong evidence linking TL to health and longevity, and its links to negative affective state are also robust, its links to positive affective states are currently less clear.

Why is biological age a plausible welfare indicator?

Ageing results from accumulated molecular damage that affects biological function at cellular, tissue and organism levels. Ageing rates vary between and within individuals of the same species, leading to the distinction between *chronological* and *biological* age. Biological age is estimated using ageing biomarkers, including those at a cellular level which reflect *cellular age* and which we focus on in this chapter (Table 12.1).

The argument for using ageing biomarkers as welfare indicators depends on evidence that affective states, which are central to animal welfare, also play a role in the biological pathways leading to ageing (Figure 12.1; see also Walker et al. 2012). The central argument is that exposure to stressors, such as physical or psychological abuse, will typically cause negative affective states, such as pain, fear and anxiety (pathway 1 in Figure 12.1). Negative affective states will in turn be associated with a physiological stress response and inflammation leading to acute somatic damage (pathway 2). If such damage is not mitigated or repaired, it will accumulate, leading to ageing (pathway 3). Therefore, how well an individual copes with the psychological and physiological impact of exposure to stressors will affect their rate of ageing. This resilience is determined by both genes and experiences associated with good physical health and positive affect.

This model predicts that individuals who are biologically old for their chronological age will typically have experienced more unmitigated negative affect, worse health and less positive affect than individuals who are biologically younger. Measurements of relative biological age should therefore reflect differences in cumulative affective experience (Bateson 2016; Bateson and Poirier 2019).

Several metrics are used to estimate biological age, some based on single biomarkers, and others on composites based on multiple measures (for examples in humans see Hastings et al. 2019; Table 12.1). In this chapter, we focus on TL, because while there are other markers

Table 12.1 Definitions of different ageing terms and examples of metrics.

Term	Definition	Example metrics
Chronological age	The amount of time elapsed since birth.	Units of time (days, months, years)
Biological age	The predicted age of an individual based on body appearance and function.[a]	Biomarkers including: • greying hair and wrinkled skin • physical frailty • cognitive impairment • decline of wound healing capacity and immune system reactivity • accumulation of senescent cells in organs • shortened lifespan
Cellular age	A measure made at the cellular level, that is used to assess the biological age of an individual. Not all cells within a tissue or organism will age at the same rate.	Biomarkers including: • telomere length • DNA methylation • mitochondrial dysfunction • DNA damage

[a] The concept of biological age is closely related to that of allostatic load, which describes the cumulative wear and tear on the body resulting from chronic stress, and which is often measured with similar sets of biomarkers.

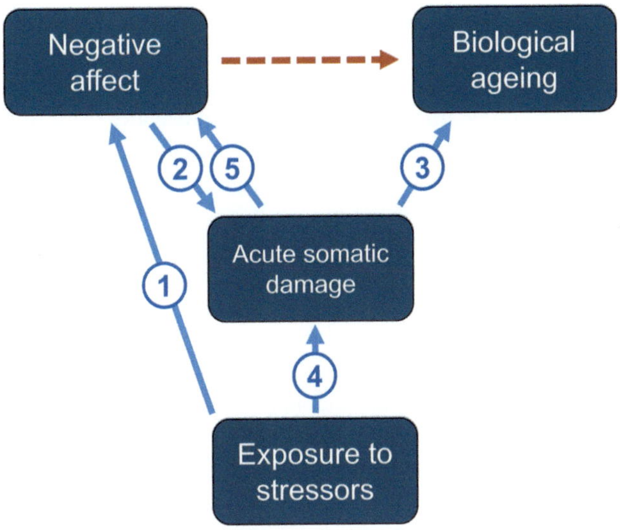

Figure 12.1 Pathways connecting exposure to stressors, negative affect, acute somatic damage and ageing. The dashed arrow indicates the association identified by Selye (1956) and the solid arrows indicate established causal relationships (see text for further explanation). Source: Adapted from Bateson and Poirier (2019).

being used to assess biological age, none of them have been studied to the same extent in either human or animal populations.

What are telomeres?

Telomeres (Greek for 'end parts') are the DNA–protein complexes found at the ends of all animal chromosomes that protect coding regions of DNA from damage (Figure 12.2). A short fragment of DNA is lost from the end of chromosomes during mitosis – the so-called 'end

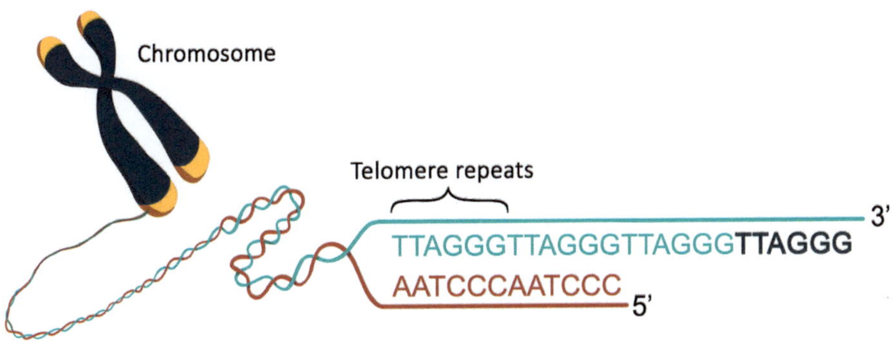

Figure 12.2 Telomeres are repetitive DNA (TTAGGG) sequences at the ends of chromosomes that are bound to specialised proteins (not depicted). Telomere length (TL) reflects the number of telomeric repeats and is measured in base pairs (bp, often expressed in kilobases, kb) or as a relative value (see section on telomere measurement methods). The telomeric DNA sequence is highly evolutionarily conserved in vertebrate species, offering the potential for TL to translate across species as a welfare indicator.

replication problem' – meaning that in the absence of repair processes, telomeres shorten with each cell division.

Telomeres can be repaired by the enzyme telomerase, but this varies between species and tissues, and over the life course. In postnatal humans and many other species, telomerase is repressed, meaning that telomeres shorten from birth onwards. In adult humans, the annual decline in TL in blood is 25–40 base pairs, which equates to 0.4–0.6%. If telomeres become critically short, this triggers a DNA damage response, cellular senescence or cell death, providing a potential causal link between telomere shortening and the loss of function characteristic of ageing. Within a species, individuals with longer telomeres stay healthy and live longer on average which has been observed for humans and other vertebrates (Kimura et al. 2008; Wilbourn et al. 2018).

Since telomeres shorten with age in most species, TL was initially believed to be a biomarker of chronological age. However, focus has shifted towards the large variation in TL typically observed within age groups and the varying rates of telomere decline among individuals. TL is now regarded as a marker of biological age.

How long telomeres are and how fast they shorten is explained both by genes and individual experience. Of relevance to welfare, many of the experiences known to affect TL in humans are likely to be associated with either positive or negative affective states (Figure 12.3). This means that there is scope for using TL as an indicator of cumulative affective experience.

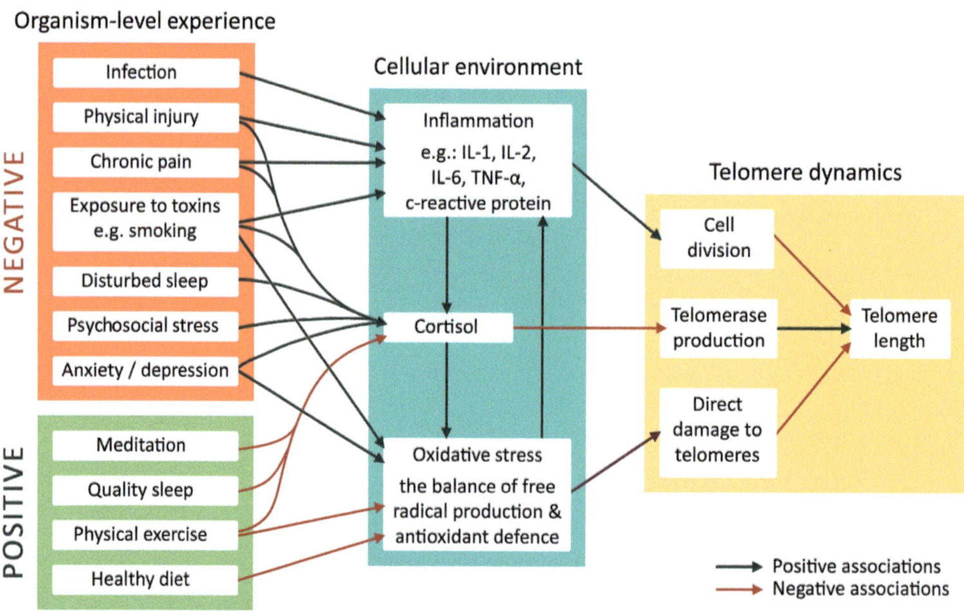

Figure 12.3 Established relationships between an individual's experiences, the cellular environment and telomere length in humans. Experiences are grouped into those likely to be associated with positive versus negative changes in affective state. As can be seen, affectively negative experiences typically shorten telomeres, whereas positive experiences retard, prevent or possibly reverse telomere shortening (see construct validity section later for details). Source: Bateson (2016) / John Wiley & Sons / CC BY 4.0.

Table 12.2 Comparison of quantitative polymerase chain reaction (qPCR) and terminal restriction fragments (TRF) methods for measuring telomere length (TL).

	qPCR methods	TRF methods
Quantity of DNA required	Low	High
Effort to establish technique	Low	High
Handling time per sample	Low	High
TL metric yielded	T/S ratio (a relative measure of mean TL)	Mean TL (in kb) and also variance in TL
Includes measurement of interstitial telomeres?*	Yes	Yes (Southern blot method) No (in-gel hybridisation method)
Provides distribution of TL?	No	Yes
Precision of measurements	Lower	Higher

kb: kilobases; T/S ratio: amount of telomeric DNA (T) relative to the amount of a chosen single-copy gene (S).
* Telomeric DNA sequences called interstitial telomeres sometimes occur within chromosomes and their abundance varies between species. These are not affected by age- or stress-related shortening but are captured with many measurement methods. Interstitial telomeres therefore have the potential to dilute the signal of TL as an indicator of cumulative affective state, rendering it less sensitive. Source: Adapted from Nussey et al. (2014).

Measurement of telomere length

All methods for measuring TL estimate the number of telomeric repeats present in a sample. A range of methods exists with differing costs and benefits (Lai et al. 2018; Nussey et al. 2014; Table 12.2). The most common methods use the quantitative polymerase chain reaction (qPCR) to measure the amount of telomeric DNA (T) relative to the amount of a chosen single-copy gene (S) and provide a relative measure of total telomeric DNA in a sample, known as the T/S ratio. Methods based on absolute measurements of terminal restriction fragments (TRF) measure TL (in kb) and provide information on the distribution of TLs in a sample. Overall, although qPCR-based methods are cheaper and simpler, TRF-based methods provide a more sensitive measure of cumulative experience (as expanded upon in the section on responsiveness and selectivity later).

Basic construct validity

Exposure to diverse stressors that cause cellular inflammation, oxidative stress and increased 'stress hormones' (see Chapter 5) accelerate telomere shortening. Furthermore, stress reduction may be associated with opposite effects (Figure 12.3). TL is therefore a good candidate biomarker of biological age, and because it seems to respond in opposite directions to positively and negatively valenced experiences, it is also an attractive potential indicator of cumulative affective experience, naturally integrating the impacts of diverse experiences into a single metric. In this section, we describe the evidence that TL lives up to this promise. We start with evidence from humans, where the affective valence of experiences can be established via self-report (Test 1 in Chapter 2).

Negative affective experience in humans is associated with shorter TL

Many human studies show that exposure to stressors is associated with either shorter TL or increased telomere shortening. The experiences studied include various types of childhood adversity, diverse psychosocial stressors and the presence of mood disorders characterised by negative affect such as anxiety, depression and post-traumatic stress disorder (reviewed by Pepper et al. 2018).

These studies typically use naturally occurring variation in stressful experiences and affective state in human populations, because it is ethically problematic to expose humans to experimental manipulations inducing chronic negative affect. Most are cross-sectional, correlating some measure of experience with a single measurement of TL per individual. Longitudinal studies, which instead correlate exposure to stressors with change in TL, provide stronger evidence that stress causes telomere shortening by reducing the likelihood that a third, unconsidered variable causes both shorter TL and an increased stress exposure (Bateson and Nettle 2018; Figure 12.4).

Evidence for associations between negative affect and shorter TL is supported by meta-analyses, which summarise effects from multiple independent studies. For example, meta-analyses report overall associations between the presence of mood disorders such as depression and shorter TL (Darrow et al. 2016; Lin et al. 2016; Ridout et al. 2016; Schutte and Malouff 2015); chronic perceived stress (as opposed to physiological stress) is also associated with shorter TL (Mathur et al. 2015). Effect sizes are small; however, a large meta-analysis of 138 cross-sectional and longitudinal studies exploring exposure to diverse stressors, including

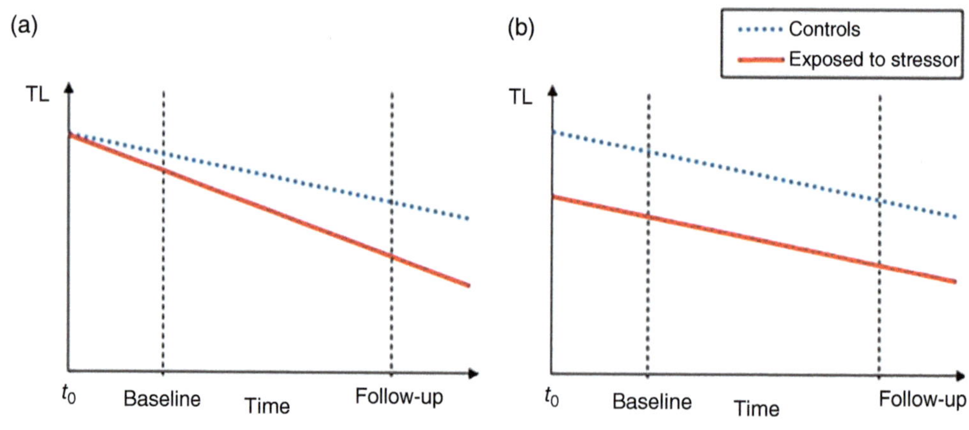

Figure 12.4 Alternative explanations for an observed cross-sectional association between on-going stress exposure and shorter TL. In both graphs, stress exposure starts at time t_0 and continues throughout for the exposed group. TL is measured at two subsequent time points (indicated by the vertical dashed lines) – a common scenario in human correlational studies. At both the baseline and follow-up measurements, individuals exposed to stress have shorter TL than controls not exposed to stress. In (a), stress exposure causes accelerated telomere shortening, whereas in (b), the difference in TL is already present at the start of stress exposure (t_0) and the rate of telomere loss is no faster in individuals exposed to stress, suggesting no causal relationship between stress exposure and telomere shortening. It is only by measuring TL at least twice and estimating the rate of TL change that scenarios (a) and (b) can be distinguished.

mood disorders, reported a summary association between TL and stress exposure of only r = −0.15 (95% CI: −0.18 to −0.11; Pepper et al. 2018).

While meta-analyses suggest small effects of single stressors on TL, if multiple sources of negative affect are present in the same individual, or if doses are higher, the effects on TL can be larger. For example, fibromyalgia patients reporting higher levels of pain had shorter TL and patients reporting both pain and depression had shorter TL than those without either (Hassett et al. 2012). Similarly, in individuals with chronic knee pain, higher reported pain frequency, intensity, duration and extent was associated with shorter TL in a dose-dependent pattern (Sibille et al. 2017). In Danish men, the number of stressful events in childhood (but not the total number over the life course) was associated with shorter adult TL (Osler et al. 2016). This latter study not only suggests additive effects of multiple forms of stress, but also supports the results of many others showing that childhood is a particularly sensitive period for the effects of stress on TL.

More positive affect in humans is associated with longer TL

Some studies indicate that possession of psychological resilience factors such as an optimistic personality or engaging in active coping strategies, such as meditation, proven to positively affect mood, are associated with longer TL, reduced telomere shortening or even telomere lengthening. For example, a meta-analysis of 12 studies reported a small association between dispositional optimism and TL of r = 0.06 (95% CI: 0.01 to 0.11). This association was increased in those participants exposed to stressful circumstances (r = 0.31, 95% CI: 0.12 to 0.47), supporting the hypothesis that an optimistic personality may buffer the effects of exposure to stressors (Schutte and Malouff 2022). Similarly, the presence of psychological resilience indicators including optimism, positive affect and active coping, which are known to moderate pain severity, was associated with longer TL in patients with and without chronic pain, when pain severity was statistically controlled for (Johnson et al. 2019).

While the above studies are correlational, there are studies that involve manipulations that typically induce more positive affect. In one such trial, participants who were randomly assigned to meditation classes had longer TL following the 6-week intervention, compared to a control group placed on a waiting list (Le Nguyen et al. 2019). However, this study has been challenged on the grounds of poor design and analysis (Friedman et al. 2020), and a systematic review concluded that better designed trials with larger sample sizes are required before definitive conclusions can be drawn on the effects of meditation on TL (Schutte et al. 2020).

A healthier lifestyle in humans is associated with longer TL

Much of the human telomere epidemiology literature focuses on associations between TL and exposures likely to affect physical health and hence Darwinian fitness (Test 3 in Chapter 2). For example, in the large meta-analysis cited earlier, physical disease, exposure to environmental hazards, poor diet, smoking and physical inactivity were all associated with shorter TL (Pepper et al. 2018). Furthermore, in a longitudinal study, performance of behaviours contributing to a healthy lifestyle including good diet, sufficient sleep and physical exercise was correlated with a reduced rate of telomere shortening over a year-long period in a dose-dependent manner (Puterman et al. 2015).

Although these studies do not measure associations between TL and affective state *per se*, there is independent evidence that a healthy lifestyle and practices such as meditation cause more positive affective state in humans (e.g. Ekinci and Sanlier 2023; Huang et al. 2020; Spector et al. 2024). The previous studies thus contribute to the validation of TL as an indicator of cumulative affective experience.

Validated indicators of affective state co-vary with TL in humans

An additional type of evidence potentially validating TL as an indicator of cumulative affective experience comes from correlations between TL and another indicator with good validity (Test 5 in Chapter 2): hippocampal grey matter volume (Bateson and Poirier 2019; Poirier et al. 2019, and see Chapter 13). As with TL, hippocampal volume decreases with age, at a rate that accelerates with stress. In a study of 1960 middle-aged people, hippocampal volume positively correlated with leukocyte TL (King et al. 2014). However, a meta-analysis combining this with four smaller studies found that TL and hippocampal volume are *non*-significantly positively correlated overall (Nilsonne et al. 2015). Only one study has compared changes in TL and hippocampal volume longitudinally in the same individuals (Staffaroni et al. 2018). This study found that greater TL shortening over 2.9 years is associated with greater hippocampal volume loss over the same period. This suggests that, in humans at least, changes in TL and hippocampal volume may reflect similar experiences.

In summary, there is evidence for associations between affective state and TL in humans, although effects are typically small. These predominantly correlational data make it impossible to establish definitively that cumulative experience *causes* changes in TL in humans. However, experimental evidence from animals addresses this problem.

Exposing animals to stressful experiences reduces TL

In many animals, exposure to stressful experiences is associated with shorter TL or faster TL shortening. This includes exposure to known punishers by which we mean stimuli or harsh environmental conditions that animals will actively work to avoid (Test 2 in Chapter 2), and likely threats to Darwinian fitness (Test 3 in Chapter 2), such as physically demanding conditions, competition, social isolation for naturally social animals and disease. For example, experimentally exposing captive adult blackbirds repeatedly to stressors comprising overcrowding, loud noise, chasing, catching, restraint and immune challenge, over a 1-year period caused reduced TL (Hau et al. 2015). Extending this result, an experimental study of nestling starlings that manipulated two types of stress (reduced quantity of food and increased begging effort) in a factorial design showed that the effects of the two types of stress were additive, with larger effects on telomere loss when both stressors were applied simultaneously (Nettle et al. 2017). A meta-analysis of 109 studies exploring exposure to diverse challenges in non-human vertebrates reported an overall association between stress exposure and TL of $r = -0.16$ (95% CI: −0.22 to −0.10; Chatelain et al. 2020). In this analysis, the association between TL and stress exposure was consistent across phylogenetic classes and life stages, although stronger in juveniles than adults, supporting the earlier suggestion from humans that childhood is a sensitive period for effects of stress on TL. As in the human meta-analysis described earlier, the association was consistently in the same direction across the different categories of stressors explored. Importantly, 133 out of 347 effects (38%) came from experimental studies. As expected, given the better control for random variation that this allows, the association between stress exposure and TL is 25% stronger in experimental studies than correlational ones. Such experimental studies strengthen the evidence that exposure to stressful experiences assumed to induce negative affect causes telomere shortening. A caveat with this study is that in many cases, there is no direct evidence that the putative stressors considered are associated with negative affective states.

In summary, animal data provide robust evidence that exposure to diverse stressors causes telomere shortening, although effect sizes are typically small. Thus, animal studies seem concordant with human ones, although the small effects have implications for the practical use of TL as a welfare indicator, especially in correlational studies (explored further in subsequent sections). We are not aware of any animal studies exploring the effects of positive affect on TL.

Responsiveness and selectivity of TL

How responsive is TL for detecting differences in experience?

In the previous section, we reviewed evidence that TL is responsive to either individual differences in affective state, or manipulations likely to produce differences in affective state. We saw evidence for graded responses, with larger doses of stress causing larger impacts on TL. However, if measurements of TL either reveal no impact of a putative negative experience, or fail to distinguish a very negative experience from a mildly negative one, then this constitutes a false-negative error – a false null. A high rate of false nulls would indicate that TL has poor **responsiveness** as an indicator of cumulative experience. In this section, we describe two major causes of false nulls in studies of the effects of stress on TL arising from low power due to TL measurement error and large individual differences in TL.

Measurements of TL lack precision. Therefore, the ability to detect differences in TL will depend on the size of the true difference relative to the precision of measurement: detecting small differences requires very precise measurement (see Figure 12.5), and a major influence on responsiveness is therefore measurement error. All methods for measuring TL suffer from this, but qPCR methods do more so than TRF (see Table 12.2; Nettle et al. 2021). Both methods require careful optimisation and quality control, and repeatability estimates for both qPCR and TRF techniques vary considerably between labs (Kärkkäinen et al. 2022).

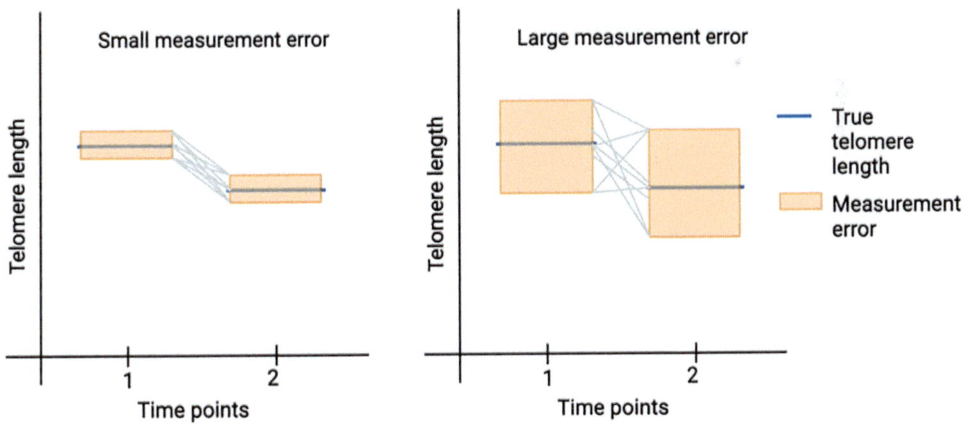

Figure 12.5 Illustration of the impact of measurement error in a longitudinal study. The same true difference in telomere length (TL) between time points 1 and 2 measured with small (left) and large (right) measurement error. While the measurements associated with the small error show clearly that TL at time point 2 is shorter than at time point 1, many different conclusions may be drawn from the measurements made with large error (false leads as well as false nulls, as we come to later). The left panel shows telomere shortening for all individuals (indicated by downward sloping grey lines), but the right panel suggests TL shortening, maintenance or even lengthening for a given individual. The cases of lengthening arise from regression to the mean (see text for details).

Measurement methods that allow quantification of the distribution of telomeres of different lengths within an individual are preferred because of evidence that stress has differential effects on telomeres of different lengths, with the longest telomeres being the most sensitive to stress-induced shortening (Andrews et al. 2021; Bauch et al. 2014).

A second major cause of false nulls is the large variation among individuals in TL. In cross-sectional studies, where TL is measured only once per subject, this causes poor responsiveness even if measurement error is minimal. In the best-case scenario, in which measurement error is zero, then given the among-individual standard deviation in adult human TL of around 650 bp, a sample of over 200 people would be required to detect a 300 bp difference in TL between two groups with a power of 0.9. For context, a 300 bp difference equates to approximately a decade of additional ageing in adult humans. Thus, even quite large sample sizes followed for many years may be too small to detect true effects, leading to false nulls.

Floor and ceiling effects are a common cause of poor responsiveness in welfare indicators; however, they are unlikely to constrain the responsiveness of TL as an indicator of cumulative affective experience. While TL has biological upper and lower limits that will vary among species, measurements at those limits are rare. Having a substantial number of very short telomeres is unlikely to occur, because it would be incompatible with basic life functions, such as the regeneration of tissue, healing of wounds and defence against pathogens. Since TL typically reduces with age, telomeres at the upper limit will also be rare and are usually restricted to cancer cells where telomerase and other mechanisms are activated to maintain TL in the face of uncontrolled cell division.

In summary, the responsiveness of TL as a welfare indicator is limited both by the precision of available measurement methodologies and the intrinsic biological variability in TL among individuals. Several other practical considerations influence the likely responsiveness of TL as a welfare indicator, but we reserve a discussion of these for the section on use in practice.

How selective is TL for detecting the impact of stressful experience?

If TL measurements suggest negative impacts of exposures that are not actually stressful, this constitutes a false lead: it appears to indicate cumulative affective experience has been negative when, in fact, the valence of experience has been neutral, or even positive (Chapter 2). A high rate of false leads would then indicate that TL has poor **selectivity** as an indicator of cumulative negative experience. Later we discuss situations where associations with TL may not be indicative of affective experience.

Returning to the model presented in Figure 12.1, it is possible that exposure to stressors might cause somatic damage directly, without the involvement of negative affect (pathway 4); for example, poor diet can directly cause inflammation. If this pathway is common, the implication is that biological age could provide false-positive information (false leads) about negative affective experience. However, if somatic damage causes negative affect (pathway 5), then an association between negative affect and ageing (and hence TL) will still be present via an indirect route. In support of this pathway, there is evidence for example, that inflammation may cause depression (Lee and Giuliani 2019), providing a mechanistic explanation for a link between poor diet and negative affect (Ekinci and Sanlier 2023).

More problematically, giving subjects what they want, which is assumed to improve affective state and welfare, can sometimes decrease TL. In humans, the excessive consumption of sugar and alcohol are both associated with shorter TL, while in animals, a greater investment in reproduction is associated with shorter TL (Beirne et al. 2015; Nussey et al. 2006; Rafie et al. 2017; Strandberg et al. 2012; Wojcicki et al. 2018). This could reflect a true-positive result mediated by the negative health effects of sugar, alcohol and reproduction (e.g. Partridge et al. 2005; pathways 4 and 5 in Figure 12.1), but it could also reflect a false lead (pathways 3 and 4). In the latter case, using TL to determine the valence of cumulative experience may lead us to deny animals some affectively positive experiences they would choose to engage in, if given the option, such as eating sugar. We revisit such potential tensions between health and affect-based views of welfare in the final section of this chapter.

A second source of possible false leads stems from how TL changes with age. In humans and many animal species (although not all: see for example Hoelzl et al. 2016 on the edible dormouse), TL shortens rapidly in early life due to many cell divisions required for growth and is either stable or declines much more slowly thereafter (Gomes et al. 2011; Gorbunova and Seluanov 2009; Swanberg et al. 2010). However, evidence suggests that on average, human and non-human primate well-being does not decline with increasing age (Weiss et al. 2012). Thus, in many species, chronological age is likely to be a non-affective variable associated with telomere shortening. This problem is addressed by controlling for chronological age either through good experimental design or statistics (see later section).

A final point worth mentioning is that the TL measurement error discussed earlier can generate false leads as well as false null results. In longitudinal studies, it is often the case that individuals with relatively long baseline TL will appear to show most telomere shortening at follow-up, while individuals with relatively short baseline TL will appear to show lengthening simply as a consequence of measurement error (see the right-hand panel of Figure 12.5). While this pattern is sometimes interpreted as reflecting a biological phenomenon of interest, an alternative interpretation, supported by simulation and analysis, is that it exemplifies a statistical artefact termed 'regression to the mean' (Bateson et al. 2019a, 2019b; Nettle et al. 2021; Verhulst et al. 2013).

Use in practice

In this section, we discuss practical questions relating to the successful use of TL as a welfare indicator and illustrate the use of TL to assess cumulative experience in animals.

What type of welfare is TL appropriate for assessing?

Evidence supports the use of TL for assessing chronic exposure to affectively negative experiences. An attraction of TL is that it has the potential to integrate different types of negative experiences arising from the physical and social environment, including food availability, exposure to scientific procedures, injury, disease and social isolation, into one measure. TL may also be sensitive to positive experiences, especially where these have a role in mitigating the negative effects of stress, although more research is needed here. Therefore, TL is particularly appropriate for measuring the cumulative impact of multidimensional exposures, where many variables are different between the treatments being compared.

Successful uses of TL in a welfare context include evaluating the impacts of high stocking density in broiler chickens (Beloor et al. 2010), hot summers (assumed to cause heat stress) in dairy cows (Seeker et al. 2021) and social isolation in naturally social African grey parrots kept as pets (Aydinonat et al. 2014). TL has also been associated with the overall habitat quality of wild animals. For example, shorter TL was reported in adult griffon vultures living in more anthropised habitats of the Iberian Peninsula characterised by higher human population density, intensive agriculture and more buildings and roads, compared to birds living in less anthropised habitats (Gangoso et al. 2021). Due to increased sensitivity in early life (see later), TL has even been successfully used in research contexts to measure quite subtle effects

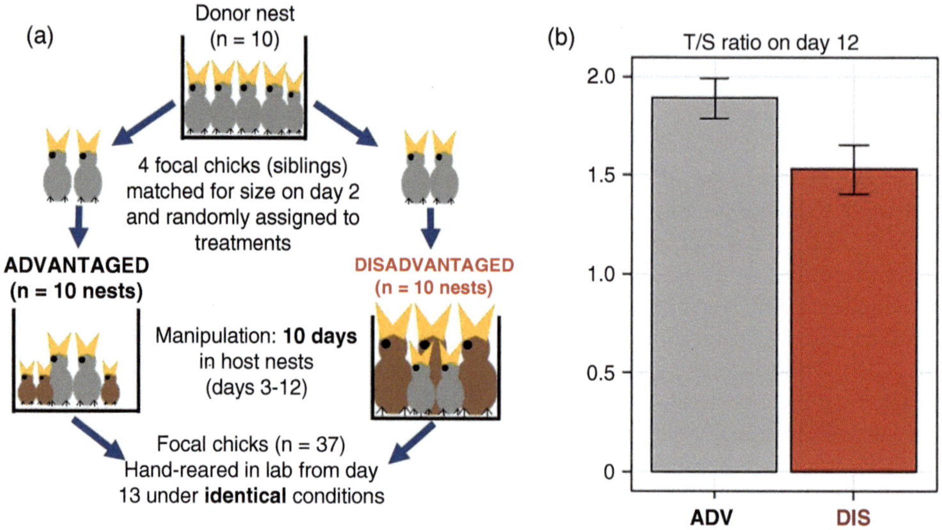

Figure 12.6 An experiment demonstrating the effects of size disadvantage on telomere loss in starling chicks. (a) Experimental design; (b) on day 12, at the end of the manipulation, T/S ratios measured in erythrocytes were significantly lower in disadvantaged chicks (DIS) compared to advantaged chicks (ADV). Source: Nettle et al. (2015) / The Royal Society / CC BY 4.0.

of stress. For example, starling chicks cross-fostered to nests where they were the smallest chicks, and thus presumably faced tougher competition for food, showed faster telomere loss than siblings cross-fostered to nests where they were the largest chicks, despite no differences in growth rates (Nettle et al. 2015; Figure 12.6). Future uses could include addressing the cumulative stress experienced by laboratory animals.

Use of TL as an indicator of cumulative experience is likely to be most successful in either large epidemiological studies or controlled experiments designed to compare effects of different treatments. However, due to the large individual differences in TL and the consequent lack of universal TL reference values comparable to those available for clinical measures such as leukocyte count or specific hormone levels, TL is unlikely to be a reliable indicator of cumulative experience in *individual* animals.

What species and life stages can TL be used in?

Most of what we know about TL is based on humans (see section on construct validity). However, telomeres are present in all vertebrates, and the telomeric DNA sequence is conserved throughout. Species do vary, however, in mean TL, the presence of interstitial telomeric DNA repeats in the middle of chromosomes, and the extent to which telomerase is active in adult tissues. Stress-induced telomere shortening could therefore be harder to detect in species with very long telomeres or a high proportion of interstitial telomere repeats (such as chickens), because a given amount of telomere loss would represent a smaller proportion of the total telomeric DNA present. Stress-induced telomere shortening could also be harder to detect in species such as the laboratory mouse and fish that retain telomerase expression as adults, because of the greater potential for telomere repair. Despite these concerns, there are several examples of stress-induced telomere loss in chickens, fish and mice (reviewed in Bateson 2016).

Furthermore, despite some species differences in how TL changes with age, rates of telomere shortening are often greater in early life when cells are dividing rapidly, followed by much slower shortening during the adult period. This means that, although TL can be assessed in adulthood, stress exposure will typically have a larger effect in early life. To illustrate, 10 days of stress was sufficient to detect effects of experimentally increased competition for food on TL in newly hatched starling chicks (Nettle et al. 2015; Figure 12.6), while 19 weeks was required to detect an effect of experimentally increased food insecurity in fully fledged juveniles, despite using a more precise TRF measurement method (Andrews et al. 2021).

What tissues can TL be measured in?

TL can be measured in nucleated cells yielding undamaged DNA that can be obtained using minimally to moderately invasive methods from most species. Since telomeres shorten fastest in rapidly dividing cells, samples for studying stress-induced telomere loss should be taken from proliferating tissues with high rates of cell division, such as blood, rather than tissues with low rates of cell division, such as skeletal muscle. In humans, most studies have used leukocytes (white blood cells). A complication with leukocytes is that a decline in mean TL can be caused by a change in the relative numbers of cell types with different TL as opposed to an actual decline in TL within a single-cell type. Since birds, fish, amphibians and reptiles differ from mammals in having nucleated erythrocytes, much less blood is required for TL measurement in these species – only a few drops for most measurement methods. Furthermore, since erythrocytes are a more homogenous population of cells than leukocytes and comprise 99% of blood cells, changes in TL measured in erythrocytes are likely to be less variable and simpler to interpret than those measured in leukocytes.

Obtaining adequate blood samples non-lethally may not always be possible. Alternative tissue sources include clipped pelvic fin tips in salmon (Pauliny et al. 2015), wing membrane biopsies in bats (Power et al. 2021) and tail tip skin biopsies in meerkats (Cram et al. 2017). Where non-invasive sampling is required, buccal cells obtained from oral swabs are an alternative proliferating cell source, used successfully to measure TL in both children and dogs (Dutra et al. 2020). Milk cells have been used in dairy cows (Iannuzzi et al. 2022).

While archived samples can sometimes be used successfully for telomere measurements, the way in which samples are collected and stored can have consequences for the quality and quantity of DNA available (see Nussey et al. 2014).

Does the timing of measurements matter?

The duration needed for a stressor to measurably affect TL will depend on its magnitude and on TL measurement precision. Assuming habituation does not occur, a given stressor will have a larger effect on TL the longer the animal has been exposed to it. All else being equal, TL measurements from studies with longer stress exposures will therefore have higher sensitivity (Aviv et al. 2006). For this reason, longitudinal studies in humans often measure changes in TL over a decade or more (e.g. Bateson et al. 2019a).

Reducing the assessment of cumulative experience to a single number measured at a single time point inevitably results in the loss of information. For example, a group of animals could go through a period of negative experience, resulting in accelerated biological ageing, followed by a period of more positive experience that slows the rate of ageing and returns biological age to average. If TL is measured only once at the end of the period of positive experience, then the period of negative experience will be missed entirely, and cumulative experience will be assessed as identical to a control group that went through neither negative nor positive experiences. More generally, depending on the timing of measurements, very different patterns of experience can result in the same final TL. Timing therefore needs to be chosen with knowledge of the research question being asked and the population studied; for example, yearly measurements at a single time point may work well for animals that are synchronised in when they experience stress (e.g. heat stress in summer), but less well where stressful experience is unsynchronised.

What type of study design is most statistically powerful and least prone to false nulls?

If the aim is to assess the impact of a putative stressor, an experimental study, in which subjects are randomly assigned to control versus stressed groups, will always be preferable, because only experiments permit definitive inferences about the causes of changes in TL. They also allow better control of known sources of variation in TL such as genetic similarity, sex, age and developmental experience. To eliminate the non-affective influences of age (discussed earlier), it is also necessary either to control for chronological age statistically, or to compare animals of the same age. Some of the clearest demonstrations of stress effects on TL have come from small experimental studies of European starlings involving fewer than 40 birds, in which age- and size-matched siblings were randomly assigned to treatments differing in competition for food (Nettle et al. 2017, 2015). In contrast, correlational studies, in which there is no manipulation of stress, require much larger sample sizes to detect effects of putative stressors on TL. For example, correlational studies of TL in dairy cows and Soay sheep involved multiple longitudinal samples from many hundreds of individuals (Froy et al. 2021; Seeker et al. 2018).

Longitudinal studies are also generally more sensitive than cross-sectional studies. In longitudinal designs, TL is measured two or more times in each subject and the difference in TL between measurements is analysed; this removes random variation due to individual differences in TL. Thus, all else being equal, these designs require fewer subjects to achieve comparable statistical power. However, what is less well appreciated is that if TL measurement error is high (as is likely with qPCR-based measurements), the power gain from performing within-subject comparisons will be offset by the loss arising from taking more readings of TL with high measurement error. Thus, the relative benefits of a longitudinal analysis will depend on the precision of the method used to measure TL. If a relatively imprecise method such as qPCR is being used to measure TL, then it could be beneficial in terms of power to use a cross-sectional design with a more subjects rather than a longitudinal design with a smaller number (Nettle et al. 2021).

In cross-sectional studies, involving only a single TL measurement for each individual, controlling for chronological age can increase power. This may therefore limit the sensitivity of TL as a welfare indicator in wild animal populations where individuals are sampled only once and information on age is unavailable. One potential solution here is to use TL from a non-dividing tissue (e.g. skeletal muscle) as a proxy for birth TL: an approach validated in a cross-sectional study of dogs of varying ages (Benetos et al. 2011).

Summary and conclusions

Evidence supports the hypothesis that TL is a valid cellular biomarker of biological age that is a moderately sensitive and specific indicator of the valence of cumulative affective experience, if used appropriately. TL is likely to be especially useful for assessing the cumulative impact of chronic negative affective states or multiple discrete negative events over relatively long periods of time. Shorter studies can be successful in very young animals due to naturally faster telomere shortening at young ages (see Figure 12.7). The main practical limitations of TL as a welfare indicator arise from the imprecision of telomere measurement methods and the large individual differences in mean TL within species. TL is therefore appropriate for use in experimental and epidemiological welfare studies with sufficiently large sample sizes, but not for clinical assessments of individuals.

A conceptual limitation of TL depends on whether one's view of welfare is entirely affective and the timescales over which inferences are to be drawn. Giving animals what they want is a common requirement for good welfare, yet paradoxically can be associated with poorer health (see also Chapter 3). For example, sugar is immediately reinforcing to humans and many other animals but is associated with subsequent negative affect due to blood–sugar dips, and in the longer term, with painful tooth decay, obesity and disease. Animals are also highly motivated to fight for mating opportunities, to defend territories and to reproduce, suggesting that these activities are reinforcing (fulfilling validation Test 2 from Chapter 2; and they are certainly good for evolutionary fitness, fulfilling validation Test 3). Reproduction is, however, physically costly and can be psychologically stressful, and individuals who invest more in reproduction typically live less long (Harshman and Zera 2007; Partridge et al. 2005). As we have seen, these preferred activities result in shorter TL. Thus, where hedonism and

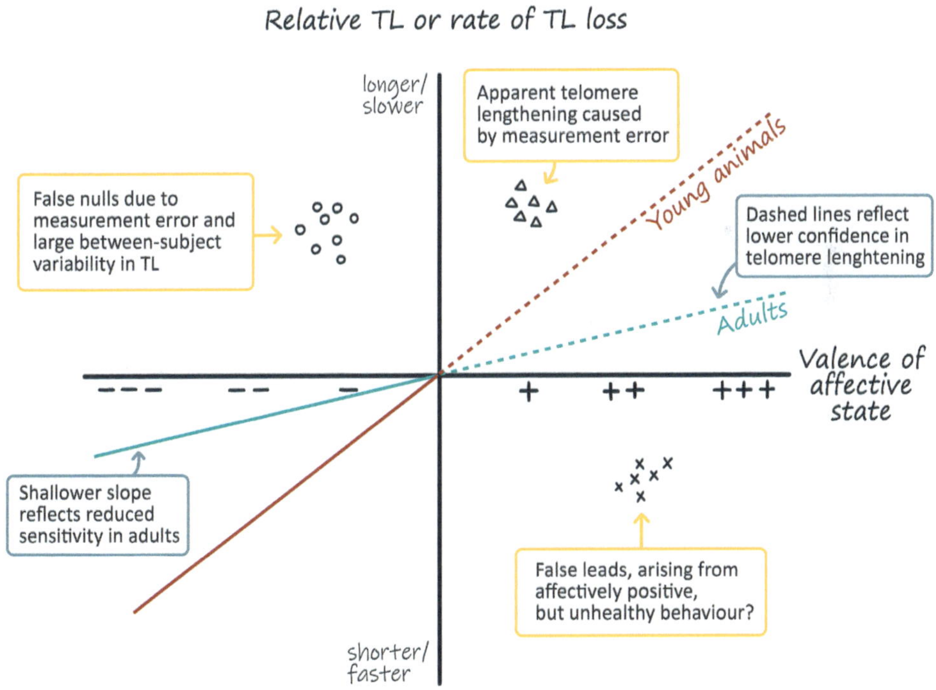

Figure 12.7 Measured telomere length (TL) correlates positively with cumulative affective valence across the range from negative to positive. The horizontal axis reflects either the intensity or duration of the cumulative affective state experienced. Although evidence supports the response of TL to stress being dose-dependent and monotonic, effects are small, especially in adult animals, reflected in the shallow slope of the blue line. Evidence for positive effects is less strong than for negative effects, reflected by the dotted lines.

health conflict, the evidence could be taken to indicate that TL is a better indicator of health than affective state. However, as we have suggested in Figure 12.1, the somatic damage associated with ill health can cause negative affect and there is accumulating evidence in humans that unhealthy lifestyles contribute to the development of anxiety and depression (e.g. Tang et al. 2024). Moreover, when it comes to cumulative affective experience – which is by definition a medium to long-term construct – we would argue that it is an open question whether the short-term rewards derived from eating sugar are offset by the long-term suffering associated with the diseases of overconsumption; and likewise, whether the short-term rewarding nature of winning a fight and mating compensates for the long-term mental toll of reproduction. It is possible that for cumulative experience, there is no conflict between affective and health-centred views of welfare.

In conclusion, TL represents a promising, reasonably well validated indicator of cumulative affective experience that is both responsive and selective enough to address some questions in animal welfare. While further validation is still required in relation to the effects of positive affect on TL, when used appropriately, TL offers an important addition to our battery of available methods for assessing animal welfare.

References

Andrews, C., Zuidersma, E., Verhulst, S., Nettle, D. and Bateson, M. (2021) *Exposure to food insecurity increases energy storage and reduces somatic maintenance in European starlings (Sturnus vulgaris). Royal Society Open Science*, 8, 211099. https://doi.org/10.1098/rsos.211099

Aviv, A., Valdes, A.M. and Spector, T.D. (2006) *Human telomere biology: pitfalls of moving from the laboratory to epidemiology. International Journal of Epidemiology*, 35(6), 1424–1429. https://doi.org/10.1093/ije/dyl169

Aydinonat, D., Penn, D.J., Smith, S., Moodley, Y., Hoelzl, F., Knauer, F. and Schwarzenberger, F. (2014) *Social isolation shortens telomeres in African grey parrots (Psittacus erithacus erithacus). PLoS One*, 9, 2–6. https://doi.org/10.1371/journal.pone.0093839

Bateson, M. (2016) *Cumulative stress in research animals: Telomere attrition as a biomarker in a welfare context? BioEssays*, 38, 201–212. https://doi.org/10.1002/bies.201500127

Bateson, M., Aviv, A., Bendix, L., Benetos, A., Ben-Shlomo, Y., Bojesen, S.E., Cooper, C., Cooper, R., Deary, I.J., Hägg, S., Harris, S.E., Kark, J.D., Kronenberg, F., Kuh, D., Labat, C., Martin-Ruiz, C.M., Meyer, C., Nordestgaard, B.G., Penninx, B.W.J.H., Pepper, G.V., Révész, D., Said, M.A., Starr, J.M., Syddall, H., Thomson, W.M., Van Der Harst, P., Whooley, M., Von Zglinicki, T., Willeit, P., Zhan, Y. and Nettle, D. (2019a) *Smoking does not accelerate leucocyte telomere attrition: A meta-analysis of 18 longitudinal cohorts. Royal Society Open Science*, 6, 190420. https://doi.org/10.1098/rsos.190420

Bateson, M., Eisenberg, D.T.A. and Nettle, D. (2019b) *Controlling for baseline telomere length biases estimates of the effect of smoking on leukocyte telomere attrition in longitudinal studies. Royal Society Open Science*, 6, 190937. https://doi.org/10.5281/zenodo.1009086

Bateson, M. and Nettle, D. (2018) *Why are there associations between telomere length and behaviour? Philosophical Transactions of the Royal Society B: Biological Sciences*, 373, 20160438. https://doi.org/10.1098/rstb.2016.0438

Bateson, M. and Poirier, C. (2019) *Can biomarkers of biological age be used to assess cumulative lifetime experience? Animal Welfare*, 28, 41–56. https://doi.org/10.71.0/09627286.28.1.041

Bauch, C., Becker, P.H. and Verhulst, S. (2014) *Within the genome, long telomeres are more informative than short telomeres with respect to fitness components in a long-lived seabird. Molecular Ecology*, 23, 300–310. https://doi.org/10.1111/mec.12602

Beirne, C., Delahay, R. and Young, A. (2015) *Sex differences in senescence: The role of intra-sexual competition in early adulthood. Proceedings of the Royal Society B Biological Sciences*, 282, 20151086. https://doi.org/10.1098/rspb.2015.1086

Beloor, J., Kang, H.K., Kim, Y.J., Subramani, V.K., Jang, I.S., Sohn, S.H. and Moon, Y.S. (2010) *The effect of stocking density on stress related genes and telomeric length in broiler chickens. Asian-Australasian Journal of Animal Sciences*, 23, 437–443.

Benetos, A., Kimura, M., Labat, C., Buchoff, G.M., Huber, S., Labat, L., Lu, X. and Aviv, A. (2011) *A model of canine leukocyte telomere dynamics. Aging Cell*, 10, 991–995. https://doi.org/10.1111/j.1474-9726.2011.00744.x

Chatelain, M., Drobniak, S.M. and Szulkin, M. (2020) *The association between stressors and telomeres in non-human vertebrates: a meta-analysis. Ecology Letters*, 23, 381–398. https://doi.org/10.1111/ele.13426

Cram, D.L., Monaghan, P., Gillespie, R. and Clutton-Brock, T. (2017) *Effects of early-life competition and maternal nutrition on telomere lengths in wild meerkats. Proceedings of the Royal Society B Biological Sciences*, 284, 20171383. https://doi.org/10.1098/rspb.2017.1383

Darrow, S.M., Verhoeven, J.E., Révész, D., Lindqvist, D., Penninx, B.W., Delucchi, K.L., Wolkowitz, O.M. and Mathews, C.A. (2016) *The association between psychiatric disorders and telomere length: A Meta-analysis involving 14,827 persons. Psychosomatic Medicine*, 78, 776–787. https://doi.org/10.1097/psy.0000000000000356

Dutra, L., Souza, F., Friberg, I., Araújo, M., Vasconcellos, A. and Young, R. (2020) *Validating the use of oral swabs for telomere length assessment in dogs. Journal of Veterinary Behavior*, 40, 16–20. https://doi.org/10.1016/j.jveb.2020.07.011

Ekinci, G.N. and Sanlier, N. (2023) *The relationship between nutrition and depression in the life process: A mini-review. Experimental Gerontology*, 172, 112072. https://doi.org/10.1016/j.exger.2022.112072

Friedman, H.L., MacDonald, D.A., Brown, N.J.L. and Coyne, J.C. (2020) *Extraordinary claims require compelling evidence: Concerns about "loving-kindness meditation slows biological aging in novices.". Psychoneuroendocrinology*, 111, 104410. https://doi.org/10.1016/j.psyneuen.2019.104410

Froy, H., Underwood, S.L., Dorrens, J., Seeker, L.A., Watt, K., Wilbourn, R.V., Pilkington, J.G., Harrington, L., Pemberton, J.M. and Nussey, D.H. (2021) *Heritable variation in telomere length predicts mortality in Soay sheep. Proceedings of the National Academy of Sciences of the United States of America*, 118, e2020563118. https://doi.org/10.1073/pnas.2020563118

Gangoso, L., Cortés-Avizanda, A., Sergiel, A., Pudifoot, B., Miranda, F., Muñoz, J., Delgado-González, A., Moleón, M., Sánchez-Zapata, J.A., Arrondo, E. and Donázar, J.A. (2021) *Avian scavengers living in anthropized landscapes have shorter telomeres and higher levels of glucocorticoid hormones. Science of the Total Environment*, 782, 146920. https://doi.org/10.1016/j.scitotenv.2021.146920

Gomes, N.M.V., Ryder, O.A., Houck, M.L., Charter, S.J., Walker, W., Forsyth, N.R., Austad, S.N., Venditti, C., Pagel, M., Shay, J.W. and Wright, W.E. (2011) *Comparative biology of mammalian telomeres: Hypotheses on ancestral states and the roles of telomeres in longevity determination. Aging Cell*, 10, 761–768. https://doi.org/10.1111/j.1474-9726.2011.00718.x

Gorbunova, V. and Seluanov, A. (2009) *Coevolution of telomerase activity and body mass in mammals: From mice to beavers. Mechanisms of Ageing and Development*, 130, 3–9. https://doi.org/10.1016/j.mad.2008.02.008

Harshman, L.G. and Zera, A.J. (2007) *The cost of reproduction: the devil in the details. Trends in Ecology and Evolution*, 22(2), 80–86. https://doi.org/10.1016/j.tree.2006.10.008

Hassett, A.L., Epel, E., Clauw, D.J., Harris, R.E., Harte, S.E., Kairys, A., Buyske, S. and Williams, D.A. (2012) *Pain is associated with short leukocyte telomere length in women with fibromyalgia. Journal of Pain*, 13, 959–969. https://doi.org/10.1016/j.jpain.2012.07.003

Hastings, W.J., Shalev, I. and Belsky, D.W. (2019) *Comparability of biological aging measures in the National Health and Nutrition Examination Study, 1999–2002. Psychoneuroendocrinology*, 106, 171–178. https://doi.org/10.1016/j.psyneuen.2019.03.012

Hau, M., Greives, T.J., Haussmann, M.F., Matlack, C., Costantini, D., Quetting, M., Adelman, J.S. and Partecke, J. (2015) *Repeated stressors in adulthood increase the rate of biological ageing. Frontiers in Zoology*, 12, 4. https://doi.org/10.1186/s12983-015-0095-z

Hoelzl, F., Smith, S., Cornils, J., Aydinonat, D., Bieber, C. and Ruf, T. (2016) *Telomeres are elongated in older individuals in a hibernating rodent, the edible dormouse (Glis glis). Scientific Reports*, **6**, 36856. https://doi.org/10.1038/srep36856

Huang, Y., Li, L., Gan, Y., Wang, C., Jiang, H., Cao, S. and Lu, Z. (2020) *Sedentary behaviors and risk of depression: a meta-analysis of prospective studies. Translational Psychiatry*, 10(1), 1–10. https://doi.org/10.1038/s41398-020-0715-z

Iannuzzi, A., Albarella, S., Parma, P., Galdiero, G., D'Anza, E., Pistucci, R., Peretti, V. and Ciotola, F. (2022) *Characterization of telomere length in Agerolese cattle breed, correlating blood and milk samples. Animal Genetics*, 53, 676–679. https://doi.org/10.1111/age.13227

Johnson, A.J., Terry, E., Bartley, E.J., Garvan, C., Cruz-Almeida, Y., Goodin, B., Glover, T.L., Staud, R., Bradley, L.A., Fillingim, R.B. and Sibille, K.T. (2019) *Resilience factors may buffer cellular aging in individuals with and without chronic knee pain. Molecular Pain*, 15, 1744806919842962. https://doi.org/10.1177/1744806919842962

Kärkkäinen, T., Briga, M., Laaksonen, T. and Stier, A. (2022) *Within-individual repeatability in telomere length: A meta-analysis in nonmammalian vertebrates. Molecular Ecology*, 31, 6339–6359. https://doi.org/10.1111/mec.16155

Kimura, M., Hjelmborg, J.V.B., Gardner, J.P., Bathum, L., Brimacombe, M., Lu, X., Christiansen, L., Vaupel, J.W., Aviv, A. and Christensen, K. (2008) *Telomere length and mortality: a study of leukocytes in elderly Danish twins. American Journal of Epidemiology*, 167, 799–806. https://doi.org/10.1093/aje/kwm380

King, K.S., Kozlitina, J., Rosenberg, R.N., Peshock, R.M., McColl, R.W. and Garcia, C.K. (2014) *Effect of leukocyte telomere length on total and regional brain volumes in a large population-based cohort. JAMA Neurology* 8591, 1–8. https://doi.org/10.1001/jamaneurol.2014.1926

Lai, T.P., Wright, W.E. and Shay, J.W. (2018) *Comparison of telomere length measurement methods. Philosophical Transactions of the Royal Society B: Biological Sciences*, 373(1741), 20160451. https://doi.org/10.1098/rstb.2016.0451

Le Nguyen, K.D., Lin, J., Algoe, S.B., Brantley, M.M., Kim, S.L., Brantley, J., Salzberg, S. and Fredrickson, B.L. (2019) *Loving-kindness meditation slows biological aging in novices: Evidence from a 12-week randomized controlled trial. Psychoneuroendocrinology*, 108, 20–27. https://doi.org/10.1016/j.psyneuen.2019.05.020

Lee, C.-H. and Giuliani, F. (2019) *The role of inflammation in depression and fatigue. Frontiers in Immunology*, 10, 1696. https://doi.org/10.3389/fimmu.2019.01696

Lin, P.Y., Huang, Y.C. and Hung, C.F. (2016) *Shortened telomere length in patients with depression: A meta-analytic study. Journal of Psychiatric Research*, 76, 84–93. https://doi.org/10.1016/j.jpsychires.2016.01.015

Mathur, M.B., Epel, E., Kind, S., Desai, M., Parks, C.G., Sandler, D.P. and Khazeni, N. (2015) *Perceived stress and telomere length: A systematic review, meta-analysis, and methodologic considerations for advancing the field. Brain, Behavior, and Immunity*, 54, 158–169. https://doi.org/10.1016/j.bbi.2016.02.002

Nettle, D., Andrews, C., Reichert, S., Bedford, T., Kolenda, C., Parker, C., Martin-Ruiz, C., Monaghan, P. and Bateson, M. (2017) *Early-life adversity accelerates cellular ageing and affects adult inflammation: Experimental evidence from the European starling. Scientific Reports*, 7, 40794. https://doi.org/10.1038/srep40794

Nettle, D., Gadalla, S.M., Lai, T.-P., Susser, E., Bateson, M. and Aviv, A. (2021) *Measurement of telomere length for longitudinal analysis: Implications of assay precision. American Journal of Epidemiology*, 190(7), 1406, kwab025–1413. https://doi.org/10.1093/aje/kwab025

Nettle, D., Monaghan, P., Gillespie, R., Brilot, B., Bedford, T. and Bateson, M. (2015) *An experimental demonstration that early-life competitive disadvantage accelerates telomere loss. Proceedings of the Royal Society B Biological Sciences*, 282, 20141610. https://doi.org/10.1098/rspb.2014.1610

Nilsonne, G., Tamm, S., Månsson, K.N.T., Åkerstedt, T. and Lekander, M. (2015) *Leukocyte telomere length and hippocampus volume: a meta-analysis. F1000 Research*, 4, 1073. https://doi.org/10.12688/f1000research.7198.1

Nussey, D.H., Baird, D., Barrett, E., Boner, W., Fairlie, J., Gemmell, N., Hartmann, N., Horn, T., Haussmann, M., Olsson, M., Turbill, C., Verhulst, S., Zahn, S. and Monaghan, P. (2014) *Measuring telomere length and telomere dynamics in evolutionary biology and ecology. Methods in Ecology and Evolution*, 5, 299–310. https://doi.org/10.1111/2041-210x.12161

Nussey, D.H., Kruuk, L.E.B., Donald, A., Fowlie, M. and Clutton-Brock, T.H. (2006) *The rate of senescence in maternal performance increases with early-life fecundity in red deer. Ecology Letters*, 9, 1342–1350. https://doi.org/10.1111/j.1461-0248.2006.00989.x

Osler, M., Bendix, L., Rask, L. and Rod, N.H. (2016) *Stressful life events and leucocyte telomere length: Do lifestyle factors, somatic and mental health, or low grade inflammation mediate this relationship? Results from a cohort of Danish men born in 1953. Brain, Behavior, and Immunity*, 58, 248–253. https://doi.org/10.1016/j.bbi.2016.07.154

Partridge, L., Gems, D. and Withers, D.J. (2005) *Sex and death: what is the connection? Cell*, 120, 461–472. https://doi.org/10.1016/j.cell.2005.01.026

Pauliny, A., Devlin, R.H., Johnsson, J.I. and Blomqvist, D. (2015) *Rapid growth accelerates telomere attrition in a transgenic fish. BMC Ecology and Evolution*, 15, 159. https://doi.org/10.1186/s12862-015-0436-8

Pepper, G.V., Bateson, M. and Nettle, D. (2018) *Telomeres as integrative markers of exposure to stress and adversity: A systematic review and meta-analysis. Royal Society Open Science*, 5, 180744. https://doi.org/10.1101/320150

Poirier, C., Bateson, M., Gualtieri, F., Armstrong, E.A., Laws, G.C., Boswell, T. and Smulders, T.V. (2019) *Validation of hippocampal biomarkers of cumulative affective experience.* ***Neuroscience*** *& Biobehavioral Reviews*, 101, 113–121. https://doi.org/10.1016/j.neubiorev.2019.03.024

Power, M.L., Power, S., Bertelsen, M.F., Jones, G. and Teeling, E.C. (2021) *Wing: A suitable nonlethal tissue type for repeatable and rapid telomere length estimates in bats. Molecular Ecology Resources*, 21, 421–432. https://doi.org/10.1111/1755-0998.13276

Puterman, E., Lin, J., Krauss, J., Blackburn, E.H. and Epel, E.S. (2015) *Determinants of telomere attrition over one year in healthy older women: Stress and health behaviors matter. Molecular Psychiatry*, 20, 529–535. https://doi.org/10.1038/mp.2014.70

Rafie, N., Golpour Hamedani, S., Barak, F., Safavi, S.M. and Miraghajani, M. (2017) *Dietary patterns, food groups and telomere length: A systematic review of current studies. European Journal of Clinical Nutrition*, 71, 151–158. https://doi.org/10.1038/ejcn.2016.149

Ridout, K.K., Ridout, S.J., Price, L.H., Sen, S. and Tyrka, A.R. (2016) *Depression and telomere length: A meta-analysis. Journal of Affective Disorders*, 191, 237–247. https://doi.org/10.1016/j.jad.2015.11.052

Schutte, N. and Malouff, J.M. (2022) *The association between optimism and telomere length: A meta-analysis. The Journal of Positive Psychology*, 17, 82–88. https://doi.org/10.1080/17439760.2020.1832249

Schutte, N.S. and Malouff, J.M. (2015) *The association between depression and leukocyte telomere length: A meta-analysis. Depression and Anxiety*, 32, 229–238. https://doi.org/10.1002/da.22351

Schutte, N.S., Malouff, J.M. and Keng, S.-L. (2020) *Meditation and telomere length: A meta-analysis. Psychology & Health*, 35, 901–915. https://doi.org/10.1080/08870446.2019.1707827

Seeker, L.A., Ilska, J.J., Psifidi, A., Wilbourn, R.V., Underwood, S.L., Fairlie, J., Holland, R., Froy, H., Salvo-, E., Bagnall, A., Whitelaw, B., Coffey, M.P., Nussey, D.H. and Banos, G. (2018) *Bovine telomere dynamics and the association between telomere length and productive lifespan. Scientific Reports*, 8, 12748. https://doi.org/10.1038/s41598-018-31185-z

Seeker, L.A., Underwood, S.L., Wilbourn, R.V., Dorrens, J., Froy, H., Holland, R., Ilska, J.J., Psifidi, A., Bagnall, A., Whitelaw, B., Coffey, M., Banos, G. and Nussey, D.H. (2021) *Telomere attrition rates are associated with weather conditions and predict productive lifespan in dairy cattle. Scientific Reports*, 11, 5589. https://doi.org/10.1038/s41598-021-84984-2

Selye, H. (1956) *The stress of life*. McGraw-Hill, New York, NY.

Sibille, K.T., Chen, H., Bartley, E.J., Riley, J., Glover, T.L., King, C.D., Zhang, H., Cruz-Almeida, Y., Goodin, B.R., Sotolongo, A., Petrov, M.E., Herbert, M., Bulls, H.W., Edberg, J.C., Staud, R., Redden, D., Bradley, L.A. and Fillingim, R.B. (2017) *Accelerated aging in adults with knee osteoarthritis pain: Consideration for frequency, intensity, time, and total pain sites. PAIN Reports*, 2, e591. https://doi.org/10.1097/pr9.0000000000000591

Spector, A., Li, Z., He, L., Badawy, Y. and Desai, R. (2024) *The effectiveness of psychosocial interventions on non-physiological symptoms of menopause: A systematic review and meta-analysis. Journal of Affective Disorders*, 352, 460–472. https://doi.org/10.1016/j.jad.2024.02.048

Staffaroni, A.M., Tosun, D., Lin, J., Elahi, F.M., Casaletto, K.B., Wynn, M.J., Patel, N., Neuhaus, J., Walters, S.M., Epel, E.S., Blackburn, E.H. and Kramer, J.H. (2018) *Telomere attrition is associated with declines in medial temporal lobe volume and white matter microstructure in functionally independent older adults. Neurobiology of Aging*, 69, 68–75. https://doi.org/10.1016/j.neurobiolaging.2018.04.021

Strandberg, T.E., Strandberg, A., Saijonmaa, O., Tilvis, R.S., Pitkälä, K.H. and Fyhrquist, F. (2012) *Association between alcohol consumption in healthy midlife and telomere length in older men. The Helsinki Businessmen Study. European Journal of Epidemiology*, 27, 815–822. https://doi.org/10.1007/s10654-012-9728-0

Swanberg, S.E., O'Hare, T.H., Robb, E.A., Robinson, C.M., Chang, H. and Delany, M.E. (2010) *Telomere biology of the chicken: A model for aging research. Experimental Gerontology*, 45, 647–654. https://doi.org/10.1016/j.exger.2010.04.002

Tang, Z., Yang, X., Tan, W., Ke, Y., Kou, C., Zhang, M., Liu, L., Zhang, Y., Li, X., Li, W. and Wang, S.-B. (2024) *Patterns of unhealthy lifestyle and their associations with depressive and anxiety symptoms among Chinese young adults: A latent class analysis. Journal of Affective Disorders*, 352, 267–277. https://doi.org/10.1016/j.jad.2024.02.055

Verhulst, S., Aviv, A., Benetos, A., Berenson, G.S. and Kark, J.D. (2013) *Do leukocyte telomere length dynamics depend on baseline telomere length? An analysis that corrects for "regression to the mean.". European Journal of Epidemiology*, 28, 859–866. https://doi.org/10.1007/s10654-013-9845-4

Walker, M., Duggan, G., Roulston, N., Van Slack, A. and Mason, G. (2012) *Negative affective states and their effects on morbidity, mortality and longevity. Animal Welfare*, 21, 497–509. https://doi.org/10.7120/09627286.21.4.497

Weiss, A., King, J.E., Inoue-Murayama, M., Matsuzawa, T. and Oswald, A.J. (2012) *Evidence for a midlife crisis in great apes consistent with the U-shape in human well-being. Proceedings of the National Academy of Sciences*, 109(49), 19949–19952. https://doi.org/10.1073/pnas.1212592109

Wilbourn, R.V., Moatt, J.P., Froy, H., Walling, C.A., Nussey, D.H. and Boonekamp, J.J. (2018) *The relationship between telomere length and mortality risk in non-model vertebrate systems: a meta-analysis. Philosophical Transactions of the Royal Society B: Biological Sciences*, 373, 20160447. https://doi.org/10.1098/rstb.2016.0447

Wojcicki, J.M., Medrano, R., Lin, J. and Epel, E. (2018) *Increased cellular aging by 3 years of age in Latino, preschool children who consume more sugar-sweetened beverages: a pilot study. Childhood Obesity* 14(3), 149–157. https://doi.org/10.1089/chi.2017.0159

Hippocampal Structural Plasticity

13

Colline Poirier

Faculty of Medical Sciences, Newcastle University, Newcastle upon Tyne, UK

Introduction

Neurons (brain cells that process external and internal information) are what give rise to affective states. In the mammalian and avian brain ('brain' hereafter), these are topographically organised into two tissues that can be distinguished macroscopically: grey and white matter. Grey matter mainly consists of neuron cell bodies and their branching dendrites, while white matter contains their myelinated axons (long extensions that connect brain regions). Several decades ago, neuroscientists discovered structural brain plasticity: that structural changes can occur in the brain with experience (e.g. motor practice like piano playing). More recently, structural brain plasticity has been discovered when individuals repeatedly experience negative or positive affective states. The brain region most consistently affected is the hippocampus (Koolschijn et al. 2009; Paquola et al. 2016). This is a bilateral brain region highly conserved across vertebrates, composed mainly of grey matter (Figure 13.1), that is involved in cognition (especially learning and memory) and emotional regulation (for instance, being key in the physiological stress response, applying negative feedback to the hypothalamic–pituitary–adrenal [HPA] axis to restore homeostasis [de Kloet et al. 1998; Jacobson and Sapolsky 1991]).

Some structural changes associated with positive or negative affective states are suspected to be adaptively functional (Suri and Vaidya 2015) for emotion regulation (as well as for learning and memory). However, exposure to the types of chronic or unpredictable aversive events that induce dysregulation of the HPA axis is often associated with *pathological* structural changes in the hippocampus, leading to stress-related disorders such as anxiety disorder, major depression and post-traumatic stress disorder (Suri and Vaidya 2015). Such stress disorders are commonly considered as non-adaptive (see Chapter 1), although some have suggested that they might have adaptive value (e.g. Kinney and Tanaka 2009; Constant et al. 2021).

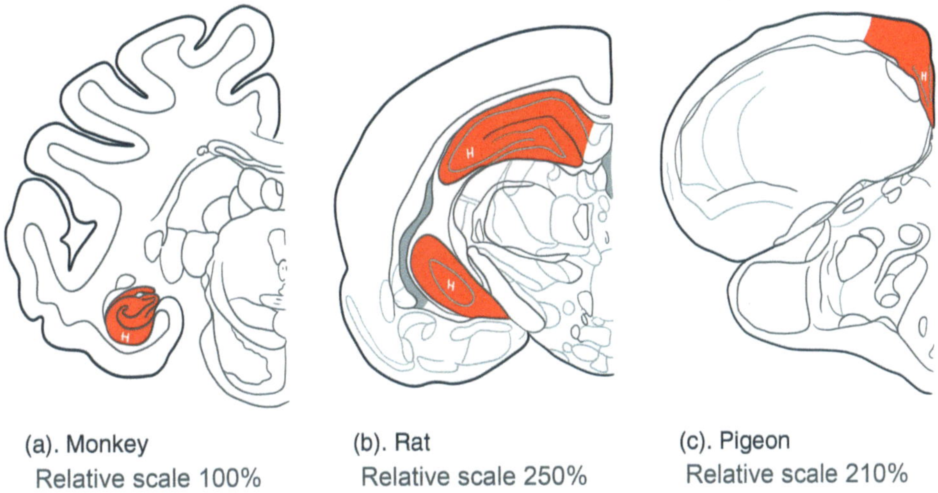

Figure 13.1 The location of the hippocampus in the forebrain of a primate, rodent and bird (shown in a transverse cross section of one hemisphere). Source: Adapted from Colombo and Broadbent (2000). Source: Colombo and Scarf (2012) / Michael Colombo / CC BY 4.0.

At the cellular level, structural plasticity involves two mechanisms. One is variation in the size and complexity of dendrites. The second, more specific to the hippocampus, is the birth of new neurons: *neurogenesis*. In the hippocampus, neurogenesis persists during the whole life of individuals (while in most other brain regions, it is restricted to early life stages). Variation in neurogenesis rate is thus a crucial mechanism underlying the ability of the hippocampus to change, an increased rate resulting in the net increase of neuron number (Boldrini et al. 2018; Chawana et al. 2014, Kempermann et al. 2018; Moreno-Jiménez et al. 2019). This cellular hippocampal plasticity is underpinned by changes at the molecular level (e.g. in specific molecules such as BDNF) that will not be reviewed here because conflicting results limit their use for welfare assessment purposes. More usefully for welfare researchers, both cellular mechanisms (changes in dendrite size and neuron number) result in macroscopic changes of grey matter density and/or hippocampal volume (Figure 13.2).

Structural changes in the hippocampus can be measured *post-mortem* using histology, or *in vivo* using magnetic resonance imaging (MRI). Despite constant technological improvements in MRI scanning, this currently has only modest spatial resolution (typically 0.5 or 1 mm in each dimension), and therefore can only measure macroscopic changes (hippocampal volume and grey matter density). However, histological approaches, combined with powerful microscopes, provide data at very high spatial resolution and so can measure hippocampal changes at the cellular level (Table 13.1), as well as being useful for hippocampal volume.

Basic construct validity

Hippocampal plasticity and affective states in humans

Many human studies in clinically healthy participants link events known to induce negative or positive affective states with respectively decreased or increased hippocampal structural measures. For instance, aversive events (e.g. bereavement, severe illness or unemployment) are

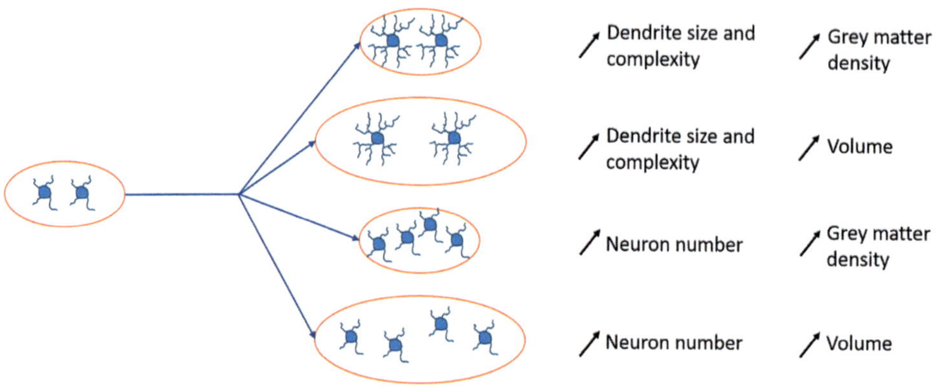

Figure 13.2 Schematic representation of different cellular mechanisms responsible for structural hippocampal plasticity and their respective macroscopic outcomes. The orange outline depicts the hippocampus (not to scale), and the blue shapes the neurons. These four scenarios can occur simultaneously. While structural changes can happen in both directions (increases and decreases), for simplicity this figure only illustrates increases.

Table 13.1 Typical outcome measures of hippocampal structural plasticity in mammals and birds.

Spatial scale	In vivo measures	Post-mortem measures
Cellular level		
Dendrite size and complexity		Histology: staining allowing visualisation of neuronal dendritic trees (e.g. Golgi impregnation procedure).
Neuron number (neurogenesis)		Histology: staining allowing visualisation of precursor cell proliferation (BrdU; Ki67) or newborn neurons (e.g. doublecortin; BrdU/calbindin; BrdU/NeuN). For protocols, see Zhao and van Praag (2020).
Macroscopic level		
Hippocampus volume	Structural MRI (T1-weighted contrast)	Histology: staining allowing visualisation of the hippocampus (e.g. thionin)
Grey matter density	Structural MRI (T1-weighted contrast)	

MRI, magnetic resonance imaging.

often associated with decreased hippocampal volume and/or reduced hippocampal grey matter (e.g. Papagni et al. 2011; Bromis et al. 2018). Conversely, longitudinal interventional studies have shown that physical exercise and mindfulness meditation, both well-established to improve affective states (Brown and Ryan 2003; Gál et al. 2021; Bourke et al. 2022; Heissel et al. 2023; Li et al. 2022), increase hippocampal volume and grey matter amount (Erickson et al. 2011; Holzel et al. 2011).

Some studies have directly assessed the link between structural changes in the hippocampus and self-reported affective states. One found a negative correlation between hippocampal volume and psychological distress in healthy adolescents (Broadhouse et al. 2019). Another revealed a positive correlation between hippocampal volume and subjective well-being and happiness in a mixed sample of clinically healthy and non-healthy participants (Van't Ent et al. 2017).

In clinical populations, several meta-analyses indicate that hippocampal volume and/or the amount of grey matter is reduced in people suffering from mood disorders such as major depression (Arnone et al. 2016) and post-traumatic stress disorder (e.g. O'Doherty et al. 2015; Bromis et al. 2018). Conversely, mood-improving treatments (antidepressant drugs, electroconvulsive therapy) increase hippocampal volume and the amount of grey matter in depressed patients (Abbott et al. 2014; Arnone et al. 2013; Frodl et al. 2008). Altogether, results from human studies suggest that structural changes in the hippocampus reflect changes in affective state valence.

Hippocampal plasticity and affective states in non-human mammals

The investigation of stress neurobiology in mammals (mostly rodents, and to a lesser extent non-human primates) has resulted in a huge body of literature that cannot be reviewed fully here. In summary, multiple stimuli or paradigms have been used to induce stress experimentally, and they have been consistently found to reduce the various measures of hippocampal plasticity (Table 13.2).

Examples of stress-inducing stimuli and paradigms include immobilisation, electric foot shocks, predator odour, social intruders, maternal care disruption, social separation and the 'unpredictable chronic stress' paradigm (during which animals are exposed to various aversive stimuli, usually twice per day, over 2–8 weeks). Beyond these stimuli/paradigms all representing threats to fitness, the evidence that these induce a deterioration in animals' affective states is physiological and behavioural (Tanapat et al. 2001; Gould et al. 1998, Rice et al. 2008; Perera et al. 2011; Pryce and Fuchs 2016). For instance, they increase levels of circulating corticosteroids, heart rate, blood pressure as well as anxiety-related and depressive-like behaviours (expressed spontaneously, or measured by classical tests developed to assess these phenotypes, e.g. open field, novelty suppressed feeding, elevated plus-maze and forced swim tests), and they decrease reward reactivity (i.e. induce anhedonia, see Chapter 9). While each type of physiological and behavioural evidence is prone to false leads, the negative impact of each stimulus on affective states is typically supported by several strands of evidence, decreasing this probability. There is therefore strong evidence that stimuli provoking negative affective states also induce reductions in hippocampal plasticity measures (see Table 13.2). In parallel, experimentally disrupting hippocampal neurogenesis in rodents and non-human primates has been shown to increase anxiety and depressive-like behaviour frequency (Revest et al. 2009; Perera et al. 2011; Snyder et al. 2011; Mohammad et al. 2016), suggesting that neurogenesis plays a causal role in these negative affective states.

Conversely, measures of hippocampal plasticity increase when animals are exposed to rewarding stimuli. Three types of stimuli have been tested: access to sexual partners, environmental

Table 13.2 Examples of aversive and rewarding stimuli that induce changes in hippocampal plasticity.

	Cellular/Macroscopic measure	Change direction	Duration	Species	Reference
Aversive stimuli					
Acute physical stress	Cellular (N)	↘	1 h cold immobilization + 30 min forced swimming	Rat	Heine et al. (2004)
Acute electrical foot shocks	Cellular (N)	↘	15 min	Rat	Malberg and Duman (2003)
Acute social stress	Cellular (N)	↘	1 h of resident-intruder test	Marmoset	Gould et al. (1998)
Acute Predator odour	Cellular (N)	↘	1 h	Rat	Tanapat et al. (2001)
Chronic social stress	Cellular (N)	↘	10 episodes per day, for 10 days	Mouse	Yap et al. (2006)
	Cellular (D)	↘	1 h of social confrontation per day, for 28 days	Tree shrew	Magarinos et al. (1996)
Chronic electrical foot shocks	Cellular (N)	↘	9 days	Rat	Malberg and Duman (2003)
Chronic physical restraint	Cellular (D)	↘	6 h per day, for 3 weeks	Rat	Watanabe et al. (1992)
	Cellular (D) and Macroscopic (Vol)	↘	6 h per day, for 3 weeks	Mouse	Kassem et al. (2013)
	Macroscopic (Vol)	↘	6 h per day, for 3 weeks	Rat	Lee et al. (2009)
	Macroscopic (Vol)	↘	2 h per day, for 3, 7, and 11 days	Rat	Rahman et al. (2016)
	Cellular (N)	↘	6 h per day, for 3 weeks	Rat	Pham et al. (2003)
Chronic Unpredictable Stress paradigm	Cellular (N)	↘	2 weeks	Rat	de Andrade et al. (2013)
	Cellular (N)	↘	3 weeks	Rat	Heine et al. (2004)
	Cellular (D)	↘	1 month	Rat	Sousa et al. (2000)
	Macroscopic (Vol)	↘	2 to 8 weeks	Rat	Luo et al. (2014)

Maternal care disruption	Cellular (D) Macroscopic (VBM)	↘ ↘	7 days 16 weeks	Mouse Macaque	Liu et al. (2016) Jackowski et al. (2011)
Social separation stress	Cellular (N)	↘	15 weeks	Macaque	Perera et al. (2011)
Rewarding stimuli					
Acute sexual experience	Cellular (N)	↗	30 min of exposure to a female	Rat	Leuner et al. (2010)
Chronic sexual experience	Cellular (N and D)	↗	1 exposure to a female per day, for 14 days	Rat	Leuner et al. (2010)
Environmental enrichment	Cellular (N) Cellular (D)	↗ ↗	40 days 1 month	Mouse Marmoset	Kempermann et al. (1997) Kozorovitskiy et al. (2005)
Voluntary exercise	Cellular (N and D) Cellular (N) Cellular (N) and Macroscopic (VBM) Macroscopic (VBM)	↗ ↗ ↗ ↗	17 days 12 days and 4 weeks 2 months 8–10 weeks	Rat Mouse Mouse Mouse	Eadie et al. (2005) van Praag et al. (1999) Biedermann et al. (2016) Fuss et al. (2014)

D, dendrite size and complexity; N, neurogenesis; Vol, total hippocampal volume; VBM, voxel-based morphometry, measuring the local amount of grey matter.

In each study, subjects were considered by the authors as healthy at the beginning of the experiment (although for animals housed in barren cages, their welfare was unlikely to be good), aversive and rewarding stimuli were imposed by the researchers, and the experimental design included a control group.

enrichment and voluntary running (see Table 13.2). While sexual activity is beneficial from a fitness point of view and well-established to be rewarding (Paredes, 2014), evidence for the rewarding nature of physical enrichment and voluntary running comes from preference and 'willingness-to-work-for' studies (Sherwin 1998; Olsson and Dahlborn 2002; Belke and Wagner 2005). These studies indicate that rodents develop place preference and are willing to work to access an enriched environment composed of a large cage with nesting material, a running wheel, a tunnel and toys. Similar results are obtained without the running wheel and with a running wheel only. As in humans, voluntary running in rodents is also associated with many positive health outcomes, including neuroprotective and antidepressant effects (reviewed in van Praag 2008). In one seminal paper, Kempermann et al. (1997) showed that exposure to such enriched environment (i.e. a large cage with nesting material, tunnels and toys and a running wheel) increased hippocampal neurogenesis in adult mice, compared to animals in a small, barren environment. Subsequent studies showed that access to enrichment without running wheels, but still containing elements that mice prefer and/or work for, was sufficient to promote neurogenesis (Brown et al. 2003). Conversely, a running wheel alone was sufficient to increase neurogenesis (van Praag et al. 1999). This effect of voluntary running in rodents has since been replicated in many studies (reviewed in van Praag 2008), including on other measures of hippocampal plasticity (e.g. increased dendritic length and amount of grey matter). As described earlier, there is widespread evidence that voluntary running is usually rewarding in rodents. However, there have been few reports that in some circumstances, voluntary running might be associated with a deterioration of affective states (increased anxiety) and yet still increase neurogenesis (Fuss et al. 2010; Onksen et al. 2012). The implication for welfare researchers of such potential false leads is discussed later in the Selectivity section.

Pharmacological studies provide further evidence that hippocampal structural changes are linked to affective states in non-human mammals (Test 4 in Chapter 2). Antidepressant drugs increase hippocampal volume and/or grey matter density (or its cellular underpinning) in non-human primates and rodents, while also reducing depressive-like behaviours (e.g. Santarelli et al. 2003; Perera et al. 2011; Snyder et al. 2011; Wu et al. 2014; Morais et al. 2017). Furthermore, experimentally disrupting hippocampal plasticity in rodent and non-human primates prevents such effects of antidepressant drugs on depressive-like behaviours (Santarelli et al. 2003; Perera et al. 2011; Wu et al. 2014): additional evidence that increased plasticity plays a causal role in the amelioration of affective states.

Finally, turning to Test 5 in Chapter 2 (i.e. checking that a new indicator reliably covaries with pre-validated affect indicators), many studies in rodent and non-human primates have documented changes in hippocampal plasticity alongside other indicators of affective states (e.g. depressive-like or anxiety-like phenotypes). Most are limited to group comparisons, which are prone to the 'ecological fallacy' (Robinson 1950): that co-variation at the group level does not necessarily means co-variation at the individual level. However, one study (MacLellan et al. 2022) recently investigated the relationship between hippocampal volume and a behavioural welfare indicator (inactive-but-awake behaviour) in individual mice, and found a negative correlation (i.e. greater hippocampal volumes were associated with less inactive-but-awake behaviour). This study provides further evidence that differences in hippocampal plasticity measures reflect variation in affective state valence in mammals.

Hippocampal plasticity and affective states in birds

Until recently, studies of hippocampal plasticity in birds have been interpreted as changes in learning and memory: in food-storing species, hippocampal structural changes were thought to facilitate memorisation of food caches, and on migrating species to support spatial navigation (for review, see Sherry and MacDougall-Shackleton 2015). However, this narrow interpretation of hippocampal plasticity supporting cognition only is challenged by a growing number of studies showing that a wide range of affect-significant conditions seem to modulate hippocampal neurogenesis in birds, as in mammals. Indeed, hippocampal neurogenesis is increased by environmental enrichment in pigeons (Melleu et al. 2016). On the other hand, it is decreased by captivity in wild-caught black-capped chickadees (Ladage et al. 2010), exposure to constant light in crows (Taufique et al. 2018), food restriction and keel bone fractures in chickens (Robertson et al. 2017; Armstrong et al. 2020) and chronic unpredictable stress in quails (Lormant et al. 2020) and chicken (Gualtieri et al. 2019). Furthermore, hippocampal neurogenesis in chickens with poor body condition, a classic welfare indicator in poultry, is lower than in those with a good body condition (Armstrong et al. 2022). Even if some of these changes were driven at least partially by changes in spatial memory demands (e.g. in the enrichment and captivity studies), the common denominator between these various factors is their presumed impact on affective states: those likely to induce more positive affect are associated with increased hippocampal neurogenesis, while those likely to induce more negative affect are associated with decreased neurogenesis.

These studies thus suggest that hippocampal neurogenesis (and other measures of hippocampal structural plasticity) might be a valid indicator of affective valence in birds. However, the current evidence is sparser than in mammals, and additional studies are required to support the validity of the hippocampal approach. The avian welfare literature would thus benefit from hippocampal plasticity studies where the affective state of the animals is experimentally manipulated (including with pharmacological drugs), and independently characterised using other affect indicators.

Validity of hippocampal structural changes as indicators of changes in emotions, moods and cumulative experience

Affective states can vary according to different time scales (see Chapter 1). Emotions are brief affective states that usually last seconds to minutes, while moods result from the integration of emotions over hours or days. Cumulative affective experience is defined as the net impact of all the events that affect the welfare of an animal over their lifetime, be it negatively, positively and/or by way of amelioration (Poirier et al. 2019, adapted from APC 2013), and can thus be conceived as resulting from the integration of emotions and moods over months or years. The question is therefore whether hippocampal plasticity is a valid indicator of affective states varying at each time scale. Evidence in humans and other mammals suggest this is the case (while this hypothesis has not been explicitly tested in birds, one would expect similar properties due to the conserved nature of the hippocampus in vertebrates).

Hippocampal changes have been observed after exposure to paradigms increasing or decreasing the valence of emotions in rodents, (e.g. a 30 min long exposure to a sexually receptive female, Leuner et al. 2010; electrical foot shocks during 15 min, Malberg and Duman 2003). Studies in animal models of mood disorders, involving exposure of the animals for weeks or months to aversive stimuli, suggest that hippocampal structural changes also validly track moods (see Table 13.2). Finally, hippocampal measures also seem to be a valid indicator of cumulative experience in humans and other mammals (see later for references). Indeed, hippocampal structural changes induced by aversive or rewarding events are sometimes transient but, in some circumstances (depending on the number, duration and frequency of the events), recovery to baseline levels can be only partial. For instance, one recovery day is sufficient to fully restore the reduced neurogenesis induced by one day of exposure to unpredictable aversive stimuli, but three weeks of recovery only leads to a partial recovery of the effect of three weeks of the same regime (Heine et al. 2004). Any exposure to an aversive stimulus before full recovery from the previous one will therefore lead to cumulative effects. Similarly, exposure to a series of rewarding stimuli at a frequency that does not allow the transient effect on the hippocampus to fully vanish will lead to cumulative effects.

Illustrating this property, hippocampal volume decreases proportionally with the number of days or weeks rats are exposed to an unpredictable stress paradigm (Luo et al. 2014; Rahman et al. 2016; Figure 13.3a); the amount of hippocampal grey matter negatively correlates with the number of aversive events healthy human participants have experienced (Papagni et al. 2011) and hippocampal volume decreases with the number of days of untreated depression over many years (Sheline et al. 2003; Figure 13.3b). Reciprocally, hippocampal volume increases with the duration of a physical exercise program in which healthy human participants are enrolled (Erickson et al. 2011); hippocampal neurogenesis increases with the duration rodents spend voluntarily running in a wheel or the total distance they voluntarily run (Bednarczyk et al. 2009; Kodali et al. 2016). There is also some evidence of a combined, net effect of positive versus negative experiences. In one study, sexual activity counteracted the effect of concomitant chronic exposure to aversive stimuli on neurogenesis in mice (Kim et al. 2013; Figure 13.3c), while similar results were found between concomitant voluntary physical activity and chronic exposure to aversive stimuli, also in mice (Castilla-Ortega et al. 2014). Collectively, these studies indicate that at least in mammals, hippocampal structural changes validly indicate changes in emotions, moods and – importantly – cumulative experience.

Responsiveness

The responsiveness of a measure reflects the probability of obtaining a false null result. Here, it means asking: what is the probability of observing no significant change in hippocampal measures, despite the affective state of the subject having changed? In this section, I first discuss whether hippocampal plasticity measures are responsive to changes in affective states across the whole range of valence (i.e. in a graph plotting how hippocampal measures change with affective state valence as illustrated in Figure 13.4, whether the slope of the function is constant and not null). I then discuss which factors influence this responsiveness (i.e. factors influencing the average slope of the function in Figure 13.4).

General responsiveness

Group comparison studies and cumulative effects over time in human and non-human mammals (as described in the construct validity section earlier) already suggest that hippocampal plasticity measures track affective state across a wide range of valence. Animal studies come from the biomedical literature, where housing conditions even in 'enriched' environments are likely to be suboptimal from a welfare point of view (Cait et al. 2022) and where the affective state of 'normal' subjects is unlikely to reach the most positive valence values. Studies demonstrating variation in hippocampal measures in healthy animals can therefore be interpreted as evidence of responsiveness across the range from mildly negative to mildly positive affective states. However, variation in healthy humans is consistent with responsiveness across the range from sub-clinical negative states to the most positive ones. Combined with studies

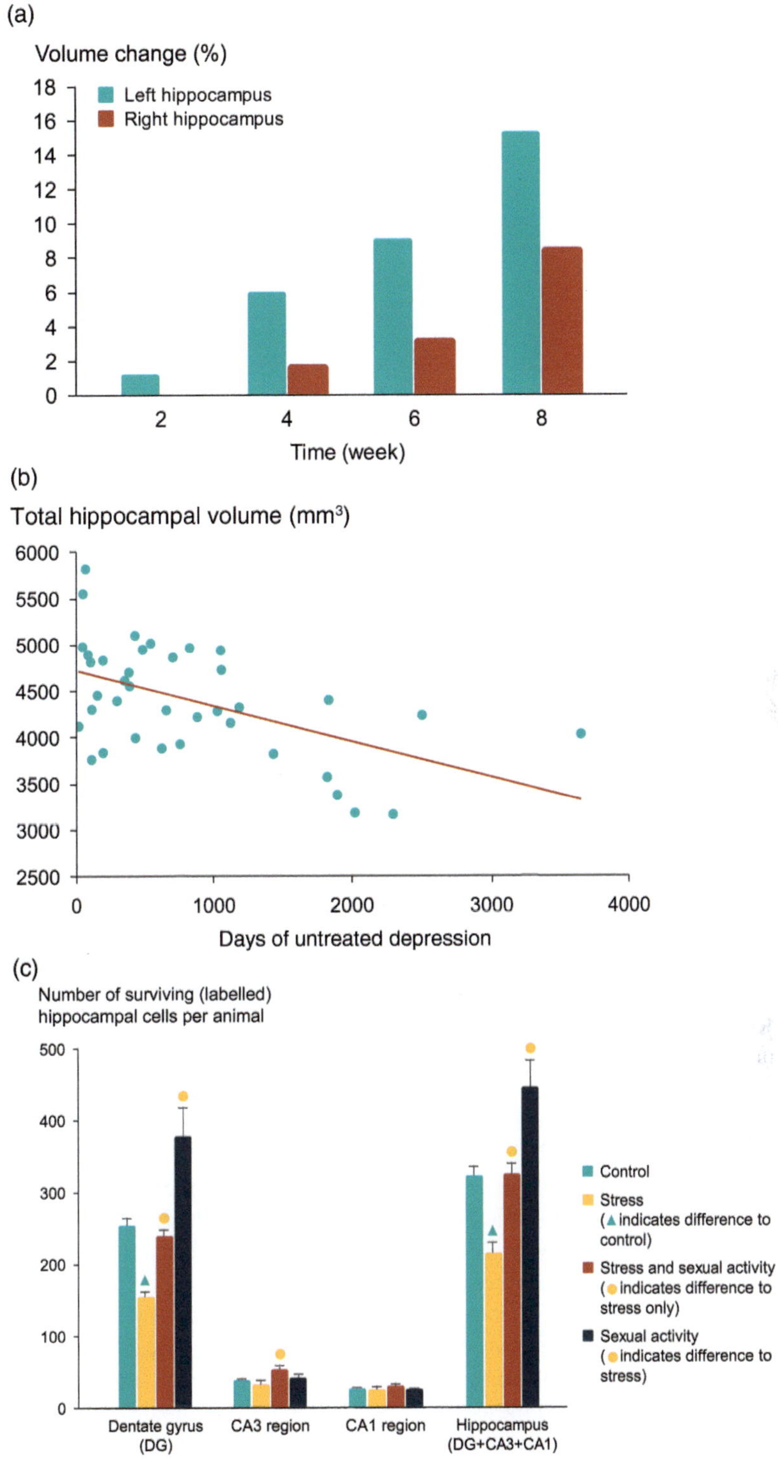

Figure 13.3 Examples where hippocampal volume or neurogenesis tracks cumulative affective experience: (a) difference to control animals in hippocampal volume in male Sprague–Dawley rats (n = 15) measured every 2 weeks in a 56-day chronic stress procedure. Source: Luo et al. (2014) / with permission of Springer Nature; (b) relationship between hippocampal volume and days of untreated depression in female outpatients (n = 38) with recurrent depression in remission. Source: Sheline et al. (2003) / with permission of American Psychiatric Association; (c) hippocampal newly born cell survival in mice (n = 12 per group) following 3 weeks of sexual activity (90 min/day) and/or repeated restraint stress (3 h/day). Source: Kim et al. (2013) / with permission of Elsevier.

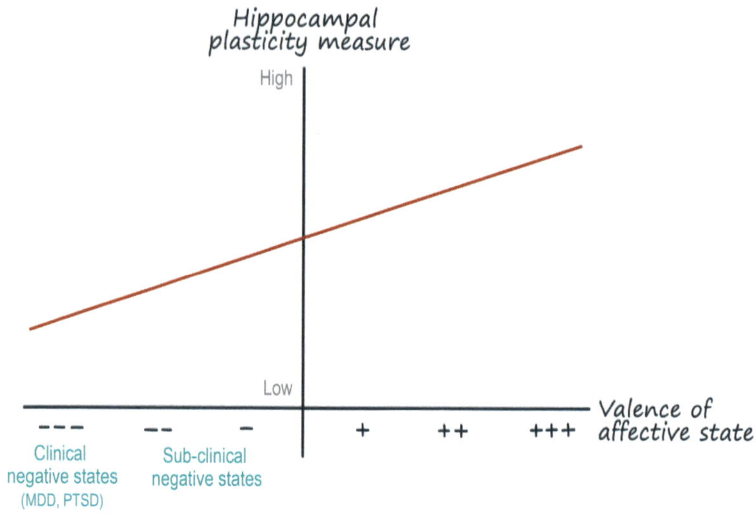

Figure 13.4 Variation of hippocampal plasticity measure according to affective state valence. The slope of the function corresponds to the responsiveness of the measure. Given the modest granularity with which current studies are able to differentiate various levels of positive and negative valence, this responsiveness appears to be similar and not null across valenced states spanning from negative to positive, resulting into a linear function (see text for justification). The slope of this linear function varies according to several factors, including the specific hippocampal measure assessed, methodological sophistication, the hippocampal sub-region examined and the age of the subjects. MDD: major depressive disorder; PTSD: post-traumatic stress disorder.

showing hippocampal variation between patients suffering from mood disorders (who arguably experience very negative affective states) and control participants, as well as a graded effect with the duration of untreated depression in depressed patients, these results suggest that hippocampal measures are responsive across the whole range of valence (in other words, the slope of the function illustrated in Figure 13.4 is never null).

One study (Van't Hent et al. 2017) has explicitly tested whether the responsiveness of the hippocampal approach is similar across the whole range of affective states (i.e. whether the slope of the function in Figure 13.4 was constant and not null and the function therefore linear). This study meta-analysed five MRI studies in human beings, summarizing findings from 724 twins and siblings. The large sample size and the liberal inclusion criteria suggest that a very wide range of affective states (from pathologically negative to very positive) were represented in this meta-analysis. Subjective well-being was measured using two scales: subjective happiness and life satisfaction. Comparing linear and non-linear models, the authors found that a linear relationship best described the association between hippocampal volume and subjective happiness. The association between hippocampal measures and the life satisfaction scale was instead best described by a non-linear relationship, the slope of the function tending to infinity for the most positive affective states (why such a 'wall effect' would occur with life satisfaction but not subjective happiness, and which of these two concepts best relates to cumulative affective experience in non-human animals is currently unknown however). This meta-analysis thus suggests that hippocampal plasticity measures are not prone to flooring or ceiling effects, instead tracking affective states in a graded way across the main range of affective state valence.

Factors affecting the responsiveness of hippocampal measures

Having established that the relationship between hippocampal plasticity measures and affective state valence is roughly linear, I now discuss the main factors influencing the slope of this linear function (i.e. the responsiveness of these measures), namely the specific measure assessed, methodological sophistication, the hippocampal sub-region examined and the age of the subjects.

The responsiveness of different hippocampal plasticity measures

The diverse measures of hippocampal plasticity differ in how responsive they are to different types of affective state (i.e. emotions vs. moods vs. cumulative experience). Methods able to quantify changes at the cellular level (neurogenesis, dendritic changes) appear sensitive to affective states of all durations, as illustrated in the construct validity section and Table 13.2 (with the shortest aversive stimulus reported to induce detectable hippocampal changes being 15 minutes). In contrast, the macroscopic measures (e.g. hippocampal volume, amount of grey matter) are responsive to changes in moods and cumulative experience (see references in construct validity section), but appear quite insensitive to short-lasting changes in emotions. Indeed, the shortest duration of aversive or rewarding stimuli reported to induce macroscopic changes is three days of exposure to a chronic physical restraint stress paradigm in rodents (consisting of complete immobilisation for 2 hours per day), with the MRI measure taken at least one hour after the end of the aversive stimulus (Rahman et al. 2016), a paradigm likely to affect animals' moods but not their emotions.

Within a similar spatial scale (i.e. cellular or macroscopic), it is however unclear whether hippocampal plasticity measures differ in their responsiveness. For instance, at the cellular level, changes in the size/complexity of dendrites, and neurogenesis, are both influenced by the same stimuli (see Table 13.2) and appear to co-occur (e.g. Eadie et al. 2005; Leuner et al. 2010), suggesting that both measures are similarly responsive. However, subtle variations in their sensitivity might still exist. Among measures of neurogenesis, it is unclear whether quantification of different stages (i.e. proliferation and survival of precursor cells, differentiation and maturation into neurons) differ in responsiveness. At the macroscopic level, we also do not know whether measures of hippocampal volume and grey matter density differ in their responsiveness to detect changes in affective states. Nevertheless, an analytical technique called voxel-based morphometry (VBM) provides an outcome measure of the local amount of grey matter (one measure per voxel, which is a three-dimensional pixel in an MRI image), which has the benefit of being sensitive to both changes (i.e. total hippocampal volume and hippocampal grey matter density). Since it is not yet possible to predict whether a putative aversive or rewarding stimulus will change hippocampal volume and/or its grey matter density, the use of VBM should reduce the probability of false null results (compared to the use of measures that only respond to changes in volume or density).

Other methodological aspects affecting responsiveness

Additional methodological factors influencing the responsiveness of hippocampal plasticity measures as affect indicators include technical differences: less measurement error will obviously lead to greater abilities to detect true effects. For instance, the use of a higher magnetic field or better-quality coils will increase the sensitivity of the MRI approach. Technical improvements occurring continuously over time, recent publications are likely to include fewer false null results than older ones (an example of such false null result being the failure to detect hippocampal atrophy in patients suffering from major depression in the early 2000s: Koolschijn et al. 2009). Additionally, as for most or even all indicators in this book, the responsiveness of hippocampal measures varies with experimental design. For example, various hippocampal properties are partially genetically determined (e.g. in humans, around 40% of the variation in hippocampal volume is of genetic origin: Sullivan et al. 2001). By eliminating genetic variability, within-subject designs (obviously limited to *in vivo* MRI approaches) therefore tend to be statistically better able to detect changes in affective state. Everything else being equal, a *post-mortem* measure implemented in a cross-sectional design (e.g. hippocampal volume measured with histology in different groups of individuals) will therefore be less sensitive and therefore more prone to false null results than an *in vivo* one used in a longitudinal design (e.g. within-subject variation in hippocampal volume measured with high-resolution MRI).

Variation in responsiveness between hemispheres and hippocampal sub-regions

Evidence supporting the validity of hippocampal plasticity measures as affect indicators comes from both left and right hippocampi. Many studies have reported effects in both (e.g. Erickson et al. 2011; Paquola et al. 2016; Van't Ent et al. 2017; see also Figure 13.3a), while others described significant effects in one hemisphere only (left hippocampus only: e.g. Wise et al. 2017; Holzel et al. 2011; right hippocampus only: e.g. Du et al. 2012; Dannlowski et al. 2012; Papagni et al. 2011). However, to claim lateralisation of an effect, one must statistically demonstrate that the effect is significantly larger in one hemisphere compared to the other one. One study, investigating the effect of chronic physical restraint on hippocampus volume in rats, did find a significant interaction between the stress condition (stressed vs. control group) and the hemisphere assessed, with the left hippocampus being more sensitive to stress than the right (Rahman et al. 2016). Whether such greater responsiveness in the left hippocampus can be replicated and generalised to other species and other measures of hippocampal plasticity remains to be determined.

Furthermore, each hippocampus can be subdivided into three sub-regions along its longitudinal axis: anterior, intermediary and posterior in primates, or ventral, intermediary and dorsal in rodents. Many studies suggest that the anterior (ventral) part of the hippocampus is more involved in affective processes than the posterior (dorsal) part (see Fanselow and Dong 2010; Moser and Moser 1998; Strange et al. 2014). It is therefore plausible that responsiveness of hippocampal markers to changes in affective states is greater in the anterior/ventral than posterior/dorsal hippocampus. Supporting this hypothesis, many studies found the effects of aversive/rewarding stimuli or effect of depression/antidepressant drugs to be localised in the anterior/ventral part of the hippocampus (e.g. Du et al. 2012; Dannlowski et al. 2012; Papagni et al. 2011; Willard et al. 2009; Wu and Hen 2014). Most of these studies fail to statistically test whether differences between sub-regions were significant, but Erickson et al. (2011) did find that the effect of voluntary exercise on human hippocampal volume was significantly greater in the anterior hippocampus than the posterior.

Variation in responsiveness according to subject age

Studies supporting the validity of the hippocampal approach have been performed in both juvenile and adult individuals of a wide range of ages, including old age. For example, the study from Erickson et al. (2011) involved healthy humans aged 55–80 years (see also Kempermann et al. (1998) for the effects of enrichment on neurogenesis in senescent mice). Collectively, these studies indicate that hippocampal plasticity is responsive to affective state across the whole lifespan of individuals, even into old age.

Nevertheless, hippocampal plasticity is at its maximum during early life, and is reduced during old age (Kempermann et al. 2018); its responsiveness to affective states may thus show a similar pattern. Consistent with this, studies documenting the adverse impact of early life stress on hippocampal plasticity vastly outnumber those on the effect of stress during adulthood. This could well be due to the reduced plasticity of the hippocampi in adulthood. However, it is important to keep in mind that the affective reaction of individuals to aversive/rewarding stimuli might also change with age: the same event might be perceived as less stressful by an adult than a juvenile. It is therefore difficult to determine whether an absence of change in hippocampal measures in adults, when such effects have been observed in juveniles, are false null results (reflecting a reduced responsiveness of hippocampal measures with age) or true null ones (reflecting an increased affective resilience with age). Similar issues are also seen for stereotypic behaviour and telomere length, as can be seen in Chapters 11 and 12 in this section of the book.

Selectivity

As seen earlier, hippocampal plasticity measures are responsive to the valence of affective states, changing in opposite directions when states become more positive or more negative. They also change in the same direction with diverse experiences of the same valence but differing arousal levels (e.g. in animals, chronic social isolation inducing a low-arousal depressive-like state, and acute electric shocks inducing a high-arousal negative affective state, see Table 13.2). Hippocampal plasticity measures are thus specific to the valence dimension of affective states. However, hippocampal plasticity is known to vary with non-affective factors (e.g. age, learning and memory processes) and is therefore not specific to affective states.

The difficulty in using hippocampal plasticity as an affect indicator lies in many of these non-affective factors not only having an additive effect on hippocampal plasticity (i.e. an effect independent of affective state, acting as 'noise' when using hippocampal plasticity in a welfare context), but also an interactive one. For instance, as seen in the previous section on responsiveness, age will influence hippocampal plasticity in the absence of any change in affective state (additive effect) but might also change stress sensitivity and therefore influence the change in affective state induced by exposure to an aversive stimulus (interactive effect). These interactive effects are currently poorly understood and often neglected in the literature. For instance, sex differences in hippocampal plasticity are typically interpreted as (non-affective) sexual dimorphism, despite a lack of controls to assess whether the affective state induced by an experimental stressor was similar in the two sexes. To increase the specificity (selectiveness) of hippocampal plasticity as an affect indicator, it is therefore important to control for non-affective factors, as they could induce false leads for welfare researchers. However, controlling statistically for these factors and considering them as noise could result in some signal linked to affective states being overlooked (if there are interactive effects between the factor and affective state), and therefore result in decreased responsiveness. Welfare researchers thus need to be aware of this potential trade-off between responsiveness (sensitivity) and selectivity (specificity) when using the hippocampal approach.

In the following, I discuss factors known to influence hippocampal plasticity, while being agnostic about whether they induce additive and/or interactive effects. Factors typically controlled for in hippocampal plasticity studies include sex, age and – for macroscopic measures only – brain size. Between-subject cross-sectional studies usually do this *a priori* by using a control group; while between-subject correlational studies (typical of human studies) and within-subject longitudinal ones, do this *a posteriori* by including these factors as additional explanatory variables in statistical models. The *a posteriori* control of the age factor requires a few words of caution: the most common statistical approach to analyse data is the general(ised) linear model, and, by definition, these models only consider linear effects (except if non-linear effects are explicitly specified). However, the relationship between hippocampal plasticity and age can depart significantly from a linear function, even during adulthood: while the general trend is a decrease of hippocampal plasticity during adulthood, transient periods of increase have been reported (Walhovd et al. 2011; Fjell et al. 2013). The exact shape might depend on the animal species and potentially the hippocampal measure (this last aspect still requires systematic comparison). A solution is to use a restrictive age range (with the cost that any effects might be limited to this specific age range), or to implement a longitudinal design with a control group (i.e. an *n-by-2* design, with n being the number of time points).

Another source of false leads in between-subject studies is variation in hippocampal plasticity of genetic origin. Longitudinal designs circumvent this problem by focusing on *within*-subject variability. When such designs are not possible (e.g. with *post-mortem* approaches), using large sample sizes will minimise the risk of false leads (based on the assumption that if experimental and control groups are big enough, genotypes associated with high- and low-hippocampal plasticity should be evenly distributed in the groups). This strategy will simultaneously minimise the risk of false nulls.

The hippocampus is, as mentioned, well established to play a crucial role in learning and memory. At the structural level, increased hippocampal plasticity is associated with enhanced cognitive abilities (e.g. Erickson et al. 2011; Wu and Hen 2014) and decreased plasticity with cognitive deficits (e.g. Rahman et al. 2016). Cognition function and affective states usually co-vary in the same direction, with the same rewarding stimuli (e.g. environmental enrichment and physical activity) enhancing cognitive function and generating more positive affect, while aversive stimuli decrease cognitive capability and make affective state more negative. In these instances, a change in hippocampal plasticity following an experimental manipulation is likely to reflect a change in both types of process: cognitive and affective ones. However, if cognitive changes occur in absence of a change in affective state, or if an improvement in cognitive abilities is concomitant with a more negative affective state, this could result in welfare researchers being misled by false leads. Such dissociation might potentially explain some results where voluntary wheel running in rodents and increased plasticity measures were found alongside increased anxiety (Fuss et al. 2010; Onksen et al. 2012, see however Garrett et al. (2012) for conflicting results): the increase in hippocampal plasticity measures observed in these studies might be due to the cognitive enhancement properties of wheel running, *despite* deteriorating affective states. Another example of possible dissociation between cognitive abilities and affective states can be found in the seasonal fluctuation of hippocampal measures observed in food-storing birds and mammals (e.g. ground squirrels and some mouse species in the wild, see Pyter et al. 2005; Burger et al. 2014): increases in hippocampal measures due to the increased cognitive load of memorising caching places might occur in the absence of change in affective states (or even a deterioration). In such instances when changes in affective states and cognitive abilities are expected to be of opposing direction, the specific location of the change within the hippocampus along its longitudinal axis might help decrease the risk for false leads: a change in the anterior/ventral sub-region is more likely to reflect a change in affective state, and a change in the posterior/dorsal one a change in cognition (Fanselow and Dong 2010; Strange et al. 2014). However, this topic requires further investigation (see for instance Wu and Hen (2014) for a more complex picture). Another, not mutually exclusive, solution is to combine the hippocampal plasticity measure with another validated indicator of affective states (even if it is more responsive to variation in arousal than valence). Here, the goal is to exclude the possibility that hippocampal measures changed without any affective state variation.

Use in practice

For welfare researchers, using the hippocampal approach might seem daunting. Box 13.1 describes recommendations for new users, and later I expand on specific points. As described in the previous sections, hippocampal measures differ in terms of their responsiveness to affective states of different durations, and strategies available to increase their selectivity will depend on experimental design (cross-sectional versus longitudinal ones), and to a lesser extent, which measure has been chosen by the experimenter. Besides their responsiveness and selectivity to affect, the choice of a measure will likely be guided by pragmatic constraints. The easiest one to implement is the *post-mortem* determination of hippocampal volume.

Box 13.1 Eleven recommendations when using the hippocampal approach

Study objective

- Be clear if you want to further validate the hippocampal approach (e.g. in a species where it has not been documented yet) or if you want to use it as a valid welfare indicator (to investigate the welfare consequence of a manipulation of interest); you cannot do both with the same dataset. Make this clear when you report your study.

Experimental design

- Be clear about the timescale of the affective state under investigation (emotions vs. moods vs. cumulative experience) and choose a relevant hippocampal measure (e.g. macroscopic measures are not sensitive to emotions).
- Adjust the sample size according to the magnitude of the expected change in affective state, the experimental design (between- vs. within-subject designs) and the specific hippocampal measure.
- When using a longitudinal approach, include a control group if possible.
- Control for the influence of learning and memory processes.
- Report the steps you have taken to maximise the responsiveness (sensitivity) and selectivity (specificity) of the welfare indicator.

Data collection and data analysis

- If you have never used the technique before, first demonstrate that you can detect the effect of a well-established welfare manipulation.
- For macroscopic measures, include total brain volume (ideally defined as total grey matter + total white matter) as a covariate in your statistical analysis.

Results description

- Report results separately for left and right hippocampus and for anterior/ventral and posterior/dorsal sub-regions.
- Ideally, report effects for both sexes separately.

Result interpretation

- If the results do not match your a priori hypothesis, before claiming that your results are false nulls or false leads, do carefully consider and discuss all the steps you have taken to maximise your hippocampal measures' responsiveness and selectivity to affect states.

Laboratory animals, when not rehomed, are usually killed at the end of any experiment, and in farm settings, heads could be collected at slaughterhouses. This approach does not necessitate specific equipment, and the basic skills required to fix, slice and stain the brain, and to delineate the highly distinguishable hippocampus, should be accessible to any biologist. In contrast, the cellular approaches will benefit from more expert skills in histology, while the VBM approach (see earlier section) requires an MRI scanner and some expert knowledge to process the data.

Choosing an adequate sample size is crucial in the success of any hippocampal approach. Beside the considerations previously mentioned about responsiveness and selectivity to affect, users should be aware that many published studies are underpowered: effect sizes reported are thus likely to be inflated, and, consequently, it seems sensible to use sample sizes slightly bigger than those reported in the relevant literature. In a welfare experiment aiming at assessing the impact of a husbandry or experimental procedure, the best approach to minimise the risk of false null results is to use benchmarking. This requires the experimenters to first demonstrate that they are able to detect the effect of a well-established aversive or rewarding stimulus expected to a have a welfare impact of similar magnitude as the one under investigation. Then, the researchers can replicate the exact same experimental protocol with the procedure of interest. Finally, the responsiveness and selectivity to changes in affective state of any hippocampal approach is likely to vary with sub-region of the hippocampus (anterior/ventral vs. posterior/dorsal) and/or the hemisphere assessed. Reporting results per sub-region and per hemisphere would therefore help gather the evidence to test these hypotheses and ultimately refine the hippocampal approach.

Conclusion

Measures of hippocampal plasticity are valid indicators of affective valence and cumulative affective experience in mammals, and appear likely to be valid in birds as well. The responsiveness of hippocampal measures is similar across the full range of affective valence (although varying according to a multitude of factors including the specific measure used and the experimental design). Changes in hippocampal measures are not specific to affective processes, but non-affective factors influencing the measures can be controlled for experimentally and/or statistically. Furthermore, measuring hippocampal volume is quite straightforward.

Despite its benefits, the hippocampal approach does suffer from several limitations. First, like several indicators in this book, it does not allow welfare researchers to determine whether an animal or group of animals is in an *absolutely* positive or negative affective state, but can only be used to detect *relative* differences in affective state (i.e. if the affective state is better or worse than a control). Second, while the apparent responsiveness of hippocampal measures to a wide range of discrete affective states (including high- and low-arousal ones) decreases the probability of obtaining false null results, it also represents a potential limitation, because solutions for improving welfare might differ according to the discrete affective state involved. In such a scenario, the hippocampal plasticity approach will not help welfare scientists identifying which specific affective state was suboptimal and needs to be targeted. Finally, some hippocampal plasticity measures are not easy to implement, requiring specific knowledge and/or equipment, and all are time consuming to collect and/or analyse. However, the strong validity of this approach, its sensitivity along the whole range of affective state valence and its value for assessing cumulative affective experience, should make these efforts worthwhile.

References

Abbott, R.A., Whear, R., Rodgers, L.R., Bethel, A., Coon, J.T., Kuyken, W., Stein, K. and Dickens, C. (2014) *Effectiveness of mindfulness-based stress reduction and mindfulness based cognitive therapy in vascular disease: A systematic review and meta-analysis of randomised controlled trials. Journal of Psychosomatic Research*, 76(5), 341–351. https://doi.org/10.1016/j.jpsychores.2014.02.012

APC (2013). Review of the assessment of cumulative severity and lifetime experience in non-human primates used in neuroscience research. *Report of the Animal Procedures Committee (APC) Primate Subcommittee Working Group* (chaired by Professor John Pickard). Available at: https://assets.publishing.service.gov.uk/media/5a7cb908e5274a2f304efaf9/cs_nhp_review_FINAL_2013_corrected.pdf (accessed February 2025)

Armstrong, E.A., Richards-Rios, P., Addison, L., Sandilands, V., Guy, J.H., Wigley, P., Boswell, T. and Smulders, T.V. (2022) *Poor body condition is associated with lower hippocampal plasticity and higher gut methanogen abundance in adult laying hens from two housing systems. Scientific Reports*, 12(1), 15505. https://doi.org/10.1038/s41598-022-18504-1

Armstrong, E.A., Rufener, C., Toscano, M.J., Eastham, J.E., Guy, J.H., Sandilands, V., Boswell, T. and Smulders, T.V. (2020) *Keel bone fractures induce a depressive-like state in laying hens. Scientific Reports*, 10(1), 3007. https://doi.org/10.1038/s41598-020-59940-1

Arnone, D., Job, D., Selvaraj, S., Abe, O., Amico, F., Cheng, Y., Colloby, S.J., O'Brien, J.T., Frodl, T., Gotlib, I.H., Ham, B.J., Kim, M.J., Koolschijn, P.C.M., Perico, C.A.M., Salvadore, G., Thomas, A.J., Van Tol, M.J., van der Wee, N.J.A., Veltman, D.J., Wagner, G. and Mcintosh, A.M. (2016) *Computational meta-analysis of statistical parametric maps in major depression. Human Brain Mapping*, 37, 1393–1404. https://doi.org/10.1002/hbm.23108

Arnone, D., McKie, S., Elliott, R., Juhasz, G., Thomas, E.J., Downey, D., Williams, S., Deakin, J.F.W. and Anderson, I.M. (2013) *State-dependent changes in hippocampal grey matter in depression. Molecular Psychiatry*, 18, 1265–1272. https://doi.org/10.1038/mp.2012.150

Bednarczyk, M.R., Aumont, A., Décary, S., Bergeron, R. and Fernandes, K.J.L. (2009) *Prolonged voluntary wheel-running stimulates neural precursors in the hippocampus and forebrain of adult CD1 mice. Hippocampus*, 19, 913–927. https://doi.org/10.1002/hipo.20621

Belke, T.W. and Wagner, J.P. (2005) *The reinforcing property and the rewarding aftereffect of wheel running in rats: A combination of two paradigms. Behavioural Processes*, 68, 165–172. https://doi.org/10.1016/j.beproc.2004.12.006

Biedermann, S., Fuss, J., Steinle, J., Auer, M.K., Dormann, C., Falfan-Melgoza, C., Ende, G., Gass, P. and Weber-Fahr, W. (2016) *The hippocampus and exercise: Histological correlates of MRI-detected volume changes. Brain Structure & Function*, 221, 1353–1363. https://doi.org/10.1007/s00429-014-0976-5

Boldrini, M., Fulmore, C.A., Tartt, A.N., Dwork, A.J. and Mann, J.J. (2018) *Human hippocampal neurogenesis persists throughout aging. Cell Stem Cell*, 22, 589–599. https://doi.org/10.1016/j.stem.2018.03.015

Bourke, M., Patten, R.K., Klamer, T.L., Klepac, B., Dash, S. and Pascoe, M.C. (2022) *The acute affective response to physical activity in people with depression: A meta-analysis. Journal of Affective Disorders*, 311, 353–363. https://doi.org/10.1016/j.jad.2022.05.089

Broadhouse, K.M., Boyes, A., Winks, N., Dokonal, T., McLoughlin, L., Parker, M., Beaudequin, D., Simcock, G., Lagopoulos, J. and Hermens, D.F. (2019) *Subcortical volume correlates of psychological distress in early adolescence. Developmental Neuroscience*, 41(3–4), 193–202. https://doi.org/10.1159/000502339

Bromis, K., Calem, M., Reinders, A.A.T.S., Williams, S.C.R. and Kempton, M.J. (2018) *Meta-analysis of 89 Structural MRI studies in posttraumatic stress disorder and comparison with major depressive disorder. American Journal of Psychiatry*, 175, 1–10. https://doi.org/10.1176/appi.ajp.2018.17111199

Brown, J., Cooper-Kuhn, C.M., Kempermann, G., Van Praag, H., Winkler, J., Gage, F.H. and Kuhn, H.G. (2003) *Enriched environment and physical activity stimulate hippocampal but not olfactory bulb neurogenesis. European Journal of Neuroscience*, 7(10), 2042–2046. https://doi.org/10.1046/j.1460-9568.2003.02647.x

Brown, K.W. and Ryan, R.M. (2003) *The benefits of being present: mindfulness and its role in psychological well-being. Journal of Personality and Social Psychology*, 84, 822–848. https://doi.org/10.1037/0022-3514.84.4.822

Burger, D.K., Gulbrandsen, T., Saucier, D.M. and Iwaniuk, A.N. (2014) *The effects of season and sex on dentate gyrus size and neurogenesis in a wild rodent, Richardson's ground squirrel (Urocitellus richardsonii). Neuroscience*, 272, 240–251. https://doi.org/10.1016/j.neuroscience.2014.04.067

Cait, J., Cait, A., Wilder Scott, R., Winder, C.B. and Masom, G.J. (2022) *Conventional laboratory housing increases morbidity and mortality in research rodents: Results of a meta-analysis. BMC Biology*, 20, 15. https://doi.org/10.1186/s12915-021-01184-0

Castilla-Ortega, E., Rosell-Valle, C., Pedraza, C., Rodríguez de Fonseca, F., Estivill-Torrús, G. and Santín, L.J. (2014) *Voluntary exercise followed by chronic stress strikingly increases mature adult-born hippocampal neurons and prevents stress-induced deficits in 'what-when-where' memory. Neurobiology of Learning and Memory*, 109, 62–73. https://doi.org/10.1016/j.nlm.2013.12.001

Chawana, R., Alagaili, A., Patzke, N., Spocter, M.A., Mohammed, O.B., Kaswera, C., Gilissen, E., Bennett, N.C., Ihunwo, A.O. and Manger, P.R. (2014) *Microbats appear to have adult hippocampal neurogenesis but post-capture stress causes a rapid decline in the number of neurons expressing doublecortin. Neuroscience*, 277, 724–733. https://doi.org/10.1016/j.neuroscience.2014.07.063

Colombo, M. and Broadbent, N. (2000) *Is the avian hippocampus a functional homologue of the mammalian hippocampus? Neuroscience and Biobehavioral Reviews*, 24, 465–484. https://doi.org/10.1016/S0149-7634(00)00016-6

Colombo, M. and Scarf, D. (2012) *Neurophysiological studies of learning and memory in pigeons. Comparative Cognition & Behavior Reviews*, 7, 23–43. https://doi.org/10.3819/ccbr.2012.70002

Constant, A., Hesp, C., Davey, C.G., Friston, K.J. and Badcock, P.B. (2021) *Why depressed mood is adaptive: A numerical proof of principle for an evolutionary systems theory of depression. Computational Psychiatry*, 5(1), 60–80. https://doi.org/10.5334/cpsy.70

Dannlowski, U., Stuhrmann, A., Beutelmann, V., Zwanzger, P., Lenzen, T., Grotegerd, D., Domschke, K., Hohoff, C., Ohrmann, P., Bauer, J., Lindner, C., Postert, C., Konrad, C., Arolt, V., Heindel, W., Suslow, T. and Kugel, H. (2012) *Limbic scars: long-term consequences of childhood maltreatment revealed by functional and structural magnetic resonance imaging. Biological Psychiatry*, 71, 286–293. https://doi.org/10.1016/j.biopsych.2011.10.021

de Andrade, J.S., Cespedes, I.C., Abrao, R.O., dos Santos, T.B., Diniz, L., Britto, L.R.G., Spadari-Bratfisch, R.-C., Ortolani, D., Melo-Thomas, L., da Silva, R.C.B. and Viana, M.B. (2013) *Chronic unpredictable mild stress alters an anxiety-related defensive response, Fos immunoreactivity and hippocampal neurogenesis. Behavioural Brain Research*, 250, 81–90. https://doi.org/10.1016/j.bbr.2013.04.031

de Kloet, E.R., Vreugdenhil, E., Oitzl, M.S. and Joëls, M. (1998) *Brain corticosteroid receptor balance in health and disease. Endocrine Reviews*, 19(3), 269–301. https://doi.org/10.1210/edrv.19.3.0331

Du, M.Y., Wu, Q.Z., Yue, Q., Li, J., Liao, Y., Kuang, W.H., Huang, X.Q., Chan, R.C., Mechelli, A. and Gong, Q.Y. (2012) *Voxelwise meta-analysis of gray matter reduction in major depressive disorder. Progress in Neuro-Psychopharmacology & Biological Psychiatry*, 36(1), 11–16. https://doi.org/10.1016/j.pnpbp.2011.09.014

Eadie, B.D., Redila, V.A. and Christie, B.R. (2005) *Voluntary exercise alters the cytoarchitecture of the adult dentate gyrus by increasing cellular proliferation, dendritic complexity, and spine density. Journal of Comparative Neurology*, 486, 39–47. https://doi.org/10.1002/cne.20493

Erickson, K.I., Voss, M.W., Prakash, R.S., Basak, C., Szabo, A., Chaddock, L., Kim, J.S., Heo, S., Alves, H., White, S.M., Wojcicki, T.R., Mailey, E., Vieira, V.J., Martin, S.A., Pence, B.D., Woods, J.A., McAuley, E. and Kramer, A.F. (2011) *Exercise training increases size of hippocampus and improves memory. Proceedings of the National Academy of Sciences of the United States of America*, 108, 3017–3022. https://doi.org/10.1073/pnas.1015950108

Fanselow, M.S. and Dong, H.W. (2010) *Are the dorsal and ventral hippocampus functionally distinct structures? Neuron*, 65, 7–19. https://doi.org/10.1016/j.neuron.2009.11.031

Fjell, A.M., Westlye, L.T., Grydeland, H., Amlien, I., Espeseth, T., Reinvang, I., Raz, N., Holland, D., Dale, A.M. and Walhovd, K.B. (2013) *Critical ages in the life course of the adult brain: Nonlinear subcortical aging. Neurobiology of Aging*, 34, 2239–2247. https://doi.org/10.1016/j.neurobiolaging.2013.04.006

Frodl, T., Jäger, M., Smajstrlova, I., Born, C., Bottlender, R., Palladino, T., Reiser, M., Möller, H. and Meisenzahl, E.M. (2008) *Effect of hippocampal and amygdala volumes on clinical outcomes in major depression: A 3-year prospective magnetic resonance imaging study. Journal of Psychiatry and Neuroscience*, 33, 423–430.

Fuss, J., Ben Abdallah, N.M., Vogt, M.A., Touma, C., Pacifici, P.G., Palme, R., Witzemann, V., Hellweg, R. and Gass, P. (2010) *Voluntary exercise induces anxiety-like behavior in adult C57BL/6J mice correlating with hippocampal neurogenesis. Hippocampus*, 20(3), 364–376. https://doi.org/10.1002/hipo.20634

Fuss, J., Biedermann, S., Falfan-Melgoza, C., Auer, M.K., Zheng, L.Z., Steinle, J., Horner, F., Sartorius, A., Ende, G., Weber-Fahr, W. and Gass, P. (2014) *Exercise boosts hippocampal volume by preventing early age-related gray matter loss. Hippocampus*, 24, 131–134. https://doi.org/10.1002/hipo.22227

Gál, É., Ştefan, S. and Cristea, I.A. (2021) *The efficacy of mindfulness meditation apps in enhancing users' well-being and mental health related outcomes: A meta-analysis of randomized controlled trials. Journal of Affective Disorders*, 279, 131–142. https://doi.org/10.1016/j.jad.2020.09.134

Garrett, L., Lie, D.C., Hrabé de Angelis, M., Wurst, W. and Hölter, S.M. (2012) *Voluntary wheel running in mice increases the rate of neurogenesis without affecting anxiety-related behaviour in single tests. BMC Neuroscience*, 13, 61. https://doi.org/10.1186/1471-2202-13-61

Gould, E., Tanapat, P., McEwen, B.S., Fluge, G. and Fuchs, E. (1998) *Proliferation of granule cell precursors in the dentate gyrus of adult monkeys is diminished by stress. Proceedings of the National Academy of Sciences of the United States of America*, 95, 3168–3171. https://doi.org/10.1073/pnas.95.6.3168

Gualtieri, F., Armstrong, E.A., Longmoor, G.K., D'Eath, R.B., Sandilands, V., Boswell, T. and Smulders, T.V. (2019) *Unpredictable chronic mild stress suppresses the incorporation of new neurons at the caudal pole of the chicken hippocampal formation. Scientific Reports*, 9(1), 7129. https://doi.org/10.1038/s41598-019-43584-x

Heine, V.M., Maslam, S., Zareno, J., Joels, M. and Lucassen, P.J. (2004) *Suppressed proliferation and apoptotic changes in the rat dentate gyrus after acute and chronic stress are reversible. European Journal of Neuroscience*, 19, 131–144. https://doi.org/10.1046/j.1460-9568.2003.03100.x

Heissel, A., Heinen, D., Brokmeier, L.L., Skarabis, N., Kangas, M., Vancampfort, D., Stubbs, B., Firth, J., Ward, P.B., Rosenbaum, S., Hallgren, M. and Schuch, F. (2023) *Exercise as medicine for depressive symptoms? A systematic review and meta-analysis with meta-regression. British Journal of Sports Medicine*, 57(16), 1049–1057. https://doi.org/10.1136/bjsports-2022-106282

Holzel, B.K., Carmody, J., Vangel, M., Congleton, C., Yerramsetti, S.M., Gard, T. and Lazar, S.W. (2011) *Mindfulness practice leads to increases in regional brain gray matter density. Psychiatry Research: Neuroimaging*, 191, 36–43. https://doi.org/10.1016/j.pscychresns.2010.08.006

Jackowski, A., Perera, T.D., Abdallah, C.G., Garrido, G., Tang, C.Y., Martinez, J., Mathew, S.J., Gorman, J.M., Rosenblum, L.A., Smith, E.L.P., Dwork, A.J., Shungu, D.C., Kaffman, A., Gelernter, J., Coplan, J.D. and Kaufman, J. (2011) *Early-life stress, corpus callosum development, hippocampal volumetrics, and anxious behavior in male nonhuman primates. Psychiatry Research: Neuroimaging*, 192, 37–44. https://doi.org/10.1016/j.pscychresns.2010.11.006

Jacobson, L. and Sapolsky, R. (1991) *The role of the hippocampus in feedback regulation of the hypothalamic-pituitary-adrenocortical axis. Endocrine Reviews*, 12, 118–134. https://doi.org/10.1210/edrv-12-2-118

Kassem, M., Lagopoulos, J., Stait-Gardner, T., Price, W.S., Chohan, T.W., Arnold, J.C., Hatton, S.N. and Bennett, M.R. (2013) *Stress-induced grey matter loss determined by MRI is primarily due to loss of dendrites and their synapses. Molecular Neurobiology*, 47, 645–661. https://doi.org/10.1007/s12035-012-8365-7

Kempermann, G., Gage, F.H., Aigner, L., Song, H., Curtis, M.A., Thuret, S., Kuhn, H.G., Jessberger, S., Frankland, P.W., Cameron, H.A., Gould, E. and Hen, R. (2018) *Human adult neurogenesis: Evidence and remaining questions. Cell Stem Cell*, 23, 25–30. https://doi.org/10.1016/j.stem.2018.04.004

Kempermann, G., Kuhn, H.G. and Gage, F.H. (1997) *More hippocampal neurons in adult mice living in an enriched environment. Nature*, 386, 493–495. https://doi.org/10.1038/386493a0

Kempermann, G., Kuhn, H.G. and Gage, F.H. (1998) *Experience-induced neurogenesis in the senescent dentate gyrus. Journal of Neuroscience*, 18(9), 3206–3212. https://doi.org/10.1523/jneurosci.18-09-03206.1998

Kim, J., Won, J., Ah, Y., Lee, D., Soo, N., Choi, Y., Ram, B., Joon, H. and Soo, J. (2013) *Sexual activity counteracts the suppressive effects of chronic stress on adult hippocampal neurogenesis and recognition memory. Brain Research*, 1538, 26–40. https://doi.org/10.1016/j.brainres.2013.09.007

Kinney, D.K. and Tanaka, M. (2009) *An evolutionary hypothesis of depression and its symptoms, adaptive value, and risk factors. The Journal of Nervous and Mental Disease*, 197(8), 561–567. https://doi.org/10.1097/nmd.0b013e3181b05fa8

Kodali, M., Megahed, T., Mishra, V., Shuai, B., Hattiangady, B. and Shetty, A.K. (2016) *Voluntary running exercise-mediated enhanced neurogenesis does not obliterate retrograde spatial memory. Journal of Neuroscience*, 36, 8112–8122. https://doi.org/10.1523/jneurosci.0766-16.2016

Koolschijn, P.C., van Haren, N.E.M., Lensvelt-Mulders, G.J., Pol, H.E.H. and Kahn, R.S. (2009) *Brain volume abnormalities in major depressive disorder: A meta-analysis of magnetic resonance imaging studies. Human Brain Mapping*, 30(11), 3719–3735. https://doi.org/10.1002/hbm.20801

Kozorovitskiy, Y., Gross, C.G., Kopil, C., Battaglia, L., McBreen, M., Stranahan, A.M. and Gould, E. (2005) *Experience induces structural and biochemical changes in the adult primate brain. Proceedings of the National Academy of Sciences of the United States of America*, 102, 17478–17482. https://doi.org/10.1073/pnas.0508817102

LaDage, L.D., Roth, T.C., Fox, R. and Pravosudov, V.V. (2010) *Ecologically relevant spatial memory use modulates hippocampal neurogenesis. Proceedings of the Royal Society of London, Series B: Biological Sciences*, 277, 1071–1079. https://doi.org/10.1098/rspb.2009.1769

Lee, T., Jarome, T.J., Li, S.-J., Kim, J.J. and Helmstetter, F.J. (2009) *Chronic stress selectively reduces hippocampal volume in rats: A longitudinal magnetic resonance imaging study. Neuroreport*, 20, 1554–1558. https://doi.org/10.1097/wnr.0b013e328332bb09

Leuner, B., Glasper, E.R. and Gould, E. (2010) *Sexual experience promotes adult neurogenesis in the hippocampus despite an initial elevation in stress hormones. PLoS One*, 5. https://doi.org/10.1371/journal.pone.0011597

Li, J., Huang, Z., Si, W. and Shao, T. (2022) *The effects of physical activity on positive emotions in children and adolescents: A systematic review and meta-analysis. International Journal of Environmental Research and Public Health*, 19(21), 14185. https://doi.org/10.3390/ijerph192114185

Liu, R., Yang, X.-D., Liao, X.-M., Xie, X.-M., Su, Y.-A., Li, J.-T., Wang, X.-D. and Si, T.-M. (2016) *Early postnatal stress suppresses the developmental trajectory of hippocampal pyramidal neurons: The role of CRHR1. Brain Structure & Function*, 221, 4525–4536. https://doi.org/10.1007/s00429-016-1182-4

Lormant, F., Ferreira, V.H.B., Meurisse, M., Lemarchand, J., Constantin, P., Morisse, M., Cornilleau, F., Parias, C., Chaillou, E., Bertin, A., Lansade, L., Leterrier, C., Lévy, F. and Calandreau, L. (2020) *Emotionality modulates the impact of chronic stress on memory and neurogenesis in birds. Scientific Reports*, 10(1), 14620. https://doi.org/10.1038/s41598-020-71680-w

Luo, Y., Cao, Z., Wang, D., Wu, L., Li, Y., Sun, W. and Zhu, Y. (2014) *Dynamic study of the hippocampal volume by structural MRI in a rat model of depression. Neurological Sciences*, 35, 1777–1783. https://doi.org/10.1007/s10072-014-1837-y

MacLellan, A., Nazal, B., Young, L. and Mason, G. (2022) *Waking inactivity as a welfare indicator in laboratory mice: Investigating postures, facial expressions and depression-like states. Royal Society Open Science*, 9(11), 221083. https://doi.org/10.1098/rsos.221083

Magariños, A.M., McEwen, B.S., Flügge, G. and Fuchs, E. (1996) *Chronic psychosocial stress causes apical dendritic atrophy of hippocampal CA3 pyramidal neurons in subordinate tree shrews. Journal of Neuroscience*, 16, 3534–3540. https://doi.org/10.1523/jneurosci.16-10-03534.1996

Malberg, J.E. and Duman, R.S. (2003) *Cell proliferation in adult hippocampus is decreased by inescapable stress: Reversal by fluoxetine treatment. Neuropsychopharmacology*, 28, 1562–1571. https://doi.org/10.1038/sj.npp.1300234

Melleu, F.F., Pinheiro, M.V., Lino-de-Oliveira, C. and Marino-Neto, J. (2016) *Defensive behaviors and prosencephalic neurogenesis in pigeons (Columba livia) are affected by environmental enrichment in adulthood. Brain Structure & Function*, 221, 2287–2301. https://doi.org/10.1007/s00429-015-1043-6

Mohammad, H., Marchisella, F., Ortega-Martinez, S., Hollos, P., Eerola, K., Komulainen, E., Kulesskaya, N., Freemantle, E., Fagerholm, V., Savontous, E., Rauvala, H., Peterson, B.D., van Praag, H. and Coffey, E.T. (2016) *JNK1 controls adult hippocampal neurogenesis and imposes cell-autonomous control of anxiety behaviour from the neurogenic niche. Molecular Psychiatry*, 23, 362–374. https://doi.org/10.1038/mp.2016.203

Morais, M., Patrício, P., Mateus-Pinheiro, A., Alves, N.D., Machado-Santos, A.R., Correia, J.S., Pereira, J., Pinto, L., Sousa, N. and Bessa, J.M. (2017) *The modulation of adult neuroplasticity is involved in the mood-improving actions of atypical antipsychotics in an animal model of depression. Translational Psychiatry*, 7, e1146. https://doi.org/10.1038/tp.2017.120

Moreno-Jiménez, E.P., Flor-garcía, M., Terreros-roncal, J., Rábano, A., Cafini, F., Pallas-bazarra, N., Ávila, J. and Llorens-martín, M. (2019) *Adult hippocampal neurogenesis is abundant in neurologically healthy subjects and drops sharply in patients with Alzheimer ' s disease. Nature Medicine*, 25, 554–560. https://doi.org/10.1038/s41591-019-0375-9

Moser, E.I. and Moser, M.-B. (1998) *Functional differentiation in the hippocampus. Hippocampus*, 8, 608–619. https://doi.org/10.1002/(sici)1098-1063(1998)8:6%3c608::aid-hipo3%3e3.0.co;2-7

O'Doherty, D.C.M., Chitty, K.M., Saddiqui, S., Bennett, M.R. and Lagopoulos, J. (2015) *A systematic review and meta-analysis of magnetic resonance imaging measurement of structural volumes in posttraumatic stress disorder. Psychiatry Research: Neuroimaging*, 232, 1–33. https://doi.org/10.1016/j.pscychresns.2015.01.002

Olsson, I.A. and Dahlborn, K. (2002) *Improving housing conditions for laboratory mice: A review of "environmental enrichment". Laboratory Animals*, 36(3), 243–270. https://doi.org/10.1258/002367702320162379

Onksen, J.L., Briand, L.A., Galante, R.J., Pack, A.I. and Blendy, J.A. (2012) *Running-induced anxiety is dependent on increases in hippocampal neurogenesis. Genes, Brain and Behavior*, 11(5), 529–538. https://doi.org/10.1111/j.1601-183x.2012.00788.x

Papagni, S.A., Benetti, S., Arulanantham, S., McCrory, E., McGuire, P. and Mechelli, A. (2011) *Effects of stressful life events on human brain structure: A longitudinal voxel-based morphometry study. Stress*, 14, 227–232. https://doi.org/10.3109/10253890.2010.522279

Paquola, C., Benett, M.R. and Lagopoulos, J. (2016) *Understanding heterogeneity in grey matter research of adults with childhood maltreatment – a meta-analysis and review. Neuroscience & Biobehavioral Reviews*, 69, 299–312. https://doi.org/10.1016/j.neubiorev.2016.08.011

Paredes, R.G. (2014) *Opioids and sexual reward. Pharmacology Biochemistry and Behavior.*, 121, 124–131. https://doi.org/10.1016/j.pbb.2013.11.004

Perera, T.D., Dwork, A.J., Keegan, K.A., Thirumangalakudi, L., Lipira, C.M., Joyce, N., Lange, C., Higley, J.D., Rosoklija, G., Hen, R., Sackeim, H.A. and Coplan, J.D. (2011) *Necessity of hippocampal neurogenesis for the therapeutic action of antidepressants in adult nonhuman primates. PLoS One*, 6, e17600. https://doi.org/10.1371/journal.pone.0017600

Pham, K., Nacher, J., Hof, P.R. and McEwen, B.S. (2003) *Repeated restraint stress suppresses neurogenesis and induces biphasic PSA-NCAM expression in the adult rat dentate gyrus. European Journal of Neuroscience*, 17, 879–886. https://doi.org/10.1046/j.1460-9568.2003.02513.x

Poirier, C., Bateson, M., Gualtieri, F., Armstrong, E.A., Laws, G.C., Boswell, T. and Smulders, T.V. (2019) *Validation of hippocampal biomarkers of cumulative affective experience. Neuroscience & Biobehavioral Reviews*, 101, 113–121. https://doi.org/10.1016/j.neubiorev.2019.03.024

Pryce, C.R. and Fuchs, E. (2016) *Chronic psychosocial stressors in adulthood: Studies in mice, rats and tree shrews. Neurobiology of Stress*, 6, 94–103. https://doi.org/10.1016/j.ynstr.2016.10.001

Pyter, L.M., Reader, B.F. and Nelson, R.J. (2005) *Short photoperiods impair spatial learning and alter hippocampal dendritic morphology in adult male white-footed mice (Peromyscus leucopus). Journal of Neuroscience*, 25(18), 4521–4526. https://doi.org/10.1523/jneurosci.0795-05.2005

Rahman, M.M., Callaghan, C.K., Kerskens, C.M., Chattarji, S. and O'Mara, S.M. (2016) *Early hippocampal volume loss as a marker of eventual memory deficits caused by repeated stress. Scientific Reports*, 6, 29127. https://doi.org/10.1038/srep29127

Revest, J.-M., Dupret, D., Koehl, M., Funk-Reiter, C., Grosjean, N., Piazza, P.-V. and Abrous, D.N. (2009) *Adult hippocampal neurogenesis is involved in anxiety-related behaviors. Molecular Psychiatry*, 14, 959–967. https://doi.org/10.1038/mp.2009.15

Rice, C.J., Sandman, C.A., Lenjavi, M.R. and Baram, T.Z. (2008) *A novel mouse model for acute and long-lasting consequences of early life stress. Endocrinology*, 149(10), 4892–4900. https://doi.org/10.1210/en.2008-0633

Robertson, B., Rathbone, L., Cirillo, G., D'Eath, R.B., Bateson, M., Boswell, T., Wilson, P.W., Dunn, I.C. and Smulders, V. (2017) *Food restriction reduces neurogenesis in the avian hippocampal formation. PLoS One*, 12, e0189158. https://doi.org/10.1371/journal.pone.0189158

Robinson, W.S. (1950) *Ecological correlations and the behavior of individuals. American Sociological Review*, 15, 351–357. https://doi.org/10.1093/ije/dyn357

Santarelli, L., Saxe, M., Gross, C., Surget, A., Battaglia, F., Dulawa, S., Weisstaub, N., Lee, J., Duman, R., Arancio, O., Belzung, C. and Hen, R. (2003) *Requirement of hippocampal neurogenesis for the behavioral effects of antidepressants. Science*, 301, 805–809. https://doi.org/10.1126/science.1083328

Sheline, Y.I., Gado, M.H. and Kraemer, H.C. (2003) *Untreated depression and hippocampal volume loss. American Journal of Psychiatry*, 160(8), 1516–1518. https://doi.org/10.1176/appi.ajp.160.8.1516

Sherry, D.F. and MacDougall-Shackleton, S.A. (2015) *Seasonal change in the avian hippocampus. Frontiers in Neuroendocrinology*, 37, 158–167. https://doi.org/10.1016/j.yfrne.2014.11.008

Sherwin, C.M. (1998) *The use and perceived importance of three resources which provide caged laboratory mice the opportunity for extended locomotion. Applied Animal Behaviour Science*, 55, 353–367. https://doi.org/10.1016/s0168-1591(97)00049-x

Snyder, J.S., Soumier, A., Brewer, M., Pickel, J. and Cameron, H.A. (2011) *Adult hippocampal neurogenesis buffers stress responses and depressive behaviour. Nature*, 476, 458–461. https://doi.org/10.1038/nature10287

Sousa, N., Lukoyanov, N.V., Madeira, M.D., Almeida, O.F.X. and Paula-Barbosa, M. (2000) *Reorganization of the morphology of hippocampal neurites and synapses after stress-induced damage correlates with behavioral improvement. Neuroscience*, 97, 253–266. https://doi.org/10.1016/s0306-4522(00)00050-6

Strange, B.A., Witter, M.P., Lein, E.S. and Moser, E.I. (2014) *Functional organization of the hippocampal longitudinal axis. Nature Reviews Neuroscience*, 15, 655–669. https://doi.org/10.1038/nrn3785

Sullivan, E.V., Pfefferbaum, A., Swan, G.E. and Carmelli, D. (2001) *Heritability of hippocampal size in elderly twin men: Equivalent influence from genes and environment. Hippocampus*, 11, 754–762. https://doi.org/10.1002/hipo.10.1

Suri, D. and Vaidya, V.A. (2015) *The adaptive and maladaptive continuum of stress responses - A hippocampal perspective. Reviews in the Neurosciences*, 26, 415–442. https://doi.org/10.1515/revneuro-2014-0083

Tanapat, P., Hastings, N., Rydel, T., Galea, L. and Gould, E. (2001) *Exposure to fox odor inhibits cell proliferation in the hippocampus of adult rats via an adrenal hormone-dependent mechanism. Journal of Comparative Neurology*, 437, 196–504. https://doi.org/10.1002/cne.1297

Taufique, S.K.T., Prabhat, A. and Kumar, V. (2018) *Constant light environment suppresses maturation and reduces complexity of newborn neuron processes in the hippocampus and caudal nidopallium of a diurnal corvid: Implication for impairment of the learning and cognitive performance. Neurobiology of Learning and Memory*, 147, 120–127. https://doi.org/10.1016/j.nlm.2017.12.001

van Praag, H. (2008) *Neurogenesis and exercise: Past and future directions. Neuromolecular Medicine*, 10, 128–140. https://doi.org/10.1007/s12017-008-8028-z

van Praag, H., Kempermann, G. and Gage, F.H. (1999) *Running increases cell proliferation and neurogenesis in the adult mouse dentate gyrus. Nature Neuroscience*, 2, 266–270. https://doi.org/10.1038/6368

Van't Ent, D., den Braber, A., Baselmans, B.M.L., Brouwer, R.M., Dolan, C.V., Hulshoff Pol, H.E., de Geus, E.J.C. and Bartels, M. (2017) *Associations between subjective well-being and subcortical brain volumes. Scientific Reports*, 7, 6957. https://doi.org/10.1038/s41598-017-07120-z

Walhovd, K.B., Westlye, L.T., Amlien, I., Espeseth, T., Reinvang, I., Raz, N., Agartz, I., Salat, D.H., Greve, D.N., Fischl, B., Dale, A.M. and Fjell, A.M. (2011) *Consistent neuroanatomical age-related volume differences across multiple samples. Neurobiology of Aging*, 32, 916–932. https://doi.org/10.1016/j.neurobiolaging.2009.05.013

Watanabe, Y., Gould, E. and McEwen, B.S. (1992) *Stress induces atrophy of apical dendrites of hippocampal CA3 pyramidal neurons. Brain Research*, 588, 341–345. https://doi.org/10.1016/0006-8993(92)91597-8

Willard, S.L., Friedman, D.P., Henkel, C.K. and Shively, C.A. (2009) *Anterior hippocampal volume is reduced in behaviorally depressed female cynomolgus macaques. Psychoneuroendocrinology*, 34, 1469–1475. https://doi.org/10.1016/j.psyneuen.2009.04.022

Wise, T., Radua, J., Via, E., Cardoner, N., Abe, O., Adams, T.M., Amico, F., Cheng, Y., Cole, J.H., de Azevedo Marques Périco, C., Dickstein, D.P., Farrow, T.F.D., Frodl, T., Wagner, G., Gotlib, I.H., Gruber, O., Ham, B.J., Job, D.E., Kempton, M.J., Kim, M.J., Koolschijn, P.C.M.P., Malhi, G.S., Mataix-Cols, D., McIntosh, A.M., Nugent, A.C., O'Brien, J.T., Pezzoli, S., Phillips, M.L., Sachdev, P.S., Salvadore, G., Selvaraj, S., Stanfield, A.C., Thomas, A.J., van Tol, M.J., van der Wee, N.J.A., Veltman, D.J., Young, A.H., Fu, C.H., Cleare, A.J. and Arnone, D. (2017) *Common and distinct patterns of grey-matter volume alteration in major depression and bipolar disorder: evidence from voxel-based meta-analysis. Molecular Psychiatry*, 22(10), 1455–1463. https://doi.org/10.1038/mp.2016.72

Wu, M.V. and Hen, R. (2014) *Functional dissociation of adult-born neurons along the dorsoventral axis of the dentate gyrus. Hippocampus*, 24, 751–761. https://doi.org/10.1002/hipo.22265

Wu, M.V., Shamy, J.L., Bedi, G., Choi, C.-W.J., Wall, M.M., Arango, V., Boldrini, M., Foltin, R.W. and Hen, R. (2014) *Impact of social status and antidepressant treatment on neurogenesis in the baboon hippocampus. Neuropsychopharmacology*, 39, 1861–1871. https://doi.org/10.1038/npp.2014.33

Yap, J.J., Takase, L.F., Kochman, L.J., Fornal, C.A., Miczek, K.A. and Jacobs, B.L. (2006) *Repeated brief social defeat episodes in mice: Effects on cell proliferation in the dentate gyrus. Behavioural Brain Research*, 172(2), 344–350. https://doi.org/10.1016/j.bbr.2006.05.027

Zhao, X. and van Praag, H. (2020) *Steps towards standardized quantification of adult neurogenesis. Nature Communications*, 11(1), 4275. https://doi.org/10.1038/s41467-020-18046-y

14 The Importance of Construct Validation, and (More) Tools for Valid Use of Indicators – *What Have We Learned and What Is Still Missing?*

Birte L. Nielsen[1], Georgia J. Mason[2], and Michael T. Mendl[3]

[1]Science for Animal Welfare, Hertfordshire, UK
[2]Department of Integrative Biology & Campbell Centre for the Study of Animal Welfare, University of Guelph, Guelph, Ontario, Canada
[3]Bristol Veterinary School, University of Bristol, Bristol, UK

We hope you have enjoyed both the content and structure of the book: the three sections covering different indicators of emotions, moods and cumulative animal welfare, and the framework set up in Chapters 1 and 2. As editors, we are very grateful to all the authors who have contributed their time and wisdom to write chapters on indicators close to their hearts.

In this final chapter, we try to draw out some common themes while avoiding repeating what you have already read. In particular, we will focus on how to not just choose but also combine indicators; the ongoing challenges of identifying absolute versus relative affective state and how to spot absent or imperfect construct validation. We end by pondering where to go from here (touching on future indicators, the role of AI, and the value of pre-registering research).

Challenges and opportunities when choosing or combining indicators

As you all know by now (and as detailed in Chapter 2), we have proposed five tests that we can use to assess the construct validity of a potential indicator of animal affect. Throughout this book, we have discussed whether different indicators

- accurately track self-reported affective states in humans (**Test 1**)
- reliably change in animals
  - exposed to stimuli that they prefer/avoid (**Test 2**)
  - exposed to ancestrally fitness-boosting or fitness-harming situations (**Test 3**)
  - given affect-modulating drugs (**Test 4**)
- reliably covary with pre-validated affect indicators (**Test 5**)

In this section, we want to highlight some of the challenges and opportunities we can encounter when choosing and combining indicators of affective states. These principles are useful for people validating individual variables, or weighting the elements of composite scores and, increasingly importantly, they provide suggested standards for people training AI to detect affective states.

When choosing indicators of animal welfare, you need to pick ones that are suitable and valid for your intended purposes, including whether the assessment is of emotions, moods or long-term cumulative affective experience. This book has thus taught us that the indicator should fit the affective timescale we are interested in. Monitoring 22 kHz calls may be a great way of assessing negative emotions in rats, for instance, but much less effective for assessing cumulative impacts over a lifetime. Performance of stereotypic behaviour, on the other hand, is not a good way to assess negative emotions but a great way to assess cumulative affective harms.

An understanding of the limitations of your indicator is also essential for making an informed choice. Does it have a threshold effect, below which it is unresponsive? Does it have ceiling effects, above which you won't be able to detect changes? Between these thresholds, can it provide a graded estimate of the intensity of affective valence and, if so, is it more suited to assessing positive or negative affective states? Knowing the strengths and weaknesses of your indicator is important, but it should come with the recognition that there is almost certainly no perfect indicator – we just need to select the best ones (or combinations) for the job. As we have learned throughout this book, even valid indicators are rarely perfect. But understanding their imperfections, including poor responsiveness (or insensitivity) and poor selectivity, can help us choose and use them well.

Irritating as it may be, the point at which we can conclude that an indicator is 'validated' is often nuanced too. The rule of thumb is '*the more times, the better*': Findings across *many* studies of a reasonably replicable relationship between affective valence and change in an indicator provide strong evidence of validation of that particular indicator. This is because each validation method relies on assumptions that might be wrong, so the more tests an indicator passes, the better. We can do this by implementing the five tests for construct validation while using manipulations which we are pretty confident will induce the predicted affective state (see Chapter 2, and many examples in the previous chapters)

One important way of addressing the limitations of individual indicators is to use several together and to select the combination used with care. The more (validated) indicators we measure and include in our assessments, the more evidence we can marshal to identify the affective state, and therefore welfare, of the animals we're interested in. But when selecting which indicators to use in combination, they need to complement each other well. From the previous chapters we have learned that some indicators may be grounded in similar physiological changes with similar properties, and therefore not independent (a bit like measuring both the volume and the weight of water). For example, we would expect cortisol and ACTH to change in a similar manner because both are involved in the activation of the HPA system. Furthermore, we would also expect cortisol and heart rate to change in a similar manner because both are associated with arousal(rather than valence). One of these (or both if you are interested in detecting arousal responses of short but different durations), should therefore be used in conjunction with a measure that can indicate the *direction of valence* (especially if you are not sure *a priori* where on the valence-axis the animal is likely to be). Examples could be specific vocalisations. That is because together, the indicators should be able to provide useful information about both affective valence and emotional intensity, noting that for some affective states such as depression, emotional intensity may be high even when arousal is low.

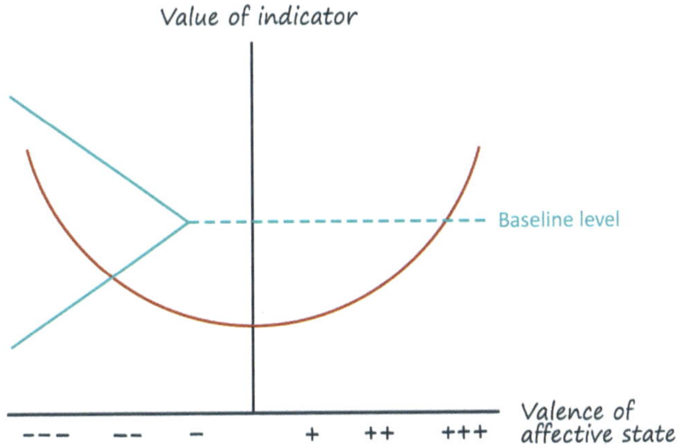

Figure 14.1 Examples of indicators that need additional information to help interpretation.

Furthermore, indicators that themselves are intrinsically rewarding or aversive may not just provide information about an animal's current affective state: they may also tell us something about their *future* state, and may even play a role in influencing this state. For example, play and stereotypic behaviour may perhaps help animals to cope (have positive consequences), while decreased telomere length, hippocampal volume, and negative judgement bias may increase a subject's vulnerability to future negative states. Other similar examples not covered in this book might include mating and drug-taking (e.g. self-administering analgesics). This adds further nuance to how we interpret these markers in welfare assessments. For example, welfare indicators that also *cause* poor welfare (like telomere length seems to; Chapter 12) should perhaps be given more weight or used differently to those that also *reflect* poor welfare but have somewhat protective consequences (e.g. like at least some abnormal repetitive behaviours; Chapter 11).

Some problematic measures may even change in both directions from a baseline level (see blue line in Figure 14.1, which echoes the last figure of Chapter 2), like body weight in times of distress. Using this type of measure as an indicator of affective state is not very useful, as it is only the occurrence of change itself that potentially indicates a shift in state, but we cannot infer which response indicates the poorest welfare. For example, as mentioned in Chapter 2, a fully grown animal who starts to gain *or* lose weight may be in a more negative state than one whose weight remains stable, but then who is the best off: an animal who gains weight, or one who loses it? It is impossible to know without further information. A similar example is anticipatory behaviour preceding a signalled reward (as mentioned in Chapter 9 on reward reactivity), in which an *increase* may be observed in short-lived or mildly aversive situations while a *decrease* is more commonly seen in longer-lived and more severe contexts. For this measure, information on the duration and severity of the context and the affective manipulation is therefore needed to enable interpretation of any anticipatory behaviour displayed. And even then, this is a rather subjective judgement call, unless we are assisted by further indicators that do not suffer from this problem. Thus, when choosing indicators of affective states, measures of this 'toppled Y' type are not useful in and of themselves.

Combining the information from multiple indicators to make an overall assessment of affective state and welfare can be problematic if their findings are inconsistent. To minimise the likelihood of inconsistent findings, we should aim to use only validated indicators based on the affective time course of interest and used in wisely selected combinations. Before any data are collected, we can therefore make informed judgements, based on the framework we have described in this book, about the likely validity and appropriateness of our chosen indicators in the context we are using them, so as to weight them accordingly when combining information to draw conclusions.

When using multiple indicators, algorithms and weightings potentially allow us to combine evidence in a quantitative manner and draw conclusions in this way. This approach has been implemented in some studies of on-farm welfare assessment, e.g. Welfare Quality® used literature-based and expert-derived weightings of the importance of different indicators to combine values from multiple markers (e.g. Botreau et al. 2007, 2009). Risks associated with this type of approach include that weightings may result in a few indicators having a major influence on any final global score. For example, measures of water access were found to account for much of the variation in complex aggregated scoring methods used for dairy cattle welfare in the Welfare Quality® project (e.g. Heath et al. 2014; de Graaf et al. 2018). Such weighting methods could therefore benefit from using the approaches outlined in this book to identify well-validated animal-based indicators *a priori*, for example those that are better estimated, more responsive, or more selective than others, and weighting them accordingly. Much work remains to be done in developing effective ways of combining validated indicators.

Absolute versus relative affective state

We have tried to link the different chapters of this book through the use of the same type of heuristic graph, where the indicator on the y-axis is plotted against affective state along the x-axis, the latter ranging from very negative to very positive. Although this may have made the various indicators' relationships with affective states easier to comprehend, it still highlights other questions mentioned in Chapter 1. In particular, how do we know where an animal is on the x-axis? Is there such a thing as 'neutral' affective state at $x = 0$, and what is the difference between increasing positive affect and reducing negative affect? Is happiness just less sadness, or is it a different beast? When a chilly hen learns to peck a button to turn on a heat lamp, for example, is she being positively reinforced by the increase in temperature (addition of a positive stimulus or reward) or negatively reinforced by the decrease in coldness (removal of an aversive stimulus)?

In Chapter 1, it was therefore sugggested that the valence state inferred from an indicator is often relative as opposed to absolute. This allows us to say, for example, that an animal with a higher indicator value is in a relatively more positive affective state than one with a lower value. However, we may wish to assess whether animals have, not just *relatively* poor welfare compared to others, but if they are in *absolute* negative states. For example, we may need to know this to judge if these states are above a certain threshold (e.g. requiring a 'severe' categorisation for research animals or triggering a euthanasia decision), while the concept of positive animal welfare (PAW; Lawrence et al. 2019; Rault et al. 2025) has the existence of measurably *absolute* positive states at its very core. Another situation where this is imperative is when we use an indicator that has a V- or U-shape on the heuristic graph (red line in Figure 14.1; echoing figures in Chapters 4 and 5). Here we need indicators that can tell us whether we are in the positive or negative half of the valence axis (such as certain signals/expressions or the presence of abnormal repetitive behaviour).

Even if we don't know absolutely where we are on the valence axis, responsive indicators can provide graded information about relative differences in affective valence. The nuance lies in the shape of the 'responsiveness' curve – some indicators may give us lovely linear gradients allowing us to say with granularity where we are (relatively or absolutely) on the valence axis, while others may be responsive in a discontinuous (e.g. binary) way providing much coarser info on location on the valence axis. Dose-response studies of how indicators change as the assumed positivity or negativity of the affective manipulation is increased or decreased are comparatively rare, but they can be very valuable in identifying indicators that tell us about relative or absolute location on the valence axis.

So, to summarize this section: When interpreting our findings, we should be aware that many indicators are likely to provide us with *relative* information about animals' affective states, rather than where in absolute terms they lie on the valence axis of our heuristic graphs. If we are confident in which half of the graph the animal is likely to be, interpretation can be facilitated by simultaneous use of indicators that operate in just the positive or negative halves of the graphs. Their combined output may then allow us to estimate where on the scale we are in absolute terms.

How to spot absent validation or questionable use

Most good, valid welfare studies will fall into two types: (i) those seeking to validate new indicators of affect, and (ii) those using already well-validated indicators to discover new things about animal well-being. However, in some studies, more questionable practices can be seen. We hope that this book has made you more able to spot when something is awry or needs more explanation to make sense. In this section, we discuss some mistakes that can be made when selecting indicators of affective state, and highlight some of the signs of improper use. Please note that we focus on *construct* validity, but please remember that internal validity (e.g. the correct use of statistics; randomising subjects; collecting data blind), and external validity (e.g. the extent to which findings are generalisable beyond the context and populations in which they are made) are of course very important too (Würbel 2017; Tuyttens et al. 2014; Mason 2023).

At first glance, a study may have sound structures, and the research goals seem clear. But on further reading, dubious assumptions may reduce the robustness and defensibility of the conclusions stated. In some of these cases, an alleged indicator is being used to investigate a potential welfare issue, but that indicator's construct validity is uncertain (thus making any conclusions uncertain). Sometimes there is no explanation why this indicator is thought to be valid (a simple omission that does not demonstrate a lack of validity, but that does make the logic hard to follow); while in other cases, loaded language is used as if it is an explanation (dangerous because it can imply greater construct validity than is defensible). Even as researchers, we can be misled by so-called *false friends*, where certain seductive words or specialised terms give rise to preconceived (but wrong) ideas about a situation. (If we provide animals with '*enrichment*', then their lives and moods must *surely* be enhanced, for example And if economists say that elasticity of demand indicates a '*luxury*', surely things that are elastically demanded cannot be that important). In Table 14.1, we therefore list a number of dubious reasons that may be given for choosing a putative indicator of affective state. Please note that just because an indicator has not (yet) been validated does not necessarily make it *in*valid, but it does make the use of it concerning and not to be relied upon. Please feel free to use Table 14.1

Table 14.1 Examples of how indicators of affective states are chosen if *not* via construct validity.

Reasons given...	Why this reasoning is problematic...
They give me highly repeatable results	Confusing validity with reliability
'Stress hormones' and 'optimistic/pessimistic judgment biases' sound convincing	Swayed by loaded language
Other people have used them and been well-cited for it	'*Everyone does it*' fallacy; Appeal to authority
People have used clinical scoring for ages	Appeal to tradition
The spline functions and other stats underlying this scoring system are so impressive, they must be right	Appeal to authority
They give me the results I expect	Potential HARKing[a]; Potential circular reasoning
They're affect-relevant in humans and so simply must be in all other species	Potential unjustified anthropomorphism
I take multiple measures and present those that give me the most plausible results	Potential HARKing[a] plus P-hacking[b] Potential circular reasoning

[a] Hypothesising After the Results are Known.
[b] Trying various statistical approaches until the desired finding appears.

as a checklist next time you go to a conference or leaf through a journal. Full disclosure: we have all been guilty of some of these ourselves – but we are trying to be better.

Inaccurate, invalid indicators can lead to the two types of error that have been referred to in all the chapters in this book: failures to detect changes in animals' affective states (false nulls) and suggesting changes in affective state that are not actually present (false leads). Both errors are important. If we fail to detect states of negative affect when conducting animal welfare research, then this will leave welfare problems unrecognised and therefore unsolved. Conversely, if we are misled by false leads, for example indications of apparent negative affect despite animals actually being in neutral or positive states, this could lead to animal care or husbandry systems being changed in ways that do not actually matter to the animals themselves (so wasting time, money and other resources) or, worse, even impose new welfare harms that were previously absent. And such problems are not unique to animal welfare research. The use of indicators of uncertain validity in animal-based biomedical research on pain and depression is one likely reason why such work has generated few successful new therapies for humans (e.g. Cobos and Portillo-Salido 2013; Garner 2014; Harro 2019; Planchez et al. 2019), in turn ensuring that '*patients will continue to suffer, and unimaginable amounts of money and animals are needlessly wasted*' (Garner 2014). Construct validation is thus essential for correct inferences about animal feelings. Without it, to quote the blunt words of Garner (2014), we have instead '*bad science*'.

Figure 14.2 The circularity and inappropriateness of using an unknown indicator to test a treatment with unknown effects on affective state.

Furthermore, in studies that seek to validate new indicators via some version of Tests 2–4, the *a priori* affective manipulations required for this are not always clearly justified, again making any conclusions uncertain. For example, uncritical or unjustified anthropomorphism (very different from legitimate arguments from homology; see e.g. McNaughton and Zangrossi 2008; Panksepp 2011a,b) may drive assumptions about what is rewarding or aversive for an animal, what situations are fitness-threatening or -harming, or whether a drug induces a particular affective state. In other cases, loaded language again might be at play, e.g. researchers providing alleged 'enrichments' without any *a priori* evidence that these are good for animal welfare. And if an unvalidated indicator is then used to assess the effectiveness of this 'enrichment', a circularity of reasoning is taking place (Figure 14.2). All this is important: without clear logic as to the validatory process, we risk there simply being no ground truth in research on animal welfare (see Chapter 2).

Uncertainty and lack of clarity can then increase the role played by preconceptions and conflicts of interest: without good construct validity, wobbly logic can lead to both real and perceived biases in data interpretation. Such biases may sometimes reflect desires to find and make sense of 'statistically significant' results. For example, in an impressive meta-analysis of 206 mouse anxiety studies involving the drug diazepam and certain behavioural tests (open-field or hole-board tests), Rosso et al. (2024) found that the rationales given for these tests were strongly related to the reported results: papers finding significantly *decreased* activity inferred a sedative effect that they claimed they had been aiming to assay, while papers finding significantly *increased* activity instead inferred reduced anxiety, which, again, they claimed they had been aiming to assay. The most likely explanation for this pattern is that the alleged meaning of each test result was tweaked *post hoc* to fit the data obtained, thus hypothesizing after the results are known (HARKing). Animal welfare scientists should obviously *not* be doing this.

Our aim in this section has been to make us all think harder about these important issues underpinning the integrity, strength and utility of our science, and which many of us, the authors of this chapter included, have almost certainly fallen prey to. Without proper construct validation, the conclusions reached about animal feelings may simply be wrong, because the indicators used are inaccurate. If '*many of the methods currently used for estimating animal welfare... are subjective, unreliable and unlikely to be accurate*' (Browning 2022), then we (and animals) are in trouble. Hopefully, this book provides tools to repair or at least improve the situation, and to also encourage increased rigour, critical thinking and a greater appreciation of best practice.

Where to go from here?

As highlighted in this chapter, the appropriate use of well-selected, validated indicators of affective state is an excellent place to start. The eleven indicators discussed in the previous chapters provide us with a good foundation to choose what measures to use when. There is still a lot of work to be done, but the methodologies outlined in Chapters 1 and 2 for construct validation of further potential indicators (some suggestions are shown in Figure 14.3) are there for everyone to use.

Some of the shortcomings identified in the previous chapters include question marks over validity in specific or wider contexts which now need further attention. And even well-validated indicators still need to be feasible for use beyond the controlled laboratory environment: they may need research to clarify how we can we use these indicators for practical welfare assessment or welfare monitoring. With increasing technological progress, for instance, we can now track and monitor individuals or groups of animals automatically, which may allow us to translate the knowledge presented here into more practical settings. One example is the use of vocalisations as indicators of affective state (see Chapter 3), as automatic detection of chicken *distress* calls has shown potential as a component of on-farm animal welfare monitoring (Mcloughlin et al. 2019; Herborn et al. 2020; Mao et al. 2022). However, although AI is sweeping through every aspect of our lives, we need to proceed with caution. Baidu, the owner of China's largest search engine, has recently filed a patent for a

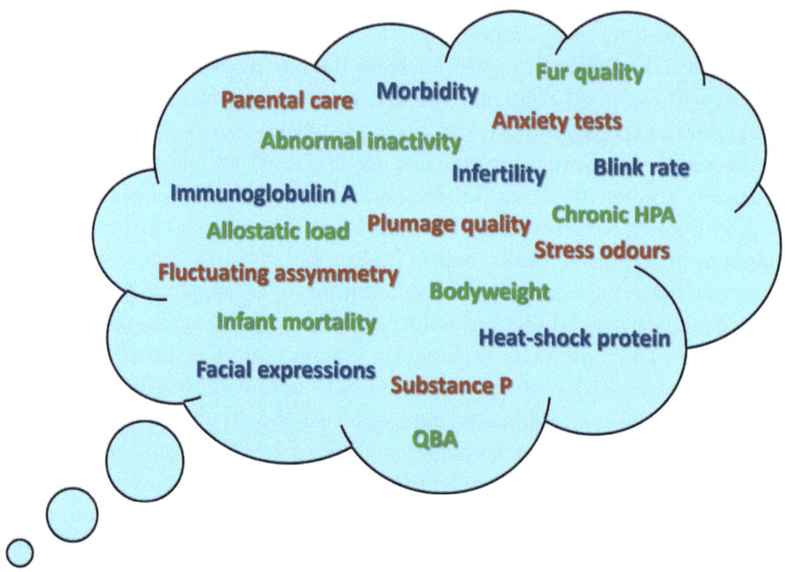

Figure 14.3 Cloud of some potential indicators of affective states for (potential) future volumes of this book.

system that collects and processes animal data including vocalisations, behaviour and physiological signals using an AI-powered analysis designed to recognise the animal's emotional state (Reuters 2025). Such a system *must* be based on proper construct validation, to avoid generating flawed information resulting in false nulls and false leads.

In Chapter 1 we introduced how operational definitions of constructs such as affective valence allow us to conceptualise them in ways that are measurable and open to empirical investigation. The use of operational definitions can also be extended to discrete emotions (e.g. fear, joy) to help the validation process. Operational definitions have been devised for fear and anxiety (Forkman et al. 2007), and we need further operationalisation of states such as sadness, guilt and (for social species) loneliness to allow us to progress with identifying and validating relevant indicators. Operationalisation and investigations of boredom have also started (Burn 2017; Hintze et al. 2020; Meagher and Robbins 2022) and, with increasing interest in the positive spectrum of animal welfare, operationalisation of states like joy and contentment is needed (Lawrence et al. 2019).

Finally, we want to draw attention to the growing practice of *pre-registering* studies before they are carried out. In brief, this involves clearly stating the assumptions, logic, and rationale for the study, its hypotheses and the predictions to be tested, the design of the study (including blinding, randomisation, sample size justification, control groups and so on), the data (variables) to be collected and the analyses to be carried out. This information can be uploaded, often using a pre-formatted template, to a public registry (e.g. the Open Science Framework). It provides evidence of what you actually set out to test in your study *before you do the work*, which in turn prevents HARKing, p-hacking, and other undesirable practices. The pre-registration document is also open to review and comment by other researchers which can help to refine the study design before it takes place.

A further step along this path is to generate a *registered report* which essentially involves writing the full Introduction and Methods for the study, including all the details highlighted above, which is then formally assessed by appointed reviewers for a journal that supports this process. The reviewing step is designed to ensure that the study's rationale, design, methods and statistics are as rigorous as possible. Once accepted, the journal is committed to publishing the resulting study as long as the registered methods are followed (e.g. Bombail et al. 2022; see also Muñoz-Tamayo et al. 2022).

Most importantly from our current perspective, *pre-registration* and the writing of *registered reports* concentrate our minds on the types of issue, including construct validation, that we have been considering here. We may not choose to follow these routes – it is likely to take some time to shift the scientific culture such that they become a standard approach – but if not, we should still think hard about the logic of our study and the validation of the constructs we are invoking. Ensuring replicability (via good experimental design – blinding, randomisation and so on – and by avoiding p-hacking), thinking about better external validity and translatability, and being more open with data and code: these are all things our community should be doing more and better if we want our work to legitimately have real, lasting impact.

Take-home message from the book

In this book we have described ways to assess animal affective states and thereby to assess their welfare. The book is founded on the notion that **affective states** (across the whole range of types) **cannot be measured directly**, but their correlates and consequences can be. We have included eleven different (types of) indicators in this volume, some better validated than others. And yes, telling what animals are feeling is hard (how could it not be?), but we can make justified inferences thanks to the five tests for construct validation (listed earlier and detailed in Chapter 2).

There is no perfect indicator of affect/welfare (and probably never will be). But by careful construct validation, and thinking formally and explicitly about responsiveness and selectivity, we can pick and interpret our tools better. This process is absolutely crucial, especially as people start delving into automation, machine learning and extensive use of AI: if we automate the assessment of invalid indicators or use naive teaching rules in the training phase of machine learning, we will make mistaken inferences about animals and then, ultimately, not help them or the people who look after them. To end, as Dawkins (2001) beautifully put it: '*Animals do not have indicators of welfare for our benefit. What they have are physiological and behavioral responses to a wide variety of situations, and it is up to us to read the signs*'.

References

Bombail, V., Brown, S.M., Marti, J.E., Meddle, S.L., Mendl, M., Robinson, E., Hammond, T.J., Nielsen, B.L., LaFollette, M.R., Vinuela-Fernandez, I., Tivey, E.K.L. and Lawrence, A.B. (2022) *Stage 1 Registered Report. Refinement of tickling protocols to improve positive animal welfare in laboratory rats. F1000 Research*, 11, 1053. https://doi.org/10.12688/f1000research.125649.2

Botreau, R., Bracke, M.B.M., Perny, P., Butterworth, A., Capdeville, J., van Reenen, C.G. and Veissier, I. (2007) *Aggregation of measures to produce an overall assessment of animal welfare. Part 2: Analysis of constraints. Animal*, 1(8), 1188–1197. https://doi.org/10.1017/s1751731107000547

Botreau, R., Veissier, I. and Perny, P. (2009) *Overall assessment of animal welfare: Strategy adopted in Welfare Quality®. Animal Welfare*, 18(4), 363–370. https://doi.org/10.1017/s0962728600000762

Browning, H. (2022) *Assessing measures of animal welfare. Biology & Philosophy*, 37, 36. https://doi.org/10.1007/s10539-022-09862-1

Burn, C. (2017) *Bestial boredom: A biological perspective on animal boredom and suggestions for its scientific investigation. Animal Behaviour*, 130, 141–151. https://doi.org/10.1016/j.anbehav.2017.06.006

Cobos, E.J. and Portillo-Salido, E. (2013) *'Bedside-to-bench' behavioral outcomes in animal models of pain: Beyond the evaluation of reflexes. Current Neuropharmacology, 11*(6), 560–591. https://doi.org/10.2174/1570159x113119990041

Dawkins, M.S. (2001) How can we recognise and assess good welfare? In Broom, D. M. (ed.) Dahlem Workshop Report 87,*Coping with Challenge: Welfare in Animals Including Humans*. Dahlem University Press, Berlin, pp. 63–67.

de Graaf, S., Ampe, B., Buijs, S., Andreasen, S.N., de Boyer Des Roches, A., van Eerdenburg, F.J.C.M., Haskell, M.J., Kirchner, M.K., Mounier, L., Radeski, M., Winckler, C., Bijttebier, J., Lauwers, L., Verbeke, W. and Tuyttens, F.A.M. (2018) *Sensitivity of the integrated Welfare Quality® scores to changing values of individual dairy cattle welfare measures. Animal Welfare*, 27(2), 157–166. https://doi.org/10.7120/09627286.27.2.157

Forkman, B., Boissy, A., Meunier-Salaün, M.-C., Canali, E. and Jones, R.B. (2007) *A critical review of fear tests used on cattle, pigs, sheep, poultry and horses. Physiology & Behavior*, 92(3), 340–374. https://doi.org/10.1016/j.physbeh.2007.03.016

Garner, J.P. (2014) *The significance of meaning: Why do over 90% of behavioral neuroscience results fail to translate to humans, and what can we do to fix it? ILAR Journal*, 55(3), 438–456. https://doi.org/10.1093/ilar/ilu047

Harro, J. (2019) *Animal models of depression: Pros and cons. Cell and Tissue Research*, 377(1), 5–20. https://doi.org/10.1007/s00441-018-2973-0

Heath, C.A.E., Browne, W.J., Mullan, S. and Main, D.C.J. (2014) *Navigating the iceberg: Reducing the number of parameters within the Welfare Quality® assessment protocol for dairy cows. Animal*, 8(12), 1978–1986. https://doi.org/10.1017/S1751731114002018

Herborn, K.A., McElligott, A.G., Mitchell, M.A., Sandilands, V., Bradshaw, B. and Asher, L. (2020) *Spectral entropy of early-life distress calls as an iceberg indicator of chicken welfare. Journal of the Royal Society Interface*, 17, 20200086. https://doi.org/10.1098/rsif.2020.0086

Hintze, S., Maulbetsch, F., Asher, L. and Winckler, C. (2020) *Doing nothing and what it looks like: Inactivity in fattening cattle. PeerJ*, 8, e9395. https://doi.org/10.7717/peerj.9395

Lawrence, A.B., Vigors, B. and Sandøe, P. (2019) *What is so positive about positive animal welfare?—a critical review of the literature. Animals*, *9*(10), 783. https://doi.org/10.3390/ani9100783

Mao, A., Giraudet, C.S.E., Liu, K., De Almeida Nolasco, I., Xie, Z., Xie, Z., Gao, Y., Theobald, J., Bhatta, D., Stewart, R. and McElligott, A.G. (2022) *Automated identification of chicken distress vocalizations using deep learning models. Journal of the Royal Society Interface*, 19, 20210921. https://doi.org/10.1098/rsif.2021.0921

Mason, G.J. (2023) *Animal welfare research is fascinating, ethical, and useful—but how can it be more rigorous? BMC Biology*, *21*(1), 302. https://doi.org/10.1186/s12915-023-01793-x

Mcloughlin, M.P., Stewart, R. and McElligott, A.G. (2019) *Automated bioacoustics: Methods in ecology and conservation and their potential for animal welfare monitoring. Journal of the Royal Society Interface*, 16, 20190225. https://doi.org/10.1098/rsif.2019.0225

McNaughton, N. and Zangrossi, H. Jr. (2008) *Theoretical approaches to the modeling of anxiety in animals. Handbook of Behavioral Neuroscience*, 17, 11–27. https://doi.org/10.1016/s1569-7339(07)00002-1

Meagher, R.K. and Robbins, J. (2022) Parallels to boredom in nonhuman animals. In Elpidorou, A. (ed.) *The Moral Psychology of Boredom1*. Rowman & Littlefield, Washington DC, USA, pp. 267–289.

Muñoz-Tamayo, R., Nielsen, B.L., Gagaoua, M., Gondret, F., Krause, E.T., Morgavi, D., Olsson, I.A.S., Pastell, M., Taghipoor, M., Tedeschi, L., Veissier, I. and Nawroth, C. (2022) *Seven steps to enhance open science practices in animal science. PNAS Nexus*, 1, pgac106. https://doi.org/10.1093/pnasnexus/pgac106

Panksepp, J. (2011a) *Cross-species affective neuroscience decoding of the primal affective experiences of humans and related animals. PLoS One*, 6(9), e21236. https://doi.org/10.1371/journal.pone.0021236

Panksepp, J. (2011b) *The basic emotional circuits of mammalian brains: Do animals have affective lives? Neuroscience & Biobehavioral Reviews*, 35(9), 1791–1804. https://doi.org/10.1016/j.neubiorev.2011.08.003

Planchez, B., Surget, A. and Belzung, C. (2019) *Animal models of major depression: Drawbacks and challenges. Journal of Neural Transmission*, 126, 1383–1408. https://doi.org/10.1007/s00702-019-02084-y

Rault, J.-L., Bateson, M., Boissy, A., Forkman, B., Grinde, B., Gygax, L., Harfeld, J.L., Hintze, S., Keeling, L.J., Kostal, L., Lawrence, A.B., Mendl, M.T., Miele, M., Newberry, R.C., Sandøe, P., Spinka, M., Taylor, A.H., Webb, L.E., Whalin, L. and Jensen, M.B. (2025) *A consensus on the definition of positive animal welfare. Biology Letters*, 21, 20240382. https://doi.org/10.1098/rsbl.2024.0382

Reuters (2025). China's Baidu looks to patent AI system to decipher animal sounds. https://www.reuters.com/business/media-telecom/chinas-baidu-looks-patent-ai-system-decipher-animal-sounds-2025-05-08/ (accessed July 2025)

Rosso, M., Herrera, A., Würbel, H. and Voelkl, B. (2024) *Evidence for HARKing in mouse behavioural tests of anxiety. Royal Society Open Science*, 11, 231744. https://doi.org/10.1098/rsos.231744

Tuyttens, F.A.M., de Graaf, S., Heerkens, J.L., Jacobs, L., Nalon, E., Ott, S., Stadig, L., Van Laer, E. and Ampe, B. (2014) *Observer bias in animal behaviour research: Can we believe what we score, if we score what we believe? Animal Behaviour*, 90, 273–280. https://doi.org/10.1016/j.anbehav.2014.02.007

Würbel, H. (2017) *More than 3Rs: The importance of scientific validity for harm-benefit analysis of animal research. Laboratory Animals*, 46, 164–166. https://doi.org/10.1038/laban.1220

Index

Page locators in **bold** indicate tables. Page locators in *italics* indicate figures. Some of the entries are covered throughout the book, in which case only the page(s) containing the first use or the definition are listed. These entries are underlined.